Epigenetic Regulation and Epigenomics

Edited by
Robert A. Meyers

Related Titles

Meyers, R.A. (ed.)

Encyclopedia of Molecular Cell Biology and Molecular Medicine

Online version: www.meyers-emcbmm.com

Sippl, W., Jung, M. (eds.)

Epigenetic Targets in Drug Discovery

ISBN: 978-3-527-32355-5

Kahl, G.

The Dictionary of Genomics, Transcriptomics and Proteomics

ISBN: 978-3-527-32073-8

Giordano, A., Macaluso (eds.)

Cancer Epigenetics

Biomolecular Therapeutics in Human Cancer

ISBN: 978-0-471-71096-7

Niculescu, M.D., Haggarty, P. (eds.)

Nutrition in Epigenetics

ISBN: 978-0-8138-1605-0

Epigenetic Regulation and Epigenomics

Advances in Molecular Biology and Medicine

Edited by
Robert A. Meyers

Volume 2

The Editor

Dr. Robert A. Meyers
Editor in Chief
RAMTECH Limited
122, Escalle Lane
Larkspur, CA 94939
USA

Cover

Simplified model of the interplay between histone modifications and small RNAs in the maintenance of pericentric heterochromatin in fission yeast (for more information see Chapter 19 "Histone Modifications", Figure 8)". Designed and drawn by Andrew Bannister and Blerta Xhemalce, The Gurdon Institute, University of Cambridge, CB2 1QN, UK.

Library of Congress Card No.: applied for

British Library Cataloguing-in-Publication Data
A catalogue record for this book is available from the British Library.

Bibliographic information published by the Deutsche Nationalbibliothek
The Deutsche Nationalbibliothek lists this publication in the Deutsche Nationalbibliografie; detailed bibliographic data are available on the Internet at <http://dnb.d-nb.de>.

Wiley-Blackwell is an imprint of John Wiley & Sons, formed by the merger of Wiley's global Scientific, Technical, and Medical business with Blackwell Publishing.

Composition Laserwords Private Limited, Chennai, India

Printing and Binding Strauss GmbH, Mörlenbach

Cover Design Adam Design, Weinheim, Germany

Printed in the Federal Republic of Germany
Printed on acid-free paper

Print ISBN: 978-3-527-32682-2

Contents

Preface and Commentary

Epigenetics is the term given to heritable traits that occur over rounds of cell division and sometimes transgenerationally, in which the mechanisms are reversible, but do not involve changes to the underlying DNA sequence. This involves regulatory systems such as DNA methylation, histone modification, nucleosome location, and noncoding RNA. The *epigenome*, meanwhile, is a parallel to the word genome, refers to the overall epigenetic state of a cell and can be considered essentially a network of chemical switches within our cells.

Our compendium is written for university undergraduates, graduate students, faculty and investigators at research institutes. There are 33 articles with a combined length of over 1100 pages and as such is the largest in depth, up to date treatment of epigenetics presently available.

Epigenetics Regulation and Epigenomics differs in content and quality from all others available in five ways 1) the overall coverage was approved by our Board, which includes 11 Nobel Prize winners; 2) the selection of each article and author was validated by several reviewers from major university research centers; 3) each article was then reviewed by peers from other universities; 4) a glossary of terms with definitions is provided at the beginning of each article and 5) the articles average 35 print pages – which provides several times the depth of other such compendia.

The content is divided into five sections of articles covering key epigenetics areas. These sections are *Analytical Methods, Basic Molecular Mechanisms, The Epigenome, Medical Applications* and *Model Organisms.*

Analytical Methods articles range from chromatin immunoprecipitation (ChIP), to tag sequencing (impacting epigenomics), DNA methylation analysis, high throughput epigenotyping by mass spectrometry and includes RNA methodologies which provide an understanding of aspects of gene regulation. The *Basic Mechanisms* section covers the cell nucleus and chromatin organization and dynamics; epigenetics of stem cells; imprinting and histone modifications and methylation; as well as epigenetic aspects of prions, twins, cloning and RNA interference and all types of regulation of gene expression. *The Epigenome* coverage includes computational epigenetics and the human epigenome. *Medical Applications* include a comprehensive article on epigenetic medicine and additional detail in several articles on the of epigenetics of cancer, the immune system and aging as well as pharmaco-epigenomics to improve cancer therapies. In fact, drugs that inhibit the DNA methyltransferases, which place methyl groups on the

DNA, are now approved for clinical use in the United States for the treatment of certain cancers. This may be the beginning of a new era of cancer treatment involving epigenetic therapy. Pharmacology and emerging clinical application of RNA is also presented in this section. *Model Organisms* range from bacteria to protozoans as well as fungi and plants.

Our team of authors and peer reviewers are located at top rated epigenetics departments at institutions including the University of Cambridge, the University of Southern California, the University of California at Los Angeles, Washington University, St. Louis, and the National Institutes of Health. The team is truly global with authors or coauthors from the U.S., Sweden, Belgium, Germany, France, the UK, Austria, Spain, Hungary, Japan, India, China, Singapore, Canada and Israel.

Our team hopes that you, the reader, will benefit from our hard work – finding the content useful in your research as well as educational. We wish to thank our Managing Editor, Sarah Mellor, as well as our Executive Editor, Gregor Cicchetti for both their advice and hard work in the course of this project.

Larkspur, California, March 2012

Robert A. Meyers
RAMTECH Limited

List of Contributors

Julia Arand
University of Saarland
Institute for Genetics/Epigenetics
Uni Campus Bld. A2.4
66123 Saarbrücken
Germany

Andrew J. Bannister
University of Cambridge
Wellcome Trust/Cancer Research UK
Gurdon Institute
Tennis Court Road
Cambridge, CB2 1QN
UK

Sailen Barik
Cleveland State University
Center for Gene Regulation in Health and Disease and Department of Biological, Geological and Environmental Sciences
College of Sciences and Health Professions
2121 Euclid Avenue
Cleveland, OH 44115
USA

Therese M. Becker
University of Sydney
Westmead Institute for Cancer Research at Westmead Millennium Institute
Westmead Hospital
Westmead
Sydney, New South Wales
Australia

Sven Beichmanis
University of Heidelberg
Kirchhoff-Institute for Physics and BioQuant Center
69120 Heidelberg
Germany

Frédéric Berger
Temasek Life Sciences Laboratory (TLL)
1 Research Link
Singapore 117604
Singapore

Yehudit Bergman
The Hebrew University Medical School
Institute for Medical Research
Israel-Canada
Department of Developmental Biology and Cancer Research
Jerusalem
Israel

Autumn Bernal
Duke University
Radiation Oncology
139 Environmental Safety DUMC
Durham, NC 27710
USA

Vira Bitko
NanoBio Corporation
2311 Green Rd
Ste A
Ann Arbor, MI 48105
USA

Erin L. Bredeweg
Oregon State University
Department of Biochemistry and Biophysics

and

Center for Genome Research and Biocomputing (CGRB)
Corvallis, OR 97331-7305
USA

Romulo Martin Brena
University of Southern California
USC Epigenome Center
Harlyne Norris Medical Research Tower
G511, 1450 Biggy Street
Los Angeles, CA 90033
USA

Michael J. Buck
State University of New York at Buffalo
Department of Biochemistry

and

The Center of Excellence in Bioinformatics and Life Sciences
701 Ellicott Street
Buffalo, NY 14203
USA

Josep Casadesús
Universidad de Sevilla
Departamento de Genética
Facultad de Biología
Apartado 1095
41080 Seville
Spain

Douglas L. Chalker
Washington University in St Louis
Biology Department
1 Brookings Drive
St Louis, MO 63130
USA

Lingyi Chen
Nankai University
The Ministry of Education
Key Laboratory of Bioactive Materials
Laboratory of Stem Cells and Developmental Biology
College of Life Sciences
94 Weijin Road
Tianjin 300071
China

Pao-Yang Chen
Department of Molecular Cell and Developmental Biology
University of California
610 Charles Young Drive East
Los Angeles, CA 90095
USA

Bart Claes
VIB
Vesalius Research Center
Herestraat 49
Box 912
3000 Leuven
Belgium

and

KU Leuven Campus Gasthuisberg
Vesalius Research Center
Herestraat 49
Box 912
3000 Leuven
Belgium

Nicole Cloonan
The University of Queensland
Queensland Centre for Medical Genomics
Institute for Molecular Bioscience
306 Carmody Road
St Lucia, Queensland 4072
Australia

Lanelle R. Connolly
Oregon State University
Department of Biochemistry and Biophysics

and

Center for Genome Research and Biocomputing (CGRB)
Corvallis, OR 97331-7305
USA

Ignacio Cota
Universidad de Sevilla
Departamento de Genética
Facultad de Biología
Apartado 1095
41080 Seville
Spain

Marion Cremer
Ludwig Maximilians University
Biocenter
Department of Biology II
82152 Martinsried
Germany

Christoph Cremer
University of Heidelberg
Kirchhoff-Institute for Physics and BioQuant Center
69120 Heidelberg
Germany

Thomas Cremer
Ludwig Maximilians University
Biocenter
Department of Biology II
82152 Martinsried
Germany

Mark A. Dawson
University of Cambridge
Wellcome Trust/Cancer Research UK
Gurdon Institute
Tennis Court Road
Cambridge, CB2 1QN
UK

and

University of Cambridge
Cambridge Institute for Medical Research
Department of Haematology
Cambridge, CB2 0XY
UK

Melvin L. DePamphilis
National Institute of Child Health and Human Development
National Institutes of Health
Building 6A
Room 3A15
9000 Rockville Pike
Bethesda, MD 20892-2753
USA

Andras Dinnyes
Szent Istvan University
Molecular Animal Biotechnology Laboratory
Hungary

and

BioTalentum Ltd
2100 Gödöllö
Hungary

Jean-Sébastien Doucet
University of Toronto
Centre for Addiction and Mental Health
Department of Pharmacology
Toronto, Ontario
Canada

Karl Ekwall
Karolinska Institutet
Department of Biosciences and Nutrition
Center for Biosciences
Novum
Huddinge
141 57 Stockholm
Sweden

Robert E. Farrell Jr.
Penn State University
Department of Biology
1031 Edgecomb Avenue
NY, PA 17403
USA

Daniel M. Fass
Broad Institute of MIT and Harvard
Stanley Center for Psychiatric Research
7 Cambridge Center
Cambridge, MA 02142
USA

Robert Feil
Centre National de la Recherche Scientific and University of Montpellier
CNRS, UMR 5535
Institute of Molecular Genetics (IGMM)
1919 route de Mende
34293 Montpellier
France

Agustin F. Fernandez
Universidad de Oviedo
Cancer Epigenetics Laboratory
Instituto Universitario de Oncología del Principado de Asturias (IUOPA)
Hospital Universitario Central de Asturias (HUCA)
Bloque Polivalente A
33006 Oviedo
Spain

Mario F. Fraga
National Center for Biotechnology (CNB) and Spanish National Research Council (CSIC)
Department of Immunology and Oncology
Cantoblanco
28049 Madrid
Spain

Michael Freitag
Oregon State University
Department of Biochemistry and Biophysics

and

Center for Genome Research and Biocomputing (CGRB)
Corvallis, OR 97331-7305
USA

Neha Garg
Devi Ahilya University
School of Biotechnology
Khandwa Road
Indore 452001
India

and

Barkatullah University
Biotechnology Department
Bhopal 462026
India

Sarika Garg
Devi Ahilya University
School of Biotechnology
Khandwa Road
Indore 452001
India

and

Max Planck Unit for Structural Molecular Biology
C/O DESY
Gebäude 25b
Notkestrasse 85
22607 Hamburg
Germany

and

Present address: University of Saskatchewan
Department of Psychiatry
Rm B45 HSB
107 Wiggins Road
Saskatoon, SK S7N 5E5
Canada

Sean M. Grimmond
The University of Queensland
Queensland Centre for Medical Genomics
Institute for Molecular Bioscience
306 Carmody Road
St Lucia
Queensland 4072
Australia

Ivo G. Gut
Centro Nacional de Analisis Genomico
C/Baldiri Reixac 4
08028 Barcelona
Spain

Thomas Haaf
Julius-Maximilians-University Würzburg
Institute of Human Genetics
Biozentrum
Am Hubland
97074 Würzburg
Germany

Martin Heß
Ludwig Maximilians University
Biocenter
Department of Biology II
82152 Martinsried
Germany

Ryutaro Hirasawa
Centre National de la Recherche Scientific and University of Montpellier
CNRS, UMR 5535
Institute of Molecular Genetics (IGMM)
1919 route de Mende
34293 Montpellier
France

Edward B. Holson
Broad Institute of MIT and Harvard
Stanley Center for Psychiatric Research
7 Cambridge Center
Cambridge, MA 02142
USA

Barbara Hübner
Ludwig Maximilians University
Biocenter
Department of Biology II
82152 Martinsried
Germany

Dean A. Jackson
University of Manchester
Faculty of Life Sciences
MIB 131 Princess Street
Manchester M1 7DN
UK

Randy Jirtle
Duke University
Radiation Oncology
139 Environmental Safety DUMC
Durham, NC 27710
USA

Kotaro J. Kaneko
National Institute of Child Health and Human Development
National Institutes of Health
Building 6A
Room 3A15
9000 Rockville Pike
Bethesda, MD 20892-2753
USA

Melissa M. Kemp
Broad Institute of MIT and Harvard
Chemical Biology
7, Cambridge Center
Cambridge, MA 02142
USA

Matthew J. Kohn
University at Albany
Department of Biomedical Sciences
School of Public Health and NYSTEM
New York State Department of Health
Empire State Plaza
Biggs Laboratory
C345 Albany, NY 12201
USA

Satya K. Kota
Centre National de la Recherche Scientific and University of Montpellier
CNRS, UMR 5535
Institute of Molecular Genetics (IGMM)
1919 route de Mende
34293 Montpellier
France

Keerthana Krishnan
The University of Queensland
Queensland Centre for Medical Genomics
Institute for Molecular Bioscience
306 Carmody Road
St Lucia
Queensland 4072
Australia

Anil Kumar
Devi Ahilya University
School of Biotechnology
Khandwa Road
Indore 452001
India

Diether Lambrechts
VIB Vesalius Research Center
Herestraat 49, Box 912
3000 Leuven
Belgium

and

KU Leuven Campus Gasthuisberg
Vesalius Research Center
Herestraat 49, Box 912
3000 Leuven
Belgium

Konstantin Lepikhov
University of Saarland
Institute for Genetics/Epigenetics
Uni Campus Bld. A2.4
66123 Saarbrücken
Germany

Rena Levin-Klein
The Hebrew University Medical School
Institute for Medical Research
Israel-Canada
Department of Developmental Biology and Cancer Research
Jerusalem
Israel

Shen Jean Lim
National University of Singapore
Department of Biochemistry
Yong Loo Lin School of Medicine
8 Medical Drive
Singapore 117597
Singapore

Lin Liu
Nankai University
The Ministry of Education
Key Laboratory of Bioactive Materials
Laboratory of Stem Cells

and

Developmental Biology
College of Life Sciences
94 Weijin Road
Tianjin 300071
China

Alexandra Lusser
Innsbruck Medical University
Division of Molecular Biology
Biocenter
Fritz-Pregl Strasse 3
6020 Innsbruck
Austria

Javier López-Garrido
Universidad de Sevilla
Departamento de Genética
Facultad de Biología
Apartado 1095
41080 Seville
Spain

Yolanda Markaki
Ludwig Maximilians University
Biocenter
Department of Biology II
82152 Martinsried
Germany

Jason A. Motl
Washington University in St Louis
Biology Department
1 Brookings Drive
St Louis, MO 63130
USA

Wesley R. Naeimi
University of Kent
Kent Fungal Group
School of Biosciences
Stacey Building
Canterbury
Kent CT2 7NJ
UK

Björn Oback
AgResearch
Ruakura Research Centre
East Street
Private Bag 3123
Hamilton
New Zealand

Matteo Pellegrini
Department of Molecular Cell and Developmental Biology
University of California
610 Charles Young Drive East
Los Angeles, CA 90095
USA

Pallavi A. Phatale
Oregon State University
Department of Biochemistry and Biophysics

and

Center for Genome Research and Biocomputing (CGRB)
Corvallis, OR 97331-7305
USA

Paolo Piatti
Innsbruck Medical University
Division of Molecular Biology
Biocenter
Fritz-Pregl Strasse 3
6020 Innsbruck
Austria

Kyle R. Pomraning
Oregon State University
Department of Biochemistry and Biophysics

and

Center for Genome Research and Biocomputing (CGRB)
Corvallis, OR 97331-7305
USA

Jason M. Rizzo
State University of New York at Buffalo
Department of Biochemistry

and

The Center of Excellence in Bioinformatics and Life Sciences
701 Ellicott Street
Buffalo, NY 14203
USA

Lothar Schermelleh
Ludwig Maximilians University
Biocenter
Department of Biology II
82152 Martinsried
Germany

Frederick A. Schroeder
Massachusetts General Hospital
Harvard Medical School
185, Cambridge Street
6th Floor
Boston, MA 02114
USA

Annie W. Shieh
Washington University in St Louis
Biology Department
1 Brookings Drive
St Louis, MO 63130
USA

Bernard Siebens
VIB Vesalius Research Center
Herestraat 49
Box 912
3000 Leuven
Belgium

and

KU Leuven Campus Gasthuisberg
Vesalius Research Center
Herestraat 49
Box 912
3000 Leuven
Belgium

Mikiko C. Siomi
Keio University School of Medicine
Department of Molecular Biology
35 Shinanomachi
Shinjuku-ku
Tokyo 160-8582
Japan

David Skaar
Duke University
Radiation Oncology
139 Environmental Safety DUMC
Durham, NC 27710
USA

Kristina M. Smith
Oregon State University
Department of Biochemistry and Biophysics

and

Center for Genome Research and Biocomputing (CGRB)
Corvallis, OR 97331-7305
USA

Jason A. Steen
The University of Queensland
Queensland Centre for Medical Genomics
Institute for Molecular Bioscience
306 Carmody Road
St Lucia
Queensland 4072
Australia

Hilmar Strickfaden
Ludwig Maximilians University
Biocenter
Department of Biology II
82152 Martinsried
Germany

Annelie Strålfors
Karolinska Institutet
Department of Biosciences and Nutrition
Center for Biosciences
Novum
Huddinge
141 57 Stockholm
Sweden

Xiuchun Cindy Tian
University of Connecticut
Department of Animal Science
Center for Regenerative Biology
Storrs, CT 06269
USA

Joo Chuan Tong
National University of Singapore
Department of Biochemistry
Yong Loo Lin School of Medicine
8 Medical Drive
Singapore 117597
Singapore

and

Institute for Infocomm Research
Data Mining Department
1 Fusionopolis Way
No. 21-01
Connexis
South Tower
Singapore 138632
Singapore

Jörg Tost
Centre National de Génotypage
CEA-Institut de Genomique
Laboratory for Epigenetics
Bâtiment G2
2 Rue Gaston Crémieux
91000 Evry
France

and

Fondation Jean Dausset – CEPH
Laboratory for Functional Genomics
27 rue Juliette Dodu 75010
Paris
France

Mick F. Tuite
University of Kent
Kent Fungal Group
School of Biosciences
Stacey Building
Canterbury
Kent CT2 7NJ
UK

Rocio G. Urdinguio
Universidad de Oviedo
Cancer Epigenetics Laboratory
Instituto Universitario de Oncología del Principado de Asturias (IUOPA)
Hospital Universitario Central de Asturias (HUCA)
Bloque Polivalente A
33006 Oviedo
Spain

Florence F. Wagner
Broad Institute of MIT and Harvard
Stanley Center for Psychiatric Research
7 Cambridge Center
Cambridge, MA 02142
USA

Joern Walter
University of Saarland
Institute for Genetics/Epigenetics
Uni Campus Bld. A2.4
66123 Saarbrücken
Germany

Qiu Wang
Duke University
Department of Chemistry
French Family Science Center
Durham, NC 27708-0354
USA

Albert H. C. Wong
University of Toronto
Centre for Addiction and Mental Health
Departments of Psychiatry and Pharmacology
Faculty of Medicine
250 College Street
Toronto
Ontario, M5T 1R8
Canada

David L. A. Wood
The University of Queensland
Queensland Centre for Medical Genomics
Institute for Molecular Bioscience
306 Carmody Road
St Lucia
Queensland 4072
Australia

Mark Wossidlo
University of Saarland
Institute for Genetics/Epigenetics
Uni Campus Bld. A2.4
66123 Saarbrücken
Germany

Blerta Xhemalce
University of Cambridge
Wellcome Trust/Cancer Research UK
Gurdon Institute
Tennis Court Road
Cambridge, CB2 1QN
UK

Anette Zeilner
Innsbruck Medical University
Division of Molecular Biology
Biocenter
Fritz-Pregl Strasse 3
6020 Innsbruck
Austria

Andreas Zunhammer
Ludwig Maximilians University
Biocenter
Department of Biology II
82152 Martinsried
Germany

Part III
The Epigenome

Epigenetic Regulation and Epigenomics: Advances in Molecular Biology and Medicine, First Edition. Edited by Robert A. Meyers.

22
Computational Epigenetics

Joo Chuan Tong[1,2] *and Shen Jean Lim*[1]
[1]*National University of Singapore, Department of Biochemistry, Yong Loo Lin School of Medicine, 8 Medical Drive, Singapore 117597, Singapore*
[2]*Institute for Infocomm Research, Data Mining Department, 1 Fusionopolis Way, No. 21-01, Connexis, South Tower, Singapore 138632, Singapore*

Epigenetic Regulation and Epigenomics: Advances in Molecular Biology and Medicine, First Edition. Edited by Robert A. Meyers.

Keywords

Bioinformatics
The application of information technology and computer science to the field of molecular biology.

Database
A collection of data that is organized so that its contents can easily be accessed, managed, and updated.

Epigenetics
The study of changes in phenotype or gene expression caused by mechanisms other than changes in the DNA sequence.

Epigenomics
The omics study of epigenetic elements.

Epigenetic informatics
The application of information technology and computer science to the field of epigenetics.

Epigenetics has recently emerged as a crucial field to study how non-gene factors may function as heritable codes for regulating gene expression. In this chapter, details are provided as to how traditional genomics, in combination with methods in computer science, mathematics, chemistry, biochemistry, and proteomics, has been applied to the large-scale analysis of heritable changes in phenotype, gene function, and gene expression that are not dependent on gene sequence.

1 Introduction

Sequencing of the genomes of human and other model organisms has generated increasingly huge volumes of data that are relevant to an understanding of natural selection, development, and evolution, the causation of disease, and the interplay between genotypes and phenotypes during development. While much progress has been made in genomic research, studies in recent years have shown that gene factors alone could not cover all aspects of heritable changes in phenotype, gene function, or gene expression. It is now known that "epigenetic" or non-gene factors can modify gene activity, either by switching off the genes or by making chromosomes difficult to unwind [1].

Epigenetic control is highly combinatorial. The haploid human genome contains approximately 23 000 genes that may be active in specific cells at specific moments. Cells control gene expression by wrapping DNA around clusters of core histone proteins to form nucleosomes, which are further organized into chromatin [2]. Gene expression patterns are moderated by changes in chromatin structure: genes are expressed when chromatin is open, and inactivated when the chromatin is condensed [3]. These dynamic chromatin states are controlled by DNA methylation, histone modifications, and interactions with nonhistone proteins. In particular, histones are subject to a large variety of post-translational modifications, such as methylation, acetylation, phosphorylation, sumoylation, and ubiquitylation [4, 5], and are functionally associated with a wide variety of processes that are continuously occurring within the cell [6]. Changes in these normal regulatory mechanisms can disrupt gene expression patterns, resulting in adverse clinical outcomes [7].

The idea of a "histone code" [8] or "epigenetic code" [9] has been proposed to describe the combinatorial nature of histone modifications that substantially extend the information potential of the genetic code. While this may serve as a valuable hypothesis for studying the heritable effects of histone modifications, many questions remain unanswered. For example, if such a "code" does in fact exist, how many distinct combinations are required for the normal functioning of a cell? Does a particular combination of DNA/histone modification always effect the same function? At present, the nature of such activities remains poorly understood. With the increasing number of new modification sites being reported each year, however, it has been postulated that "... nearly every histone residue that is accessible to solvent may be a target for post-translational modification" [2] and "... provide, in combination, an almost infinite source of variability that can be used for signal transduction" [6].

Such forms of complexity require extraordinary efforts for systematic analysis. Several large-scale mapping initiatives are now in place, such as those created by the Alliance for the Human Epigenome and Disease (AHEAD) Task Force [10], the ENCyclopedia of DNA Elements (ENCODE) Project Consortium [11], the Human Epigenome Project (HEP) Consortium [12], and the High-throughput Epigenetic Regulatory Organisation in Chromatin (HEROIC) Project Consortium (http://www.heroic-ip.eu). The huge quantity of experimental data generated by these and other projects requires appropriate bioinformatics infrastructure spanning general and specialist databases, basic bioinformatics tools, and sophisticated algorithms for management and detailed analysis. In this chapter, a survey is provided of the major tools and resources that have been developed in this rapidly growing field.

2 Data Sources

Molecular database efforts have kept pace with the rapid rate at which epigenetic data and related information are being generated. These repositories are valuable sources of information to support basic and applied research, by virtue of curation, annotation, new data linkages, cross-referencing, and other novel approaches. Some of these resources are described in the following subsections.

2.1 DNA Methylation Databases

2.1.1 Meth DB (http://www.methdb.net) [13]

This searchable database, hosted at Institut de Génétique Humaine, France, contains experimentally validated information on 20 236 methylation content data and 6312 methylation patterns for more than 48 species, 1511 individuals, 198 tissues and cell lines, and 79 phenotypes (last updated in September 2009). Each entry contains the following fields: species name, sex, tissue, phenotype, experimental technique, DNA methylation type, gene/locus, analyzed sequence/array, expression level, 5mC content environment, experimental diagrams, and literature reference. MethDB also provides links to the National Center for Biotechnology Information (NCBI) Taxonomy browser [14], PubMed, GenBank, and Online Mendelian Inheritance in Man (OMIM) [15], wherever possible. The information in MethDB has been updated regularly. An online data submission system is available for investigators to upload their experimental data to share with the research community.

2.1.2 MethPrimerDB (http://medgen.ugent.be/methprimerdb/) [16]

This public repository, compiled and hosted at Ghent University of Belgium, was developed for the storage and retrieval of validated PCR-based methylation assays. It currently (July 2010) contains 259 primer sets derived from human, mouse, and rat. Database records can be searched by gene symbol, nucleotide sequence, analytical method used, Entrez Gene [14] or methPrimerDB identifier, and submitter's name. Each entry contains the source organism name, official gene symbols, alias gene symbols, analysis methods, primer sequences, submitter information, and links to Entrez Gene, MethDB [13], and RTPrimerDB [17] and PubMed reference abstracts, whichever is available. In addition, the database is integrated with a sequence similarity search tool termed *methBLAST* [16], that is used to evaluate oligonucleotide sequence similarities by querying against *in silico* bisulfate-modified genome sequences.

2.1.3 MethyLogiX (www.methylogix.com/genetics/database.shtml.htm) [18]

This database, from Sequenom GmbH, Germany, specializes in DNA methylation data in late-onset Alzheimer's disease (LOAD). The information is derived from the matrix-assisted laser desorption/ionization time-of-flight (MALDI-TOF) mass spectrometry analyses of post-mortem brain samples and lymphocytes across 12 potential Alzheimer's susceptibility loci. DNA methylation data specific to each gene, including sample ID, patient age, patient gender, diagnosis (disease or control), tissue, and expression levels, are presented. The contents of MethyLogiX is a supplement to Schumacher and coworker's literature report on age-specific epigenetic drift in LOAD [18].

2.1.4 The Krembil Family Epigenetics Laboratory Databases (http://www.epigenomics.ca)

In 2002, the Krembil Foundation sponsored the establishment of a laboratory that works in collaboration with the Foundation of the Centre for Addiction and Mental Health (CAMH) to study if, and how, epigenetic factors can predispose and/or cause severe human diseases. This study has led to the development of a website that records information on DNA methylation data of human chromosomes 21 and 22, germline DNA methylation variations, and DNA methylation profiles in monozygotic and dizygotic twins. The DNA methylation maps are derived from the high-throughput profiling of human chromosomes 21 and 22 in eight individuals, using tiling microarrays consisting of over 340 000 oligonucleotide probe pairs [19]. DNA methylation variation information within the germlines of normal males is derived using a 12 198-feature CpG island microarray [20]. Both datasets can be browsed using the University of California, Santa Cruz (UCSC) Genome Browser [21]. DNA methylation profiles in monozygotic and dizygotic twins are presented in a table with the probe ID, position, link to UCSC Genome Browser, [21] and the intra-class correlation coefficient (ICC) values in blood, buccal, and gut.

2.2 Cancer Methylation Databases

2.2.1 PubMeth (www.pubmeth.org) [22]

This database was developed at Ghent University, Belgium. The contents are derived from text-mining of Medline/PubMed abstracts, in combination with manual reviewing and annotation of preselected abstracts. PubMeth provides access to more than 5000 records of genes that are reportedly methylated in various cancer types, collected from over 1000 literature sources. Each record contains information about the source of publication, the gene, as well as the cancer type and subtypes if specified. The number of primary cancer samples where methylation is analyzed, as well as the number of analyzed cell lines and the number of normal tissues, are also included. Other information includes the type of detection technologies, evidence sentence, as well as links to GeneCards [23] and Database of Transcriptional Start Sites (DBTSS) [24] whichever is available. Online data submission is also provided.

2.3 Histone Databases

2.3.1 The Histone Database (http://genome.nhgri.nih.gov/histones/) [25]

The Histone or Histone Sequence Database was developed by the National Human Genome Research Institute, National Institutes of Health, USA. This is a searchable collection of histones and histone fold-containing proteins derived from Swiss-Prot, PIR, PDB, GenBank, EMBL Nucleotide Sequence Database, DDBJ, and the Protein Research Foundation (PRF) (http://www.prf.or.jp). The database contains annotated alignments of full-length nonredundant sequence sets, sequence sets in redundant, and nonredundant FASTA format, as well as partial sequence information. It also provides summaries of the latest data on solved histone fold structures, post-translational modifications of histones, and the human histone gene complement. Links to PDB, the Molecular Modelling Database (MMDB) [26] and NCBI's Entrez molecular structure viewer Cn3D [27] are also provided, whichever is available. The last update in March 2007 saw a collection of 254 redundant sequences from histone H1, 383 from histone H2, 311 from histone H2B, 1043 from histone H3, and 198 from histone H4, derived from more than 857 species that were available as of October 1999.

2.4
Chromatin Databases

2.4.1 ChromDB (http://www.chromdb.org) [28]

The ChromDB database, compiled and hosted at the University of Arizona, USA is a repository for chromatin-related proteins, including RNAi-associated proteins for a broad range of organisms. Three types of information are included in the database: genomic-based; transcript-based; and NCBI Reference Sequence (RefSeq) [29]-based data. Genome-based sequences are limited to plant, algal, and diatom genomes. Information on animal and fungal model organisms are available as transcript-based sequences, and are derived from NCBI RefSeq collection. The database also contains integrated tools such as Basic Local Alignment Search Tool (BLAST) [30], Exon Viewer, Pfam [31] Domain Viewer, and SMART [32] Domain Viewer to facilitate sequence analysis. Each entry records the following information: formal name; ChromDB ID; taxonomy name (linked to NCBI's taxonomy browser) and lineage; protein group; ChromDB model type; sequence; transcript view diagram; splice model status; and organism-specific links to external websites where available.

2.4.2 CREMOFAC (http://www.jncasr.ac.in/cremofac/) [33]

This searchable database, hosted at the Jawaharlal Nehru Centre for Advanced Scientific Research, India, is dedicated for ATP and non-ATP-dependent chromatin-remodeling factors. The database currently (July 2010) stores 64 types of remodeling factors from 49 different organisms reported in literature, with 1725 redundant remodeling factor sequences and 720 nonredundant sequences. Information found in the database includes gene, protein, promoter, and isoform protein sequences, and protein domain images. It also provides detailed information on the chromatin-remodeling factors found in human, mouse, and rat, extracted from NCBI, Ensembl [34], Mouse [35], and Rat [36] Genome Database. These data are categorized into classes or families based on functionality, and include ISWI, Swi/Snf2, CHD or Mi-2, bromodomain chromatin modifiers, and other ARID chromatin-modifying proteins. In addition, phylogeny trees, chromatin-remodeling pathway diagrams extracted from Biocarta database (http://www.biocarta .com/) and links to PubMed are also provided, whichever is available.

2.5
Gene Expression Databases

2.5.1 Gene Expression Omnibus (GEO) (www.ncbi.nlm.nih.gov/geo/) [37]

The NCBI (GEO) is a public repository for gene expression data derived from single- and dual-channel microarray-based experiments that measures mRNA, miRNA, genomic DNA, and protein abundance, as well as non-array techniques such as serial analysis of gene expression (SAGE) and mass spectrometry peptide profiling, among others. GEO records are linked to other NCBI resources such as PubMed, GenBank, UniGene, MapViewer, and OMIM, wherever possible. Precomputed interactive hierarchical cluster heat map images are available on each record, to facilitate analysis of coordinated regulated genes within the database. The database holds information from over 10 000 experiments comprising 300 000 samples, 16 billion individual abundance measurements, for over 500 organisms, submitted by 5000 laboratories from around the world.

2.5.2 Gene Expression Nervous System ATlas (GENSAT) (http://www.gensat .org/) [38]

The NCBI GENSAT at Rockefeller University, USA is a searchable collection of pictorial gene expression maps of the brain and spinal cord of the mouse. The database currently (July 2010) contains limited gene expression data for epigenetic factors: one for demethylases; eight for methyltransferases; four for deacetylases; and two for phosphorylases.

2.5.3 HugeIndex (http://www.hugeindex.org/) [39]

The Human Gene Expression Index (HugeIndex) records gene expression data on normal human tissues from high-density oligonucleotide arrays. It contains the results of 59 gene expression experiments on 19 human tissues. Interactive scatter plots are provided to allow a user to compare results between tissues and individual experiments.

2.5.4 COXPRESdb (http://coxpresdb.hgc.jp) [40]

The Co-Expressed Gene Database (COXPRESdb) provides information on coexpressed gene networks for the estimation of gene functions, gene regulation, and/or protein–protein interactions in human and mouse. The database contains four types of coexpressed gene network: (1) highly coexpressed genes; (2) genes sharing the same GO annotation; (3) genes expressed in the same tissue; and (4) user-defined gene sets.

2.6 Other Data Sources

Many more publicly accessible databases that contain epigenetic-related information exist. Specific examples include the Integrated Resource of Protein Domains and Functional Sites (InterPro) [41], PROSITE [42], Protein Analysis through Evolutionary Relationships (PANTHER) [43], the Restriction Enzyme Database (REBASE) [44], and Biomolecular Interaction Network Database (BIND) [45].

3 Computational Tools

Computational methods of sequence analysis, data mining, molecular interactions, and molecular interaction networks are used routinely to support epigenome mapping initiatives, such as chromatin immunoprecipitation (ChIP)-on-chip [46], ChIP-Seq [47], and bisulfite sequencing [48].

ChIP-on-chip: This is a microarray-based platform that combines ChIP with microarray technology (chip) to investigate DNA–protein interactions on a genome-wide basis [46]. Computational tools for ChIP-on-chip analysis are primarily focused on identifying ChIP enrichment sites, which is useful for inferring sites of direct DNA–protein interaction. Specific examples include chromatin immunoprecipitation on tiled arrays (ChIPOTle) [49], TileMap [50], and Ringo [51].

ChIP-Seq: This is a variant of ChIP-on-chip that uses high-throughput DNA sequencing to detect differences between sample and control DNA [47]. It offers important advantages over ChIP-on-chip, including minimal data processing and allowing analysis to be made directly from sequence read counts [52]. However, the method requires the accurate mapping of short sequence reads to the reference genome. As such, algorithms such as BLASTN [53] and BLAT [54] that can identify regions of similarity between sequences, and those that can detect short-read assembly, such as QPALMA [55] and AMOScmp [56], are useful for this approach.

Bisulfite sequencing: Bisulfite sequencing [46] involves the use of bisulfite treatment of DNA to determine its cytosine methylation patterns. Computational tools that focus on bisulfite sequencing are commonly used to quantify cytosine methylation levels [57], to estimate the effectiveness of bisulfite treatment [57], and to visualize the results [58]. Collectively, the developed algorithms enable the analysis of DNA methylation patterns of different tissue types, and also the genome-wide comparison of histone modification sites identified by various epigenome mapping initiatives [59].

4
Computational Analysis of DNA Methylation

DNA methylation plays an integral role in the regulation of genomic stability and cellular plasticity. It is essential for normal cell development, and is associated with numerous fundamental processes that include genomic imprinting [60], X-chromosome inactivation [61], the maintenance of repetitive elements [62], and carcinogenesis [63]. The statistical method of support vector machine (SVM) has been applied to predict the methylation landscape of human brain DNA, and to map the entire genomic methylation patterns for all 22 human autosomes [64]. Attempts to predict protein methylation sites have also been reported. These include procedures based on SVM, hidden Markov model (HMM), artificial neural network (ANN), naïve Bayes, logistic regression, K-nearest neighbors, and decision trees [65, 66]. However, the implementation of such systems is difficult due to the lack of experimental data for model construction. As such, available systems are primarily focused on arginine and lysine methylations, as their mechanisms are currently the best understood and the training data most readily available.

5
Computational Analysis of Histone Modifications

Histones are the main protein components of chromatin. They act as spools around which the DNA is wound, and play an important role in DNA packaging, chromosome stabilization, and gene expression. Histone proteins are subject to a wide variety of post-translational modifications, including methylation, acetylation, phosphorylation, sumoylation, and ubiquitylation [4, 5]. Covalent modifications of the histone proteins may affect chromosome function via two distinct mechanisms [67]. First, they may alter its electrostatic properties, resulting in a change in the histone structure or its DNA-binding activity. Second, they may generate binding surfaces for protein recognition modules, and help engage specific functional complexes to their relevant sites of action. Computational methods that analyzes chromatin structures, are therefore, particularly useful for identifying activating and repressive histone modification events.

The development of machine-learning algorithms for locating histone-occupied as well as acetylation, methylation, and phosphorylation positions in DNA sequences have been widely reported [59, 68]. An example is the use of HMMs to infer the states of histone modification changes at each genomic position, based on ChIP fragment counts [69]. The use of wavelet analysis, combined with HMMs, for discovering activating and repressive histone modifications using ChIP-on-chip datasets has also been reported [70]. These algorithms allow the screening of histone marks in large sets of protein sequences, such as those encoded by the complete genomes of higher complexity organisms. To understand the interplay between various histone modifications, including methylation and acetylation, Schübeler and colleagues [71] performed a genome-wide chromatin structure analysis in the *Drosophila* genome. These studies revealed the existence of a binary pattern of histone modifications among euchromatic genes, with active genes being hyperacetylated at H3/4 and hypermethylated at H3, and inactive genes being hypomethylated and deacetylated at the same locations. Roh and coworkers [72] reported the genome-wide mapping of diacetylation of histone H3 at Lys9 and Lys14 in resting and activated human T cells. Roh's group showed that this form of chromatin modification is correlated with active gene promoters and with regulatory elements associated with gene expression. In a follow-up study, they extended their investigations to the genome-wide screening of conserved and nonconserved enhancers by histone acetylation patterns [73].

6
Computational Analysis of Cancer Epigenetics

Cancer is a class of disease characterized by the breakdown of DNA methylation and histone-modification patterns, the aberrant expression of miRNAs, and the aberrant dysregulation of various epigenetic machinery proteins [74]. Currently, several initiatives are under way to identify novel methylation patterns that correlate with the progression to malignancy. One such initiative is the EU-funded CancerDip Consortium, which focuses on identifying methylation and epigenetic patterns in different tumor types [75]. A genome-wide analysis of Methyl-DNA immunoprecipitation (MeDIP) assay data from colon (Caco-2) and prostate cancer (PC3), as well as several tumor cell lines, has shown that tumor-specific methylated genes can be classified into distinct functional categories, that they possess common sequence motifs in their promoters, and that they occur in clusters on chromosomes [76].

Abnormal DNA methylation within the CpG islands represents one of the most frequent forms of alterations in cancers.

Various studies have shown that entire CpG islands may become aberrantly methylated in cancer, and that this is mechanistically linked to histone methylation [77, 78]. An analysis of interindividual stability and variations of DNA methylation profiles among healthy individuals using linear regression models and the EpiGRAPH web service (http://epigraph.mpi-inf.mpg.de/WebGRAPH/) have shown that CpG islands may act collectively as emergent and bistable epigenetic switches for maintaining a CpG-island-wide "on" or "off" state [79]. Data from CpG islands have also been used to build computational systems for tumor class prediction. For example, Olek and colleagues [80] constructed SVM models to recognize the difference between T- and B-cell leukemias and CD19+ B cells and CD4+ T cells obtained from healthy donors, using a set of selected CpG sites.

Genomic imprinting is a genetic phenomenon that results in preferential gene expression in a parent-of-origin specific manner. A loss of genomic imprinting may cause human cancers, although the exact mechanism by which imprinting operates remains unknown [81]. Yang and Lee [82] developed a workflow to measure allele-specific gene expression quantitatively, so as to facilitate the identification of sequence motifs that are associated with imprinted genes. These authors reported an accuracy of 98% (sensitivity = 92%, specificity = 99%) for their developed system, which was trained using a dataset of 24 imprinted genes and 128 nonimprinted genes. Other computational models have also been reported for analyzing epigenetic marks in other cancer subtypes. These include the use of SVMs to differentiate acute lymphoblastic leukemia from acute myeloid leukemia [83], and the use of Manhattan distance and average linkage algorithms to analyze human colorectal tumors [84].

7 Computational Analysis of Stem Cell Epigenetics

Stem cells are unspecialized cells with the ability to renew themselves through mitotic cell division, or to undergo differentiation into more specialized cell types [85]. In the mammalian system, two classes of stem cell are available: (1) embryonic stem (ES) cells, which can differentiate into all cell types except the extraembryonic tissue; and (2) adult stem cells, which are responsible for replenishing specialized cells and regenerating damaged tissues. Various studies have shown that DNA methyltransferases [86] and Polycomb group response elements (PREs)/Trithorax group response elements (TREs) [87] possess epigenetic signatures that are important for the differentiation of both human ES cells and germline stem cells. Recently acquired data have indicated that cancer has a common basis that is grounded in a polyclonal epigenetic disruption of stem/progenitor cells [88]. By unraveling the nature of epigenetic alternations, it is hoped that this will lead to improved culture and differentiation technologies, as well as new therapeutic agents that can be used to directly manipulate stem cells in patients.

The computational analysis of epigenetic marks in stem cells is at its early stages. A genome-wide prediction of transcription factor-binding sites in mouse ES cells has been reported [89], to capture the characteristic patterns of transcription factor- binding motif occurrences and the histone profiles associated with regulatory elements, such as promoters and enhancers. Analyses of upregulated and downregulated gene clusters could improve the present understanding of exogenous control on ES cell state in human. Stanford and colleagues [90] have analyzed temporal expression microarray data obtained from ES cells after the initiation of commitment, and integrated these data with known genome-wide transcription factor binding. These studies demonstrated a repressive model of ES cell maintenance, and helped to define the regulatory balance that is needed to maintain the ES cell state. Ringrose and coworkers [91] performed an analysis of PRE/TREs in the *Drosophila melanogaster* genome, and defined the sequence criteria that distinguish PRE/TREs from non-PRE/TREs. By using a series of weighted motifs, these authors were able to identify 167 candidate PRE/TRE sequences, which map to genes involved in development and cell proliferation. Position-specific matrices for predicting cis-regulatory elements have also been developed, and used to study PRE/TREs in *D. melanogaster* [92].

8 Conclusion

In this chapter, a survey has been conducted of how computational methods enable the high-throughput analysis of epigenetic and related information. Propelled by an increasingly powerful technology, bioinformatics are today an essential component for modern epigenetic research. The main challenges in the field include the way in which experimental data are being processed and harmonized across various experiments from different research groups, and how the different analytical tools, with their varying levels of complexity, are being integrated. The integration of different bioinformatics tools and resources, combined with advances in computational infrastructures, might allow a more sophisticated analysis at multiple levels of complexity, from the subcellular molecular level, to the cellular and systems levels, and beyond.

References

1 Tost, J. (2008) *Epigenetics*, Horizon Scientific Press, Norwich.

2 Peterson, C.L., Laniel, M.A. (2004) Histones and histone modifications. *Curr. Biol.*, **14**, R546–R551.

3 Rodenhiser, D., Mann, M. (2006) Epigenetics and human disease: translating basic biology into clinical applications. *Can. Med. Assoc. J.*, **174**, 341–348.

4 Jenuwein, T., Allis, C.D. (2001) Translating the histone code. *Science*, **293**, 1074–1080.

5 Nathan, D., Sterner, D.E., Berger, S.L. (2003) Histone modifications: now summoning sumoylation. *Proc. Natl Acad. Sci. USA*, **100**, 13118–13120.

6 Turner, B.M. (2007) Defining an epigenetic code. *Nat. Cell Biol.*, **9**, 2–6.

7 Feinberg, A.P., Tycko, B. (2004) The history of cancer epigenetics. *Nat. Rev. Cancer*, **4**, 143–153.

8 Margueron, R., Trojer, P., Reinberg, D. (2005) The key to development: interpreting the histone code? *Curr. Opin. Genet. Dev.*, **15**, 163–176.

9 Nightingale, K.P., O'Neill, L.P., Turner, B.M. (2006) Histone modifications: signalling receptors and potential elements of a heritable epigenetic code. *Curr. Opin. Genet. Dev.*, **16**, 125–136.

10 Jones, P.A., Martienssen, R. (2005) A blueprint for a human epigenome project: the AACR human epigenome workshop. *Cancer Res.*, **65**, 11241–11246.

11 ENCODE Project Consortium (2004) The ENCODE (ENCyclopedia of DNA Elements) project. *Science*, **306**, 636–640.

12 Rakyan, V.K., Hildmann, T., Novik, K.L., Lewin, J., Tost, J., Cox, A.V., Andrews, T.D., Howe, K.L., Otto, T., Olek, A., Fischer, J., Gut, I.G., Berlin, K., Beck, S. (2004) DNA methylation profiling of the human major histocompatibility complex: a pilot study for

the human epigenome project. *PLoS Biol.*, **2**, e405.

13 Negre, V., Grunau, C. (2006) The MethDB DAS server: adding an epigenetic information layer to the human genome. *Epigenetics*, **1**, 101–105.

14 Wheeler, D.L., Barrett, T., Benson, D.A., Bryant, S.H., Canese, K., Chetvernin, V., Church, D.M., DiCuccio, M., Edgar, R., Federhen, S., Feolo, M., Geer, L.Y., Helmberg, W., Kapustin, Y., Khovayko, O., Landsman, D., Lipman, D.J., Madden, T.L., Maglott, D.R., Miller, V., Ostell, J., Pruitt, K.D., Schuler, G.D., Shumway, M., Sequeira, E., Sherry, S.T., Sirotkin, K., Souvorov, A., Starchenko, G., Tatusov, R.L., Tatusova, T.A., Wagner, L., Yaschenko, E. (2008) Database resources of the National Center for Biotechnology Information. *Nucleic Acids Res.*, **36**, D13–D21.

15 Hamosh, A., Scott, A.F., Amberger, J., Bocchini, C., Valle, D., McKusick, V.A. (2002) Online Mendelian Inheritance in Man (OMIM), a knowledgebase of human genes and genetic disorders. *Nucleic Acids Res.*, **30**, 52–55.

16 Pattyn, F., Hoebeeck, J., Robbrecht, P., Michels, E., De Paepe, A., Bottu, G., Coornaert, D., Herzog, R., Speleman, F., Vandesompele, J. (2006) methBLAST and methPrimerDB: web-tools for PCR based methylation analysis. *BMC Bioinform.*, **7**, 496.

17 Lefever, S., Vandesompele, J., Speleman, F., Pattyn, F. (2009) RTPrimerDB: the portal for real-time PCR primers and probes. *Nucleic Acids Res.*, **37**, D942–D945.

18 Wang, S.C., Oelze, B., Schumacher, A. (2008) Age-specific epigenetic drift in late-onset Alzheimer's disease. *PLoS ONE*, **3**, e2698.

19 Schumacher, A., Kapranov, P., Kaminsky, Z., Flanagan, J., Assadzadeh, A., Yau, P., Virtanen, C., Winegarden, N., Cheng, J., Gingeras, T., Petronis, A. (2006) Microarray-based DNA methylation profiling: technology and applications. *Nucleic Acids Res.*, **34**, 528–542.

20 Flanagan, J.M., Popendikyte, V., Pozdniakovaite, N., Sobolev, M., Assadzadeh, A., Schumacher, A., Zangeneh, M., Lau, L., Virtanen, C., Wang, S.C., Petronis, A. (2006) Intra- and interindividual epigenetic variation in human germ cells. *Am. J. Hum. Genet.*, **79**, 67–84.

21 Kuhn, R.M., Karolchik, D., Zweig, A.S., Fujita, P.A., Diekhans, M., Smith, K.E., Rosenbloom, K.R., Raney, B.J., Pohl, A., Pheasant, M., Meyer, L.R., Learned, K., Hsu, F., Hillman-Jackson, J., Harte, R.A., Giardine, B., Dreszer, T.R., Clawson, H., Barber, G.P., Haussler, D., Kent, W.J. (2008) The UCSC Genome Browser Database: update 2010. *Nucleic Acids Res.*, **38**, D613–D619.

22 Ongenaert, M., Van Neste, L., De Meyer, T., Menschaert, G., Bekaert, S., Criekinge, W.V. (2008) PubMeth: a cancer methylation database combining text-mining and expert annotation. *Nucleic Acids Res.*, **36**, D842–D846.

23 Safran, M., Solomon, I., Shmueli, O., Lapidot, M., Shen-Orr, S., Adato, A., Ben-Dor, U., Esterman, N., Rosen, N., Peter, I., Olender, T., Chalifa-Caspi, V., Lancet, D. (2002) GeneCards 2002: towards a complete, object-oriented, human gene compendium. *Bioinformatics*, **18**, 1542–1543.

24 Wakaguri, H., Yamashita, R., Suzuki, Y., Sugano, S., Nakai, K. (2008) DBTSS: database of transcription start sites, progress report 2008. *Nucleic Acids Res.*, **36**, D97–D101.

25 Sullivan, S., Sink, D.W., Trout, K.L., Makalowska, I., Taylor, P.M., Baxevanis, A.D., Landsman, D. (2002) The Histone database. *Nucleic Acids Res.*, **30**, 341–342.

26 Wang, Y., Addess, K.J., Chen, J., Geer, L.Y., He, J., He, S., Lu, S., Madej, T., Marchler-Bauer, A., Thiessen, P.A., Zhang, N., Bryant, S.H. (2007) MMDB: annotating protein sequences with Entrez's 3D-structure database. *Nucleic Acids Res.*, **35**, D298–D300.

27 Wang, Y., Geer, L.Y., Chappey, C., Kans, J.A., Bryant, S.H. (2000) Cn3D: sequence and structure views for Entrez. *Trends Biochem. Sci.*, **25**, 300–302.

28 Gendler, K., Paulsen, T., Napoli, C. (2008) ChromDB: the chromatin database. *Nucleic Acids Res.*, **36**, D298–D302.

29 Pruitt, K.D., Tatusova, T., Maglott, D.R. (2007) NCBI reference sequences (RefSeq): a curated non-redundant sequence database of genomes, transcripts and proteins. *Nucleic Acids Res.*, **35**, D61–D65.

30 Altschul, S.F., Madden, T.L., Schaffer, A.A., Zhang, Z., Miller, W., Lipman, D.J. (1997) Gapped BLAST and PSI-BLAST: a new generation of protein database search programs. *Nucleic Acids Res.*, **25**, 3389–3402.

31 Finn, R.D., Tate, J., Mistry, J., Coggill, P.C., Sammut, S.J., Hotz, H.R., Ceric, G., Forslund, K., Eddy, S.R., Sonnhammer, E.L.L., Bateman, A. (2008) The Pfam protein families database. *Nucleic Acids Res.*, **36**, D281–D288.

32 Letunic, I., Doerks, T., Bork, P. (2008) SMART 6: recent updates and new developments. *Nucleic Acids Res.*, **37**, D229–D232.

33 Shipra, A., Chetan, K., Rao, M.R. (2006) CREMOFAC – a database of chromatin remodeling factors. *Bioinformatics*, **22**, 2940–2944.

34 Hubbard, T., Barker, D., Birney, E., Cameron, G., Chen, Y., Clark, L., Cox, T., Cuff, J., Curwen, V., Down, T., Durbin, R., Eyras, E., Gilbert, J., Hammond, M., Huminiecki, L., Kasprzyk, A., Lehvaslaiho, H., Lijnzaad, P., Melsopp, C., Mongin, E., Pettett, R., Pocock, M., Potter, S., Rust, A., Schmidt, E., Searle, S., Slater, G., Smith, J., Spooner, W., Stabenau, A., Stalker, J., Stupka, E., Ureta-Vidal, A., Vastrik, I., Clamp, M. (2002) The Ensembl genome database project. *Nucleic Acids Res.*, **30**, 38–41.

35 Bult, C.J., Eppig, J.T., Kadin, J.A., Richardson, J.E., Blake, J.A., and the Mouse Genome Database Group (2008) The Mouse Genome Database (MGD): mouse biology and model systems. *Nucleic Acids Res.*, **36**, D724–D728.

36 Dwinell, M.R., Worthey, E.A., Shimoyama, M., Bakir-Gungor, B., DePons, J., Laulederkind, S., Lowry, T., Nigram, R., Petri, V., Smith, J., Stoddard, A., Twigger, S.N., Jacob, H.J., and the RGD Team (2009) The Rat Genome Database 2009: variation, ontologies and pathways. *Nucleic Acids Res.*, **37**, D744–D749.

37 Barrett, T., Troup, D.B., Wilhite, S.E., Ledoux, P., Rudnev, D., Evangelista, C., Kim, I.F., Soboleva, A., Tomashevsky, M., Marshall, K.A., Phillippy, K.H., Sherman, P.M., Muertter, R.N., Edgar, R. (2008) NCBI GEO: archive for high-throughput functional genomic data. *Nucleic Acids Res.*, **37**, D885–D890.

38 Heintz, N. (2004) Gene expression nervous system atlas (GENSAT). *Nat. Neurosci.*, **7**, 483.

39 Haverty, P.M., Weng, Z., Best, N.L., Auerbach, K.R., Hsiao, L.L., Jensen, R.V., Gullans, S.R. (2002) HugeIndex: a database with visualization tools for high-density oligonucleotide array data from normal human tissues. *Nucleic Acids Res.*, **30**, 214–217.

40 Obayashi, T., Hayashi, S., Shibaoka, M., Saeki, M., Ohta, H., Kinoshita, K. (2008) COXPRESdb: a database of coexpressed gene networks in mammals. *Nucleic Acids Res.*, **36**, D77–D82.

41 Hunter, S., Apweiler, R., Attwood, T.K., Bairoch, A., Bateman, A., Binns, D., Bork, P., Das, U., Daugherty, L., Duguenne, L., Finn, R.D., Gough, J., Haft, D., Hulo, N., Kahn, D., Kelly, E., Laugraud, A., Letunic, I., Lonsdale, D., Lopez, R., Madera, M., Maslen, J., McAnulla, C., McDowall, J., Mistry, J., Mitchell, A., Mulder, N., Natale, D., Orengo, C., Quinn, A.F., Selengut, J.D., Sigrist, C.J., Thimma, M., Thomas, P.D., Valentin, F., Wilson, D., Wu, C.H., Yeats, C. (2009) InterPro: the integrative protein signature database. *Nucleic Acids Res.*, **37**, D211–D215.

42 Hulo, N., Bairoch, A., Bulliard, V., Cerutti, L., De Castro, E., Langendijk-Genevaux, P.S., Pagni, M., Sigrist, C.J. (2006) The PROSITE database. *Nucleic Acids Res.*, **34**, D227–D230.

43 Mi, H., Guo, N., Kejariwal, A., Thomas, P.D. (2007) PANTHER version 6: protein sequence and function evolution data with expanded representation of biological pathways. *Nucleic Acids Res.*, **35**, D247–D252.

44 Roberts, R.J., Vincze, T., Posfai, J., Macelis, D. (2005) REBASE--restriction enzymes and DNA methyltransferases. *Nucleic Acids Res.*, **33**, D230–D232.

45 Gilbert, D. (2005) Biomolecular interaction network database. *Brief. Bioinform.*, **6**, 194–198.

46 Buck, M.J., Lieb, J.D. (2004) ChIP-chip: considerations for the design, analysis, and application of genome-wide chromatin immunoprecipitation experiments. *Genomics*, **83**, 349–360.

47 Mikkelsen, T.S., Ku, M., Jaffe, D.B., Issac, B., Lieberman, E., Giannoukos, G., Alvarez, P., Brockman, W., Kim, T.K., Koche, R.P., Lee, W., Mendenhall, E., O'Donovan, A., Presser, A., Russ, C., Xie, X., Meissner, A., Wernig,

M., Jaenisch, R., Nusbaum, C., Lander, E.S., Berstein, B.E. (2007) Genome-wide maps of chromatin state in pluripotent and lineage-committed cells. *Nature*, **448**, 553–560.

48 Hajkova, P., el-Maarri, O., Engemann, S., Oswald, J., Olek, A., Walter, J. (2002) DNA-methylation analysis by the bisulfite-assisted genomic sequencing method. *Methods Mol. Biol.*, **200**, 143–154.

49 Buck, M.J., Nobel, A.B., Lieb, J.D. (2005) ChIPOTle: a user-friendly tool for the analysis of ChIP-chip data. *Genome Biol.*, **6**, R97.

50 Ji, H., Wong, W.H. (2005) TileMap: create chromosomal map of tiling array hybridizations. *Bioinformatics*, **21**, 3629–3636.

51 Toedling, J., Sklyar, O., Huber, W. (2007) Ringo – an R/Bioconductor package for analyzing ChIP-chip readouts. *BMC Bioinformatics*, **8**, 221.

52 Barski, A., Cuddapah, S., Cui, K., Roh, T.Y., Schones, D.E., Wang, Z., Wei, G., Chepelev, I., Zhao, K. (2007) High-resolution profiling of histone methylations in the human genome. *Cell*, **129**, 823–837.

53 Altschul, S.F., Gish, W., Miller, W., Myers, E.W., Lipman, D.J. (1990) Basic local alignment search tool. *J. Mol. Biol.*, **215**, 403–410.

54 Kent, W.J. (2002) BLAT – the BLAST-like alignment tool. *Genome Res.*, **12**, 656–664.

55 De Bona, F., Ossowski, S., Schneeberger, K., Ratsch, G. (2008) Optimal spliced alignments of short sequence reads. *Bioinformatics*, **24**, i174–i180.

56 Pop, M., Phillippy, A., Delcher, A.L., Salzberg, S.L. (2004) Comparative genome assembly. *Brief. Bioinform.*, **5**, 237–248.

57 Lewin, J., Schmitt, A.O., Adorján, P., Hildmann, T., Piepenbrock, C. (2004) Quantitative DNA methylation analysis based on four-dye trace data from direct sequencing of PCR amplicates. *Bioinformatics*, **20**, 3005–3012.

58 Boyer, L.A., Lee, T.I., Cole, M.F., Johnstone, S.E., Levine, S.S., Zucker, J.P., Guenther, M.G., Kumar, R.M., Murray, H.L., Jenner, R.G., Gifford, D.K., Melton, D.A., Jaenisch, R., Young, R.A. (2005) Core transcriptional regulatory circuitry in human embryonic stem cells. *Cell*, **122**, 947–956.

59 Xu, H., Wei, C.L., Lin, F., Sung, W.K. (2008) An HMM approach to genome-wide identification of differential histone modification sites from ChIP-seq data. *Bioinformatics*, **24**, 2344–2349.

60 Li, E., Beard, C., Jaenisch, R. (1993) Role for DNA methylation in genomic imprinting. *Nature*, **366**, 362–365.

61 Kaslow, D.C., Migeon, B.R. (1987) DNA methylation stabilizes X chromosome inactivation in eutherians but not in marsupials: evidence for multistep maintenance of mammalian X dosage compensation. *Proc. Natl Acad. Sci. USA*, **84**, 6210–6214.

62 Liang, G., Chan, M.F., Tomigahara, Y., Tsai, Y.C., Gonzales, F.A., Li, E., Laird, P.W., Jones, P.A. (2002) Cooperativity between DNA methyltransferases in the maintenance methylation of repetitive elements. *Mol. Cell. Biol.*, **22**, 480–491.

63 Jones, P.A. (2002) DNA methylation and cancer. *Oncogene*, **21**, 5358–5360.

64 Das, R., Dimitrova, N., Xuan, Z., Rollins, R.A., Haghighi, F., Edwards, J.R., Ju, J., Bestor, T.H., Zhang, M.Q. (2006) Computational prediction of methylation status in human genomic sequences. *Proc. Natl Acad. Sci. USA*, **103**, 10713–10716.

65 Bhasin, M., Reinherz, E.L., Reche, P.A. (2005) Prediction of CpG methylation using a support vector machine. *FEBS Lett.*, **579**, 4302–4308.

66 Chen, H., Xue, Y., Huang, N., Yao, X., Sun, Z. (2006) MeMo: a web tool for prediction of protein methylation modifications. *Nucleic Acids Res.*, **34**, W249–W253.

67 Iizuka, M., Smith, M.M. (2003) Functional consequences of histone modifications. *Curr. Opin. Genet. Dev.*, **13**, 154–160.

68 Won, K.J., Chepelev, I., Ren, B., Wang, W. (2008) Prediction of regulatory elements in mammalian genomes using chromatin signatures. *BMC Bioinformatics*, **9**, 547.

69 Miranda-Saavedra, D., Barton, G.J. (2007) Classification and functional annotation of eukaryotic protein kinases. *Proteins*, **68**, 893–914.

70 Thurman, R.E., Day, N., Noble, W.S., Stamatoyannopoulos, J.A. (2007) Identification of higher-order functional domains in the human ENCODE regions. *Genome Res.*, **17**, 917–927.

71 Schubeler, D., MacAlpine, D.M., Scalzo, D., Wirbelauer, C., Kooperberg,

C., van Leeuwen, F., Gottschling, D.E., O'Neill, L.P., Turner, B.M., Delrow, J., Bell, S.P., Groudine, M. (2004) The histone modification pattern of active genes revealed through genome-wide chromatin analysis of a higher eukaryote. *Genes Dev.*, **18**, 1263–1271.

72 Roh, T.Y., Cuddapah, S., Zhao, K. (2005) Active chromatin domains are defined by acetylation islands revealed by genome-wide mapping. *Genes Dev.*, **19**, 542–552.

73 Roh, T.Y., Wei, G., Farrell, C.M., Zhao, K. (2007) Genome-wide prediction of conserved and nonconserved enhancers by histone acetylation patterns. *Genome Res.*, **17**, 74–81.

74 Halusková, J. (2010) Epigenetic studies in human diseases. *Folia Biol.*, **56**, 83–96.

75 Jacinto, F.V., Ballestar, E., Esteller, M. (2008) Methyl-DNA immunoprecipitation (MeDIP): hunting down the DNA methylation. *Biotechniques*, **44**, 35–43.

76 Keshet, I., Schlesinger, Y., Farkash, S., Rand, E., Hecht, M., Segal, E., Pikarski, E., Young, R.A., Niveleau, A., Cedar, H., Simon, I. (2006) Evidence for an instructive mechanism of *de novo* methylation in cancer cells. *Nat. Genet.*, **38**, 149–153.

77 Laird, P.W. (2005) Cancer epigenetics. *Hum. Mol. Genet.*, **14**, R65–R76.

78 Eden, E., Lipson, D., Yogev, S., Yakhini, Z. (2007) Discovering motifs in ranked lists of DNA sequences. *PLoS Comput. Biol.*, **3**, e39.

79 Bock, C., Walter, J., Paulsen, M., Lengauer, T. (2008) Inter-individual variation of DNA methylation and its implications for large-scale epigenome mapping. *Nucleic Acids Res.*, **36**, e55.

80 Adorjan, P., Distler, J., Lipscher, E., Model, F., Muller, J., Pelet, C., Braun, A., Florl, A.R., Gutig, D., Grabs, G., Howe, A., Kursar, M., Lesche, R., Leu, E., Lewin, A., Maler, S., Muller, V., Otto, T., Scholz, C., Schulz, W.A., Selfert, H.H., Schwope, I., Ziebarth, H., Berlin, K., Piepenbrock, C., Olek, A. (2002) Tumour class prediction and discovery by microarray-based DNA methylation analysis. *Nucleic Acids Res.*, **30**, e21.

81 Rainier, S. (1993) Relaxation of imprinted genes in human cancer. *Nature*, **362**, 747–749.

82 Yang, H.H., Lee, M.P. (2004) Application of bioinformatics in cancer epigenetics. *Ann. N. Y. Acad. Sci.*, **1020**, 67–76.

83 Model, F., Adorjan, P., Olek, A., Piepenbrock, C. (2001) Feature selection for DNA methylation based cancer classification. *Bioinformatics*, **17**, S157–S164.

84 Weisenberger, D.J., Siegmund, K.D., Campan, M., Young, J., Long, T.I., Faasse, M.A., Kang, G.H., Widschwendlter, M., Weener, D., Buchanan, D., Koh, H., Simms, L., Barker, M., Leggett, B., Levine, J., Kim, M., French, A.J., Thibodeau, S.N., Jass, J., Haile, R., Laird, P.W. (2006) CpG island methylator phenotype underlies sporadic microsatellite instability and is tightly associated with BRAF mutation in colorectal cancer. *Nat. Genet.*, **38**, 787–793.

85 Stojkovic, M., Lako, M., Strachan, T., Murdoch, A. (2004) Derivation, growth and applications of human embryonic stem cells. *Reproduction*, **128**, 259–267.

86 Bibikova, M., Chudin, E., Wu, B., Zhou, L., Garcia, E.W., Liu, Y., Shin, S., Plaia, T.W., Auerbach, J.M., Arking, D.E., Gonzalez, R., Crook, J., Davidson, B., Schulz, T.C., Robins, A., Khanna, A., Sartipy, P., Hyllner, J., Vanguri, P., Savant-Bhonsale, S., Smith, A.K., Chakravarti, A., Maitra, A., Rao, M., Barker, D.L., Loring, J.F., Fan, J.B. (2006) Human embryonic stem cells have a unique epigenetic signature. *Genome Res.*, **16**, 1075–1083.

87 Boyer, L.A., Plath, K., Zeitlinger, J., Brambrink, T., Medeiros, L.A., Lee, T.I., Levine, S.S., Wernig, M., Tajonar, A., Ray, M.K., Bell, G.W., Otte, A.P., Vidal, M., Gifford, D.K., Young, R.A., Jaenisch, R. (2006) Polycomb complexes repress developmental regulators in murine embryonic stem cells. *Nature*, **441**, 349–353.

88 Feinberg, A.P., Ohlsson, R., Henikoff, S. (2006) The epigenetic progenitor origin of human cancer. *Nat. Rev. Genet.*, **7**, 21–33.

89 Won, K.J., Ren, B., Wang, W. (2010) Genome-wide prediction of transcription factor binding sites using an integrated model. *Genome Biol.*, **11**, R7.

90 Walker, E., Ohishi, M., Davey, R.E., Zhang, W., Cassar, P.A., Tanaka, T.S., Der, S.D., Morris, Q., Hughes, T.R., Zandstra, P.W., Stanford, W.L. (2007) Prediction and testing of novel transcriptional networks

regulating embryonic stem cell self-renewal and commitment. *Cell Stem Cell*, **1**, 71–86.

91 Ringrose, L., Rehmsmeier, M., Dura, J.M., Paro, R. (2003) Genome-wide prediction of Polycomb/Trithorax response elements in *Drosophila melanogaster*. *Dev. Cell*, **5**, 759–771.

92 Fiedler, T., Rehmsmeier, M. (2006) jPREdictor: a versatile tool for the prediction of cis-regulatory elements. *Nucleic Acids Res.*, **34**, W546–W550.

23
The Human Epigenome

Romulo Martin Brena
University of Southern California, USC Epigenome Center,
Harlyne Norris Medical Research Tower, G511, 1450 Biggy Street,
Los Angeles, CA 90033, USA

Epigenetic Regulation and Epigenomics: Advances in Molecular Biology and Medicine, First Edition. Edited by Robert A. Meyers.

Keywords

Major groove of DNA
A 22 Armstrong-wide groove in the DNA double helix where proteins, such as transcription and other regulatory factors, make contact with the DNA.

CpG islands
Sequence regions at least 200 bp in length with a GC content greater than 50% and an observed/expected CpG ratio greater than 60%. Most CpG islands are generally methylation-free in somatic tissues, with the exception of the inactive X chromosome. Over 50% of human genes are associated with a CpG island.

Monozygotic twins
Twins that develop from a single zygote that splits to form two embryos (also known as "identical" twins).

Lipotropes
Nutrients that regenerate or supply methyl groups, including include folate, choline methionine, and vitamin B_{12}.

Imprinted locus
A locus with monoallelic expression determined by the parental origin of the allele.

Epigenetic mechanisms are responsible for the transmission of information that is "layered onto" the DNA from one cell division to the next. That is, epigenetic information is not contained in the DNA sequence itself, but it is nonetheless essential for normal development, for maintaining the overall integrity of the genome, and for modulating gene dosage via processes such as imprinting and X-chromosome inactivation in females. Epigenetic modifications are reversible, which makes them an attractive therapeutic target for cancer and other diseases. DNA methylation is affected by nutrition and by environmental stimuli, which lends the epigenome a remarkable level of plasticity. DNA methylation is profoundly disrupted in cancer, and several techniques have been developed to analyze the cancer epigenome both globally and at the single gene level. Importantly, DNA methylation has been shown to serve as a biomarker. A large body of research is currently under way in the hope of identifying sequences that could lead to clinical applications. It should also be noted that DNA methylation inhibitors have been used in the successful treatment of myelodysplastic syndrome in human patients. This opens a promising avenue for the clinical treatment of solid tumors in the future.

1 Introduction

Epigenetics is currently defined as "... information heritable during cell division other than the DNA sequence itself." [1]. Originally, the term was coined by Waddington and used to describe the sequential changes undergone by an organism during development, dictated by a defined genome under the influence of the environment [2]. As opposed to the irreversible nature of genetic modifications, epigenetic events leave the original DNA sequence intact. Epigenetic regulation is the description of these sequential changes and includes all of the transcriptional regulatory processes that are crucial in maintaining cellular differentiation and cell-type identity within a given cell lineage [3].

2 DNA Methylation

Cytosine methylation is the most common base modification in the *eukaryotic genome,* and is defined as the "addition of a methyl group to the carbon 5 position of the cytosine ring to form 5-methyl-cytosine [4]." 5-Methyl-cytosine is primarily found in the context of 5′-CpG-3′ dinucleotides (~3–8% of all cytosines in the genome), and it occurs symmetrically on both strands of DNA [5]. Cytosine methylation, however, has also been described in the context of 5′-CpNpG-3′ and 5′-CpCpWpGpG-3′ sequences [6–9]. The addition of the methyl group is catalyzed by a family of DNA methyltransferases (DNMTs) which employ *S*-adenosyl-methionine (SAM) as the methyl donor [10, 11]. Methyl groups attached to the 5-carbon position of cytosine protrude into the major groove of DNA, where they are accessible to methylation-sensitive transcription factors and methyl-binding proteins [12, 13]. After DNA synthesis, the two newly synthesized strands of DNA undergo DNA methylation via the action of DnmtI [14], which copies the DNA methylation pattern from the parental onto the daughter DNA strands. *De novo* DNA methylation results from the action of a different class of DNA methyltransferases, Dnmt3a and Dnmt3b. These enzymes are abundantly expressed in embryonic stem cells, and are downregulated upon differentiation [15, 16]. During development, Dnmt3a and Dnmt3b exhibit nonoverlapping functions, with Dnmt3b specifically required for the methylation of centromeric satellite repeats [17].

5-Methyl-cytosine has a relatively high propensity to deaminate spontaneously to thymine; thus, CpG dinucleotides are under-represented in the human genome [18]. Importantly, the methylation status and distribution of CpG sites in the human genome is not random; approximately 80% of all CpGs are located in repetitive sequences and centromeric repeat regions of chromosomes, and are heavily methylated [19]. The remaining 20% are found preferentially in 0.5–5.0 kb sequence stretches that occur at average intervals of 100 kb [20]. These sequence stretches – termed "CpG islands" – are generally methylation-free in somatic tissues, with the exception of the X chromosome and, to a large extent, have been maintained throughout evolution. Importantly, 50–60% of human genes are associated with a CpG island [21–23]. The functional importance of CpG islands derives from the fact that changes in their DNA methylation status are generally associated with changes in the expression of their associated gene.

Genes associated with methylated CpG islands tend to be downregulated or silenced [24–27]. Because of its potential to abrogate gene expression, DNA methylation has been proposed as one of the two hits of Knudson's two-hit hypothesis for oncogenic transformation [28].

3 DNA Methylation in Normal Development

DNA methylation is essential for normal development, chromosome stability, maintenance of gene expression, and proper telomere length [17, 29–40]. Genetic knockout of *Dnmt1*, *Dnmt3a*, or *Dnmt3b* in the mouse embryo results in embryonic or perinatal lethality, underscoring the essential role of DNA methylation in normal developmental processes [17, 36]. Although, in both human and mouse, DNA methylation patterns are first established during gametogenesis, the genetic material contributed by each of the gametes undergoes profound changes after fertilization. A recent report has indicated that the paternal genome is actively demethylated in mitotically active zygotes [41]; this active demethylation phase is then followed by a passive and selective loss of DNA methylation that continues until the morula stage [42, 43]. DNA methylation patterns are then re-established after implantation and maintained through somatic cell divisions [44]. Interestingly, amidst the sweeping changes that occur during embryonic development, the methylation status of imprinted genes remains unchanged [43, 45].

Normal DNA methylation patterns may vary among individuals [46, 47], potentially stemming from environmental exposure, stochastic methylation events, or trans-generational inheritance [48–50]. The importance of inter-individual epigenomic variance has been postulated to influence the development of disease and the time of disease onset [51]. An example of this phenomenon is the onset of psychiatric disorders, such as schizophrenia and bipolar disorder in monozygotic twins. In some instances, only one member of the twin pair develops the pathology, while in others the time of disease onset between the twins may differ by years or even decades. Most importantly, however, is the fact that molecular studies have failed to identify a genetic component that may account for this phenotypic discordance [51].

4 Nutrition and DNA Methylation

In several studies, attention has been focused on the connection between nutrition and DNA methylation. Of particular interest is the role played by a number of nutrients directly involved in either supplying or regenerating methyl groups. Since methyl groups are labile, a chronic deficiency in methyl-supplying nutrients can result in a change in the ratio of SAM to *S*-adenosylhomocystein (SAH), concomitant with a reduction in the cellular potential for methylation reactions (including DNA methylation) [52]. Nutrients that either regenerate or supply methyl groups fall into the category of lipotropes, and include folate, choline, methionine, and vitamin B_{12}. Riboflavin and vitamin B_6 might also contribute to the modulation of DNA methylation processes, as both of these nutrients are integral components in 1-carbon metabolism [53]. Studies in which rodents were subjected to diets deficient in different combinations of folate, choline, methionine and vitamin B_{12} were able to show a reduction in the SAM : SAH ratio in those animals. Furthermore, DNA hypomethylation could be detected at the genomic level not only in specific tissues but also at specific loci [54–57]. Taken together, these results suggest that the mechanism regulating the epigenome can be influenced by environmental factors. Moreover, the modulation exerted by environmental factors on the epigenome can potentially contribute and/or trigger the development or onset of disease. In light of this evidence, high-resolution mapping of the methylome – ideally at single CpG dinucleotide resolution – may provide a new avenue for understanding disease susceptibility factors that could be used to detect at-risk individuals.

5 Epigenetic Crosstalk: DNA Methylation and Histone Modifications

DNA methylation is not the only regulatory mechanism that comprises the epigenome; rather, histone modifications have been the subject of intense investigation for many years, and have been defined as "epigenetic modifiers." Eight histone proteins, two each of H2A, H2B, H3, and H4, along with 146 bp of DNA, comprise a single nucleosome. The interaction among neighboring nucleosomes can be altered by the complex combination of covalent post-translational modifications (PTMs) on the histone tails, which may in fact represent a "histone code." Different types of histone modifications include phosphorylation, acetylation, mono-, di-, and tri-methylation, ubiquitination, ADP ribosylation, deimination, proline isomerization, and sumoylation. These modifications may directly alter protein–histone interactions, or indirectly influence protein–histone, protein–DNA and histone–DNA interactions by attracting other proteins that bind specifically to modified histones. The enzymes responsible for these modifications, and for their reversal, have significant specificity for the type of mark, the particular amino acid, and the position of the amino acid in the histone tail. Histone modifications can be very dynamic in nature, changing rapidly in response to stimuli. Today, the mapping individual histone modifications genome-wide is possible with chromatin immunoprecipitation applied to tiling arrays, although the resolution is not yet at the single nucleosome level and depends heavily on the antibody that recognizes the modification of interest [58]. Because of the complexity of histone marks on a given nucleosome, new tools and approaches for testing the functional significance of individual modifications will be particularly useful, such as the synthesis of nucleosomes with pure, single modifications added *in vitro* [59, 60]. The interaction between, and interdependence of, DNA methylation and histone modifications has been the subject of many studies, particularly of cancer [61–65]. Alterations in the pattern and overall amount of each histone modification have also been reported in human cancers and cancer cell lines [66]. For example, H3K27 trimethylation in promoters has been reported in association with gene silencing. This and other silencing marks may co-occur with aberrant DNA methylation and function synergistically in gene silencing. Importantly, H3K27 trimethylation has been observed in the absence of aberrant DNA methylation. Experimental models using cancer cell lines have suggested a relative order of silencing events involving both histones and DNA methylation, but this may be gene- and/or cell-type dependent. More globally, two characteristic changes of histone modifications in cancer are a decrease in acetylation of Lys16 and trimethylation of Lys20 on histone H4, in large part from repetitive portions of the genome and in association with hypomethylation of these DNA sequences [67].

6 Genome-Wide DNA Methylation Analyses

Analyzing the human genome for changes in DNA methylation is a challenging endeavor. A majority of the approximately 29 million CpG dinucleotides in the haploid genome are located in ubiquitous repetitive sequences common to all chromosomes, which hampers determination of the precise genomic location where many DNA methylation changes occur [68, 69]. In addition, gene-associated CpG islands encompass a minor fraction of all CpG sites, and consequently their hypermethylation has only a limited effect on global 5-methylcytosine levels in cancer cell DNA [70]. However, as changes in CpG island methylation can abrogate gene expression [71], the identification of aberrant CpG island methylation often – but not always – leads to the identification of genes for which expression is affected during, or because of, the tumorigenic process.

The first method to emerge as a genome-wide screen for CpG island methylation, restriction landmark genomic scanning (RLGS), was originally described in 1991 [72, 73]. In RLGS, the genomic DNA is digested with rare-cutting methylation-sensitive restriction enzymes such as NotI and AscI. The recognition sequences for these enzymes occur preferentially in CpG islands [74, 75], effectively creating a bias toward the assessment of DNA methylation in gene promoters. Importantly, NotI and AscI recognition sequences rarely occur within the same island, effectively doubling the number of CpG islands interrogated for DNA methylation in any given assay [76]. Following digestion, the DNA is radiolabeled and subjected to two-dimensional (2-D) gel electrophoresis. DNA methylation is detected as the absence of a radiolabeled fragment, which stems from the failure of the enzymes to digest a methylated DNA substrate. The main strengths of RLGS are that PCR and hybridization are not part of the protocol, allowing for the quantitative representation of DNA methylation levels and a notably low false-positive rate relative to most other global methods for detecting DNA methylation. Additionally, a priori knowledge of the sequence is not required [77], making RLGS an excellent discovery tool [27, 78–82]. One disadvantage of RLGS is that is limited to the number of NotI and AscI sites in the human genome that fall within the well-resolved region of the profile, although in practice the combinatorial analysis of both enzymes can assess the methylation status of up to 4100 landmarks.

The success of the Human Genome Project [83] helped to stimulate the development of newer methods for genome analysis, which were then adapted for DNA methylation analyses, ranging from single genes, intermediate range and high-throughput [84, 85] to more complete methylome coverage (array-based methods, next-generation sequencing) [86–93]. Arrays originally designed for the analysis of DNA alterations have been adapted for DNA methylation analysis. A main advantage of array platforms is their potential to increase the number of CpGs analyzed, and the technically advanced state of array analysis in general. Critical parameters for methylation arrays for the analysis of human cancers include effective resolution, methylome coverage (total number of CpGs analyzed), reproducibility, the ability to distinguish copy number, and DNA methylation events and accurate validation through an independent method.

Differential methylation hybridization – the first array method developed to identify novel methylated targets in the cancer genome [87] – has served as a basis for many newer-generation array methods. In this assay, DNA is first digested with MseI, an enzyme that cuts preferentially outside of CpG islands, and then ligated to linker primers. The ligated DNA is subsequently digested with up to two methylation sensitive restriction enzymes, such as BstUI, HhaI, or HpaII. As these enzymes are 4-bp restriction endonucleases, their recognition sequence is ubiquitous in GC-rich regions, such as CpG islands. Following a second round of enzymatic digestion, the DNA is then amplified by polymerase chain reaction (PCR), using the ligated linkers as primer binding sites. The detection of DNA methylation is accomplished by fluorescently labeling the PCR product from a test sample, such as tumor DNA, and co-hybridizing it with the PCR products derived from a control sample, such as normal tissue DNA. Aberrantly methylated fragments are refractory to the methylation-sensitive restriction endonuclease digestion, and this results in the generation of PCR products. On the other hand, an unmethylated fragment would be digested, preventing PCR amplification. Therefore, the comparison of signal intensities derived from the test and control samples following hybridization to CpG island arrays provides a profile of sequences that are methylated in one sample, but not in the other. A possible drawback of most methylation array methods is the need to use potentially unfaithful linked ligation and linked PCR amplification, which is prone to false positives. Nevertheless, massive improvements in oligonucleotide arrays – particularly for allelic DNA methylation analysis – hold the promise of even greater methylome coverage to methylation array-based methods in the future [86, 90, 91, 93, 94].

Bacterial artificial chromosome (BAC) arrays have also been introduced successfully as a means of high-throughput DNA methylation analysis [86, 95], and competing tiling path arrays are currently available [96]. In one application with BAC arrays, genomic DNA is digested with a rare-cutting methylation-sensitive restriction enzyme, the digested sites are filled-in with biotin, and any unmethylated fragments are selected on streptavidin beads and then co-hybridized to the BAC array with a second reference genome. In contrast to other array methods, ligation and PCR are not used in this protocol. The use of rare-cutting restriction enzymes ensures that most BACs will contain only a single site or a single cluster of sites, allowing single-CpG-effective resolution and accurate validation. Tiling path BAC arrays can be easily adapted for use with different restriction enzymes to significantly increase the number of analyzable CpGs. However, genome coverage using restriction enzymes is limited by the presence of their recognition sequence in the target of interest.

The particular combination of array and methylation-sensitive detection reagents is also critical for tumor methyl ome analysis. These reagents include methylation-sensitive restriction enzymes, 5-methylcytosine antibody, methylated

DNA-binding protein columns, or bisulfite-based methylation detection. Bisulfite is a chemical that allows for the conversion of cytosine to uracil, while leaving 5-methylcytosine unconverted [97]. This method is a staple of single gene analysis and the high-throughput analysis of small sets of genes [98, 99] although, owing to the significantly reduced sequence complexity of DNA after bisulfite treatment, its use for array applications has been more limited [84, 100, 101]. DNA selected through methyl-binding protein columns or by 5-methylcytosine antibody immunoprecipitation has also been applied to microarrays [93, 102–106]. The effective resolution of DNA methylation using either method is dependent in part on the average DNA fragment size after random shearing (generally 500 bp to 1 kb). It is not yet clear how many CpG residues are needed for productive DNA-antibody binding to occur, or whether the antibody has a significant sequence bias. An advantage of this approach is that it is not limited to specific sequences as are restriction enzyme-based approaches. The 5-methylcytosine antibody protocol has been used successfully to map the methylome of *Arabidopsis thaliana* [105, 106], with results largely confirmed by shot-gun bisulfite sequencing of the same genome [107]. This approach has also been applied to human cancer cell lines [102, 103].

Methylation-sensitive restriction enzymes, whether rare or common cutters, can in theory provide single-CpG-effective resolution. In practice, however, common cutters – even when applied to oligonucleotide arrays – will not yield single-CpG resolution because up to 10 oligonucleotides spanning multiple common cutter sites are averaged into one value. Additionally, because protocols using common cutters require ligation and PCR [87, 89, 91], the distance and sequence between sites precludes a large proportion of these sites from analysis, reducing genome coverage. The restriction enzyme McrBc has also been tested for methylation detection [103, 108], although the resolution of methylation events is undefined owing to the unusual recognition site of this enzyme (two methylated CpGs separated by 40–3000 bp of nonspecific sequence).

An innovative large-scale SAGE-like sequencing method has also been employed for the DNA methylation analysis of breast cancer and the surrounding stroma cells [109]. Gene expression arrays can also be used to identify the DNA methylation-related silencing of genes by focusing on silent genes that are reactivated in tumor cell lines exposed to a DNA-demethylating agent [110–113].

Reduced representation bisulfite sequencing, which is a large scale genome-wide shot-gun sequencing approach [114], has been used successfully to investigate the loss of DNA methylation in DNMT[1^{kd}, $3a^{-/-}$, $3b^{-/-}$] embryonic stem cells. An advantage of this method is that it is amenable to gene discovery without preselecting targets, although sites exhibiting heterogeneous methylation might be confounding when represented by only a single sequence read. Substantially increasing the depth of sequencing might mitigate this limitation, however. Moreover, as clone libraries can be constructed the system can be automated to maximize efficiency.

Human epigenome projects of normal human cells have taken a standard sequencing-based bisulfite strategy, which gives a single-CpG resolution of methylation status [96, 115, 116]. While these projects are not designed primarily

Tab. 1 Common DNA methylation techniques, detailing required DNA amounts, specimen treatment, CpG coverage, and throughput.

Technique	*Specimen treatment*	*DNA amount*	*CpGs analyzed*	*High-throughput*
HPLC	Total hydrolysis	<1 μg	N/A	Yes
RLGS	Enzyme digestion	>1 μg	>2000	No
BAC arrays	Enzyme digestion	>400 ng	Variable	Yes
DMH	Enzyme digestion	>300 ng	12000	Yes
GoldenGate	Bisulfite conversion	1 μg	1536	Yes
Infinium	Bisulfite conversion	1 μg	27578	Yes
MS-SNuPE	Bisulfite conversion	250 ng[a]	1	Possible
MSP	Bisulfite conversion	250 ng[a]	5–15	Yes
Pyrosequencing	Bisulfite conversion	250 ng[a]	5–10	Yes
MethyLight	Bisulfite conversion	250 ng[a]	5–10	Yes
MassARRAY	Bisulfite conversion	500 ng	5–30	Possible
COBRA	Bisulfite conversion	250 ng[a]	2–5	No
Bio-COBRA	Bisulfite conversion	250 ng[a]	2–5	No
MIRA	Anti-5-methyl-C antibody	>1μg	Variable	No
MeDIP	Anti-5-methyl-C antibody	>1 μg	Variable	No
RRBS	Bisulfite conversion	>1 μg	Variable	No
WGSBS	Bisulfite conversion	>1 μg	Variable	No

[a]Bisulfite treatment of genomic DNA can be performed with variable amounts of starting material (1 μg is standard). A sample of 250 ng is provided as a reference, since bisulfite reactions utilizing this amount of starting material have been successfully performed. HPLC, high-performance liquid chromatography; RLGS, restriction landmark genomic scanning; BAC, bacterial artificial chromosome; DMH, differential methylation hybridization; MS-SNuPE, methylation-sensitive single nucleotide primer extension; COBRA, combined bisulfite restriction analysis; Bio-COBRA, Bioanalyzer-based combined bisulfite restriction analysis; MIRA, methylated CpG island recovery assay; MeDIP, methylated DNA immunoprecipitation; RRBS, reduced representation bisulfite sequencing; WGSBS, whole-genome shotgun bisulfite sequencing.

to determine the DNA methylation status of 29 million CpGs, the efforts to date have been immense and impressive, and have included different cell types, as well as interindividual and interspecies comparisons. A combination of either bisulfite, the 5-methylcytosine antibody, methyl-binding protein columns or restriction enzymes with next-generation sequencing also holds great promise, and these and other studies are currently adding to whole new disciplines within epigenetic research, including population epigenetics and comparative epigenetics. In addition to the main goals of these projects, the data will also be of substantial value for comparison with cancer methylome data, whether from arrays or from sequencing bisulfite-converted DNA (Table 1).

7 Computational Analysis of the Methylome

Aberrant DNA methylation exhibits tumor-type specific patterns [74]. However, it is unclear how these patterns are established and why a large number of CpG islands seem to be refractory to DNA methylation, while others are aberrantly methylated at high frequency [78, 102, 117–119]. A functional explanation for this observation could be that all CpG islands may be equally susceptible to DNA methylation, but that only a fraction is detected in tumors because of selection pressures. This hypothesis – though probably true for some genes – is unlikely to explain the mechanism responsible for the aberrant methylation of all CpG island-associated genes. Sequence-based rules derived from cancer cell methylation data have also been explored as a way to predict the pattern of aberrant DNA methylation in cancer genome-wide [120–123]. These studies have identified consensus sequences, proximity to repetitive elements and chromosomal location as potential factors influencing, or perhaps determining, the likelihood that a CpG island might become aberrantly methylated. If the sequence context in which a CpG island is located influences its likelihood of becoming aberrantly methylated, then the convergence of different computational analyses is likely to find commonalities that could help explain this phenomenon. An important goal in these investigations will be to distinguish sequence rules that predict pan-cancer DNA methylation from those that predict tumor-type-specific DNA methylation, as these rules could be mutually exclusive. An intriguing and particularly striking association between a subset of genes susceptible to aberrant promoter methylation in adult human cancers and a subset of genes occupied or marked by polycomb group proteins in human embryonic stem cells has been reported independently by three groups [124–126]. These and earlier studies [127, 128] offer important new insights into the possible mechanism by which certain genes might be susceptible to DNA methylation in cancer, as well as epigenetic support for the theory that human tumors arise from tissue stem cells. A comparison of the sequences associated with polycomb group protein occupancy, and those derived from the computational analysis of methylation-prone and methylation-resistant loci described above, might be particularly revealing.

8
DNA Methylation in Cancer

Most of the current evidence linking DNA methylation with regulation of gene expression and disease stems from human cancers. Significant changes in genome-wide DNA methylation have been observed in cultured cancer cells and primary human tumors [74, 80]. Such changes include global hypomethylation of centromeric repeats and repetitive sequences and gene-specific hypermethylation of CpG islands. DNA hypomethylation has been associated with chromosomal instability, resulting in increased mutation rates and abnormal gene expression [29, 129, 130].

In general, DNA hypermethylation of gene-associated CpG islands results in either a downregulation or complete abrogation of gene expression, indicating that aberrant DNA methylation could serve a similar function to genetic abnormalities, such as inactivating mutations or deletions in the disease state [71]. The results of numerous studies have indicated that several gene classes – such as adhesion molecules, inhibitors of angiogenesis, DNA repair, cell-cycle regulators and metastasis suppressors, among others – are frequently hypermethylated in primary human tumors [131–137].

As opposed to the irreversible essence of genetic alterations that result in gene silencing, the importance of understanding the mechanism involved in the epigenetic abrogation of gene expression lies in the reversible nature of epigenetic processes. Thus, a number of "epigenetic therapies" geared toward reversing aberrant epigenetic events in malignant cells have been developed. Most of these therapies rely on the use of two classic inhibitors of DNA methylation, namely 5-azacytidine and 5-aza-2′-deoxycytidine, both of which were originally synthesized as cytotoxic agents [138, 139]. These molecules both act as potent inhibitors of DNA methylation and exert their action through a variety of mechanisms. One mechanism is via incorporation of the agent into the DNA during S phase; this results in the trapping of DNMTs through the formation of a covalent bond between the catalytic site of the enzyme and the pyrimidine ring of the azanucleoside. Following the completion of each cell cycle, and concomitant to the depletion of DNMTs from the cellular environment, heritable DNA demethylation is observed in cells treated with either of these agents [140–144]. It has also been reported that both 5-azacytidine and 5-aza-2′-deoxycytidine can induce a rapid degradation of DNMT1 via the proteasomal pathway, even in the absence of DNA replication [145].

Despite the fact that, when used in high concentrations, azanucleosides exhibit high cytotoxicity, promising reports have emerged from clinical trials in which low doses of these agents administered during 3- to 10-day courses have been effective in treating some myelodysplastic syndromes and leukemias [141, 146, 147].

Recent reports have underscored the commonality of the epigenetic changes observed in cancer with those present in aging cells in normal tissues [71, 148, 149]. Consequently, a hypothesis has emerged, proposing that age-related methylation may act as a precursor for malignant transformation, thus helping to explain the age-dependent increase in cancer risk [150].

9
DNA Methylation as a Biomarker

Given the role of aberrant DNA methylation in cancer initiation and progression, much effort has been directed towards the development of strategies which could facilitate early cancer detection. It is now clear that DNA methylation is an early event in tumor development, as indicated by reports where aberrant hypermethylated sites could be detected in seemingly normal epithelia from patients years before the overt development of cancer [151]. Thus, the use of DNA methylation as a biomarker might prove to be a useful tool, not only for an early diagnosis but also for the detection and assessment of high-risk individuals. The importance of early detection is clearly evident, as the five-year survival rate for patients with breast, prostate, or colon cancer – for which screening tests are available – is four- to sixfold higher than that for lung cancer patients, for which no early detection protocol is currently implemented [152].

In order for a biomarker to be clinically applicable, it must be specific, sensitive, and detectable in specimens obtained through minimally invasive procedures. Promising results have already been obtained, since aberrantly methylated CpG islands have been detected in DNA samples derived from urine, serum, sputum, and the stools of cancer patients [153]. Importantly, it should also be noted that changes in DNA methylation also occur in normal epithelia. Consequently, extensive investigations are currently under way to identify tumor-specific DNA methylation events that afford enough sensitivity and specificity to be utilized as biomarkers. Another obstacle to be overcome in this respect is the fact that tumor DNA is present only in minimal amounts in body fluids, which means that exquisitely sensitive techniques will be required to detect and analyze tumor-derived DNA.

Currently, a wide array of techniques is available to measure DNA methylation, both genome-wide and at the single gene level. In general, genome-wide techniques for DNA methylation analysis require large amounts of DNA, which makes them unsuitable for the analysis of biomarkers. Nonetheless, these techniques have been applied successfully to uncover novel tumor suppressor genes and to monitor global changes in DNA methylation in health and disease [74, 80, 132, 154].

10
Epigenetic Response to Cancer Therapy

Aberrant DNA methylation of particular CpG islands may also alter the response of a cancer cell to therapeutic agents, or serve as a clinically useful marker of clinical outcome. For example, normal expression of the DNA repair gene *O*-6-methylguanine DNA methyltransferase (MGMT) is associated with resistance to therapy, whereas aberrant DNA methylation of the MGMT 5′ CpG island, and presumable MGMT silencing [155–157], is associated with significantly improved antitumor response of alkylating agents, such as temozolomide [158, 159]. In contrast, cisplatin-resistant cancer cells can be sensitized by relieving repressive histone H3K27 methylation and DNA methylation, presumably by reactivating silenced tumor suppressors and modulators of cisplatin response [160]. Efforts directed at identifying DNA methylation-based markers for the early detection of tumors and predicting tumor response to therapy are under way in research laboratories worldwide [27, 112, 161, 162]. Assays are currently available to detect aberrant DNA methylation in minute samples that are obtained with minimally invasive procedures and are likely to contain tumor cells and tumor DNA shed from a primary tumor mass [163, 164]. In contrast, the loss of DNA methylation from normally methylated promoters of the MAGEA gene family, followed by MAGEA gene activation, may elicit the production of anti-MAGEA antibodies, which are detectable in the blood of patients with melanoma and other cancers [165].

11
Concluding Remarks

In recent years, the study of epigenetic alterations in the human genome has taken center stage in an effort to better understand the molecular basis of human disease beyond the well-documented realm of genetic events. DNA methylation analysis at both global and gene-specific levels has helped to shed light on gene function, and has also uncovered a large number of genes, the expression of which is abolished primarily through epigenetic mechanisms in diseases, such as cancer.

The fact that epigenetic changes are reversible also opens a new spectrum of potential treatment options which may lead to the amelioration, or even elimination, of the disease phenotype.

Today, DNA methylation data can be generated using different approaches, many of which are well-established and have served as important tools for epigenetic analysis. As yet, however, no single technique can provide an unambiguous approach to DNA methylation data harvesting.

Finally, it is important to emphasize the critical role of DNA methylation assays as tools for assessing the efficacy and safety of DNA demethylating agents, as these – at least potentially – may in time be developed into standard regiments for cancer therapy. Recently, drugs such as decitabine have shown promising results in clinical trials focused on the treatment of both solid and liquid tumors. However, due to the nonspecific nature of such nucleotide analogs, it is critical to monitor their effect not only on neoplastic cells but also on normal tissues, in order to ensure that no long-term damage is inflicted on unaffected targets.

Currently, a large body of evidence exists indicating that not all possible DNA methylation targets in the human genome are affected equally in the disease state. Whilst the biological mechanism behind these observations is not fully understood, the situation might involve selection pressure or an intrinsic difference in sequence susceptibility to aberrant epigenetic changes. Thus, the use of sensitive assays to monitor DNA methylation changes will surely play a key role in the development and implementation of new therapies aimed at modulating the epigenome.

References

1 Feinberg, A.P. (2007) Phenotypic plasticity and the epigenetics of human disease. *Nature*, **447**(7143), 433–440.

2 Van Speybroeck, L. (2002) From epigenesis to epigenetics: the case of C. H. Waddington. *Ann. N. Y. Acad. Sci.*, **981**, 61–81.

3 Bernstein, B.E., Mikkelsen, T.S., Xie, X., Kamal, M., Huebert, D.J., Cuff, J., Fry, B., Meissner, A., Wernig, M., Plath, K., Jaenisch, R., Wagschal, A., Feil, R., Schreiber, S.L., Lander, E.S. (2006) A bivalent chromatin structure marks key developmental genes in embryonic stem cells. *Cell*, **125**(2), 315–326.

4 Christman, J.K. (1982) Separation of major and minor deoxyribonucleoside monophosphates by reverse-phase high-performance liquid chromatography: a simple method applicable to quantitation of methylated nucleotides in DNA. *Anal. Biochem.*, **119**(1), 38–48.

5 Bird, A.P. (1986) CpG-rich islands and the function of DNA methylation. *Nature*, **321**(6067), 209–213.

6 Franchina, M., Kay, P.H. (2000) Evidence that cytosine residues within 5′-CCTGG-3′ pentanucleotides can be methylated in human DNA independently of the methylating system that modifies 5′-CG-3′ dinucleotides. *DNA Cell Biol.*, **19**(9), 521–526.

7 Malone, C.S., Miner, M.D., Doerr, J.R., Jackson, J.P., Jacobsen, S.E., Wall, R., Teitell, M. (2001) CmC(A/T)GG DNA methylation in mature B-cell lymphoma gene silencing. *Proc. Natl Acad. Sci. USA*, **98**(18), 10404–10409.

8 Clark, S.J., Harrison, J., Frommer, M. (1995) CpNpG methylation in mammalian cells. *Nat. Genet.*, **10**(1), 20–27.

9 Ramsahoye, B.H., Biniszkiewicz, D., Lyko, F., Clark, V., Bird, A.P., Jaenisch, R. (2000) Non-CpG methylation is prevalent in embryonic stem cells and may be mediated by DNA methyltransferase 3a. *Proc. Natl Acad. Sci. USA*, **97**(10), 5237–5242.

10 Chiang, P.K., Gordon, R.K., Tal, J., Zeng, G.C., Doctor, B.P., Pardhasaradhi, K., McCann, P.P. (1996) S-Adenosylmethionine and methylation. *FASEB J.*, **10**(4), 471–480.

11 Schmitt, F., Oakeley, E.J., Jost, J.P. (1997) Antibiotics induce genome-wide hypermethylation in cultured *Nicotiana tabacum* plants. *J. Biol. Chem.*, **272**(3), 1534–1540.
12 Bell, A.C., Felsenfeld, G. (2000) Methylation of a CTCF-dependent boundary controls imprinted expression of the Igf2 gene. *Nature*, **405**(6785), 482–485.
13 Jorgensen, H.F., Bird, A. (2002) MeCP2 and other methyl-CpG binding proteins. *Ment. Retard. Dev. Disabil. Res. Rev.*, **8**(2), 87–93.
14 Hermann, A., Goyal, R., Jeltsch, A. (2004) The Dnmt1 DNA-(cytosine-C5)-methyltransferase methylates DNA processively with high preference for hemimethylated target sites. *J. Biol. Chem.*, **279**(46), 48350–48359.
15 Okano, M., Bell, D.W., Haber, D.A., Li E. (1999) DNA methyltransferases Dnmt3a and Dnmt3b are essential for *de novo* methylation and mammalian development. *Cell*, **99**(3), 247–257.
16 Okano, M., Xie, S.P., Li, E. (1998) Cloning and characterization of a family of novel mammalian DNA (cytosine-5) methyltransferases. *Nat. Genet.*, **19**(3), 219–220.
17 Okano, M., Takebayashi, S., Okumura, K., Li, E. (1999) Assignment of cytosine-5 DNA methyltransferases Dnmt3a and Dnmt3b to mouse chromosome bands 12A2-A3 and 2H1 by in situ hybridization. *Cytogenet. Cell Genet.*, **86**(3-4), 333–334.
18 Egger, G., Liang, G., Aparicio, A., Jones, P.A. (2004) Epigenetics in human disease and prospects for epigenetic therapy. *Nature*, **429**(6990), 457–463.
19 Herman, J.G., Baylin, S.B. (2003) Gene silencing in cancer in association with promoter hypermethylation. *N. Engl. J. Med.*, **349**(21), 2042–2054.
20 Colot, V., Rossignol, J.L. (1999) Eukaryotic DNA methylation as an evolutionary device. *BioEssays*, **21**(5), 402–411.
21 Gardiner-Garden, M., Frommer, M. (1987) CpG islands in vertebrate genomes. *J. Mol. Biol.*, **196**(2), 261–282.
22 Larsen, F., Gundersen, G., Lopez, R., Prydz, H. (1992) CpG islands as gene markers in the human genome. *Genomics*, **13**(4), 1095–1107.
23 Takai, D., Jones, P.A. (2002) Comprehensive analysis of CpG islands in human chromosomes 21 and 22. *Proc. Natl Acad. Sci. USA*, **99**(6), 3740–3745.
24 Bird, A. (2002) DNA methylation patterns and epigenetic memory. *Genes Dev.*, **16**(1), 6–21.
25 Jaenisch, R., Bird, A. (2003) Epigenetic regulation of gene expression: how the genome integrates intrinsic and environmental signals. *Nat. Genet.*, **33**(Suppl.), 245–254.
26 Dai, C., Holland, E.C. (2003) Astrocyte differentiation states and glioma formation. *Cancer J.*, **9**(2), 72–81.
27 Brena, R.M., Morrison, C., Liyanarachchi, S., Jarjoura, D., Davuluri, R.V., Otterson, G.A., Reisman, D., Glaros, S., Rush, L.J., Plass, C. (2007) Aberrant DNA methylation of OLIG1, a novel prognostic factor in non-small cell lung cancer. *PLoS Med.*, **4**, e108.
28 Jones, P.A., Laird, P.W. (1999) Cancer epigenetics comes of age. *Nat. Genet.*, **21**(2), 163–167.
29 Chen, R.Z., Pettersson, U., Beard, C., Jackson-Grusby, L., Jaenisch, R. (1998) DNA hypomethylation leads to elevated mutation rates. *Nature*, **395**(6697), 89–93.
30 Costello, J.F. (2003) DNA methylation in brain development and gliomagenesis. *Front. Biosci.*, **8**, S175–S184.
31 Fan, G.P., Beard, C., Chen, R.Z., Csankovszki, G., Sun, Y., Siniaia, M., Biniszkiewicz, D., Bates, B., Lee, P.P., Kuhn, R., Trumpp, A., Poon, C.S., Wilson, C.B., Jaenisch, R. (2001) DNA hypomethylation perturbs the function and survival of CNS neurons in postnatal animals. *J. Neurosci.*, **21**(3), 788–797.
32 Gonzalo, S., Jaco, I., Fraga, M.F., Chen, T., Li, E., Esteller, M., Blasco, M.A. (2006) DNA methyltransferases control telomere length and telomere recombination in mammalian cells. *Nat. Cell Biol.*, **8**(4), 416–424.
33 Hansen, R.S., Wijmenga, C., Luo, P., Stanek, A.M., Canfield, T.K., Weemaes, C.M.R., Gartler, S.M. (1999) The DNMT3B DNA methyltransferase gene is mutated in the ICF immunodeficiency syndrome. *Proc. Natl Acad. Sci. USA*, **96**(25), 14412–14417.
34 Kawai, J., Hirotsune, S., Hirose, K., Fushiki, S., Watanabe, S., Hayashizaki, Y. (1993) Methylation profiles of genomic DNA of mouse developmental brain detected by restriction landmark genomic

scanning (Rlgs) method. *Nucleic Acids Res.*, **21**(24), 5604–5608.
35 Kazazian, H.H., Moran, J.V. (1998) The impact of L1 retrotransposons on the human genome. *Nat. Genet.*, **19**(1), 19–24.
36 Li, E., Bestor, T.H., Jaenisch, R. (1992) Targeted mutation of the DNA methyltransferase gene results in embryonic lethality. *Cell*, **69**(6), 915–926.
37 Maraschio, P., Zuffardi, O., Dalla Fior, T., Tiepolo, L. (1988) Immunodeficiency, centromeric heterochromatin instability of chromosomes 1, 9, and 16, and facial anomalies: the ICF syndrome. *J. Med. Genet.*, **25**(3), 173–180.
38 Takizawa, T., Nakashima, K., Namihira, M., Ochiai, W., Uemura, A., Yanagisawa, M., Fujita, N., Nakao, M., Taga, T. (2001) DNA methylation is a critical cell-intrinsic determinant of astrocyte differentiation in the fetal brain. *Dev. Cell*, **1**(6), 749–758.
39 Trasler, J.M., Trasler, D.G., Bestor, T.H., Li, E., Ghibu, F. (1996) DNA methyltransferase in normal and Dnmtn/Dnmtn mouse embryos. *Dev. Dyn.*, **206**(3), 239–247.
40 Xu, G.L., Bestor, T.H., Bourc'his, D., Hsieh, C.L., Tommerup, N., Bugge, M., Hulten, M., Qu, X.Y., Russo, J.J., Viegas-Pequignot, E. (1999) Chromosome instability and immunodeficiency syndrome caused by mutations in a DNA methyltransferase gene. *Nature*, **402**(6758), 187–191.
41 Hemberger, M., Dean, W., Reik, W. (2009) Epigenetic dynamics of stem cells and cell lineage commitment: digging Waddington's canal. *Nat. Rev. Mol. Cell Biol.*, **10**(8), 526–537.
42 Santos, F., Hendrich, B., Reik, W., Dean, W. (2002) Dynamic reprogramming of DNA methylation in the early mouse embryo. *Dev. Biol.*, **241**(1), 172–182.
43 Wood, A.J., Oakey, R.J. (2006) Genomic imprinting in mammals: emerging themes and established theories. *PLoS Genet.*, **2**(11), e147.
44 Gaudet, F., Hodgson, J.G., Eden, A., Jackson-Grusby, L., Dausman, J., Gray, J.W., Leonhardt, H., Jaenisch, R. (2003) Induction of tumors in mice by genomic hypomethylation. *Science*, **300**(5618), 489–492.
45 Tremblay, K.D., Duran, K.L., Bartolomei, M.S. (1997) A 5′ 2-kilobase-pair region of the imprinted mouse H19 gene exhibits exclusive paternal methylation throughout development. *Mol. Cell Biol.*, **17**(8), 4322–4329.
46 Fraga, M.F., Ballestar, E., Paz, M.F., Ropero, S., Setien, F., Ballestar, M.L., Heine-Suner, D., Cigudosa, J.C., Urioste, M., Benitez, J., Boix-Chornet, M., Sanchez-Aguilera, A., Ling, C., Carlsson, E., Poulsen, P., Vaag, A., Stephan, Z., Spector, T.D., Wu, Y.Z., Plass, C., Esteller, M. (2005) Epigenetic differences arise during the lifetime of monozygotic twins. *Proc. Natl Acad. Sci. USA*, **102**(30), 10604–10609.
47 Sandovici, I., Kassovska-Bratinova, S., Loredo-Osti, J.C., Leppert, M., Suarez, A., Stewart, R., Bautista, F.D., Schiraldi, M., Sapienza, C. (2005) Interindividual variability and parent of origin DNA methylation differences at specific human Alu elements. *Hum. Mol. Genet.*, **14**(15), 2135–2143.
48 Wong, A.H., Gottesman, I.I., Petronis, A. (2005) Phenotypic differences in genetically identical organisms: the epigenetic perspective. *Hum. Mol. Genet.*, **14**(Spec. No. 1), R11–R18.
49 Gartner, K. (1990) A third component causing random variability beside environment and genotype. A reason for the limited success of a 30-year-long effort to standardize laboratory animals? *Lab. Anim.*, **24**(1), 71–77.
50 Morgan, D.K., Whitelaw, E. (2008) The case for transgenerational epigenetic inheritance in humans. *Mamm. Genome*, **19**(6), 394–397.
51 Cardno, A.G., Rijsdijk, F.V., Sham, P.C., Murray, R.M., McGuffin, P. (2002) A twin study of genetic relationships between psychotic symptoms. *Am. J. Psychiatry*, **159**(4), 539–545.
52 Cantoni, G.L. (1985) The role of *S*-adenosylhomocysteine in the biological utilization of *S*-adenosylmethionine. *Prog. Clin. Biol. Res.*, **198**, 47–65.
53 Yi, P., Melnyk, S., Pogribny, M., Pogribny, I.P., Hine, R.J., James, S.J. (2000) Increase in plasma homocysteine associated with parallel increases in plasma *S*-adenosylhomocysteine and lymphocyte DNA hypomethylation. *J. Biol. Chem.*, **275**(38), 29318–29323.

54 Wainfan, E., Poirier, L.A. (1992) Methyl groups in carcinogenesis: effects on DNA methylation and gene expression. *Cancer Res.*, **52**(Suppl. 7), 2071s–2077s.

55 Pogribny, I.P., Basnakian, A.G., Miller, B.J., Lopatina, N.G., Poirier, L.A., James, S.J. (1995) Breaks in genomic DNA and within the p53 gene are associated with hypomethylation in livers of folate/methyl-deficient rats. *Cancer Res.*, **55**(9), 1894–1901.

56 Pogribny, I.P., James, S.J., Jernigan, S., Pogribna, M. (2004) Genomic hypomethylation is specific for preneoplastic liver in folate/methyl deficient rats and does not occur in non-target tissues. *Mutat. Res.*, **548**(1-2), 53–59.

57 Shivapurkar, N., Poirier, L.A. (1983) Tissue levels of *S*-adenosylmethionine and *S*-adenosylhomocysteine in rats fed methyl-deficient, amino acid-defined diets for one to five weeks. *Carcinogenesis*, **4**(8), 1051–1057.

58 Bernstein, B.E., Meissner, A., Lander, E.S. (2007) The mammalian epigenome. *Cell*, **128**(4), 669–681.

59 Shogren-Knaak, M., Ishii, H., Sun, J.M., Pazin, M.J., Davie, J.R., Peterson, C.L. (2006) Histone H4-K16 acetylation controls chromatin structure and protein interactions. *Science*, **311**(5762), 844–847.

60 Sjoblom, T., Jones, S., Wood, L.D., Parsons, D.W., Lin, J., Barber, T.D., Mandelker, D., Leary, R.J., Ptak, J., Silliman, N., Szabo, S., Buckhaults, P., Farrell, C., Meeh, P., Markowitz, S.D., Willis, J., Dawson, D., Willson, J.K., Gazdar, A.F., Hartigan, J., Wu, L., Liu, C., Parmigiani, G., Park, B.H., Bachman, K.E., Papadopoulos, N., Vogelstein, B., Kinzler, K.W., Velculescu, V.E. (2006) The consensus coding sequences of human breast and colorectal cancers. *Science*, **314**(5797), 268–274.

61 Cameron, E.E., Bachman, K.E., Myohanen, S., Herman, J.G., Baylin, S.B. (1999) Synergy of demethylation and histone deacetylase inhibition in the re-expression of genes silenced in cancer. *Nat. Genet.*, **21**(1), 103–107.

62 Frigola, J., Song, J., Stirzaker, C., Hinshelwood, R.A., Peinado, M.A., Clark, S.J. (2006) Epigenetic remodeling in colorectal cancer results in coordinate gene suppression across an entire chromosome band. *Nat. Genet.*, **38**(5), 540–549.

63 Jones, P.A., Baylin, S.B. (2007) The epigenomics of cancer. *Cell*, **128**(4), 683–692.

64 Millar, D.S., Paul, C.L., Molloy, P.L., Clark, S.J. (2000) A distinct sequence (ATAAA)(n) separates methylated and unmethylated domains at the 5′-end of the GSTP1 CpG island. *J. Biol. Chem.*, **275**(32), 24893–24899.

65 Song, J.Z., Stirzaker, C., Harrison, J., Melki, J.R., Clark, S.J. (2002) Hypermethylation trigger of the glutathione-*S*-transferase gene (GSTP1) in prostate cancer cells. *Oncogene*, **21**(7), 1048–1061.

66 Jones, P.A. (2005) Overview of cancer epigenetics. *Semin. Hematol.*, **42**(3, Suppl. 2), S3–S8.

67 Fraga, M.F., Ballestar, E., Villar-Garea, A., Boix-Chornet, M., Espada, J., Schotta, G., Bonaldi, T., Haydon, C., Ropero, S., Petrie, K., Iyer, N.G., Perez-Rosado, A., Calvo, E., Lopez, J.A., Cano, A., Calasanz, M.J., Colomer, D., Piris, M.A., Ahn, N., Imhof, A., Caldas, C., Jenuwein, T., Esteller, M. (2005) Loss of acetylation at Lys16 and trimethylation at Lys20 of histone H4 is a common hallmark of human cancer. *Nat. Genet.*, **37**(4), 391–400.

68 Kochanek, S., Renz, D., Doerfler, W. (1993) DNA methylation in the Alu sequences of diploid and haploid primary human cells. *EMBO J.*, **12**(3), 1141–1151.

69 Rein, T., DePamphilis, M.L., Zorbas, H. (1998) Identifying 5-methylcytosine and related modifications in DNA genomes. *Nucleic Acids Res.*, **26**(10), 2255–2264.

70 Di Croce, L., Raker, V.A., Corsaro, M., Fazi, F., Fanelli, M., Faretta, M., Fuks, F., Lo Coco, F., Kouzarides, T., Nervi, C., Minucci, S., Pelicci, P.G. (2002) Methyltransferase recruitment and DNA hypermethylation of target promoters by an oncogenic transcription factor. *Science*, **295**(5557), 1079–1082.

71 Jones, P.A., Baylin, S.B. (2002) The fundamental role of epigenetic events in cancer. *Nat. Rev. Genet.*, **3**(6), 415–428.

72 Hatada, I., Hayashizaki, Y., Hirotsune, S., Komatsubara, H., Mukai, T. (1991) A genomic scanning method for higher organisms using restriction sites as landmarks. *Proc. Natl Acad. Sci. USA*, **88**(21), 9523–9527.

73 Rush, L.J., Plass, C. (2002) Restriction landmark genomic scanning for DNA methylation in cancer: past, present, and future applications. *Anal. Biochem.*, **307**(2), 191–201.

74 Costello, J.F., Fruhwald, M.C., Smiraglia, D.J., Rush, L.J., Robertson, G.P., Gao, X., Wright, F.A., Feramisco, J.D., Peltomaki, P., Lang, J.C., Schuller, D.E., Yu, L., Bloomfield, C.D., Caligiuri, M.A., Yates, A., Nishikawa, R., Su Huang, H., Petrelli, N.J., Zhang, X., O'Dorisio, M.S., Held, W.A., Cavenee, W.K., Plass, C. (2000) Aberrant CpG-island methylation has non-random and tumour-type-specific patterns. *Nat. Genet.*, **24**(2), 132–138.

75 Liang, G.N., Robertson, K.D., Talmadge, C., Sumegi, J., Jones, P.A. (2000) The gene for a novel transmembrane protein containing epidermal growth factor and follistatin domains is frequently hypermethylated in human tumor cells. *Cancer Res.*, **60**(17), 4907–4912.

76 Dai, Z., Weichenhan, D., Wu, Y.Z., Hall, J.L., Rush, L.J., Smith, L.T., Raval, A., Yu, L., Kroll, D., Muehlisch, J., Fruhwald, M.C., de Jong, P., Catanese, J., Davuluri, R.V., Smiraglia, D.J., Plass, C. (2002) An AscI boundary library for the studies of genetic and epigenetic alterations in CpG islands. *Genome Res.*, **12**(10), 1591–1598.

77 Smiraglia, D.J., Fruhwald, M.C., Costello, J.F., McCormick, S.P., Dai, Z., Peltomaki, P., O'Dorisio, M.S., Cavenee, W.K., Plass, C. (1999) A new tool for the rapid cloning of amplified and hypermethylated human DNA sequences from restriction landmark genome scanning gels. *Genomics*, **58**(3), 254–262.

78 Dai, Z., Lakshmanan, R.R., Zhu, W.G., Smiraglia, D.J., Rush, L.J., Fruhwald, M.C., Brena, R.M., Li, B., Wright, F.A., Ross, P., Otterson, G.A., Plass, C. (2001) Global methylation profiling of lung cancer identifies novel methylated genes. *Neoplasia*, **3**(4), 314–323.

79 Kuromitsu, J., Kataoka, H., Yamashita, H., Muramatsu, M., Furuichi, Y., Sekine, T., Hayashizaki, Y. (1995) Reproducible alterations of DNA methylation at a specific population of CpG islands during blast formation of peripheral blood lymphocytes. *DNA Res.*, **2**(6), 263–267.

80 Smiraglia, D.J., Rush, L.J., Fruhwald, M.C., Dai, Z., Held, W.A., Costello, J.F., Lang, J.C., Eng, C., Li, B., Wright, F.A., Caligiuri, M.A., Plass, C. (2001) Excessive CpG island hypermethylation in cancer cell lines versus primary human malignancies. *Hum. Mol. Genet.*, **10**(13), 1413–1419.

81 Yoshikawa, H., de la Monte, S., Nagai, H., Wands, J.R., Matsubara, K., Fujiyama, A. (1996) Chromosomal assignment of human genomic NotI restriction fragments in a two-dimensional electrophoresis profile. *Genomics*, **31**(1), 28–35.

82 Smith, L.T., Lin, M., Brena, R.M., Lang, J.C., Schuller, D.E., Otterson, G.A., Morrison, C.D., Smiraglia, D.J., Plass, C. (2006) Epigenetic regulation of the tumor suppressor gene TCF21 on 6q23-q24 in lung and head and neck cancer. *Proc. Natl Acad. Sci. USA*, **103**(4), 982–987.

83 Venter, J.C., Adams, M.D., Myers, E.W., Li, P.W., Mural, R.J., Sutton, G.G., Smith, H.O., Yandell, M., Evans, C.A., Holt, R.A., Gocayne, J.D., Amanatides, P., Ballew, R.M., Huson, D.H., Wortman, J.R., Zhang, Q., Kodira, C.D., Zheng, X.H., Chen, L., Skupski, M., Subramanian, G., Thomas, P.D., Zhang, J., Gabor Miklos, G.L., Nelson, C., Broder, S., Clark, A.G., Nadeau, J., McKusick, V.A., Zinder, N., Levine, A.J., Roberts, R.J., Simon, M., Slayman, C., Hunkapiller, M., Bolanos, R., Delcher, A., Dew, I., Fasulo, D., Flanigan, M., Florea, L., Halpern, A., Hannenhalli, S., Kravitz, S., Levy, S., Mobarry, C., Reinert, K., Remington, K., Abu-Threideh, J., Beasley, E., Biddick, K., Bonazzi, V., Brandon, R., Cargill, M., Chandramouliswaran, I., Charlab, R., Chaturvedi, K., Deng, Z., Di Francesco, V., Dunn, P., Eilbeck, K., Evangelista, C., Gabrielian, A.E., Gan, W., Ge, W., Gong, F., Gu, Z., Guan, P., Heiman, T.J., Higgins, M.E., Ji, R.R., Ke, Z., Ketchum, K.A., Lai, Z., Lei, Y., Li, Z., Li, J., Liang, Y., Lin, X., Lu, F., Merkulov, G.V., Milshina, N., Moore, H.M., Naik, A.K., Narayan, V.A., Neelam, B., Nusskern, D., Rusch, D.B., Salzberg, S., Shao, W., Shue, B., Sun, J., Wang, Z., Wang, A., Wang, X., Wang, J., Wei, M., Wides, R., Xiao, C., Yan, C., Yao, A., Ye, J., Zhan, M., Zhang, W., Zhang, H., Zhao, Q., Zheng, L., Zhong, F., Zhong, W., Zhu, S., Zhao, S., Gilbert, D., Baumhueter, S., Spier, G.,

Carter, C., Cravchik, A., Woodage, T., Ali, F., An, H., Awe, A., Baldwin, D., Baden, H., Barnstead, M., Barrow, I., Beeson, K., Busam, D., Carver, A., Center, A., Cheng, M.L., Curry, L., Danaher, S., Davenport, L., Desilets, R., Dietz, S., Dodson, K., Doup, L., Ferriera, S., Garg, N., Gluecksmann, A., Hart, B., Haynes, J., Haynes, C., Heiner, C., Hladun, S., Hostin, D., Houck, J., Howland, T., Ibegwam, C., Johnson, J., Kalush, F., Kline, L., Koduru, S., Love, A., Mann, F., May, D., McCawley, S., McIntosh, T., McMullen, I., Moy, M., Moy, L., Murphy, B., Nelson, K., Pfannkoch, C., Pratts, E., Puri, V., Qureshi, H., Reardon, M., Rodriguez, R., Rogers, Y.H., Romblad, D., Ruhfel, B., Scott, R., Sitter, C., Smallwood, M., Stewart, E., Strong, R., Suh, E., Thomas, R., Tint, N.N., Tse, S., Vech, C., Wang, G., Wetter, J., Williams, S., Williams, M., Windsor, S., Winn-Deen, E., Wolfe, K., Zaveri, J., Zaveri, K., Abril, J.F., Guigo, R., Campbell, M.J., Sjolander, K.V., Karlak, B., Kejariwal, A., Mi, H., Lazareva, B., Hatton, T., Narechania, A., Diemer, K., Muruganujan, A., Guo, N., Sato, S., Bafna, V., Istrail, S., Lippert, R., Schwartz, R., Walenz, B., Yooseph, S., Allen, D., Basu, A., Baxendale, J., Blick, L., Caminha, M., Carnes-Stine, J., Caulk, P., Chiang, Y.H., Coyne, M., Dahlke, C., Mays, A., Dombroski, M., Donnelly, M., Ely, D., Esparham, S., Fosler, C., Gire, H., Glanowski, S., Glasser, K., Glodek, A., Gorokhov, M., Graham, K., Gropman, B., Harris, M., Heil, J., Henderson, S., Hoover, J., Jennings, D., Jordan, C., Jordan, J., Kasha, J., Kagan, L., Kraft, C., Levitsky, A., Lewis, M., Liu, X., Lopez, J., Ma, D., Majoros, W., McDaniel, J., Murphy, S., Newman, M., Nguyen, T., Nguyen, N., Nodell, M., Pan, S., Peck, J., Peterson, M., Rowe, W., Sanders, R., Scott, J., Simpson, M., Smith, T., Sprague, A., Stockwell, T., Turner, R., Venter, E., Wang, M., Wen, M., Wu, D., Wu, M., Xia, A., Zandieh, A., Zhu, X. (2001) The sequence of the human genome. *Science*, **291**(5507), 1304–1351.

84 Bibikova, M., Lin, Z., Zhou, L., Chudin, E., Garcia, E.W., Wu, B., Doucet, D., Thomas, N.J., Wang, Y., Vollmer, E., Goldmann, T., Seifart, C., Jiang, W., Barker, D.L., Chee, M.S., Floros, J., Fan, J.B. (2006) High-throughput DNA methylation profiling using universal bead arrays. *Genome Res.*, **16**(3), 383–393.

85 Ehrich, M., Nelson, M.R., Stanssens, P., Zabeau, M., Liloglou, T., Xinarianos, G., Cantor, C.R., Field, J.K., van den Boom, D. (2005) Quantitative high-throughput analysis of DNA methylation patterns by base-specific cleavage and mass spectrometry. *Proc. Natl Acad. Sci. USA*, **102**(44), 15785–15790.

86 Ching, T.T., Maunakea, A.K., Jun, P., Hong, C., Zardo, G., Pinkel, D., Albertson, D.G., Fridlyand, J., Mao, J.H., Shchors, K., Weiss, W.A., Costello, J.F. (2005) Epigenome analyses using BAC microarrays identify evolutionary conservation of tissue-specific methylation of SHANK3. *Nat. Genet.*, **37**(6), 645–651.

87 Huang, T.H., Laux, D.E., Hamlin, B.C., Tran, P., Tran, H., Lubahn, D.B. (1997) Identification of DNA methylation markers for human breast carcinomas using the methylation-sensitive restriction fingerprinting technique. *Cancer Res.*, **57**(6), 1030–1034.

88 Ishkanian, A.S., Malloff, C.A., Watson, S.K., DeLeeuw, R.J., Chi, B., Coe, B.P., Snijders, A., Albertson, D.G., Pinkel, D., Marra, M.A., Ling, V., MacAulay, C., Lam, W.L. (2004) A tiling resolution DNA microarray with complete coverage of the human genome. *Nat. Genet.*, **36**(3), 299–303.

89 Khulan, B., Thompson, R.F., Ye, K., Fazzari, M.J., Suzuki, M., Stasiek, E., Figueroa, M.E., Glass, J.L., Chen, Q., Montagna, C., Hatchwell, E., Selzer, R.R., Richmond, T.A., Green, R.D., Melnick, A., Greally, J.M. (2006) Comparative isoschizomer profiling of cytosine methylation: the HELP assay. *Genome Res.*, **16**(8), 1046–1055.

90 Misawa, A., Inoue, J., Sugino, Y., Hosoi, H., Sugimoto, T., Hosoda, F., Ohki, M., Imoto, I., Inazawa, J. (2005) Methylation-associated silencing of the nuclear receptor 1I2 gene in advanced-type neuroblastomas, identified by bacterial artificial chromosome array-based methylated CpG island amplification. *Cancer Res.*, **65**(22), 10233–10242.

91 Schumacher, A., Kapranov, P., Kaminsky, Z., Flanagan, J., Assadzadeh, A., Yau, P., Virtanen, C., Winegarden, N., Cheng,

J., Gingeras, T., Petronis, A. (2006) Microarray-based DNA methylation profiling: technology and applications. *Nucleic Acids Res.*, **34**(2), 528–542.

92 Wang, Y., Hayakawa, J., Long, F., Yu, Q., Cho, A.H., Rondeau, G., Welsh, J., Mittal, S., De Belle, I., Adamson, E., McClelland, M., Mercola, D. (2005) "Promoter array" studies identify cohorts of genes directly regulated by methylation, copy number change, or transcription factor binding in human cancer cells. *Ann. N. Y. Acad. Sci.*, **1058**, 162–185.

93 Weber, M., Davies, J.J., Wittig, D., Oakeley, E.J., Haase, M., Lam, W.L., Schubeler, D. (2005) Chromosome-wide and promoter-specific analyses identify sites of differential DNA methylation in normal and transformed human cells. *Nat. Genet.*, **37**(8), 853–862.

94 Hellman, A., Chess, A. (2007) Gene body-specific methylation on the active X chromosome. *Science*, **315**, 1141–1143.

95 Jones, P.A., Martienssen, R. (2005) A blueprint for a Human Epigenome Project: the AACR Human Epigenome Workshop. *Cancer Res.*, **65**(24), 11241–11246.

96 Jeltsch, A., Walter, J., Reinhardt, R., Platzer, M. (2006) German human methylome project started. *Cancer Res.*, **66**(14), 7378.

97 Frommer, M., McDonald, L.E., Millar, D.S., Collis, C.M., Watt, F., Grigg, G.W., Molloy, P.L., Paul, C.L. (1992) A genomic sequencing protocol that yields a positive display of 5-methylcytosine residues in individual DNA strands. *Proc. Natl Acad. Sci. USA*, **89**(5), 1827–1831.

98 Herman, J.G., Graff, J.R., Myohanen, S., Nelkin, B.D., Baylin, S.B. (1996) Methylation-specific PCR: a novel PCR assay for methylation status of CpG islands. *Proc. Natl Acad. Sci. USA*, **93**(18), 9821–9826.

99 Laird, P.W., Jackson-Grusby, L., Fazeli, A., Dickinson, S.L., Jung, W.E., Li, E., Weinberg, R.A., Jaenisch, R. (1995) Suppression of intestinal neoplasia by DNA hypomethylation. *Cell*, **81**(2), 197–205.

100 Adorjan, P., Distler, J., Lipscher, E., Model, F., Muller, J., Pelet, C., Braun, A., Florl, A.R., Gutig, D., Grabs, G., Howe, A., Kursar, M., Lesche, R., Leu, E., Lewin, A., Maier, S., Muller, V., Otto, T., Scholz, C., Schulz, W.A., Seifert, H.H., Schwope, I., Ziebarth, H., Berlin, K., Piepenbrock, C., Olek, A. (2002) Tumour class prediction and discovery by microarray-based DNA methylation analysis. *Nucleic Acids Res.*, **30**(5), e21.

101 Yan, P.S., Wei, S.H., Huang, T.H. (2004) Methylation-specific oligonucleotide microarray. *Methods Mol. Biol.*, **287**, 251–260.

102 Keshet, I., Schlesinger, Y., Farkash, S., Rand, E., Hecht, M., Segal, E., Pikarski, E., Young, R.A., Niveleau, A., Cedar, H., Simon, I. (2006) Evidence for an instructive mechanism of de novo methylation in cancer cells. *Nat. Genet.*, **38**(2), 149–153.

103 Novak, P., Jensen, T., Oshiro, M.M., Wozniak, R.J., Nouzova, M., Watts, G.S., Klimecki, W.T., Kim, C., Futscher, B.W. (2006) Epigenetic inactivation of the HOXA gene cluster in breast cancer. *Cancer Res.*, **66**(22), 10664–10670.

104 Rauch, T., Li, H., Wu, X., Pfeifer, G.P. (2006) MIRA-assisted microarray analysis, a new technology for the determination of DNA methylation patterns, identifies frequent methylation of homeodomain-containing genes in lung cancer cells. *Cancer Res.*, **66**(16), 7939–7947.

105 Zhang, X., Yazaki, J., Sundaresan, A., Cokus, S., Chan, S.W., Chen, H., Henderson, I.R., Shinn, P., Pellegrini, M., Jacobsen, S.E., Ecker, J.R. (2006) Genome-wide high-resolution mapping and functional analysis of DNA methylation in *Arabidopsis*. *Cell*, **126**(6), 1189–1201.

106 Zilberman, D., Gehring, M., Tran, R.K., Ballinger, T., Henikoff, S. (2007) Genome-wide analysis of *Arabidopsis thaliana* DNA methylation uncovers an interdependence between methylation and transcription. *Nat. Genet.*, **39**(1), 61–69.

107 Cokus, S.J., Feng, S., Zhang, X., Chen, Z., Merriman, B., Haudenschild, C.D., Pradhan, S., Nelson, S.F., Pellegrini, M., Jacobsen, S.E. (2008) Shotgun bisulphite sequencing of the *Arabidopsis* genome reveals DNA methylation patterning. *Nature*, **452**(7184), 215–219.

108 Lippman, Z., Gendrel, A.V., Black, M., Vaughn, M.W., Dedhia, N., McCombie, W.R., Lavine, K., Mittal, V., May, B., Kasschau, K.D., Carrington, J.C., Doerge, R.W., Colot, V., Martienssen, R. (2004)

Role of transposable elements in heterochromatin and epigenetic control. *Nature*, **430**(6998), 471–476.

109 Hong, C., Moorefield, K.S., Jun, P., Aldape, K.D., Kharbanda, S., Phillips, H.S., Costello, J.F. (2007) Epigenome scans and cancer genome sequencing converge on WNK2, a kinase-independent suppressor of cell growth. *Proc. Natl Acad. Sci. USA*, **104**(26), 10974–10979.

110 Karpf, A.R., Jones, D.A. (2002) Reactivating the expression of methylation silenced genes in human cancer. *Oncogene*, **21**(35), 5496–5503.

111 Karpf, A.R., Peterson, P.W., Rawlins, J.T., Dalley, B.K., Yang, Q., Albertsen, H., Jones, D.A. (1999) Inhibition of DNA methyltransferase stimulates the expression of signal transducer and activator of transcription 1, 2, and 3 genes in colon tumor cells. *Proc. Natl Acad. Sci. USA*, **96**(24), 14007–14012.

112 Shames, D.S., Girard, L., Gao, B., Sato, M., Lewis, C.M., Shivapurkar, N., Jiang, A., Perou, C.M., Kim, Y.H., Pollack, J.R.,, Fong, K.M., Lam, C.L., Wong, M., Shyr, Y., Nanda, R., Olopade, O.I., Gerald, W., Euhus, D.M., Shay, J.W., Gazdar, A.F., Minna, J.D. (2006) A genome-wide screen for promoter methylation in lung cancer identifies novel methylation markers for multiple malignancies. *PLoS Med.*, **3**(12), e486.

113 Suzuki, H., Gabrielson, E., Chen, W., Anbazhagan, R., van Engeland, M., Weijenberg, M.P., Herman, J.G., Baylin, S.B. (2002) A genomic screen for genes upregulated by demethylation and histone deacetylase inhibition in human colorectal cancer. *Nat. Genet.*, **31**(2), 141–149.

114 Meissner, A., Gnirke, A., Bell, G.W., Ramsahoye, B., Lander, E.S., Jaenisch, R. (2005) Reduced representation bisulfite sequencing for comparative high-resolution DNA methylation analysis. *Nucleic Acids Res.*, **33**(18), 5868–5877.

115 Eckhardt, F., Lewin, J., Cortese, R., Rakyan, V.K., Attwood, J., Burger, M., Burton, J., Cox, T.V., Davies, R., Down, T.A., Haefliger, C., Horton, R., Howe, K., Jackson, D.K., Kunde, J., Koenig, C., Liddle, J., Niblett, D., Otto, T., Pettett, R., Seemann, S., Thompson, C., West, T., Rogers, J., Olek, A., Berlin, K., Beck, S. (2006) DNA methylation profiling of human chromosomes 6, 20 and 22. *Nat. Genet.*, **38**(12), 1378–1385.

116 Rakyan, V.K., Hildmann, T., Novik, K.L., Lewin, J., Tost, J., Cox, A.V., Andrews, T.D., Howe, K.L., Otto, T., Olek, A., Fischer, J., Gut, I.G., Berlin, K., Beck, S. (2004) DNA methylation profiling of the human major histocompatibility complex: a pilot study for the human epigenome project. *PLoS Biol.*, **2**(12), e405.

117 Raval, A., Lucas, D.M., Matkovic, J.J., Bennett, K.L., Liyanarachchi, S., Young, D.C., Rassenti, L., Kipps, T.J., Grever, M.R., Byrd, J.C., Plass, C. (2005) TWIST2 demonstrates differential methylation in immunoglobulin variable heavy chain mutated and unmutated chronic lymphocytic leukemia. *J. Clin. Oncol.*, **23**(17), 3877–3885.

118 Rush, L.J., Dai, Z.Y., Smiraglia, D.J., Gao, X., Wright, F.A., Fruhwald, M., Costello, J.F., Held, W.A., Yu, L., Krahe, R., Kolitz, J.E., Bloomfield, C.D., Caligiuri, M.A., Plass, C. (2001) Novel methylation targets in de novo acute myeloid leukemia with prevalence of chromosome 11 loci. *Blood*, **97**(10), 3226–3233.

119 Yan, P.S., Chen, C.M., Shi, H.D., Rahmatpanah, F., Wei, S.H., Caldwell, C.W., Huang, T.H.M. (2001) Dissecting complex epigenetic alterations in breast cancer using CpG island microarrays. *Cancer Res.*, **61**(23), 8375–8380.

120 Bock, C., Paulsen, M., Tierling, S., Mikeska, T., Lengauer, T., Walter, J. (2006) CpG island methylation in human lymphocytes is highly correlated with DNA sequence, repeats, and predicted DNA structure. *PLoS Genet.*, **2**(3), e26.

121 Fang, F., Fan, S., Zhang, X., Zhang, M.Q. (2006) Predicting methylation status of CpG islands in the human brain. *Bioinformatics*, **22**(18), 2204–2209.

122 Feltus, F.A., Lee, E.K., Costello, J.F., Plass, C., Vertino, P.M. (2003) Predicting aberrant CpG island methylation. *Proc. Natl Acad. Sci. USA*, **100**(21), 12253–12258.

123 Feltus, F.A., Lee, E.K., Costello, J.F., Plass, C., Vertino, P.M. (2006) DNA motifs associated with aberrant CpG island methylation. *Genomics*, **87**(5), 572–579.

124 Ohm, J.E., McGarvey, K.M., Yu, X., Cheng, L., Schuebel, K.E., Cope, L., Mohammad,

H.P., Chen, W., Daniel, V.C., Yu, W., Berman, D.M., Jenuwein, T., Pruitt, K., Sharkis, S.J., Watkins, D.N., Herman, J.G., Baylin, S.B. (2007) A stem cell-like chromatin pattern may predispose tumor suppressor genes to DNA hypermethylation and heritable silencing. *Nat. Genet.*, **39**(2), 237–242.

125 Schlesinger, Y., Straussman, R., Keshet, I., Farkash, S., Hecht, M., Zimmerman, J., Eden, E., Yakhini, Z., Ben-Shushan, E., Reubinoff, B.E., Bergman, Y., Simon, I., Cedar, H. (2007) Polycomb-mediated methylation on Lys27 of histone H3 pre-marks genes for de novo methylation in cancer. *Nat. Genet.*, **39**(2), 232–236.

126 Widschwendter, M., Fiegl, H., Egle, D., Mueller-Holzner, E., Spizzo, G., Marth, C., Weisenberger, D.J., Campan, M., Young, J., Jacobs, I., Laird, P.W. (2007) Epigenetic stem cell signature in cancer. *Nat. Genet.*, **39**(2), 157–158.

127 Reynolds, P.A., Sigaroudinia, M., Zardo, G., Wilson, M.B., Benton, G.M., Miller, C.J., Hong, C., Fridlyand, J., Costello, J.F., Tlsty, T.D. (2006) Tumor suppressor p16INK4A regulates polycomb-mediated DNA hypermethylation in human mammary epithelial cells. *J. Biol. Chem.*, **281**(34), 24790–24802.

128 Vire, E., Brenner, C., Deplus, R., Blanchon, L., Fraga, M., Didelot, C.,, Morey, L., Van Eynde, A., Bernard, D., Vanderwinden, J.M., Bollen, M., Esteller, M., Di Croce, L., de Launoit, Y., Fuks, F. (2006) The Polycomb group protein EZH2 directly controls DNA methylation. *Nature*, **439**(7078), 871–874.

129 Herman, J.G., Latif, F., Weng, Y., Lerman, M.I., Zbar, B., Liu, S., Samid, D., Duan, D.S., Gnarra, J.R., Linehan, W.M. (1994) Silencing of the VHL tumor-suppressor gene by DNA methylation in renal carcinoma. *Proc. Natl Acad. Sci. USA*, **91**(21), 9700–9704.

130 Momparler, R.L., Eliopoulos, N., Ayoub, J. (2000) Evaluation of an inhibitor of DNA methylation, 5-aza-2′-deoxycytidine, for the treatment of lung cancer and the future role of gene therapy. *Adv. Exp. Med. Biol.*, **465**, 433–446.

131 Dai, Z., Zhu, W.G., Morrison, C.D., Brena, R.M., Smiraglia, D.J., Raval, A., Wu, Y.Z., Rush, L.J., Ross, P., Molina, J.R., Otterson, G.A., Plass, C. (2003) A comprehensive search for DNA amplification in lung cancer identifies inhibitors of apoptosis cIAP1 and cIAP2 as candidate oncogenes. *Hum. Mol. Genet.*, **12**(7), 791–801.

132 Esteller, M. (2003) Cancer epigenetics: DNA methylation and chromatin alterations in human cancer. *Adv. Exp. Med. Biol.*, **532**, 39–49.

133 Han, S.Y., Iliopoulos, D., Druck, T., Guler, G., Grubbs, C.J., Pereira, M., Zhang, Z., You, M., Lubet, R.A., Fong, L.Y., Huebner, K. (2004) CpG methylation in the Fhit regulatory region: relation to Fhit expression in murine tumors. *Oncogene*, **23**(22), 3990–3998.

134 Kim, H., Kwon, Y.M., Kim, J.S., Lee, H., Park, J.H., Shim, Y.M., Han, J., Park, J., Kim, D.H. (2004) Tumor-specific methylation in bronchial lavage for the early detection of non-small-cell lung cancer. *J. Clin. Oncol.*, **22**(12), 2363–2370.

135 Kim, J.S., Lee, H., Kim, H., Shim, Y.M., Han, J., Park, J., Kim, D.H. (2004) Promoter methylation of retinoic acid receptor beta 2 and the development of second primary lung cancers in non-small-cell lung cancer. *J. Clin. Oncol.*, **22**(17), 3443–3450.

136 Maruyama, R., Sugio, K., Yoshino, I., Maehara, Y., Gazdar, A.F. (2004) Hypermethylation of FHIT as a prognostic marker in nonsmall cell lung carcinoma. *Cancer*, **100**(7), 1472–1477.

137 Sathyanarayana, U.G., Padar, A., Huang, C.X., Suzuki, M., Shigematsu, H., Bekele, B.N., Gazdar, A.F. (2003) Aberrant promoter methylation and silencing of laminin-5-encoding genes in breast carcinoma. *Clin. Cancer Res.*, **9**(17), 6389–6394.

138 Sorm, F., Piskala, A., Cihak, A., Vesely, J. (1964) 5-Azacytidine, a new, highly effective cancerostatic. *Experientia*, **20**(4), 202–203.

139 Jones, P.A., Taylor, S.M. (1980) Cellular differentiation, cytidine analogs and DNA methylation. *Cell*, **20**(1), 85–93.

140 Goffin, J., Eisenhauer, E. (2002) DNA methyltransferase inhibitors-state of the art. *Ann. Oncol.*, **13**(11), 1699–1716.

141 Issa, J.P., Garcia-Manero, G., Giles, F.J., Mannari, R., Thomas, D., Faderl, S., Bayar, E., Lyons, J., Rosenfeld, C.S., Cortes, J., Kantarjian, H.M. (2004) Phase

1 study of low-dose prolonged exposure schedules of the hypomethylating agent 5-aza-2′-deoxycytidine (decitabine) in hematopoietic malignancies. *Blood*, **103**(5), 1635–1640.

142 Yan, L., Nass, S.J., Smith, D., Nelson, W.G., Herman, J.G., Davidson, N.E. (2003) Specific inhibition of DNMT1 by antisense oligonucleotides induces re-expression of estrogen receptor-alpha (ER) in ER-negative human breast cancer cell lines. *Cancer Biol. Ther.*, **2**(5), 552–556.

143 Chuang, J.C., Yoo, C.B., Kwan, J.M., Li, T.W., Liang, G., Yang, A.S., Jones, P.A. (2005) Comparison of biological effects of non-nucleoside DNA methylation inhibitors versus 5-aza-2′-deoxycytidine. *Mol. Cancer Ther.*, **4**(10), 1515–1520.

144 Juttermann, R., Li, E., Jaenisch, R. (1994) Toxicity of 5-aza-2′-deoxycytidine to mammalian cells is mediated primarily by covalent trapping of DNA methyltransferase rather than DNA demethylation. *Proc. Natl Acad. Sci. USA*, **91**(25), 11797–11801.

145 Ghoshal, K., Datta, J., Majumder, S., Bai, S., Kutay, H., Motiwala, T., Jacob, S.T. (2005) 5-Aza-deoxycytidine induces selective degradation of DNA methyltransferase 1 by a proteasomal pathway that requires the KEN box, bromo-adjacent homology domain, and nuclear localization signal. *Mol. Cell. Biol.*, **25**(11), 4727–4741.

146 Byrd, J.C., Stilgenbauer, S., Flinn, I.W. (2004) Chronic lymphocytic leukemia. *Hematology (Am. Soc. Hematol. Educ. Program)*, **1**, 163–183.

147 Lubbert, M. (2000) DNA methylation inhibitors in the treatment of leukemias, myelodysplastic syndromes and hemoglobinopathies: clinical results and possible mechanisms of action. *Curr. Top. Microbiol. Immunol.*, **249**, 135–164.

148 Issa, J.P. (2000) CpG-island methylation in aging and cancer. *Curr. Top. Microbiol. Immunol.*, **249**, 101–118.

149 Richardson, B. (2003) Impact of aging on DNA methylation. *Ageing Res. Rev.*, **2**(3), 245–261.

150 Ahuja, N., Issa, J.P. (2000) Aging, methylation and cancer. *Histol. Histopathol.*, **15**(3), 835–842.

151 Issa, J.P., Ahuja, N., Toyota, M., Bronner, M.P., Brentnall, T.A. (2001) Accelerated age-related CpG island methylation in ulcerative colitis. *Cancer Res.*, **61**(9), 3573–3577.

152 Belinsky, S.A. (2004) Gene-promoter hypermethylation as a biomarker in lung cancer. *Nat. Rev. Cancer*, **4**(9), 707–717.

153 Sidransky, D. (2002) Emerging molecular markers of cancer. *Nat. Rev. Cancer*, **2**(3), 210–219.

154 Oakeley, E.J., Schmitt, F., Jost, J.P. (1999) Quantification of 5-methylcytosine in DNA by the chloroacetaldehyde reaction. *Biotechniques*, **27**(4), 744–746, 748–750, 752.

155 Costello, J.F., Futscher, B.W., Kroes, R.A., Pieper, R.O. (1994) Methylation-related chromatin structure is associated with exclusion of transcription factors from and suppressed expression of the *O*-6-methylguanine DNA methyltransferase gene in human glioma cell lines. *Mol. Cell. Biol.*, **14**(10), 6515–6521.

156 Costello, J.F., Futscher B.W., Tano, K., Graunke, D.M., Pieper, R.O. (1994) Graded methylation in the promoter and body of the *O*-6-methylguanine DNA methyltransferase (MGMT) gene correlates with MGMT expression in human glioma cells. *J. Biol. Chem.*, **269**(25), 17228–17237.

157 Harris, L.C., Remack, J.S., Brent, T.P. (1994) In vitro methylation of the human *O*-6-methylguanine-DNA methyltransferase promoter reduces transcription. *Biochim. Biophys. Acta*, **1217**(2), 141–146.

158 Esteller, M., Garcia-Foncillas, J., Andion, E., Goodman, S.N., Hidalgo, O.F., Vanaclocha, V., Baylin, S.B., Herman, J.G. (2000) Inactivation of the DNA-repair gene MGMT and the clinical response of gliomas to alkylating agents. *N. Engl. J. Med.*, **343**(19), 1350–1354.

159 Hegi, M.E., Diserens, A.C., Gorlia, T., Hamou, M.F., de Tribolet, N., Weller, M., Kros, J.M., Hainfellner, J.A., Mason, W., Mariani, L., Bromberg, J.E., Hau, P., Mirimanoff, R.O., Cairncross, J.G., Janzer, R.C., Stupp, R. (2005) MGMT gene silencing and benefit from temozolomide in glioblastoma. *N. Engl. J. Med.*, **352**(10), 997–1003.

160 Abbosh, P.H., Montgomery, J.S., Starkey, J.A., Novotny, M., Zuhowski, E.G., Egorin, M.J., Moseman, A.P., Golas, A., Brannon, K.M., Balch, C., Huang, T.H., Nephew, K.P. (2006) Dominant-negative histone H3 lysine 27 mutant derepresses silenced tumor suppressor genes and reverses the drug-resistant phenotype in cancer cells. *Cancer Res.*, **66**(11), 5582–5591.

161 Brena, R.M., Plass, C., Costello, J.F. (2006) Mining methylation for early detection of common cancers. *PLoS Med.*, **3**(12), e479.

162 Cui, H., Cruz-Correa, M., Giardiello, F.M., Hutcheon, D.F., Kafonek, D.R., Brandenburg, S., Wu, Y., He, X., Powe, N.R., Feinberg, A.P. (2003) Loss of IGF2 imprinting: a potential marker of colorectal cancer risk. *Science*, **299**(5613), 1753–1755.

163 Cairns, P., Esteller, M., Herman, J.G., Schoenberg, M., Jeronimo, C., Sanchez-Cespedes, M., Chow, N.H., Grasso, M., Wu, L., Westra, W.B., Sidransky, D. (2001) Molecular detection of prostate cancer in urine by GSTP1 hypermethylation. *Clin. Cancer Res.*, **7**(9), 2727–2730.

164 Krassenstein, R., Sauter, E., Dulaimi, E., Battagli, C., Ehya, H., Klein-Szanto, A., Cairns, P. (2004) Detection of breast cancer in nipple aspirate fluid by CpG island hypermethylation. *Clin. Cancer Res.*, **10**(1, Pt 1), 28–32.

165 Chen, Y.T., Stockert, E., Chen, Y., Garin-Chesa, P., Rettig, W.J., van der Bruggen, P., Boon, T., Old, L.J. (1994) Identification of the MAGE-1 gene product by monoclonal and polyclonal antibodies. *Proc. Natl Acad. Sci. USA*, **91**(3), 1004–1008.

24
Methylomes

Pao-Yang Chen and Matteo Pellegrini
Department of Molecular Cell and Developmental Biology, University of California, 610 Charles Young Drive East, Los Angeles, CA, 90095, USA

Epigenetic Regulation and Epigenomics: Advances in Molecular Biology and Medicine, First Edition. Edited by Robert A. Meyers.

Keywords

DNA methylation
This occurs predominantly on cytosines at the carbon 5 position, yielding 5-methylcytosines (5meC or mC). It is catalyzed by DNA methyltransferases, is the most stable of all epigenetic modifications and was first discovered in 1948.

CpG islands
Sequences frequently associated with promoters and enriched (compared to the genome average) in cytosine–guanosine (CpG) dinucleotides. They are often found to be regions in the genome that are subjected to epigenetic modulation.

DNA methyltransferase (DNMT)
DNA methylation is mediated by the DNA methyltransferase (DNMT) family of enzymes, that catalyze the transfer of a methyl group from *S*-adenosyl methionine to DNA. The methylation patterns are initially established by *de novo* DNMTs in early embryonic development, and maintained by maintenance DNMTs.

Hydroxymethylation
Hydroxylation in methylcytosine produces 5-hydroxymethylcytosine (5hmC). Studies *in vitro* and in cultured cells have shown that human tet oncogene 1 (TET1) is capable of hydrolyzing 5meC to produce 5hmC in DNA. Bisulfite-based sequencing methods cannot distinguish between methylation and hydroxymethylation; thus, data already deposited in public databases contains a mixture of both marks. It has been shown recently that single-molecule, real-time (SMRT) sequencing is able to distinguish 5meC from 5hmC. It has also been reported that 5hmC is associated with enhancers and gene bodies in human embryonic stem cells.

Promoter methylation
Most gene promoters are associated with CpG islands, and usually are unmethylated. Promoter methylation is often associated with gene silencing.

Gene body methylation
When methylation occurs within the body of the gene, it may facilitate transcription by preventing spurious transcription initiations. In diseases that lead to aberrant methylation, genes can sometimes be demethylated, allowing transcription to be initiated at several incorrect sites.

Repetitive sequences

Repetitive sequences such as transposable elements are usually hypermethylated, preventing chromosomal instability, translocations, and gene disruption.

1 An Introduction to Methylomes

The methylation of DNA cytosine residues, which was first discovered in 1948 [1–3], is a common epigenetic mark in many eukaryotes, and is often found in the sequence context CG or CHG (H = A, T, C). When located at gene promoters, DNA methylation is usually a repressive mark. On the other hand, DNA methylation is increased within actively transcribed genes in plants and mammals. DNA methylation has been found to be an important factor in the regulation of embryonic development, disease progression, and aging. In many cancers, there is a dysregulation of DNA methylation leading to a loss of global methylation, hypermethylated promoters in tumor suppressor genes, and hypomethylated transposons. In monozygotic twins (MZ) differences between their methylation profiles are found to increase with age, and some of the differentially methylated loci are associated with age-related diseases. The current revolution in sequencing technology has enabled the generation of single-base-pair resolution whole-genome DNA methylation profiling, that is, the methylome. This allows, for the first time, DNA methylation and its interactions with other epigenetic and genetic factors to be studied systematically.

2 Technology

Over the past decade there has been a revolution in DNA methylation analysis technology, with analyses that previously were restricted to specific loci now being performed on a genome-scale such that entire methylomes can be characterized, at single-base-pair resolution. In this chapter, several of the current technologies employed for genome profiling are introduced. Moreover, these high-resolution, genome-wide DNA methylation profiling techniques have been used to characterize the methylome of several model organisms, including *Arabidopsis thaliana*, mouse, rice, and human [4–11].

These profiling methods can be broadly categorized into three main approaches, according to the methylation-dependent treatment of the DNA, namely *affinity enrichment*, *enzyme digestion*, and *bisulfite conversion* [6, 12]. The basic principles of DNA methylation profiling have been reviewed by Laird [6].

Typically, methylcytosines cannot be distinguished from unmethylated cytosine by using hybridization-based methods. In addition, as the DNA methyltransferases (DNMTs) are not present during polymerase chain reaction (PCR) or in biological cloning systems, DNA methylation information will be erased during amplification. Consequently, methylation-dependent treatments of the DNA before amplification or hybridization are generally employed in sequence-specific DNA methylation analysis techniques. Following the treatment of genomic DNA with one of these three methylation-dependent steps, either microarrays or high-throughput DNA sequencing can be used to reveal the location of the methylcytosine residues. Hence, depending on the profiling methods utilized, a variety of analytical steps have been developed for determining DNA methylation patterns and profiles [1, 12–14].

2.1 Affinity Enrichment-Based Methods

The affinity enrichment of methylated regions employs antibodies that are specific for methylcytosines (in the context of denatured DNA, this might involve methylated DNA immunoprecipitation; MeDIP) or by using methyl-binding domain (MBD) proteins with an affinity for methylated native genomic DNA, followed by microarray hybridization or next-generation sequencing. Affinity enrichment-based methods have been identified as powerful tools for the comprehensive profiling of DNA methylation. While the affinity purification of methylated DNA was first demonstrated with the methyl-binding protein 2 (MeCP2) [15], affinity enrichment followed by microarray has also been used for profiling methylomes in plants [16], mouse [17], and human [18]:

- **Methylated DNA immunoprecipitation (MeDIP):** This technique involves the immunoprecipitation of the methylated fraction of a genomic DNA sample with a monoclonal antibody against methylcytosine [14], followed either by hybridization of the immunoprecipitated fraction against the input fraction on a microarray [19, 20] or next-generation sequencing (MeDIP-seq) [21]. Subsequent MeDIP data were found to be enriched for highly methylated and high-CpG density regions that are often subjected to epigenetic modulation [22]. The main drawback of this technique is the production of tiling arrays, but it may also be confounded by cross-hybridization signals. In most genomes, relatively heavily methylated regions of repeat DNA have neither been sequenced nor assembled, and are thus often missing from microarray designs. As a result, microarray technology is now being replaced by high-throughput short-read sequencing as the method of choice [13]. Typically, MeDIP-seq will enrich for sequences containing methylcytosines so as to create moderate resolution methylation profiles; however, unlike MeDIP-chip, MeDIP-seq is not limited to detecting methylation at loci tiled by oligonucleotides on a microarray.
- **MethylCap-Seq:** This method involves the capture of methylated DNA using the MBD domain of MeCP2, and a subsequent next-generation sequencing of the eluted DNA. The elution of the captured methylated DNA is accomplished using a salt gradient, which stratifies the genome into fractions with different CpG densities [23].
- **MBD-Seq:** This employs the MBD2 protein methyl-CpG binding domain to enrich for methylated double-stranded DNA fragments [24].

Affinity enrichment-based methods are more tolerant of DNA impurity and integrity, but many require substantial quantities of input genomic DNA in order to produce sufficient output of enriched DNA [6]. These methods allow for a rapid and efficient genome-wide assessment of DNA methylation, but they do not yield information regarding individual CpG dinucleotides. Moreover, they require substantial experimental (sequencing input) or statistical/bioinformatic adjustment for varying CpG densities at different regions of the genome [25]. Windows containing from a few hundreds to thousands of base pairs are used to test the enrichment of methylation signals, and this limits its resolution. Such windows may include CpGs that are not directly covered by a read [21]. It is difficult to estimate the percentage methylation of a window; rather, these are typically assigned as positive and negative methylation groups.

A major advantage of affinity-based methods, however, is that they are relatively inexpensive compared to whole-genome bisulfite sequencing techniques.

2.2 Enzyme Digestion-Based Methods

Restriction enzyme digestion employs methylcytosine-sensitive enzymes to digest DNA, followed by size fractionation and hybridization to custom microarrays. A variety of restriction enzyme-based methods have been described, using combinations of restriction enzymes that either do not restrict methylated DNA or enzymes that only restrict methylated DNA (e.g., restriction enzyme McrBC). These approaches have been improved recently with respect to sensitivity and quantification, however, and now offer the advantage of a larger genomic coverage than the traditional bisulfite-based methods [1, 26]. They may also offer single-nucleotide resolution at sites cut by the specific enzyme.

- **HpaII tiny fragment enrichment by ligation-mediated PCR(HELP):** In this approach, DNA is first digested in parallel with MspI and HpaII (which is resistant to DNA methylation). The HpaII and MspI products are then amplified by ligation-mediated PCR, and hybridized using separate fluorochromes to a customized array [18, 27]. As a result, HELP assays mainly detect methylated CpG sites.
- **Restriction enzyme (McrBC):** Here, the DNA is digested with McrBC, which cleaves half of the methylated DNA in the genome and most methylated CpG islands [28]. As a consequence, the digested fractions will be enriched for unmethylated CpG sites.

In general, enzyme digestion-based methods tend to require DNA of high purity, quantity, and integrity; they also employ a variety of enzymes and are able to detect both methylated and unmethylated DNAs. In some cases, methylation may also be detected in repeat sequences. Whilst, in general, these methods are less expensive than the other approaches discussed here, the main drawbacks are that it is essential to achieve a complete digestion, and that there is a detection bias towards regions that contain the restriction site of the enzyme. Methylation-sensitive restriction enzyme-based methods are able to resolve methylation differences in low-CpG density regions (e.g., HELP), whereas McrBC-based and affinity-based methods provide better results with high-CG density regions [6, 18].

2.3 Bisulfite Conversion-Based Methods

2.3.1 Bisulfite Sequencing

Bisulfite sequencing can be used to determine the methylation status of cytosines, at single-nucleotide resolution. Briefly, single-stranded DNA is treated with bisulfite; this causes the cytosines to be sulfanated, while the methylated cytosines are unaffected. The cytosine is then deaminated and desulfanated to uracil [29]. The converted DNA is amplified using PCR with appropriate primer pairs, after which the PCR products are directly sequenced and aligned with unconverted DNA, so as to reveal which cytosines were methylated. This method effectively converts an epigenetic state into a genetic polymorphism, which then can be detected using PCR, microarrays, or direct sequencing.

Previously, traditional methods could only determine DNA methylation for specific loci by using Sanger sequencing of bisulfite-converted and PCR-amplified genomic DNA fragments. However, by combining bisulfite sequencing with the next-generation sequencing (BS-seq or MethylC-seq), it is now possible to profile genome-wide DNA methylation at single-base resolution. BS-seq is considered to be the "gold standard" for determining the DNA methylome, and has been used recently for profiling a variety of organisms, including humans [4, 7, 8, 30, 31].

2.3.2 GoldenGate

Illumina has adapted its GoldenGate BeadArray (a high-throughput, single-nucleotide polymorphism-genotyping system) to interrogate DNA methylation in genomic DNA samples [32]. For this, a multiplexed methylation-specific primer extension of bisulfite-converted DNA, at up to 1536 different CG sites in 96 samples, is performed using primers that are specific for methylated and unmethylated sequences at each site. The primers for the two different methylation states are labeled with different fluorescent dyes, and the products hybridized to bead arrays containing approximately 30 beads per CG site.

2.3.3 Infinium HumanMethylation27

The Infinium platform incorporates a whole-genome amplification step after bisulfite conversion, which is followed by fragmentation and hybridization of the sample to methylation-specific DNA oligomers that are linked to individual bead types. Each bead type corresponds to a specific DNA CpG site and methylation state. The current implementation of the Illumina Infinium assay for DNA methylation analysis (known as the *HumanMethylation27 DNA Analysis BeadChip*) interrogates 27 578 CpG sites (0.1% of human CG sites) from 14 495 protein-coding gene promoters and 110 microRNA gene promoters [33]. Unfortunately, the probes of the Infinium assay cover only a small percentage of all CpGs in the genome, and are located preferentially in unmethylated promoter regions [34].

Although bisulfite conversion-based DNA methylation profiling on arrays is not well-suited to the *de novo* analysis of DNA methylation profiles, it is suited to high-throughput validation or follow-up studies of a limited number of CpG sites in hundreds to thousands of samples.

2.3.4 **Bisulfite Padlock Probe**

The bisulfite padlock probe (BSPP) is a targeted method that isolates selected locations for methylation profiling [35, 36]. The padlock probes are 100-nucleotide DNA fragments designed to hybridize to genomic DNA targets. When the gap between the two hybridized, locus-specific arms of a padlock probe have been polymerized and ligated to form a circular strand of DNA, the circles generated can be amplified using the common "backbone" sequence that connects the two arms. This enables tens of thousands of probes to be used within a single reaction, and the resultant libraries are then analyzed using massively parallel sequencing. Whilst this method generates methylation profiles at single base resolution, sequence-specific biases due to the use of bisulfite-converted DNA may adversely affect the amplification of these libraries.

2.3.5 **Whole-Genome Bisulfite Sequencing**

Whole-genome bisulfite sequencing (BS-seq or MethylC-seq), which is considered the "gold standard" for determining DNA methylomes [7, 8], involves bisulfite conversion followed by high-throughput sequencing. Typically, BS-seq is capable of determining the state of virtually all cytosines in the genome. Two protocols have been developed for constructing bisulfite-converted libraries:

- The protocol of Cokus *et al.* [7] employs two amplification steps. The first amplification generates both forward and reverse bisulfite-converted sequences, ligated with DNA adapters of DpnI restriction sites. These sequences are then digested with DpnI restriction enzymes, which results in 5 bp sequence tags on the bisulfite-converted sequences. Two patterns of tags are created, based on the forward (+FW) and reverse (−FW) directions of the bisulfite-converted sequences. After ligation with standard Illumina adaptors and a second amplification step, four types of bisulfite-converted reads are generated: forward and reverse reads from Watson (+FW, +RC) and Crick stands (−FW, −RC), respectively. These tags are essential to reduce the ambiguity of certain classes of reads.
- The protocol of Lister *et al.* [8] is used to generate bisulfite libraries using premethylated adapters, and in this case no tags are present and all reads are forward (+FW or −FW).

The traditional aligners such as BLAT [37], SOAP [38], and Bowtie [39] are not suitable for aligning bisulfite-converted reads, because they do not explicitly account for the conversion of cytosines in the reads. A few newly developed aligners, such as BS-Seeker [40], BSMAP [41], RMAP [42], and PASH [43], have been designed explicitly for mapping bisulfite-converted reads. Thus, by comparing the genomic Cs with the aligned reads, it is possible to determine the methylation level for each cytosine.

2.3.6 **Reduced Representation Bisulfite Sequencing**

Reduced representation bisulfite sequencing (RRBS) employs an enzymatic digestion to reduce genome complexity, followed by bisulfite treatment and next-generation sequencing [44, 45]. The use of MspI, TaqI, or BglII methylation-insensitive restriction enzyme enriches for CG-rich regions of the genome. Subsequent computational analysis has revealed to digest the mouse genomic DNA with MspI, and then to select 40- to 220-bp fragments, would cover 4.8% of all CpG sites which intersect 90% of CpG islands; this would represent a 47-fold enrichment of CpG island sequences in the resultant library [1, 46].

Bisulfite treatment not only requires DNA denaturation before treatment but also causes substantial DNA degradation, with further purification needed to remove the sodium bisulfite. For these reasons, the input DNA for many bisulfite-based methods may be of low purity and integrity [6]. Yet, two artifacts may arise from the BS-seq approach: (i) an incomplete bisulfite conversion of unmethylated cytosines to uracil; and (ii) a bias caused by the fact that methylated and unmethylated sequences do not always undergo PCR amplification at the same efficiency following bisulfite treatment [47]. Incomplete conversion occurs mainly because bisulfite attacks only cytosines in single-stranded DNA. In areas of the genome with a high GC content, the DNA may not be denatured completely, which would result in patches of unmodified cytosines. For example, in the study of the first *Arabidopsis* methylome [7], bisulfite reads with three or more methylated CHH sites in a row were considered false positives, and discarded from further analysis.

2.4 Comparison of Methods

The choice of profiling methods may be influenced by several factors, including the number of samples, the quality and quantity of DNA, as well as the desired coverage, data resolution, cost, and size of genome. It is also necessary to take into account the organism that is being studied: an array-based analysis requires that a suitable array for the species of interest is available, whereas sequence-based analyses are generally applicable to any species for which a reference genome exists [6, 12].

Bock *et al.* [34] performed a quantitative comparison of genome-wide DNA methylation mapping techniques, ranging from affinity-based methods (MeDIP-seq, MethylCap-seq) to bisulfite conversion-based methods (Infinium HumanMethylation27, RRBS). Subsequently, it was found that MeDIP-seq and MethylCap-seq could each distinguish between methylated and unmethylated regions almost as precisely as RRBS, but were less accurate for quantifying the DNA methylation levels in partially methylated genomic regions. In terms of genome coverage, MeDIP-seq and MethylCap-seq provided a broad coverage of the genome, whereas the RRBS and Infinium arrays were restricted to CpG islands and promoter regions.

One major advantage of sequencing-based methods over microarrays is their ability to interrogate CpGs in repetitive elements. Approximately 45% of the human genome is derived from transposable elements (TEs), a major driving force in the evolution of mammalian gene regulation, with almost half of all CpGs falling within these repetitive regions [48]. The extent to which different sequencing-based methods will interrogate repeats is therefore of considerable interest [21], and consequently some comparisons among profiling methods, followed by sequencing, were conducted to reveal any differences [1, 14, 21]. Overall, a high degree of concordance among the methods was observed in the different comparisons, the key metrics of which were presented by Beck *et al.* [2].

In addition, when Harris *et al.* compared MeDIP-seq, MBD-seq, MethylC-seq, and RRBS [21], within the two affinity enrichment-based methods they showed first, that MeDIP-seq enrichment occurred preferentially in regions with a low CpG density, while MBD-seq enriched for regions with a high CpG density. Second, the MeDIP-seq signal was seen to increase in regions with high non-CpG cytosine methylation, whereas MBD-seq did not show this trend; this suggested that different biases were present in the two enrichment methods. Both results were also confirmed by Li *et al.* [14] whereby, for the two bisulfite conversion-based methods, RRBS was shown to generate a significant coverage of CpGs in CpG islands, whereas BS-seq offered a greater CpG coverage genome-wide. This difference suggested that RRBS would be the method of choice if CpG islands were the main focus of a study.

The main strengths of bisulfite-based methods include a single-base resolution and an ability to quantify methylation levels. Affinity enrichment-based methods are generally less expensive than bisulfite-based methods, albeit at reduced resolution. A second potential advantage of the enrichment methods is that all four nucleotides are retained, as no conversion takes place, but this increases the rate of uniquely mappable sequence reads. Unfortunately, however, affinity enrichment-based methods do not allow the precise quantification of methylation levels.

Recently, interest in the study of hydroxymethylation has increased, with various studies being targeted at an understanding of hydroxymethylation and its possible role in DNA demethylation. Unfortunately, the bisulfite-based sequencing technique is unable to detect which cytosines are hydroxymethylated, as these are indistinguishable from 5-methylcytosine [49]. However, the recently developed single-molecule real-time (SMRT) sequencing approach has been shown capable of profiling N6-methyladenine (mA), 5-methylcytosine (mC), and 5-hydroxymethylcytosine (hmC) in real time [50].

2.5 Aligning Bisulfite-Converted Reads

Whole-genome bisulfite sequencing (BS-seq) provides unbiased genome-wide coverage, single-base resolution, and quantitative methylation measurements. However, as the bisulfite-converted reads are modified from the original genomic sequences, their alignment against reference genome will require customized aligners. Such bisulfite aligners typically implement a three-letter alignment, with some differences in their methods for selecting best hits. The way in which bisulfite-converted reads are mapped to reference genomes, using three-letter alignments as implemented in BS Seeker [40], is described in the following section.

2.5.1 Three-Letter Alignment Algorithm

BS Seeker employs Bowtie [39] for mapping the reads generated from either of the two bisulfite conversion library protocols described in Sect. 2.5. In the case of Cokus *et al.*'s protocol, it generates a forward read (+FW) from the Watson strand, the reverse complement (+RC) of +FW, a forward read (−FW) from the Crick strand, and the reverse complement (−RC) of −FW. BS-Seeker first converts all Cs to Ts on FW reads and both strands of the reference genome, so that the subsequent mapping is performed using only three letters, A, T, G. Similarly, G/A conversion is performed on RC reads and both strands of the reverse complement of the reference genome. Bowtie is then used to map the C → T-converted FW reads to the C → T-converted Watson and Crick strands, and the G → A-converted RC reads to the two G → A-converted reverse complements of the Watson and Crick strands. During each of the four runs of Bowtie, the mapped positions for each read are recorded. When all runs of Bowtie are complete, only unique alignments are retained. Here, "unique alignments" are defined as those that have no other hits with-the same or fewer mismatches in the three-letter alignment (between the converted read and the converted genomic sequence). Finally, the number of mismatches is calculated. For this, a read T that aligns to a genomic C is considered a match, while a read C that aligns to a genomic T is considered a mismatch. Post-processing removes low-quality mappings based on the number of mismatches.

As Lister *et al.*'s protocol generates only +FW and −FW reads, aligning these reads is simpler and, consequently, Bowtie is run only twice. The strategy of converting the reference sequences by treating all Cs as Ts (and Gs as As) has also been used [4, 46].

Other bisulfite ready-alignment programs have also been developed, including BSMAP [41], MAQ [51], RMAP [42, 52], and Bismark [53]. BSMAP enumerates all possible combinations of C → T conversion in the BS reads to identify the uniquely mapping position with the least mismatches on the reference genome. The bisulfite mapping in RMAP uses Wildcard matching for mapping Ts. MAQ also has a methylation alignment mode, and assigns non-unique reads randomly to one of the best-matching positions. Bismark extends BS Seeker to paired end mapping. These aligners were compared to BS Seeker by mapping synthetic reads generated from a genome with known methylation levels [40]. Thus, BSMAP was found to be significantly slower than the other aligners. MAQ's strategy of randomly assigning one of the best-matching positions for non-unique reads, results in a lower accuracy and biased estimates of the methylation rates. Although RMAP and BS-Seeker perform equally when mapping reads from Lister *et al.*'s protocol, when Cokus *et al.*'s protocol is used BS-Seeker outperforms RMAP. Among these aligners, BS-Seeker was found to be the fastest when mapping to large genomes, such as human.

2.6 Downstream Data Analyses

In order to interpret whole-genome DNA methylation profiles it is necessary to generate informative statistics, which include the generation of average methylation levels across genomic features (e.g., metaplots), and the patterns of differential methylation across methylomes. How these procedures are applied to BS-seq data is discussed in the following subsection.

When an incomplete bisulfite conversion occurs, the unmethylated Cs fail to be converted to Ts and this results in false positive methylated cytosine calls. In order to measure the quality of bisulfite sequencing, bisulfite conversion rates are often calculated (i.e., the ratio of the number of converted cytosines to the total number of reads at each site). Unmethylated cytosines include both converted cytosines and false positive methylated cytosines. The rate of false positive methylated cytosines in a data set can be determined by examining the reads that map to known unmethylated regions (e.g., the unmethylated chloroplast genome, or regions that are determined to be unmethylated using by traditional bisulfite sequencing). In an effort to reduce false positive calls, reads that include three methylated CHH sites are usually removed, as these may have arisen due to a lack of conversion in reads

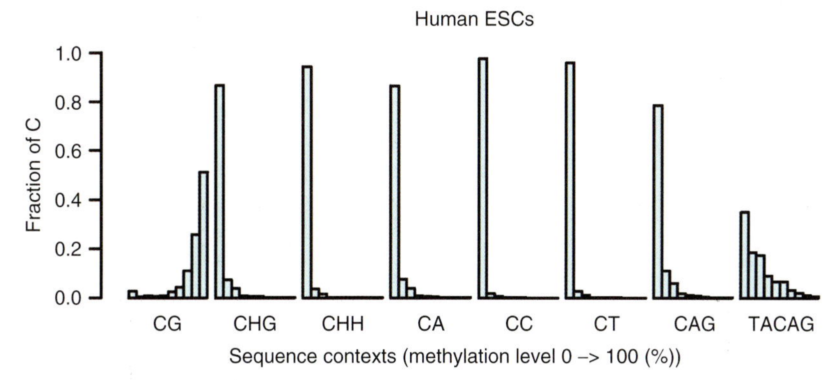

Fig. 1 Histograms of methylation levels per sequence context. Cytosines are binned into 10 groups according to their methylation levels from 0% (unmethylated/lowly methylated) to 100% (highly methylated). ESC, embryonic stem cell.

that form secondary structures. It is estimated that the overall bisulfite conversion rate in human methylomes is 98–99% [9].

A next step in analyzing methylomes involves the computation of global methylation levels (i.e., average percent methylation). In BS-seq data, methylation levels at a cytosine can be estimated from the ratio of the number of methylated reads over the sum of methylated and unmethylated reads. The global methylation level is the average (or median) methylation level among all cytosines covered by the BS-seq data. A filter of minimum coverage may be imposed to avoid biased estimates from sites with low coverage. Due to the sequence specificity of methyltransferases, cytosines within different sequence contexts have distinct methylation levels. In the *Arabidopsis* methylome, the observed methylation levels are 24% for CG sites, 6.7% for CHG sites and 1.7% for CHH sites [7]. The distinct distributions of methylcytosines in CG and non-CG sites observed in human embryonic stem (ES) cells are shown in Fig. 1.

Global methylation levels have been found to vary between cells types and different stages of development. In the mouse, methylation levels in fetal tissue, ES cells and sperm were found to be between 73% and

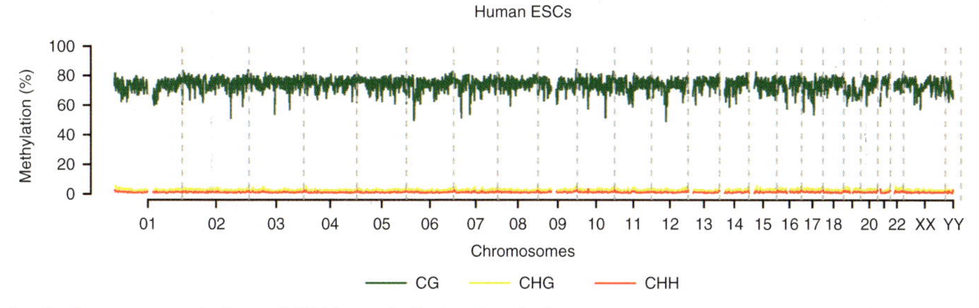

Fig. 2 A chromosomal view of DNA methylation levels in human embryonic stem cells (ESCs). Methylation levels are plotted across chromosomes in the three sequence contexts of CG (green), CHG (yellow), and CHH (red).

85%, whereas those in placenta were lower, at 42%. Both male and female primordial germ cells are known to undergo waves of demethylation during their development, and were observed to have methylation levels of 16.3% and 7.8%, respectively [5]. Cancer tissues are also often hypomethylated [54] and, indeed, global methylation levels may be used as a marker in the diagnosis of cancer. For the global analysis of data sets it is also helpful to generate chromosomal profiles of DNA methylation, which capture global trends

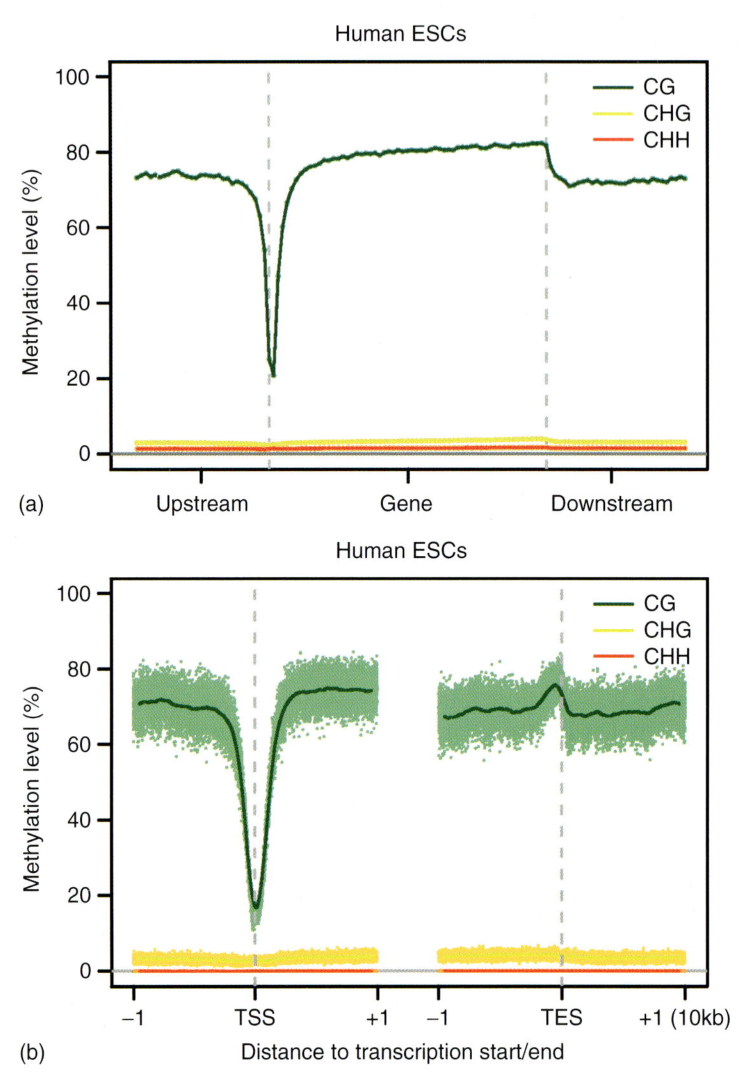

Fig. 3 Metagene plots of DNA methylation levels. (a) A stretch plot of methylation levels from upstream to downstream of coding genes; (b) A linear plot showing methylation levels 10 kbp upstream and downstream of the transcription start sites (TSSs) and transcription end sites (TESs). ESC, embryonic stem cell.

of methylation across chromosomes in megabase windows (Fig. 2).

Methylation levels are also found to vary in a consistent fashion across certain genomic features. Plots that capture these trends (referred to as *"metaplots"*) can highlight average methylation trends across protein-coding genes, pseudogenes, transposons, or exons. These plots are useful for understanding the changes of methylation levels upstream and downstream of these features. For example, in order to generate a metaplot of genes, the transcription start and end sites for selected genes are fixed, while the upstream, body, and downstream regions are binned into a specified number of windows (e.g., 100 bins); the average methylation level is then calculated for each window. A metagene plot summarizes the average methylation level per window, and is usually plotted in the upstream to downstream direction.

Promoter hypomethylation is often correlated with gene expression. In mouse data, for example, methylation levels fell from 80% to 40% in the promoter region, were highly methylated in the gene, and showed a slight decrease in the downstream region as shown in Fig. 5c from the original paper [30]). In these metaplots, the genes are stretched in order to align all of their transcription start and end sites; thus, the plot represents the methylation level in a position that is scaled by the distance between the start and end sites, rather than the physical distance (see Fig. 3a for a metagene stretch plot of human ES cells). Alternatively, a linear metaplot reveals the methylation level in absolute distance (base pairs) from the start sites or the end sites, see Fig. 3b for an example of the metagene linear plot. A linear metaplot is usually plotted from a few kilobases upstream and downstream from the sites of interest, and average methylation levels are calculated per base across all genes. Linear metaplots are helpful when the changes of methylation occur in only a few bases, and are in similar locations across all test genes. In contrast, metaplots can be used to depict methylation levels across various genomic features, and are useful when comparing the methylation patterns between mutants (e.g., methylation mutants) and wild-type, or between disease samples and normal samples.

The visualization of BS-seq data at individual sites requires the use of a genome browser, such as the UCSC genome browser (http://genome.ucsc.edu) or the Anno-J browser (http://www.annoj.org). To visualize individual sites, the methylation data are first converted to a methylation track in a wiggle format (for the UCSC browser) and the track is uploaded to the browser to be analyzed along with other annotation tracks (Fig. 4a and b).

3 Applications

3.1 The First Arabidopsis Methylome Using BS-Seq

In 2008, Cokus *et al.* reported details of the first methylome of an organism, *Arabidopsis thaliana*, by combining a bisulfite treatment of genomic DNA with next-generation sequencing, using the Illumina 1G Genome Analyzer sequencing technology (BS-seq) [7]. In this case, 2.6 billion nucleotides were uniquely mapped, covering 93% of all theoretically mappable cytosines, and reaching on average 10-fold coverage per strand. Two months later,

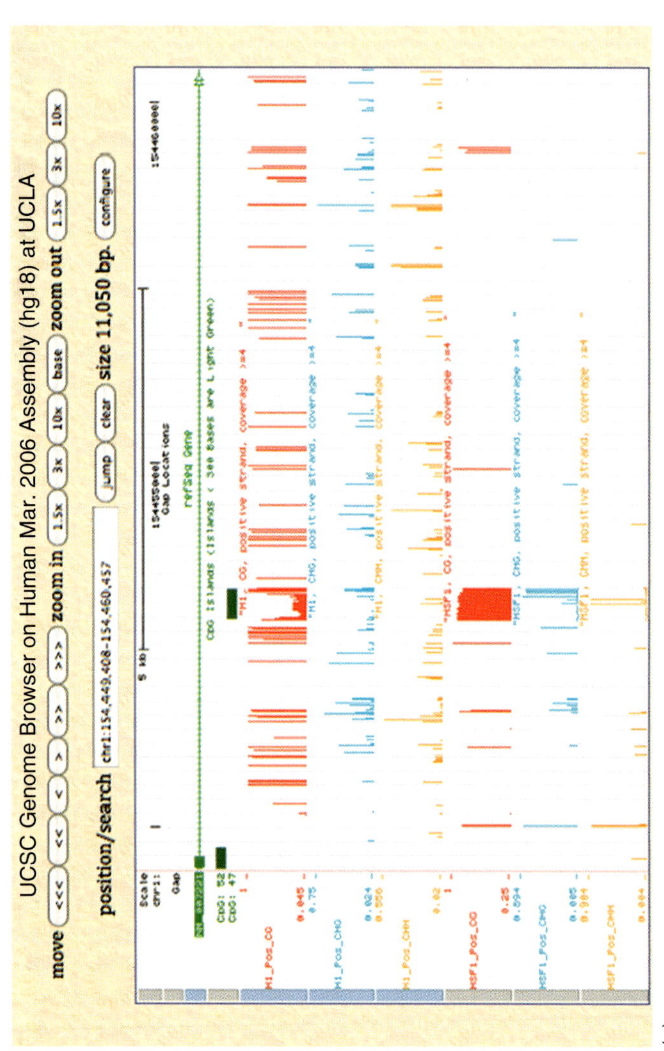
UCSC Genome Browser on Human Mar. 2006 Assembly (hg18) at UCLA
move <<< << < > >> >>> zoom in 1.5x 3x 10x base zoom out 1.5x 3x 10x
position/search chr1:154,449,408-154,460,457 jump clear size 11,050 bp. configure

(a)

(b)

Fig. 4 Snapshots of methylation tracks from genome browsers. (a) Methylation tracks of CG, CHG, and CHH from two human ES cell lines, H1 and HSF1, show differential methylation at the CpG islands in *pmf1*; (b) Multiple tracks showing gene expression (mRNA-seq), DNA methylation (BS-seq), and histone modifications (chip-seq) in one snapshot.

Lister *et al.* used a premethylated adaptor protocol (MethylC-seq) to generate an *Arabidopsis* methylome with 8X coverage [8]. Lister *et al.* reported several advantages of BS-seq compared to the traditional microarray-based methods, including an ability to analyze repetitive sequences that were difficult to study with microarrays (as they may exceed the dynamic detection range, or cross-hybridize), and allowing the determination methylation at single-base resolution. As a consequence of the increased resolution, it was now possible to study the sequence context-specific methylation of cytosines.

In the *Arabidopsis* methylome, DNA methylation is observed preferentially at CG sites. The distribution of CG methylation is bimodal; most CG sites are either highly methylated or unmethylated. In contrast, the methylation at methylated CHG and CHH sites is more often found at low levels, which are distributed uniformly between full and no methylation. These differences suggest that these methylation patterns are regulated by different methyltransferases. The chromosomal view of methylation levels shows a rise in pericentromeric regions that are repeat-rich. The metaplot of average methylation trends around genes reveals that CG methylation is low in transcription start and end sites, but high in the gene bodies. Within genes, DNA methylation is found to be higher in exons than in introns, and tends to have a slight bias for the 3′ end of genes [30]. Within repeat regions, the methylation level of all cytosines is generally higher than their surrounding regions.

The single-base resolution methylomes allows an autocorrelation analysis to be conducted of the methylation between the two cytosines and their distance. A 10-base periodicity between two CHH sites (the length of one helical DNA turn) was detected; this coincided with a similar finding in mammals that DNMT 3a forms a tetramer with DNA methyltransferase 3-like protein (Dnmt3L), leading to the formation of two active sites that can methylate two CG sequences spaced, 8 to 10 nucleotides apart. The autocorrelation analysis also revealed a period of 167 nucleotides at CHG sites, which is also the average spacing of nucleosomes in plant chromatin. Recently, the analysis of autocorrelation trends in both *Arabidopsis* and human data, also identified a 10-base periodicity in the DNA methylation status of nucleosome-bound DNA, and showed that the nucleosomal DNA was more highly methylated than flanking DNA [10].

Both, Cokus *et al.* and Lister *et al.* compared the methylation profiles between wild-type plants and a variety of methyltransferase mutants, in order to examine the mechanisms of methylation maintenance, establishment, and demethylation. The plant DNA methyltransferase enzymes, MET1, CMT3, and DRM1/DRM2, are responsible for CG, CHG, and CHH methylation, respectively. Hence, combinations of these enzymes were knocked down to reveal changes of methylation levels in cytosines with specific genome contexts. These results suggested that different DNA methyltransferases acted redundantly, and helped to explain the viability of these mutants. In contrast, the met1 cmt3 drm1 drm2 quadruple mutant caused embryonic lethality.

The correlation between smRNA (small interfering RNA; siRNA) and DNA-methylated regions can also be studied in methylomes. Genomic regions that generate smRNAs were highly methylated at CG, CHG, and CHH sites, as might be expect from the present understanding of RNA-directed DNA methylation pathways. It was also found that, at a subset of genomic loci, DNA methylation and smRNAs act in a self-reinforcing positive feedback loop; while smRNAs are found to direct over one-third of the DNA methylation, DNA methylation at smRNA-generating loci can also effect an increase in the production of smRNAs [8].

The studies of *Arabidopsis* methylomes provide the first examples of both the comprehensive analysis of individual methylation profiles, and the integration with other data such as smRNA and mRNA.

3.2 Human Methylomes from Embryonic Stem Cells

3.2.1 The First Look at the Human Methylome

Although several genome-wide studies of mammalian DNA methylation have been conducted, they have been limited by low resolution [18, 55], sequence-specific bias [35, 36], or complexity reduction approaches that have allowed only a small fraction of the genome to be analyzed [1, 45, 46]. Subsequently, in 2009, by using the BS-seq technique Lister *et al.* reported details of the first human methylome from the human embryonic stem cell (hESC) line H1 and fetal fibroblasts.

Details of the second and third human methylomes were also reported a few months later. In this case, Laurent *et al.* investigated differential methylation patterns between hESC line H9 and a fibroblastic differentiated derivative of the hESCs [9], while Chodavarapu *et al.* generated BS-seq data from the hESC line HSF1 to examine the relationship between nucleosome positioning and DNA methylation [10]. In addition, Chen *et al.* carried out a comprehensive comparison of these three methylomes, H1, HSF1, and H9 [56], by studying conserved non-CG methylation, the relationship between allele-specific transcription and allele-specific methylation, and DNA methylation at transcription factor binding sites.

The coverage of BS-seq data in the H1 and H9 methylomes is 14X (14-fold) and 9X per stand, respectively. The validation of the H9 methylome data processing and methylation calling strategy was carried out by comparing the BS-seq data with data from an independent array-based analysis (e.g., Illumina Infinium Human-Methylation27 BeadChip microarray) generated from the same cell preparations. The HumanMethylation27 array interrogates 0.1% of total CpG sites in the genome, and 0.01% of the total number of cytosines covered by the H9 methylome.

The detailed analysis of the hESC methylome, as well as the comparison with differentiated cells, has been of great interest for stem cell research, notably in improving the present understanding of epigenetic reprogramming and differentiation. Thus, a brief summary is now provided of the results of these pioneering studies with human methylomes.

In human methylomes, the methylation levels of CG dinucleotides show a bimodal distribution, similar to that found in *Arabidopsis* [7]. In the H1 genome, 77% of methylated CG (mCG) sites were 80–100% methylated, whereas 85% of mCHG and mCHH sites were between 10% and 40% methylated. A chromosomal view revealed large variations throughout the chromosome, with subtelomeric regions of the chromosomes frequently showing higher DNA methylation densities. This was consistent with the observation that DNA methylation is involved in the control of telomere length and recombination [57]. In promoter regions, both methylomes showed that the CG methylation level is anti-correlated with gene expression. Within genes, non-CG methylation levels correlate positively with transcription, while gene-body methylation has been hypothesized to suppress the spurious initiation of transcription within active genes in *Arabidopsis* [16] (a similar function may exist in mammals [35, 58]). It has been documented in previous studies that DNA methylation at transcription factor binding sites (TFBS) may interfere with the ability of some DNA-binding proteins to interact with their target sequences [59, 60]. Yet, by combining Chip-seq and BS-seq data, a decrease in methylation levels was observed at the binding sites of several transcription factors related to cell pluripotency, including NANOG, SOX2, KLF4, and OCT4 [4].

One of the most interesting findings from hESC methylomes was a confirmation of the existence of non-CG methylation, which disappears upon induction of differentiation of the ES cells, but is restored in induced pluripotent stem cells (iPSCs). Approximately 25% of the methylated cytosines identified in hESC H1 are in non-CG contexts (non-CG sites are much less methylated than CG sites). Typically, CA is found to be the most methylated non-CG dinucleotide, and shows similar methylation patterns as CG dinucleotides in promoters and genes; methylated CA sites are found across the genome and gene regions, but are reduced at promoters [9]. The absence of mCHG and mCHH methylation in differentiated cells (IMR90) coincided with a significantly lower transcript abundance of the *de novo* DNMTs DNMT3A and DNMT3B, and also the associated DNMT3L in IMR90 cells, which were reported to mediate non-CG methylation in mouse ES cells [61, 62].

In the *Arabidopsis* methylome [7], both CG and CHG methylation are symmetrical (i.e., the methylation status of the first cytosine is correlated with that of the second cytosine on the opposite strand). In human, this symmetry is found at CG sites, but is less evident at non-CG sites.

Further analysis of methylation sequence contexts reveals that TACAG sites are strongly enriched in methylated cytosines, and are conserved among hESC lines [56]. In addition, sharp spikes of CG methylation levels are observed at splice sites, and are probably influenced by the donor/acceptor sequence context around the splice junctions. The intron–exon boundaries also appear to be marked by gradients in chromatin features, including nucleosomes [63] and the H3K36me3 histone mark [64]; this suggests that the coupling of transcription and splicing may be regulated by DNA methylation, as well as by other epigenetic marks. Currently, the correlation between various histone modifications and DNA methylation is an active research topic. It is hoped that the integration of Chip-seq from histone protein binding and BS-seq data [65] may shed light on the complicated relationships between DNA methylation and chromatin modifications.

A comparison of methylomes between hESCs and differentiated cells has revealed that the former have generally higher global methylation levels and significant fractions of methylated non-CG sites. This pattern may represent an epigenetically primed state in hESCs that is followed during the early phases of differentiation by an increase in methylation of a subset of genes, in the context of a general reduction of global methylation. Differentiation-associated differential methylation profiles were observed for developmentally regulated genes. Consequently, by comparing undifferentiated and differentiated cells, dynamic DNA methylation is observed to be associated closely with changes in gene expression during differentiation. Even though the global level of methylation decreased with differentiation, almost 80% of differentially methylated regions showed an increased methylation with differentiation. Many key pluripotency and differentiation-associated genes were found in these regions, which suggests that preferential DNA methylation may affect the transcription of these genes during the course of differentiation.

3.2.2 Vertical Comparison between ES Cells and iPSCs

The iPSCs offer immense potential for regenerative medicine and studies of disease and development [66]. The reprogramming process is not a genetic transformation, but rather epigenomic in nature. Although, in a recent study, minimal differences were reported in chromatin structure and gene expression between hESCs and iPSCs [67], a growing number of reports have proposed key epigenomic differences between these cells [36, 68–71]. In order to fully characterize such differences, and to understand how complete and variable the re-establishment of ES cell-like DNA methylation patterns are throughout the entire genome, two large-scale studies were conducted to provide extensive comparisons of methylomes between multiple hESC lines and iPSC lines.

Bock *et al.* established genome-wide reference maps of DNA methylation by using the RRBS technique and gene expression for 20 previously derived hESC lines and 12 human iPSC lines [72]. The study results suggested that ES cells and iPSCs should not be regarded as one or two well-defined populations, but rather as two partially overlapping clouds with inherent variability among both ES cell and iPSC lines. As cell-line-specific variation in DNA methylation and gene expression is observed among ES cell lines, any epigenetic similarity to ES cell lines is unlikely to be a sufficient indicator of an iPSC line's utility for a specific application. However, it was possible to develop an iPSC classifier that uses methylation profiles and gene expression for the classification of ES cell or iPSC lines.

Lister *et al.* performed BS-seq on five human iPSC lines, along with methylomes of ES cells, somatic cells, and differentiated iPSCs and ES cells [11]. On a genome scale, the DNA methylomes of ES cells and iPSCs were similar to one another, and distinct from the primary somatic cell lines. The iPSCs showed significant reprogramming variability, including somatic memory and aberrant reprogramming of DNA methylation. In addition, the iPSCs shared megabase-scale differentially methylated regions proximal to centromeres and telomeres that displayed an incomplete reprogramming of non-CG methylation, as well as differences in CG methylation and histone modifications.

3.3 Phylogenetically Diverse Methylomes

With the advance of sequencing techniques, BS-seq is today used widely for the generation of methylomes of many organisms of interest. The conservation and divergence of methylomes have been analyzed extensively in a variety of reports by comparing methylation profiles across fungi, plants, invertebrates, and vertebrates [30, 31, 58]. When Xiang *et al.* analyzed the methylome of silkworm[73], an economically important model insect, they reported some interesting methylation patterns that differed from other studied organisms, such as *Arabidopsis* and human. Likewise, Feng *et al.* reported BS-seq data from eight eukaryotic organisms, including *A. thaliana*, *Oryza sativa* (rice), *Populus trichocarpa* (poplar), *Chlamydomonas reinhardtii* (green algae), *Ciona intestinalis* (sea squirt), *Apis mellifera* (honey bee), *Danio rerio* (zebrafish), and *Mus musculus* (mouse). Zemach *et al.* analyzed methylomes from 17 eukaryotic genomes of three plants (*O. sativa*, *Selaginella moellendorffii*, *Physcomitrella patens*), from two green algae (*Chlorella* sp. NC64A and *Volvox carteri*), from seven animals/insects, including *Tetraodon nigroviridis* (puffer fish), *Tribolium castaneum* (flour beetle), *Drosophila melanogaster*, *C. intestinalis*, *A. mellifera*, *Bombyx mori* (silkworm), and *Nematostella vectensis*), and five fungi, namely *Phycomyces blakesleeanus*, *Coprinopsis cinerea*, *Laccaria bicolor*, *Postia placenta*, and *Uncinocarpus reesii*.

3.3.1 Global Methylation Patterns

Among these methylomes, the DNA methylation landscape was found to be either continuous along the genome (e.g., human), or to show "mosaic" methylation patterns, with series of heavily methylated DNA domains interspersed with domains that were weakly methylated (e.g., *Arabidopsis*). Mosaic methylation patterns were mainly observed in plants, fungi, invertebrates, whereas vertebrates were often continuously methylated.

The vertebrate genomes were globally methylated, except at CpG islands that were mostly unmethylated. Other genomic elements such as TEs, genes, and intergenic regions were predominantly methylated. However, the global DNA methylation pattern seen in vertebrates was not conserved across all eukaryotes. For example, *Saccharomyces cerevisiae* (yeast) and *Caenorhabditis elegans* (worm) have no recognizable DNMT-like genes and are devoid of DNA methylation. *Tribolium castaneum* (flour beetle) adults and *D. melanogaster* embryos also do not have detectable DNA methylation of the nuclear genome. In *Bombyx mori* (silkworm), only 0.11% of the genomic cytosines were methylated, all of which occurred at CG dinucleotides. Most of the methylated cytosines in silkworm had intermediate methylation levels, which contrasted with the usual bimodal distribution observed in *Arabidopsis*, mouse, and human.

As most animals show a mosaic methylation, it remains unclear as to how the evolutionary transition from mosaic to global methylation in vertebrates evolved, although it may be connected to the development of the innate immune system [58].

3.3.2 Gene Methylation

Gene methylation is commonly observed in plants and animals, where the 5′ and 3′ ends of genes are significantly less methylated than the inner portions, and preferentially in exons. The methylation of genes may inhibit transcriptional elongation [74]. In some organisms (e.g., *Ciona* sp. and honey bee), gene methylation appears to be the main source of genomic methylcytosines.

DNA methylation in insects is usually found at very low levels, and mostly at CG sites. For example, *C. intestinalis*, honeybee, silkworm, and anemone genes have similar methylation patterns as are found in plants and fish (e.g., puffer fish), with the highest methylation levels found in moderately expressed genes. A similar association of methylation levels with transcriptional activity is also found in mammals. In the silkworm, the promoter methylation was not associated with gene expression, as in other plants and mammals, which suggested that a different regulatory mechanism may operate in insects.

The silkworm analysis also showed a significant excess of methylated genomic loci matching smRNA within genes, but depleted within TEs. This pattern contrasted with observations in plants, where highly methylated genomic loci matching smRNAs were rarely found in genes, but were prevalent in TEs and other repeats [7]. In plants, smRNA-directed DNA-methylation that targets repetitive DNA plays an important role in TE silencing [75], which explains why smRNAs in TEs are highly methylated. In silkworms, the prevalence of genomic loci of smRNA in genes and their dense CG methylation implies that smRNAs may play a role in gene body CG methylation. Through an analysis of Gene Ontology functions, it was found that methylation may contribute to maintaining the relatively high expression of genes that are essential for biosynthetic processes in the silk glands.

In contrast, methylation in fungal genes, such as *Phycomyces blakesleeanus*, *C. cinerea*, *L. bicolor*, and *P. placenta*, has different patterns. Rather, it is concentrated in transcriptionally silent, repetitive loci, whereas active genes are generally unmethylated. However, unlike the other fungi, *U. reesii* exhibits methylation of active genes.

Plants also have methylation in their genes. In rice, methylation in the genes has a convex relationship with transcription, whereby modestly expressed genes are most likely to be methylated whereas genes at either transcriptional extreme are least likely to be methylated. The two early diverging land plants, *S. moellendorffii* and *P. patens*, have minimal methylation in the genes, whereas in the green alga *Chlorella* sp. NC64A, the genes are methylated virtually without exception.

Although, in evolutionary terms, plants and animals diverged about 1.6 billion years ago, the above-described evidence suggests that similar patterns of DNA methylation in the bodies of active genes are present in both groups. This implies that gene body methylation reflects a primary and ancestral function of DNA methylation. Furthermore, gene body methylation is conserved with a clear preference for exons in most organisms, which suggests that exon methylation might also be an ancestral condition. The recent finding that exons are enriched in nucleosomes relative to introns [63, 76] has led to speculation that nucleosomes might act to guide DNA methyltransferases, resulting in exon methylation.

3.3.3 **Transposable Elements (TEs) Methylation**

There is strong evidence for the targeting of DNA methylation to repetitive elements in fungi and plants, though it is unclear if the same process applies to invertebrate animals. Vertebrate transposons are methylated in most genomes, but it is unclear whether this is due to specific targeting [58].

In the three flowering plants (rice, *Arabidopsis*, and poplar), and the two early diverging land plants (*P. patens* and *S. moellendorffii*) DNA methylation is highly enriched in repetitive DNA and transposons in all three sequence contexts (CG, CHG, and CHH). Moreover, the green algae *Volvox* and *Chlamydomonas* also display preferential CG methylation of TEs that is most likely regulated by a different mechanism than that of flowering plants.

In fungi, DNA methylation appears to be exclusively found in repetitive sequences. The methylation in fungi (e.g., *Phycomyces blakesleeanus, Coprinopsis cinerea, L. bicolor, P. placenta, Uncinocarpus reesii,* and *Neurospora crassa*) is likely to be used to silence TEs and other repeats.

TEs methylation in invertebrates is less evident. In *Ciona intestinalis,* honeybee, silk moth, and anemone, the TEs are hypomethylated while the genes are hypermethylated; this suggests that the genes are more likely the main targets of DNA methylation in invertebrates. Thus, there is little evidence that methylation in invertebrates either inhibits transcription or silences TEs.

Among the sequenced methylomes, the presence of non-CG methylation is variable although, when present, it is always found at lower levels than CG methylation. In flowering plants, the absence of CHG and CHH methylation in gene bodies demonstrates that genes and transposons are differentially targeted. Interestingly, *Chlamydomonas* has the most unusual pattern of methylation, with non-CG methylation enriched in exons of genes rather than in repeats and transposons.

4 Future Directions

4.1 Pacific Bioscience Direct Readout of DNA Methylation

Next-generation sequencing techniques consist of various strategies that rely on a combination of template preparation, sequencing, and imaging, followed by alignment and assembly methods to interpret the data. Currently, two methods are used for template preparation, namely clonally amplified templates originating from single DNA molecules, and single DNA molecule templates [77].

Clonal amplification results in a population of identical templates, each of which has undergone the sequencing reaction. Subsequently, imaging captures the consensus signal from the nucleotides added to the identical templates. During the addition, a potential source of error may occur when strands lag in the sequencing cycle, or when multiple nucleotides are added in a cycle (i.e., leading-strand dephasing). These factors limit the length of reads generated from clonal amplification. Currently, the Illumina Genome Analyzer is one of the most popular sequencers. It uses the clonally amplified template method, coupled with the four-colour cyclic reversible termination (CRT) method. Genome analysis of these data reveals an underrepresentation of AT-rich and GC-rich regions, which is most likely due to amplification bias during template preparation.

In contrast, Pacific Bioscience is currently developing single-molecule real-time (SMRT) sequencing, a new technique which requires much less DNA material and no PCR [78]. SMRT sequencing is able to generate reads of 1000bp, and offers a direct detection of DNA methylation in real time. This new technique should be available shortly, and should greatly improve the present ability to precisely and efficiently measure methylomes.

4.1.1 SMRT Sequencing

SMRT data are generated from a DNA polymerase that performs an uninterrupted template-directed synthesis, using four distinguishable

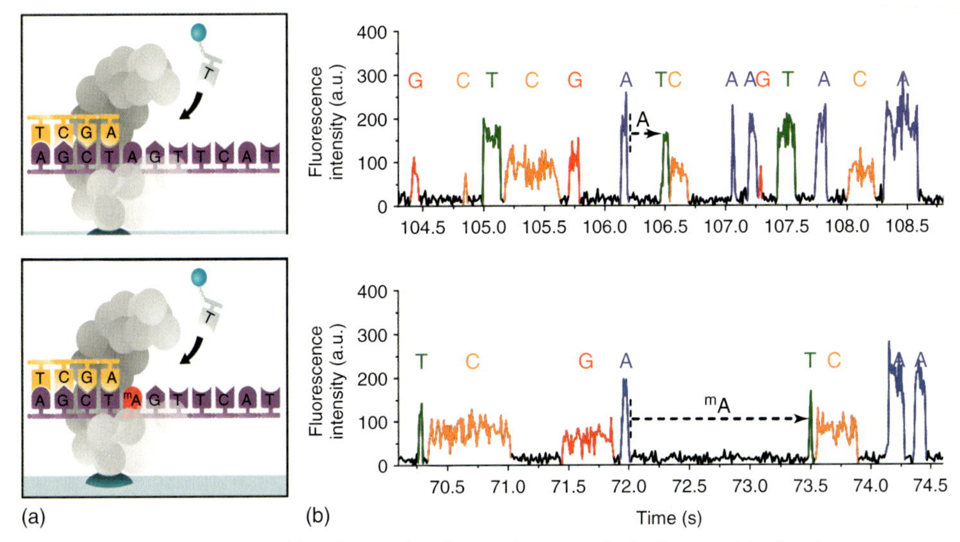

Fig. 5 Kinetic detection enables the study of modified bases. (a) Cartoons of polymerase synthesis of DNA strands containing a methylated (upper panel) or unmethylated (lower panel) adenine; (b) In this example, the IPD (shown as dashed arrows) before incorporation of the thymine is about fivefold larger for mA in the template compared to adenine. Illustration courtesy of Pacific Biosciences.

fluorescently labeled deoxyribonucleoside triphosphates (dNTPs) [78]. For this, spatially distributed single polymerase molecules are attached to the solid support, to which a primed template molecule is bound. The enzymatic incorporation into a growing DNA strand is detected with zero-mode waveguide nanostructure arrays from the binding of correctly basepaired (cognate) phospholinked dNTPs in the active site of the polymerase. The conjugation of fluorophores to the terminal phosphate moiety of the dNTPs allows for a continuous observation of DNA synthesis over thousands of bases. The data report directly on polymerase dynamics in real time, revealing distinct polymerization states and pause sites corresponding to DNA secondary structure. The error rate is estimated to be approximately 17% by sequencing known sequence [77, 78]. Typical sources of error include multiple nucleotide additions in any given cycle, or a lack of signal due to the incorporation of dark nucleotides or probes. Repeat sequencing of the same template molecule over 15 times or more has been shown to improve the base calling accuracy to more than 99%.

4.1.2 Direct Detection of DNA Methylation

In SMRT sequencing, DNA polymerases catalyze the incorporation of fluorescently labeled nucleotides into complementary nucleic acid strands. The arrival times and durations of the resulting fluorescence pulses yield information concerning the polymerase kinetics, and allow the direct detection of modified nucleotides in the DNA template, including mA, mC, and hmC [50, 79, 80] (see Fig. 5). Fluorescence pulses in SMRT sequencing are characterized not only by their emission spectra, but also by their duration and by the interval between successive pulses. These metrics are referred to respectively as *pulse width* and interpulse duration (IPD). The pulse width is a function of all kinetic steps after nucleotide binding up to fluorophore release, while the IPD is determined by the kinetics of nucleotide binding and polymerase translocation. As SMRT sequencing polymerase synthesis rates are sensitive to DNA primary and secondary structures, it is possible to discriminate how the various modifications might affect the polymerase kinetics.

SMRT has several major advantages over BS-seq; notably, it detects DNA methylation in real time without bisulfite conversion. SMRT is also able to discriminate between mC and hmC, whereas BS-seq cannot [79]. The read length of BS-seq is typically short, whereas SMRT can generate 1 kb reads, allowing more coverage in repetitive genomic sequences. SMRT is also less biased in sequencing as it does not require PCR. Moreover, SMRT can potentially detect other epigenetic modifications, or DNA damage that results in changes of fluorescence pulses. Unfortunately, Nonetheless, at the time of writing this chapter, the cost of generating a genome-wide methylation profile using SMRT is considerably higher than sequencing using Illumina technology.

4.2
Perspectives of Methylome Studies

DNA methylation has been shown to have a close relationship with transcription, and yet it adapts dynamically to environmental changes; methylomes, however, are constantly modified across developmental stages and tissues. Since its development in 2008, the whole-genome bisulfite sequencing technique (BS-seq) technique has been applied to important areas of biomedical research, including regenerative medicine, cancer research, nutrition intake, and aging studies. Each of these research areas might have significant impacts on human health; for example, by comparing the methylomes of twins at different ages, genes or regions could be identified that would be differentially methylated as a function of age, thus shedding new light on aging-related diseases and, ultimately, their associated genes.

Today, trans-generational studies of genome-wide methylation profiles are also becoming feasible. Whereas genetic mutations typically accumulate over hundreds of generations, organisms must adapt rapidly to their changing environments in order to improve and/or maintain their survival. Trans-generational studies of methylomes may reveal how epigenetic markers would impact on the regulation of genes, and how they might be passed from parents to offspring, thus allowing the organism to accumulate heritable adaptations to their environments, independently of any genetic changes In addition, horizontal comparisons of methylomes across organisms might provide a new dimension in evolutionary studies, as the conservation and divergence of methylation patterning will surely shed light on the diversity of epigenetic marks across organisms.

Currently, whole-genome bisulfite sequencing enables biological questions to be tackled that, just a few years ago, would have been impossible to resolve. It is anticipated that these techniques will allow important questions to be answered with regards to the diversity of epigenetic marks in populations, their association with clinically important phenotypes, and their heritability across generations.

References

1 Beck, S., Rakyan, V.K. (2008) The methylome: approaches for global DNA methylation profiling. *Trends Genet.*, **24**, 231–237.

2 Beck, S. (2010) Taking the measure of the methylome. *Nat. Biotechnol.*, **28**, 1026–1028.

3 Hotchkiss, R.D. (1948) The quantitative separation of purines, pyrimidines, and nucleosides by paper chromatography. *J. Biol. Chem.*, **175**, 315–332.

4 Lister, R., Pelizzola, M., Dowen, R.H., Hawkins, R.D., Hon, G., Tonti-Filippini, J., Nery, J.R., Lee, L., Ye, Z., Ngo, Q.M., Edsall, L., Antosiewicz-Bourget, J., Stewart, R., Ruotti, V., Millar, A.H., Thomson, J.A., Ren, B., Ecker, J.R. (2009) Human DNA methylomes at base resolution show widespread epigenomic differences. *Nature*, **462**, 315–322.

5 Popp, C., Dean, W., Feng, S.H., Cokus, S.J., Andrews, S., Pellegrini, M., Jacobsen, S.E., Reik, W. (2010) Genome-wide erasure of DNA methylation in mouse primordial germ cells is affected by AID deficiency. *Nature*, **463**, 1101–1126.

6 Laird, P.W. (2010) Principles and challenges of genome-wide DNA methylation analysis. *Nat. Rev. Genet.*, **11**, 191–203.

7 Cokus, S.J., Feng, S., Zhang, X., Chen, Z., Merriman, B., Haudenschild, C.D., Pradhan, S., Nelson, S.F., Pellegrini, M., Jacobsen, S.E. (2008) Shotgun bisulphite sequencing of the *Arabidopsis* genome reveals DNA methylation patterning. *Nature*, **452**, 215–219.

8 Lister, R., O'Malley, R.C., Tonti-Filippini, J., Gregory, B.D., Berry, C.C., Millar, A.H., Ecker, J.R. (2008) Highly integrated single-base resolution maps of the epigenome in *Arabidopsis*. *Cell*, **133**, 523–536.

9 Laurent, L., Wong, E., Li, G., Huynh, T., Tsirigos, A., Ong, C.T., Low, H.M., Kin Sung, K.W., Rigoutsos, I., Loring, J. (2010) Dynamic changes in the human methylome during differentiation. *Genome Res.*, **20**, 320–331.

10 Chodavarapu, R.K., Feng, S., Bernatavichute, Y.V., Chen, P.Y., Stroud, H., Yu, Y., Hetzel, J.A., Kuo, F., Kim, J., Cokus, S.J., Casero, D., Bernal, M., Huijser, P., Clark, A.T., Kramer, U., Merchant, S.S., Zhang, X., Jacobsen, S.E., Pellegrini, M. (2010) Relationship between nucleosome positioning and DNA methylation. *Nature*, **466**, 388–392.

11 Lister, R., Pelizzola, M., Kida, Y.S., Hawkins, R.D., Nery, J.R., Hon, G., Antosiewicz-Bourget, J., O'Malley, R., Castanon, R., Klugman, S., Downes, M., Yu, R., Stewart, R., Ren, B., Thomson, J.A., Evans, R.M., Ecker, J.R. (2011) Hotspots of aberrant epigenomic reprogramming in human induced pluripotent stem cells. *Nature.*, **471**, 68–73.

12 Pelizzola, M., Ecker, J.R. (2010) The DNA methylome. *FEBS Lett.*, **585**, 1994–2000.

13 Pomraning, K.R., Smith, K.M., Freitag, M. (2009) Genome-wide high throughput analysis of DNA methylation in eukaryotes. *Methods*, **47**, 142–150.

14 Li, N., Ye, M., Li, Y., Yan, Z., Butcher, L.M., Sun, J., Han, X., Chen, Q., Zhang, X., Wang, J. (2010) Whole genome DNA methylation analysis based on high throughput sequencing technology. *Methods*, **52**, 203–212.

15 Cross, S.H., Charlton, J.A., Nan, X., Bird, A.P. (1994) Purification of CpG islands using a methylated DNA binding column. *Nat. Genet.*, **6**, 236–244.

16 Zhang, X., Yazaki, J., Sundaresan, A., Cokus, S., Chan, S.W., Chen, H., Henderson, I.R., Shinn, P., Pellegrini, M., Jacobsen, S.E., Ecker, J.R. (2006) Genome-wide high-resolution mapping and functional analysis of DNA methylation in *Arabidopsis*. *Cell*, **126**, 1189–1201.

17 Farthing, C.R., Ficz, G., Ng, R.K., Chan, C.F., Andrews, S., Dean, W., Hemberger, M., Reik, W. (2008) Global mapping of DNA methylation in mouse promoters reveals epigenetic reprogramming of pluripotency genes. *PLoS Genet.*, **4**, e1000116.

18 Irizarry, R.A., Ladd-Acosta, C., Carvalho, B., Wu, H., Brandenburg, S.A., Jeddeloh, J.A., Wen, B., Feinberg, A.P. (2008) Comprehensive high-throughput arrays for relative methylation (CHARM). *Genome Res.*, **18**, 780–790.

19 Weber, M., Davies, J.J., Wittig, D., Oakeley, E.J., Haase, M., Lam, W.L., Schubeler, D. (2005) Chromosome-wide and promoter-specific analyses identify sites

of differential DNA methylation in normal and transformed human cells. *Nat. Genet.*, **37**, 853–862.

20 Jacinto, F.V., Ballestar, E., Esteller, M. (2008) Methyl-DNA immunoprecipitation (MeDIP): hunting down the DNA methylome. *Biotechniques*, **44**, 35, 37, 39 (passim).

21 Harris, R.A., Wang, T., Coarfa, C., Nagarajan, R.P., Hong, C., Downey, S.L., Johnson, B.E., Fouse, S.D., Delaney, A., Zhao, Y., Olshen, A., Ballinger, T., Zhou, X., Forsberg, K.J., Gu, J., Echipare, L., O'Geen, H., Lister, R., Pelizzola, M., Xi, Y., Epstein, C.B., Bernstein, B.E., Hawkins, R.D., Ren, B., Chung, W.Y., Gu, H., Bock, C., Gnirke, A., Zhang, M.Q., Haussler, D., Ecker, J.R., Li, W., Farnham, P.J., Waterland, R.A., Meissner, A., Marra, M.A., Hirst, M., Milosavljevic, A., Costello, J.F. (2010) Comparison of sequencing-based methods to profile DNA methylation and identification of monoallelic epigenetic modifications. *Nat. Biotechnol.*, **28**, 1097–1105.

22 Bird, A.P. (1986) CpG-rich islands and the function of DNA methylation. *Nature*, **321**, 209–213.

23 Brinkman, A.B., Simmer, F., Ma, K., Kaan, A., Zhu, J., Stunnenberg, H.G. (2010) Whole-genome DNA methylation profiling using MethylCap-seq. *Methods*, **52**, 232–236.

24 Serre, D., Lee, B.H., Ting, A.H. (2010) MBD-isolated genome sequencing provides a high-throughput and comprehensive survey of DNA methylation in the human genome. *Nucleic Acids Res.*, **38**, 391–399.

25 Down, T.A., Rakyan, V.K., Turner, D.J., Flicek, P., Li, H., Kulesha, E., Graf, S., Johnson, N., Herrero, J., Tomazou, E.M., Thorne, N.P., Backdahl, L., Herberth, M., Howe, K.L., Jackson, D.K., Miretti, M.M., Marioni, J.C., Birney, E., Hubbard, T.J., Durbin, R., Tavare, S., Beck, S. (2008) A Bayesian deconvolution strategy for immunoprecipitation-based DNA methylome analysis. *Nat. Biotechnol.*, **26**, 779–785.

26 Schumacher, A., Kapranov, P., Kaminsky, Z., Flanagan, J., Assadzadeh, A., Yau, P., Virtanen, C., Winegarden, N., Cheng, J., Gingeras, T., Petronis, A. (2006) Microarray-based DNA methylation profiling: technology and applications. *Nucleic Acids Res.*, **34**, 528–542.

27 Khulan, B., Thompson, R.F., Ye, K., Fazzari, M.J., Suzuki, M., Stasiek, E., Figueroa, M.E., Glass, J.L., Chen, Q., Montagna, C., Hatchwell, E., Selzer, R.R., Richmond, T.A., Green, R.D., Melnick, A., Greally, J.M. (2006) Comparative isoschizomer profiling of cytosine methylation: the HELP assay. *Genome Res.*, **16**, 1046–1055.

28 Sutherland, E., Coe, L., Raleigh, E.A. (1992) McrBC: a multisubunit GTP-dependent restriction endonuclease. *J. Mol. Biol.*, **225**, 327–348.

29 Frommer, M., McDonald, L.E., Millar, D.S., Collis, C.M., Watt, F., Grigg, G.W., Molloy, P.L., Paul, C.L. (1992) A genomic sequencing protocol that yields a positive display of 5-methylcytosine residues in individual DNA strands. *Proc. Natl Acad. Sci. USA*, **89**, 1827–1831.

30 Feng, S., Cokus, S.J., Zhang, X., Chen, P.Y., Bostick, M., Goll, M.G., Hetzel, J., Jain, J., Strauss, S.H., Halpern, M.E., Ukomadu, C., Sadler, K.C., Pradhan, S., Pellegrini, M., Jacobsen, S.E. (2010) Conservation and divergence of methylation patterning in plants and animals. *Proc. Natl Acad. Sci. USA*, **107**, 8689–8694.

31 Zemach, A., McDaniel, I.E., Silva, P., Zilberman, D. (2010) Genome-wide evolutionary analysis of eukaryotic DNA methylation. *Science*, **328**, 916–919.

32 Bibikova, M., Fan, J.B. (2009) GoldenGate assay for DNA methylation profiling. *Methods Mol. Biol.*, **507**, 149–163.

33 Bibikova, M., Le, J., Barnes, B., Saedinia-Melnyk, S., Zhou, L.X., Shen, R., Gunderson, K.L. (2009) Genome-wide DNA methylation profiling using Infinium® assay. *Epigenomics*, **1**, 177–200.

34 Bock, C., Tomazou, E.M., Brinkman, A.B., Muller, F., Simmer, F., Gu, H., Jager, N., Gnirke, A., Stunnenberg, H.G., Meissner, A. (2010) Quantitative comparison of genome-wide DNA methylation mapping technologies. *Nat. Biotechnol.*, **28**, 1106–1114.

35 Ball, M.P., Li, J.B., Gao, Y., Lee, J.H., LeProust, E.M., Park, I.H., Xie, B., Daley, G.Q., Church, G.M. (2009) Targeted and genome-scale strategies reveal gene-body methylation signatures in human cells. *Nat. Biotechnol.*, **27**, 361–368.

36 Deng, J., Shoemaker, R., Xie, B., Gore, A., LeProust, E.M., Antosiewicz-Bourget, J.,

Egli, D., Maherali, N., Park, I.H., Yu, J., Daley, G.Q., Eggan, K., Hochedlinger, K., Thomson, J., Wang, W., Gao, Y., Zhang, K. (2009) Targeted bisulfite sequencing reveals changes in DNA methylation associated with nuclear reprogramming. *Nat. Biotechnol.*, **27**, 353–360.

37 Kent, W.J. (2002) BLAT – the BLAST-like alignment tool. *Genome Res.*, **12**, 656–664.

38 Li, R., Li, Y., Kristiansen, K., Wang, J. (2008) SOAP: short oligonucleotide alignment program. *Bioinformatics*, **24**, 713–714.

39 Langmead, B., Trapnell, C., Pop, M., Salzberg, S.L. (2009) Ultrafast and memory-efficient alignment of short DNA sequences to the human genome. *Genome Biol.*, **10**, R25.

40 Chen, P.Y., Cokus, S.J., Pellegrini, M. (2010) BS Seeker: precise mapping for bisulfite sequencing. *BMC Bioinformatics*, **11**, 203.

41 Xi, Y., Li, W. (2009) BSMAP: whole genome bisulfite sequence MAPping program. *BMC Bioinformatics*, **10**, 232.

42 Smith, A.D., Chung, W., Hodges, E., Kendall, J., Hannon, G., Hicks, J., Xuan, Z., Zhang, M.Q. (2009) Updates to the RMAP short-read mapping software. *Bioinformatics*, **25**, 2841–2842.

43 Coarfa, C., Yu, F., Miller, C.A., Chen, Z., Harris, R.A., Milosavljevic, A. (2010) Pash 3.0: a versatile software package for read mapping and integrative analysis of genomic and epigenomic variation using massively parallel DNA sequencing. *BMC Bioinformatics*, **11**, 572.

44 Meissner, A., Gnirke, A., Bell, G.W., Ramsahoye, B., Lander, E.S., Jaenisch, R. (2005) Reduced representation bisulfite sequencing for comparative high-resolution DNA methylation analysis. *Nucleic Acids Res.*, **33**, 5868–5877.

45 Smith, Z.D., Gu, H., Bock, C., Gnirke, A., Meissner, A. (2009) High-throughput bisulfite sequencing in mammalian genomes. *Methods*, **48**, 226–232.

46 Meissner, A., Mikkelsen, T.S., Gu, H., Wernig, M., Hanna, J., Sivachenko, A., Zhang, X., Bernstein, B.E., Nusbaum, C., Jaffe, D.B., Gnirke, A., Jaenisch, R., Lander, E.S. (2008) Genome-scale DNA methylation maps of pluripotent and differentiated cells. *Nature*, **454**, 766–770.

47 Warnecke, P.M., Stirzaker, C., Song, J., Grunau, C., Melki, J.R., Clark, S.J. (2002) Identification and resolution of artifacts in bisulfite sequencing. *Methods*, **27**, 101–107.

48 Kunarso, G., Chia, N.Y., Jeyakani, J., Hwang, C., Lu, X.Y., Chan, Y.S., Ng, H.H., Bourque, G. (2010) Transposable elements have rewired the core regulatory network of human embryonic stem cells. *Nat. Genet.*, **42**, 631–635.

49 Wu, S.C., Zhang, Y. (2010) Active DNA demethylation: many roads lead to Rome. *Nat. Rev. Mol. Cell Biol.*, **11**, 607–620.

50 Flusberg, B.A., Webster, D.R., Lee, J.H., Travers, K.J., Olivares, E.C., Clark, T.A., Korlach, J., Turner, S.W. (2010) Direct detection of DNA methylation during single-molecule, real-time sequencing. *Nat. Methods*, **7**, 461–465.

51 Li, H., Ruan, J., Durbin, R. (2008) Mapping short DNA sequencing reads and calling variants using mapping quality scores. *Genome Res.*, **18**, 1851–1858.

52 Smith, A.D., Xuan, Z.Y., Zhang, M.Q. (2008) Using quality scores and longer reads improves accuracy of Solexa read mapping. *BMC Bioinformatics*, **9**, 128.

53 Krueger, F., Andrews, S.R. (2011) Bismark: a flexible aligner and methylation caller for Bisulfite-Seq applications. *Bioinformatics*, **27**, 1571–1572.

54 Goelz, S.E., Vogelstein, B., Hamilton, S.R., Feinberg, A.P. (1985) Hypomethylation of DNA from benign and malignant human colon neoplasms. *Science*, **228**, 187–190.

55 Rauch, T.A., Wu, X., Zhong, X., Riggs, A.D., Pfeifer, G.P. (2009) A human B cell methylome at 100-base pair resolution. *Proc. Natl Acad. Sci. USA*, **106**, 671–678.

56 Chen, P.Y., Feng, S., Joo, J.W., Jacobsen, S.E., Pellegrini, M. (2011) A comparative analysis of DNA methylation across human embryonic stem cell lines. *Genome Biol.*, **12**, R62.

57 Gonzalo, S., Jaco, I., Fraga, M.F., Chen, T., Li, E., Esteller, M., Blasco, M.A. (2006) DNA methyltransferases control telomere length and telomere recombination in mammalian cells. *Nat. Cell Biol.*, **8**, 416–424.

58 Suzuki, M.M., Bird, A. (2008) DNA methylation landscapes: provocative insights from epigenomics. *Nat. Rev. Genet.*, **9**, 465–476.

59 Bell, A.C., Felsenfeld, G. (2000) Methylation of a CTCF-dependent boundary controls imprinted expression of the Igf2 gene. *Nature*, **405**, 482–485.

60 Kitazawa, S., Kitazawa, R., Maeda, S. (1999) Transcriptional regulation of rat cyclin D1 gene by CpG methylation status in promoter region. *J. Biol. Chem.*, **274**, 28787–28793.
61 Aoki, A., Suetake, I., Miyagawa, J., Fujio, T., Chijiwa, T., Sasaki, H., Tajima, S. (2001) Enzymatic properties of de novo-type mouse DNA (cytosine-5) methyltransferases. *Nucleic Acids Res.*, **29**, 3506–3512.
62 Ramsahoye, B.H., Biniszkiewicz, D., Lyko, F., Clark, V., Bird, A.P., Jaenisch, R. (2000) Non-CpG methylation is prevalent in embryonic stem cells and may be mediated by DNA methyltransferase 3a. *Proc. Natl Acad. Sci. USA*, **97**, 5237–5242.
63 Schwartz, S., Meshorer, E., Ast, G. (2009) Chromatin organization marks exon-intron structure. *Nat. Struct. Mol. Biol.*, **16**, 990–995.
64 Kolasinska-Zwierz, P., Down, T., Latorre, I., Liu, T., Liu, X.S., Ahringer, J. (2009) Differential chromatin marking of introns and expressed exons by H3K36me3. *Nat. Genet.*, **41**, 376–381.
65 Hawkins, R.D., Hon, G.C., Lee, L.K., Ngo, Q., Lister, R., Pelizzola, M., Edsall, L.E., Kuan, S., Luu, Y., Klugman, S., Antosiewicz-Bourget, J., Ye, Z., Espinoza, C., Agarwahl, S., Shen, L., Ruotti, V., Wang, W., Stewart, R., Thomson, J.A., Ecker, J.R., Ren, B. (2010) Distinct epigenomic landscapes of pluripotent and lineage-committed human cells. *Cell Stem Cell*, **6**, 479–491.
66 Yamanaka, S. (2009) A fresh look at iPS cells. *Cell*, **137**, 13–17.
67 Guenther, M.G., Frampton, G.M., Soldner, F., Hockemeyer, D., Mitalipova, M., Jaenisch, R., Young, R.A. (2010) Chromatin structure and gene expression programs of human embryonic and induced pluripotent stem cells. *Cell Stem Cell*, **7**, 249–257.
68 Doi, A., Park, I.H., Wen, B., Murakami, P., Aryee, M.J., Irizarry, R., Herb, B., Ladd-Acosta, C., Rho, J., Loewer, S., Miller, J., Schlaeger, T., Daley, G.Q., Feinberg, A.P. (2009) Differential methylation of tissue- and cancer-specific CpG island shores distinguishes human induced pluripotent stem cells, embryonic stem cells and fibroblasts. *Nat. Genet.*, **41**, 1350–1353.
69 Kim, K., Doi, A., Wen, B., Ng, K., Zhao, R., Cahan, P., Kim, J., Aryee, M.J., Ji, H., Ehrlich, L.I., Yabuuchi, A., Takeuchi, A., Cunniff, K.C., Hongguang, H., McKinney-Freeman, S., Naveiras, O., Yoon, T.J., Irizarry, R.A., Jung, N., Seita, J., Hanna, J., Murakami, P., Jaenisch, R., Weissleder, R., Orkin, S.H., Weissman, I.L., Feinberg, A.P., Daley, G.Q. (2010) Epigenetic memory in induced pluripotent stem cells. *Nature*, **467**, 285–290.
70 Stadtfeld, M., Apostolou, E., Akutsu, H., Fukuda, A., Follett, P., Natesan, S., Kono, T., Shioda, T., Hochedlinger, K. (2010) Aberrant silencing of imprinted genes on chromosome 12qF1 in mouse induced pluripotent stem cells. *Nature*, **465**, 175–181.
71 Hu, B.Y., Weick, J.P., Yu, J., Ma, L.X., Zhang, X.Q., Thomson, J.A., Zhang, S.C. (2010) Neural differentiation of human induced pluripotent stem cells follows developmental principles but with variable potency. *Proc. Natl Acad. Sci. USA*, **107**, 4335–4340.
72 Bock, C., Kiskinis, E., Verstappen, G., Gu, H., Boulting, G., Smith, Z.D., Ziller, M., Croft, G.F., Amoroso, M.W., Oakley, D.H., Gnirke, A., Eggan, K., Meissner, A. (2011) Reference maps of human ES and iPS cell variation enable high-throughput characterization of pluripotent cell lines. *Cell*, **144**, 439–452.
73 Xiang, H., Zhu, J., Chen, Q., Dai, F., Li, X., Li, M., Zhang, H., Zhang, G., Li, D., Dong, Y., Zhao, L., Lin, Y., Cheng, D., Yu, J., Sun, J., Zhou, X., Ma, K., He, Y., Zhao, Y., Guo, S., Ye, M., Guo, G., Li, Y., Li, R., Zhang, X., Ma, L., Kristiansen, K., Guo, Q., Jiang, J., Beck, S., Xia, Q., Wang, W., Wang, J. (2010) Single base-resolution methylome of the silkworm reveals a sparse epigenomic map. *Nat. Biotechnol.*, **28**, 516–520.
74 Portela, A., Esteller, M. (2010) Epigenetic modifications and human disease. *Nat. Biotechnol.*, **28**, 1057–1068.
75 Zhang, X. (2008) The epigenetic landscape of plants. *Science*, **320**, 489–492.
76 Tilgner, H., Nikolaou, C., Althammer, S., Sammeth, M., Beato, M., Valcarcel, J., Guigo, R. (2009) Nucleosome positioning as a determinant of exon recognition. *Nat. Struct. Mol. Biol.*, **16**, 996–U124.
77 Metzker, M.L. (2010) Sequencing technologies – the next generation. *Nat. Rev. Genet.*, **11**, 31–46.
78 Eid, J., Fehr, A., Gray, J., Luong, K., Lyle, J., Otto, G., Peluso, P., Rank, D., Baybayan, P., Bettman, B., Bibillo, A., Bjornson, K., Chaudhuri, B., Christians, F., Cicero, R.,

Clark, S., Dalal, R., Dewinter, A., Dixon, J., Foquet, M., Gaertner, A., Hardenbol, P., Heiner, C., Hester, K., Holden, D., Kearns, G., Kong, X., Kuse, R., Lacroix, Y., Lin, S., Lundquist, P., Ma, C., Marks, P., Maxham, M., Murphy, D., Park, I., Pham, T., Phillips, M., Roy, J., Sebra, R., Shen, G., Sorenson, J., Tomaney, A., Travers, K., Trulson, M., Vieceli, J., Wegener, J., Wu, D., Yang, A., Zaccarin, D., Zhao, P., Zhong, F., Korlach, J., Turner, S. (2009) Real-time DNA sequencing from single polymerase molecules. *Science*, **323**, 133–138.

79 Huang, Y., Pastor, W.A., Shen, Y., Tahiliani, M., Liu, D.R., Rao, A. (2010) The behaviour of 5-hydroxymethylcytosine in bisulfite sequencing. *PLoS ONE*, **5**, e8888.

80 Tahiliani, M., Koh, K.P., Shen, Y., Pastor, W.A., Bandukwala, H., Brudno, Y., Agarwal, S., Iyer, L.M., Liu, D.R., Aravind, L., Rao, L. (2009) Conversion of 5-methylcytosine to 5-hydroxymethylcytosine in mammalian DNA by MLL partner TET1. *Science*, **324**, 930–935.

Part IV
Medical Applications

25
Emerging Clinical Applications and Pharmacology of RNA

Sailen Barik[1] *and Vira Bitko*[2]
[1]*Cleveland State University, Center for Gene Regulation in Health and Disease and Department of Biological, Geological and Environmental Sciences, College of Sciences and Health Professions, 2121 Euclid Avenue, Cleveland, OH 44115, USA*
[2]*NanoBio Corporation, 2311 Green Rd, Ste A, Ann Arbor, MI 48105, USA*

Epigenetic Regulation and Epigenomics: Advances in Molecular Biology and Medicine, First Edition. Edited by Robert A. Meyers.

Keywords

Aptamer
A short single-stranded nucleic acid selected from combinatorial libraries to bind specific target molecules through a protocol termed

SELEX
(Systematic Evolution of Ligands by EXponential enrichment).

CCR, CXCR
Cellular receptors for cytokine families that are named on the basis of the two key cysteine (C) residues that are either next to each other (CC) or separated by another amino acid, X (CXC).

IFN (Interferon)
A family of cellular proteins, originally discovered as interfering with virus growth (hence the name). Now also known to be important in many cellular functions, including correct cell growth.

LNA (Locked Nucleic Acid)
Often referred to as

inaccessible RNA
LNA contains modified ribose sugars with a bridge connecting the 2′-O and 4″-C which enhances base stacking, increasing the thermal stability (melting temperature).

Pharmacokinetics
The action and metabolism of pharmaceuticals (drugs) in the body, including the processes of absorption, biotransformation, distribution to tissues, duration of action, and elimination.

PNA (Peptide Nucleic Acid)
A nucleic acid analog in which the entire phosphate sugar backbone has been replaced by an uncharged polyamide backbone with the side groups, purine and pyrimidine bases, found in biological nucleic acids.

Ribozyme
Catalytic RNA, capable of cleaving target RNA of specific sequence.

RNA drug
Different types of RNA molecules (aptamers, ribozyme, siRNA, etc.) that can be used for therapeutic and/or diagnostic purposes.

RNA interference (RNAi)
A novel mechanism induced by double-stranded RNA that leads to post-transcriptional gene silencing by sequence-specific degradation or translational inhibition of specific transcripts.

Spiegelmer
Biostable aptamers that are synthesized from L-nucleotides (the mirror image of natural nucleotides), and therefore, cannot be degraded by naturally occurring nucleases.

In all living cells, the RNA acts as an important intermediate of genetic information transfer between DNA and protein. RNA is generally single-stranded, and a flexible and versatile biopolymer that incorporates many pharmacologically desirable traits of DNA and protein without some of their disadvantages. It is water-soluble and nontoxic, can be produced by chemical synthesis as well as recombinant cloning, faithfully replicates the parental DNA sequence, can be converted to DNA or protein, possesses enzymatic self-cleaving activity, and makes specific RNA–RNA, RNA–DNA, and RNA–protein interactions. The biologically remarkable forms and functions of RNA, such as antisense RNA, ribozymes, RNA decoys, aptamers, small RNA, and their role in RNA interference and epigenetic chromosomal regulation, are all rooted in this fundamentally unique combination of biochemical and molecular traits. It is only recently that some of these properties of RNA have been exploited to interfere with or repair dysfunctional or harmful nucleic acids or proteins, and to induce therapeutic gene products in a variety of pathological syndromes or infectious diseases. First, second, and third generations of RNA drugs have produced promising results, raising new hopes in bringing RNA therapeutics from the bench to the bedside. In this chapter, the basic features and discoveries of RNA that are relevant to its clinical usage are described, and details provided of some emerging clinical applications of RNA.

1
Clinically Relevant Features of RNA

1.1
RNA in Central Dogma

The core of the so-called Central Dogma of molecular biology (Fig. 1) charts the genetic information flow in all living cells: DNA makes RNA makes protein. DNA is copied into RNA in "transcription," and RNA is deciphered to protein in "translation" [1]. To this is added the fact that DNA and RNA genomes are duplicated in "replication," and in retroviruses, the viral RNA genome can be copied in either the "retro" (hence the name of these viruses) or "reverse" direction to produce DNA in a process aptly termed "*reverse transcription*." DNA replication is catalyzed by DNA polymerase, and transcription by RNA polymerase (RNAP). The enzyme that copies RNA into RNA is sometimes referred to as "replicase" or "transcriptase" (as the name "RNAP" has already been taken by its DNA-dependent counterpart); in this chapter, however, the recent trend will be followed to call the enzyme RdRP (RNA-dependent RNA polymerase). Lastly, the reverse transcriptase (RT) catalyzes the conversion of RNA to DNA. Relevant aspects of the Central Dogma are detailed below.

DNA – the stored but otherwise useless genetic information in all cells – is double-stranded and, in comparison to RNA and protein, is relatively rigid and inflexible. Practically all naturally occurring DNA, except for the very small types (e.g., plasmids), are intimately complexed with proteins (such as histones in mammalian cells) to produce the chromosomes. The principal types of RNA transcribed from the DNA in a cell are listed in Table 1. Each class of RNA is transcribed by one of three types of RNAP that often requires the assistance of transcription factors, many of which are in turn regulated by physiological signals. The messenger RNA (mRNA) holds the distinction of being the only type of RNA that is translated into protein; hence its name as the "messenger" of genetic information. The majority of mammalian mRNAs are produced as pre-mRNAs containing "exons" and "introns"; subsequently, RNA splicing (see Sect. 1.4) removes the introns and joins the exons to produce the processed mRNA, ready for translation [2–4]. By virtue of its protein-coding nature, the mRNA has evolved to be more labile and subject to a relatively greater temporal, developmental, and tissue-specific regulation than the ribosomal RNA (rRNA) and transfer RNA (tRNA). While the replication and transcription of DNA genomes occur in the nucleus, translation occurs in the cytoplasm. The replication and transcription of RNA genomes, such as those of viruses, is also a cytoplasmic event, with few exceptions.

RNAP initiates transcription at a specific sequence on the DNA, known as a *promoter*, and copies only one strand of the DNA. Various conventions have been defined to describe the two nucleic acid strands, depending on the focus of the discussion. The preference is to call the RNA the "sense" strand, because this is the first step in the Central Dogma where a molecule that "makes sense" (to the translational machinery) is produced from the information encrypted in the DNA. The complementary strand of the DNA is, therefore, "antisense" (AS) (Fig. 1). In other words, RNAP copies the AS strand of the DNA as template to produce the RNA. To rephrase, the RNA sequence is identical

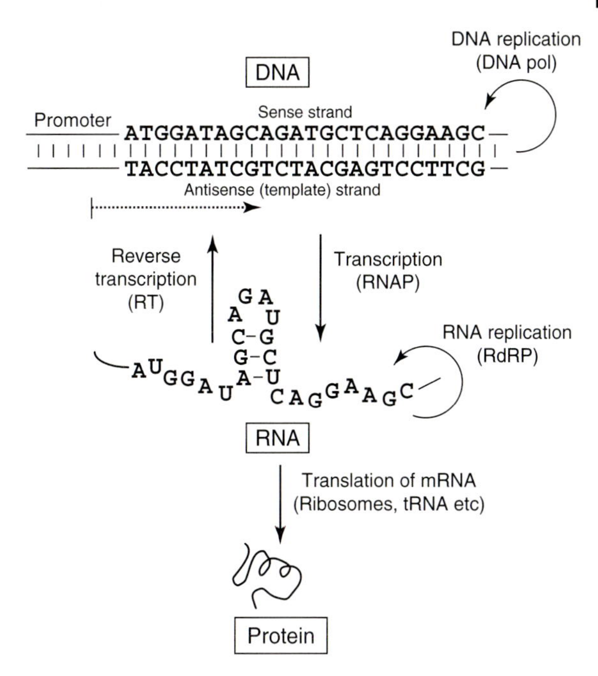

Fig. 1 The central dogma of molecular biology (see Sect. 1.1). This figure is not meant to serve as a comprehensive diagram of all aspect of replication, transcription and translation, but only to provide minimal relevant information needed to understand the remainder of the chapter. RNA splicing is omitted for simplicity, but is detailed in Sect. 1.4. The mRNA is shown without its 5′ cap structure and the 3′ poly(A) tail. An oversimplified RNA hairpin is shown. Nucleotides are paired by hydrogen bonds, indicated by small lines. The most important feature to note is that the RNA is located between DNA and protein, forms local base-pairs that resemble DNA, and at the same time folds and bends like proteins.

to the sense strand of the DNA (except for the T to U change), and complementary to the AS strand. In describing viral RNA genomes, virologists often follow another nomenclature, whereby the mRNA sense RNA is referred to as the "*positive*" or "*plus*" strand, and its complementary RNA is called the "*negative*" or "*minus*" strand. As most cellular functions – both structural and enzymatic – are performed by proteins, the overwhelming majority of conventional drugs directly target proteins. The majority of the emerging RNA drugs, in contrast, target the mRNA that produces the proteins, and thus function at a higher level in the Central Dogma.

Translation – the final step in Central Dogma – is a far more complex process that requires the participation of ribosomes, tRNA, and a slew of translational factors; hence, there is no such entity as "RNA-dependent protein polymerase." Protein is also the ultimate, dead-end product of the Central Dogma, and thus, cannot be reverse-translated. Finally, proteins do not replicate. It should be noted that RNA is the intermediate between the DNA and the protein, and shares the convenient features of both. In short, it has the nucleic acid property of DNA and the flexibility of protein; this point is elaborated further in the following sections.

Tab. 1 The principal types of RNA produced in cells.[a]

Type of RNA	*Property*	*Transcribed by*
mRNA	Messenger RNA: codes for proteins	RNAP II
rRNA	Ribosomal RNA: forms the basic structure of the ribosome and catalyzes translation	RNAP I
tRNA	Transfer RNA: functions as adaptors between amino acids and mRNA during translation	RNAP III
snRNA	Small nuclear RNAs: function in a variety of nuclear processes, including splicing	RNAP III
snoRNA	Small nucleolar RNA: used to process and chemically modify rRNAs	
siRNA	Short interfering RNA: double-stranded RNA that eventually degrades target RNA in the RNA interference (RNAi) pathway; important in RNA therapeutics	RNAP II
miRNA	microRNA: related to siRNA in biogenesis. These and other small noncoding RNAs (ncRNAs) function in diverse cellular processes, including translation inhibition, X-chromosome inactivation, telomere synthesis, and centromere silencing. Important in RNA therapeutics	RNAP II
piRNA	Piwi-interacting RNA, slightly larger than miRNA. piRNA is the newest and largest class of noncoding RNA. They bind to Piwi proteins and are important in spermatogenesis. Biogenesis is unclear; rasiRNA is a subspecies of piRNA	Exact mechanism of piRNA biogenesis currently unclear

[a] Intermediates and precursors are not described. For example, pri-miRNA is first processed by the nuclear RNase, Drosha, to produce pre-miRNA, which is further processed by Dicer in the cytoplasm to generate miRNA (Fig. 6). When first discovered, miRNA was termed "small temporal RNA" (stRNA).

1.2
Transcription and Replication of RNA Genomes

RNA genomes in Nature are found almost exclusively in RNA viruses that constitute a large group of pathogens infecting practically all species, including human [5]. Viral RNA genomes can be linear or circular, single- or double-stranded, segmented or nonsegmented, while the single-stranded genomes can be positive or negative sense or even ambisense (a mixture of the two). They can vary in size, and range from a little over a hundred nucleotides (as in hepatitis delta virus and plant viroids) to around 30 000 nt (as in coronoviruses that cause the severe acute respiratory syndrome; SARS). Viruses, by definition, are obligatory parasites, and must infect host cells in order to multiply. Due to their small size, viral genomes lack the capacity to code for the highly complex translation machinery (Fig. 1); all viruses, therefore, must utilize the translational apparatus of the host. However, animal cells in general – and human cells in particular – lack any enzymatic activity that can copy RNA templates, thus compelling all RNA viruses to encode such activities of their own. While the retroviral RNA genomes encode the RT that copies the genomic RNA into complementary DNA (cDNA) to be integrated into the host chromosomal DNA, the nonretroviral RNA genomes encode RdRP [6], which transcribes and replicates the genome exclusively in the cytoplasm. The functional viral RdRP is usually a multisubunit complex consisting of a large polymerase subunit and one or more accessory subunits. The RNA genome of positive-strand RNA viruses is essentially a sense-strand mRNA that is translated immediately after infection. The AS genome of negative-strand RNA viruses, in contrast, must first be transcribed by the associated RdRP activity to produce mRNAs that are then translated to produce new viral proteins. Many RNA genomes – especially negative-strand genomes – are tightly wrapped by the nucleocapsid protein (N), producing the N-RNA nucleoprotein complex, which is the biological template for the RdRP. The mRNAs, in contrast, are naked. Thus, the genomic RNA is resistant to various onslaughts including prospective drugs, while thc mRNA is a therapeutic target [7].

In contrast to the DNA genomes, RNA genomes are also highly mutable, mainly because the RdRP lacks a proofreading activity and thus is highly error-prone [8]. As a result, even the purest, clonally purified, RNA virus preparation may contain a mixed population of genomes of various sequences [9], leading to the concept of "quasi-species" [8]. This has two important implications. First, it allows the RNA virus to mutate quickly and produce "best-fit" mutants in a rapidly changing environment, thus enhancing its chances of survival. By the same token, however, it also creates major hurdles in designing vaccines and other antiviral strategies against RNA viruses. Indeed, in recurring RNA viral epidemics such as the annual episodes of human flu (caused by influenza virus), fresh vaccines are made against the new strain as the antigenic makeup of the new virus may be sufficiently different from those of the previous years, thus making the past vaccines ineffective.

1.3 Noncanonical Base Pairs in RNA

Although RNA is described as single-stranded in comparison to DNA, in reality more than half of the nucleotides in the typical RNA participate in base-pairing. Several motifs with various combinations of double-stranded helix, bulge, loop, and are commonly found in RNA secondary structures [10]. In small viral RNA genomes such as viroids, virusoids, satellite RNA, and human hepatitis delta virus RNA, essentially all of the bases are paired such that the RNA is practically all double-stranded and highly folded. While the Watson–Crick A:T and G:C base pairs are universal in all DNA, RNAs may contain many modified bases in addition to the standard A, C, G, and U, and the flexibility of the RNA chain also allows various noncanonical base pairs [11]. The most prevalent noncanonical RNA base pairs with the standard bases are GU, GA, and AC. The tRNAs often contain the modified base inosine (I), created post-transcriptionally, in the first wobble position of the anticodon; as I can pair with A, C, or U, this allows more promiscuous codon–anticodon pairing, such that fewer different tRNAs are required. As discussed later, such modified bases and alternate pairing rules may be used in designing better RNA drugs.

1.4 RNA Splicing

Typical mammalian genes are organized on the DNA in exon–intron structure, with an average of about nine exons per gene. Transcription of the DNA produces pre-mRNA, from which the introns are removed in a process known as *splicing*. The process is catalyzed by a large ribonucleoprotein complex, called the "*spliceosome*," containing various RNA of the snRNA family (Table 1) and multiple proteins [12, 13]. The spliceosome recognizes sequence features of the intron–exon boundaries (splice junctions) and performs the actual cutting and joining of the RNA segments. The fundamental principles of how alternate splicing can be regulated by an RNA drug and specific splice variants or faulty RNA can be destroyed or repaired by catalytic RNA drugs are summarized in the following sections.

1.4.1 Alternate RNA Splicing: Regulation by Antisense RNA

Alternate splicing occurs when the introns of a particular pre-mRNA are spliced in more than one way, generating multiple mature mRNA species – and hence multiple proteins – from a single gene (Fig. 2). More than half of all human gene transcripts are estimated to undergo alternate splicing, yielding an enormous variety of mRNAs out of a relatively small repertoire of genes [14]. As in everything else in biology, alternate splicing is highly regulated, and its pattern can be specific to tissue, development stage, physiological condition, or disease.

A representative example of one type of alternate splicing and its exploitation in therapeutics is shown in Fig. 2. Apoptosis (referred to as "programmed cell death") is regulated by the differential expression of a large number of proteins by alternative RNA splicing, some of which may even have opposite apoptotic functions. For example, the Bcl-x pre-mRNA generates two alternative splice variants, namely Bcl-x(Long) (Bcl-xL) and Bcl-x(Short) (Bcl-xS). While the Bcl-xL protein is anti-apoptotic, Bcl-xS is pro-apoptotic and sensitizes cancer cells to chemotherapeutic agents. Elevated levels of Bcl-xL generally correlate with a decreased cellular sensitivity towards chemotherapy. In a recent approach [15], an AS nucleic acid targeted towards a complementary alternative splice site of Bcl-x pre-mRNA shifted splicing from Bcl-xL to Bcl-xS, thus lowering the apoptotic threshold of prostate cancer cells and increasing the efficacy of chemotherapeutic drugs. Thus, it is conceivable that a correctly designed AS RNA could selectively alter the splicing pattern of a pre-mRNA to produce a therapeutic effect.

In summary, an AS RNA, by binding to a specific sequence, can alter the structure and function of the target RNA, which has obvious clinical implications. As discussed in Sects 1.4.2 and 1.4.3, ribozymes and siRNAs (Sect. 1.5) cannot inhibit site-specific splicing but can be used to destroy a specific splice variant by targeting a sequence that is unique to the variant.

1.4.2 *Trans*-Cleaving Ribozymes: Destruction of Undesired RNA

Ribozymes are RNA molecules that catalyze a chemical reaction – that is, the scission of a covalent bond [16]. Historically, the first catalytic RNA was described in a group I intron from *Tetrahymena thermophila* that could undergo self-cleavage in *cis*, which was followed by the discovery of *trans*-cleaving RNA that could act on

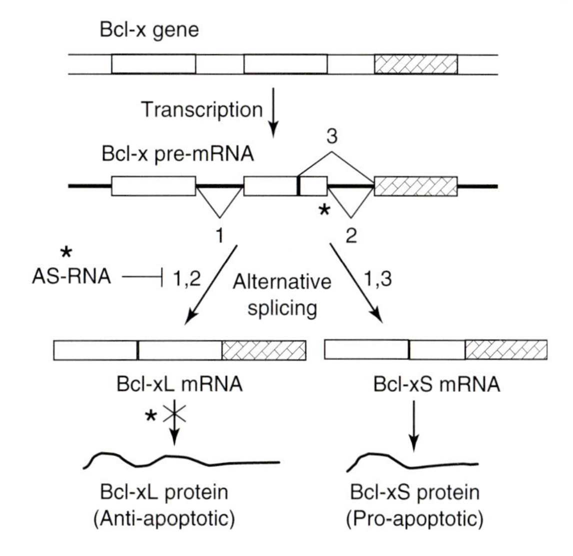

Fig. 2 Regulation of alternative splicing by antisense RNA drug [15]. The three possible splicing events on the Bcl-x pre-mRNA are labeled 1, 2, and 3. A combination of 1 and 2 splicing yields the long Bcl-xL mRNA and protein, while 1 and 3 splicing yields the short Bcl-xS mRNA and protein. The antisense RNA (AS-RNA, denoted by an asterisk) inhibits splicing 2, thus abrogating the synthesis of the long protein, shifting the balance toward the short protein, and promoting the death of cancer cells.

other RNA substrates [17–21]. It was soon recognized that the latter class could be used as "catalytic AS" tools in gene therapy, whereas the purely AS RNA described above could only inhibit target function but did not cause loss of the target or alteration of the target sequence. By far the most favored *trans*-cleaving ribozyme in clinical applications are the hammerhead ribozymes [22], which share a typical secondary structure of the catalytic core composed of three helices and variable, as well as invariant, nucleotides at specific positions (Fig. 3a). Whereas, helix II is formed intramolecularly, helices I and III are composed of paired hammerhead and substrate sequences. Cleavage occurs 3' to nucleotide H (Fig. 3a) of the target, and efficient cleavage requires a U to the 5′ side of this H to pair with a critical A residue (indicated by an asterisk in Fig. 3a) in the hammerhead. The variable nucleotides (N in Fig. 3a) in these helices allow virtually any RNA sequence to be targeted. The repertoire of target sequences can be further expanded by taking advantage of noncanonical RNA base-pairing and modified bases (see Sect. 1.3). For example, replacement of the asterisked A (in Fig. 3a) with a more promiscuous base, inosine (I), resulted in a hammerhead ribozyme that could cleave at a CH consensus via I:C pairing. In summary, a synthetically engineered hammerhead ribozyme can act as "molecular scissors," destroying unwanted RNA of specific sequence such as a viral RNA, a splice variant, or a mutant RNA associated with a pathological condition such as cancer. As the ribozymes function catalytically, a single ribozyme molecule can be reused many times to cleave a large number of substrate molecules (Fig. 3b), thus enhancing the therapeutic effect.

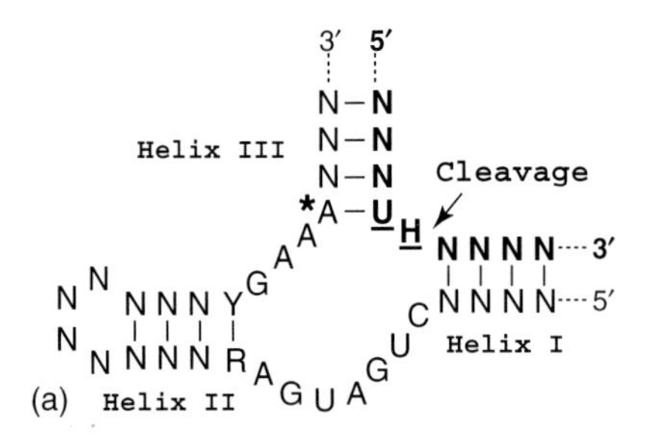

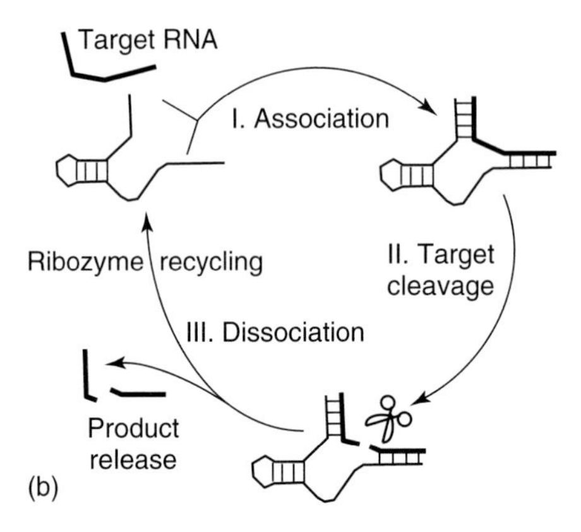

Fig. 3 (a) Secondary structure of a *trans*-cleaving hammerhead ribozyme with bound substrate. The ribozyme is in plain letters; the substrate RNA is in bold. N = any nucleotide; Y = pyrimidine (U or C); R = purine (A or G); H = any nucleotide except G. Note that A:C and G:U base pairs are permissible in all RNA (see Sect. 1.3), including ribozymes. The hammerhead ribozyme, associated with the substrate RNA, is depicted to form the three structurally and functionally important helices (I, II, III). In clinical applications, exogenously used hammerheads are generally 35–40 nt long, and the intramolecular helix II is formed by 4 bp; however, minimized hammerheads with faster cleavage rates have been described that contain only 2 bp in this helix. The cleavage site in the substrate RNA is as shown. The required UH dinucleotide is underlined; this U can be substituted by a C if the pairing A (asterisked) is changed to I; (b) The target (thick line) cleavage cycle of a ribozyme (thin line). The dissociation step (product release) is the slowest and hence rate-limiting; however, once dissociated from the cleaved target, the ribozyme is recycled as a true catalyst to carry out multiple rounds of cleavage.

1.4.3 *Trans*-Splicing Ribozymes: Repair of Defective RNA

Once defective genetic information has been transcribed into RNA, it is still possible to use RNA therapeutics to mend the defective RNA sequence without altering the DNA gene. Major strategies have been developed that exemplify the use of *trans*-splicing ribozymes in RNA repair [23, 24].

The first method (Fig. 4) employs the originally discovered self-splicing group I intron from *Tetrahymena*, which mediates *trans*-splicing of an exon attached to its 3′ end onto a targeted 5′ exon RNA that is a separate RNA molecule. The repair of *lacZ* transcripts in *Escherichia coli* and in cultured mammalian cells using shortened versions of this ribozyme has been demonstrated [25]. In such studies, the ribozyme, which is engineered to contain the wild-type sequence, base-pairs with the mutant RNA using an internal guide sequence, cleaves off the mutant segment, and then ligates the wild-type sequence to the cleaved product (Fig. 4). In a variation of this theme, a "twin ribozyme" is created by tandem duplication of a hairpin ribozyme, such that four extra nucleotides are added to the replaced RNA segment. If this is an mRNA, then the resultant repaired mRNA would contain an extra amino acid followed by a frame-shift, which may have a therapeutic value.

The second, related method is named spliceosome-mediated RNA *trans*-splicing (SMaRT) [26]. This does not require an exogenous ribozyme, because it uses the spliceosome to catalyze a *trans*-splicing reaction between the target pre-mRNA and pre-*trans*-splicing mRNA (PTMM) (Fig. 5). It should be noted that the PTM is not a ribozyme; rather, its *trans*-splicing simply competes with the natural *cis*-splicing and prevents the generation of the mutant mRNA.

In an interesting diagnostic application, "half-ribozyme" ligase molecules have been designed that are activated upon binding to their RNA targets. The catalytic nature of these enzymes leads to multiple rounds of ligation and hence to signal

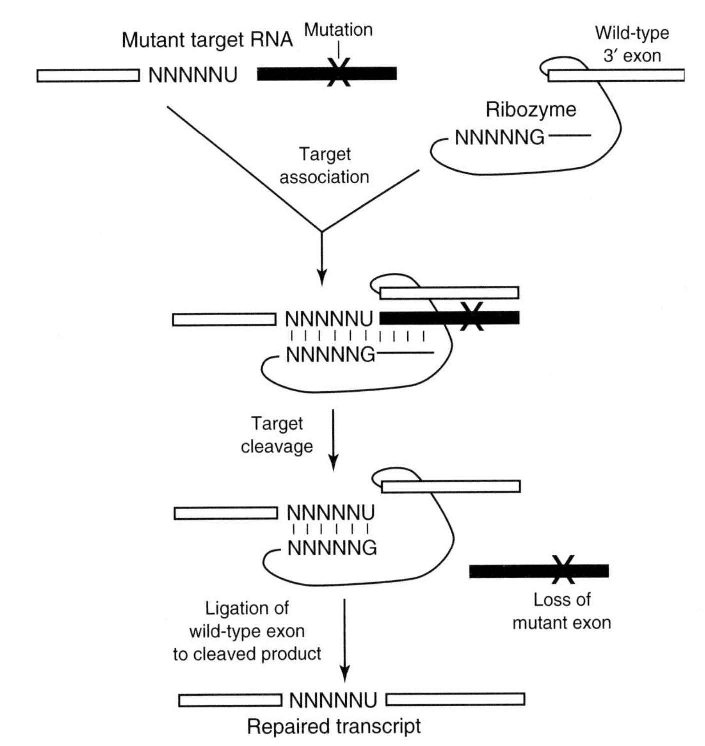

Fig. 4 Ribozyme-mediated repair of RNA. The mutant exon of the target RNA with the mutation X is shown in black. The *trans*-splicing ribozyme, introduced into the cell, brings in the wild-type exon (white). The N_5G stretch and some extra nucleotides acts as the internal guide sequence to direct *trans*-cleavage that results in a loss of the mutant exon and ligation of the wild-type exon to produce the repaired transcript (bottom).

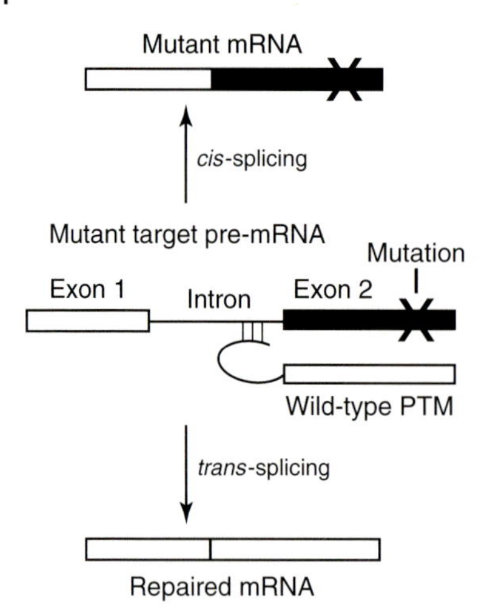

Fig. 5 Spliceosome-mediated *trans*-splicing repair of mutant RNA by SMART (see Sect. 1.4.3). The mutant exon with the mutation X is shown in black, and its wild-type counterpart in the PTM in white.

amplification, allowing the detection of attomolar (10^{-18}) quantities of the target, such as viral RNA in clinical samples.

1.5 RNA Interference: siRNA and miRNA

A relatively recent discovery – the phenomenon of RNA interference (RNAi) and the various aspects of its mechanism and application – have captivated the imagination of the biological community [27, 28]. A literature search through the PubMed site of NCBI (http://www.ncbi.nlm.nih.gov/) revealed that a total of 303 RNAi-related reports were made in just one month (September, 2011), at the rate of about 10 per day, underscoring the feverish activity in this field. The RNAi pathway is activated by double-stranded RNA (dsRNA) that, in principle, can form whenever sense RNA meets AS. In Nature, segments of many viral genomes are transcribed in both orientations, generating both sense (positive) and AS (negative) RNA in the infected cell, which may then base-pair to form dsRNA. In the laboratory, dsRNA can be synthesized chemically or by recombinant technology and then introduced into the cell, tissue, or animal [7, 28–30]. The mechanism of the RNAi pathway is briefly illustrated in Fig. 6, whereby a long dsRNA is precisely processed by Dicer, a member of the RNase III superfamily, to produce 21- to 28 nt-long dsRNA fragments with 3′-overhangs, called *siRNA*. This step is bypassed by chemically synthesized siRNAs that are introduced directly into the cell by transfection or electroporation. The siRNA is incorporated into a multiprotein complex called RNA-induced silencing complex (RISC) [31], followed by an unwinding of the strands by a helicase activity. The unwinding is asymmetric and initiates from the energetically easier terminus – that is, the one with the higher A–U base-pair content [32, 33]. The activated RISC with the AS strand engages its complementary target RNA. A RISC-associated RNase activity, mainly due to Argonaute-2, then cleaves the target RNA 10 nt from the 5′ end of the siRNA strand, resulting in knockdown or silencing of the corresponding gene function [31].

A related class of naturally occurring single-stranded cellular RNA, named microRNA (miRNA), is also 21–24 nt long [34, 35]. The miRNAs are transcribed from specialized endogenous genes as an approximately 70 nt-long hairpin precursor. The biogenesis and function of siRNA and miRNA are highly similar [36], although the full extent of overlap between the two pathways is yet to be determined. Dicer is

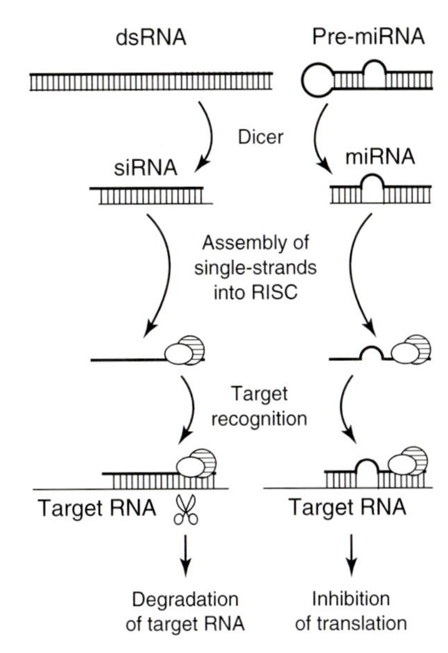

Fig. 6 RNA interference (RNAi). Only the major steps are shown. In contrast to siRNA, the miRNA starts as a hairpin and eventually inhibits the translation of an mRNA with an imperfect match, although exceptions are known in that some miRNAs do degrade the target. The protein subunits of RISC are indicated by the spheres, but their exact number remains unclear. It is likely that the siRNA-RISC and miRNA-RISC assemble from common as well as unique subunits.

involved in the processing of both siRNA and miRNA from their respective precursors (Fig. 6), although different Dicers may be preferred by siRNA and miRNA. In general, it appears that siRNA and miRNA that are perfect or a near-perfect match with their target degrades the target RNA, whereas those with mismatches repress translation instead (although a few exceptions are known) [37]. RISC is an exclusively cytoplasmic entity; consequently, siRNA and miRNA are only useful against cytoplasmic targets such as mRNA, and not nuclear pre-mRNA. Viral genomic RNAs that are either highly structured or covered with protein are also resistant to RNAi.

The exact number of miRNA in the human cell is still unclear [38], but the current estimate runs in the hundreds (http://www.mirbase.org). Accumulating evidence indicates that they regulate a variety of normal cellular pathways, including differentiation and development [34, 35]. Due to their remarkable efficiency and selectivity, both siRNA and miRNA have emerged as major tools for knocking down gene expression in basic, as well as clinical, applications.

1.6
RNA as a Protein Antagonist: SELEX, Aptamers, and Spiegelmers

The variability of nucleotide sequence and the ability to fold allow single-stranded DNA and RNA to attain a wide range of structures that can interact with high affinity and specificity with other biomolecules [39]. In principle, starting with a large library of DNA or RNA sequences (ca. 10^{15}), it is possible to select specific molecules or "aptamers" [40–43] to bind to practically any target, such as a protein, by using a combinatorial and iterative Darwinian-type *in vitro* evolution process that has been termed SELEX ("systematic evolution of ligands by exponential enrichment") [44] (Fig. 7). The aptamers are composed of a central randomized (degenerate) region flanked by fixed primer binding sites on either side for polymerase chain reaction (PCR)-based amplification. Whereas, AS RNA and siRNA (Sect. 1.5) can target other RNAs only

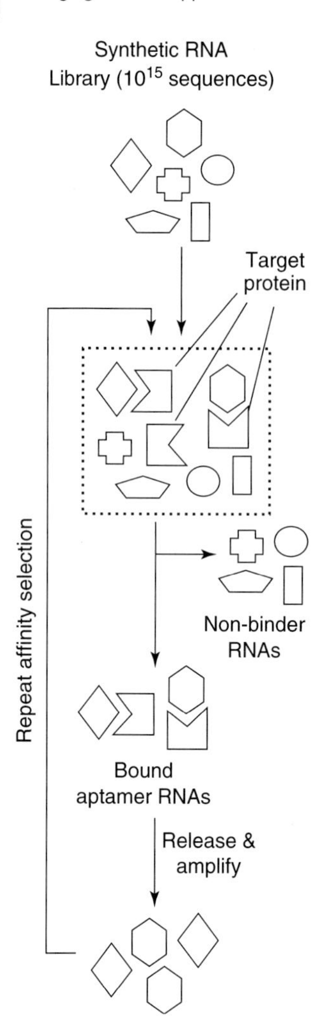

Fig. 7 RNA SELEX and aptamer. In this schematic diagram the degenerate RNA molecules are depicted as pieces of various sizes and shapes, some of which fit the immobilized target protein (in gray). The unbound RNAs are discarded, and the bound aptamers are eluted and amplified by reverse transcription-PCR. They can be further screened through additional cycles of binding and reamplification until the desired strength and specificity of binding is achieved.

by Watson–Crick base-pairing, the aptamers in contrast function by forming discrete structures that bind specific targets. Conceptually, aptamers are similar to antibodies in terms of specificity, but have several advantages over antibodies, especially stability and chemical synthesis (as elaborated in Sect. 1.9). Aptamers possess affinity in the low nanomolar to picomolar range, and can discriminate between targets that are 96% identical in sequence. RNA aptamers derived from *in vitro* selections (SELEX) may be used inside cells for at least four purposes: (i) to antagonize cellular proteins; (ii) as decoys to natural RNA-binding proteins; (iii) as regulatory modules to govern gene expression; and (iv) to antagonize nonprotein targets.

Photoaptamers are defined as aptamers with photocrosslinking functionality. Photocrosslinking never caused any degradation in specificity and, in fact, led to a significant increase in aptamer specificity in some cases. Thus, photoaptamers add a second dimension of specificity, as they recognize both the complex shape and charge distribution of their targets and the presence of specific amino acid residues at specific sites.

Spiegelmers (German "Spiegel" means mirror) are mirror-image aptamers made from L-nucleotides [45]. The L-enantiomers act just like standard aptamers, but have the added advantage of being highly biostable due to their resistance to nucleases.

1.7 Polyamide ("Peptide") Nucleic Acid (PNA)

Polyamide ("Peptide") Nucleic Acid (PNA) is a synthetic nucleic acid analog in which the entire phosphate sugar backbone is replaced by an uncharged polyamide (peptide-like) backbone that is nuclease-resistant [46, 47]. The side groups are still the natural purine and pyrimidine bases and in a similar configuration as in natural single-stranded nucleic acids. Thus, PNAs possess the base-pairing capability and specificity of nucleic acids and hence, can function as AS but are more stable and bind more strongly and specifically to their DNA or RNA targets.

A unique property of some PNA sequences is to function as gene-specific silencers. These PNAs bind to dsDNA by an invasion mechanism, whereby two PNA molecules form a triplex structure with the cDNA target sequence, while the other strand of the DNA duplex is displaced into a single-stranded loop. Although this type of binding is generally limited to pyrimidine-rich sequences, the resulting PNA–DNA complex is extremely stable. If this occurs at the promoter or enhancer sequences, the binding of RNAP or specific transcription factors can be abrogated in *cis*, leading to the silencing of specific genes.

1.8 Locked Nucleic Acid (LNA)

Often referred to as *inaccessible RNA*, a Locked Nucleic Acid (LNA) contains modified RNA nucleotides in which the ribose moiety is modified with an extra bridge connecting the 2′ oxygen and the 4′ carbon [48]. The bridge locks the ribose in the 3′-endo (North) conformation (hence the name), as often found in the A-form of DNA or RNA. LNA nucleotides can be mixed with DNA or RNA bases at the desired positions in the oligonucleotide. The locked ribose conformation enhances base stacking and backbone preorganization, which in turn significantly increases the thermal stability (melting temperature) of the LNA when bound to its target and, hence, the specificity of binding. LNA nucleotide, if used in moderation, improves the potency and specificity of siRNA in target knockdown, and is currently being used widely in this role and in the specific quantification of miRNA.

Tab. 2 The major advantages of RNA as a pharmaceutical agent.

Relevant property	*DNA*	*RNA*	*Protein*	*Small molecule*
Chemical synthesis and modification	Yes	Yes	Difficult	Some
Biochemical synthesis	Yes	Yes	Yes	Some
Replication	Yes	Yes	No	No
Sequence-specific interaction with DNA or RNA	Yes	Yes	Some	No
Specific interaction with proteins	No	Yes	Some	Some
Stability without refrigeration	Yes	Yes	No	Some
Intricately folded structure	No	Yes	Yes	No
Catalytic activity	No	Yes	Yes	No
Cloning possible	Yes	Yes	Yes	No
Easy to isolate and purify	Yes	Yes	No	No
Safe for autoimmunity	Yes	Yes	Some	Yes
Nontoxic	Yes	Yes	No	Some
Tissue-specific targeting	Yes	Yes	No	No
Can repair mutant gene or disable expression	Yes	Yes	No	No
Relatively low cost	Yes	Yes	No	No

1.9 Summary: Pharmaceutical Advantages of RNA

The relevant advantages of RNA over other forms of traditional and prospective pharmaceuticals are summarized in Table 2. Proteins contain 20 different amino acids, each with its unique chemical and physical property, whereas RNA (or DNA), in contrast, is composed of only four nucleotides, thus reducing the complexity. The costs of chemical synthesis of the three biological macromolecules of the Central Dogma are currently in the order: Protein > RNA > DNA. The recombinant expression of protein is an uncertain prospect, due to a variety of factors such as mRNA and protein stability and processing, translational regulation, and codon bias. Finally, the biological activity of many proteins requires post-translation modifications such as glycosylation and phosphorylation that do not occur in bacterial expression systems. The purification of proteins in active form can be a significant hurdle due to insolubility and incorrect folding. A drawback of DNA is the ability of its CpG motifs to activate Toll-like receptors (TLRs) and thus to induce a general innate immune response, which in fact underlies the apparent therapeutic effects of several successful AS DNA molecules [49, 50]. As noted above, RNA can also trigger cellular innate immune response via specific TLRs and mitochondrial pathways, although this can be minimized through correct RNA design.

Small molecules currently constitute the largest class of drugs, and owe their origin to two major sources: natural products from folklore; and structure-based drug design. The former source is subject to the whims of serendipity and Nature's bounty, while the latter source requires expensive cutting-edge technology. In the past, the synthesis of small-molecule drugs by chemical means has been the norm of the pharmaceutical industry; however, the chemical reactions are often challenging and need to be revised for each new compound or a new substitution. RNA can be prepared rather easily *in vitro*, as well as *in vivo*, by transcription from cloned cDNA. Small RNA molecules (e.g., siRNA) are better produced chemically; indeed, it can be expected that the associated cost will diminish with the development of innovative chemical and process engineering and increased market volumes.

It is highly likely that ribozymes and siRNAs will become the major RNA weapon of the future in the war against pathogenic RNA [51–53]. There are multiple advantages to targeting RNA instead of proteins, including sequence-specific binding, a greater number of sites accessible for interaction, selective inhibition, exploitation of multivalent interaction, and a more facile attack on drug resistance. The link between RNA and genomics and bioinformatics also allows for increased discovery and specificity testing.

2 Emerging Clinical Applications of RNA

2.1 Ribozymes against Infectious Agents

The ribozyme has been consistently recognized as a potentially sequence-specific therapeutic approach [51]. One of the first obvious viral targets for ribozyme therapy has been HIV-1 (human immunodeficiency virus, the causative agent of AIDS), in part due to the extraordinary death toll of the disease and the limited efficacy and toxicity of the multidrug cocktails targeting the viral RT. Various sequences in the HIV genome can be targeted for ribozyme-mediated therapy [54–58]. The two types of ribozyme (hairpin and hammerhead) show similar antiviral efficacies in cell culture in terms of their ability to inhibit the HIV replication, and have also fared well in clinical trials [59]. In representative studies, autologous lymphocytes were transduced with a hairpin ribozyme that cleaves the U5 region of HIV-1 RNA, and the resulting cell population consisting of transduced and nontransduced cells was infused. When tested in peripheral blood, the ribozyme-containing cells indeed exhibited a preferential survival over non-ribozyme-containing cells. Similarly, hammerhead ribozymes (Rz1, Rz2) targeted to a conserved region of the HIV *tat* gene [59, 60] were transduced into a population of syngeneic $CD4^+$ T lymphocytes, with the resultant cells being introduced into the respective HIV-positive identical twin [61]. When four patients were infused with syngeneic lymphocytes, subsequent PCR analysis demonstrated expression and viral inhibition for up to four years post-infusion (the latest time point examined). Moreover, these procedures were well tolerated by patients, with no serious adverse events. Together, these results indicate that the infusion of gene-altered, activated T cells in HIV-infected patients is safe, and that the transduced cells persist for long intervals. The results also suggest that ribozyme-transduced cells may possess a survival advantage *in vivo*.

An alternative approach to ribozyme gene therapy would be to transfer the ribozyme-expressing constructs into hematopoietic stem cells of HIV-1-infected individuals. This might serve as a potent therapeutic approach to provide mature cells arising from transduced stem cells that were resistant to the destructive events associated with HIV-1 infection. In order to determine the feasibility of gene therapy for AIDS patients, peripheral blood $CD34^+$ cells were isolated from HIV-1-infected individuals and transduced with retroviral vectors containing three different anti-HIV-1-genes [62]: the RNA binding domain of the Rev-responsive element (RRE decoy) (L-RRE-neo); a double hammerhead ribozyme vector targeted to cleave the Tat and Rev transcripts (L-TR/TAT-neo); and a *trans*-dominant mutant of Rev (M10) (L-M10-SN). As a control, a vector mediating only neomycin resistance was used. After three days of transduction on allogeneic stroma in the presence of stem cell factors interleukin-6 (IL-6) and IL-3, the cultures were neomycin-selected and then challenged with HIV-1. Compared to the control cultures, the L-RRE-neo-, L-TR/TAT-neo-, and L-M10-SN-transduced cultures displayed an up to 1000-fold inhibition of HIV-1 replication after the viral challenge. The study results showed that anti-HIV-1 genes can be introduced into $CD34^+$ cells from individuals already infected with HIV-1, and that they strongly inhibit HIV-1 replication *in vivo*.

Herpes simplex virus 1 (HSV-1) and human cytomegalovirus (HCMV) are pervasive herpes viruses that cause severe morbidity or mortality in neonates and immunocompromised individuals. In particular, HSV-1 is the causative agent for cold sores and encephalitis in newborns, while HCMV is one of the leading causes of birth defects in the US, and is accountable for more than 50% of deaths in organ transplant patients. HCMV-mediated retinitis is also the leading cause of AIDS-related blindness. In recent studies, ribozymes were shown to efficiently cleave the mRNAs of both viruses and to effectively reduce viral growth in cultured cells [63–65].

Targeting of the mRNA encoding the major transcriptional activator ICP4 (Infected Cell Protein 4) of HSV-1 – an immediate early (IE) gene which is essential for the expression of most of the viral early and late genes – caused about an 80% reduction in ICP4 and a 1000-fold decrease in viral growth. Meanwhile, other IE gene (e.g., α47 and ICP27) expressions were not affected, which showed that the ribozyme is highly specific in targeting the desired mRNA. Ribozyme expression was not cytotoxic, and the expressing cells were indistinguishable from the parental cells in terms of cell growth and viability for up to two months. Similar results were also reported when the overlapping region of HCMV IE gene 1 and 2 was targeted by the ribozyme. In HCMV-infected human cells, this ribozyme reduced IE1/IE2 expression by 85%, and inhibited HCMV growth by 150-fold. A reduction in IE1/IE2 expression levels also resulted in the down-regulation of other early and late gene products, such as US2, UL44, gB, and gH, whereas those transcripts that are not regulated by IE1/IE2 (e.g., 5 kb RNA and UL36 mRNA) were unaffected. RNAi also showed promise against gammaherpesviruses [66].

Hepatitis C virus (HCV) infection may lead to chronic hepatitis, cirrhosis, and hepatocellular carcinoma. The therapy of chronic HCV infection has been greatly improved with the combined use of ribavirin and α-interferon and, recently, with PEG (polyethylene glycol)-interferons. However, over half of the patients did not accrue any lasting benefits from these therapies. Ribozyme strategies to target certain HCV sequences have been proposed as anti-HCV treatments [67, 68]. In particular, the 5′ noncoding region of the HCV RNA is highly conserved and plays an essential role in translation initiation, by functioning as an internal ribozyme entry site. In several studies, both endogenously expressed and exogenously delivered ribozymes targeted against this region inhibited HCV replication. The extrapolation of animal data to humans was eventually abandoned, however, due to the possibility of toxic effects.

Urogenital human papillomavirus (HPV) infections are the most common sexually transmitted viral disease among women. Recently, a self-processing triple-ribozyme cassette, consisting of two *cis*-acting hammerhead ribozymes flanking an internal, *trans*-acting hammerhead (ITRz), was tested against HPV-11 E6/E7 mRNA, whereupon significant reductions (80–90%) in infection were achieved for both [69]. A similar ribozyme, designed against hepatitis B virus (HBV), caused a >80% reduction in viral liver DNA over a two-week period in a transgenic mouse model.

2.2
Ribozymes against Cellular Disorders

Recently, ribozymes have been demonstrated as a highly effective treatment against a number of cellular disorders, especially cancer. Point mutations in the K-*ras* gene are observed at a high incidence in human pancreatic carcinomas. The anti-K-*ras* ribozyme against codon 12 of the mutant K-*ras* mRNA, when expressed via a recombinant adenoviral vector, suppressed tumor growth and promoted regression [70, 71]. It was also possible to accomplish an efficient reversion of the malignant phenotype in human pancreatic tumors with K-*ras* gene mutation. Similarly, a clinically significant fraction of sickle cell hemoglobin could be repaired by *trans*-slicing ribozymes [72, 73]. In an interesting screening project, a randomized ribozyme pool was used in a mouse model to identify novel metastasis-related genes [74].

It is well known that tumor growth beyond a few cubic millimeters does not occur without the induction of a new vascular supply network, a process known as *angiogenesis*, and consequently the inhibition of angiogenesis has been heralded as a potential approach to cancer therapy. To date, vascular endothelial growth factor (VEGF) is the best-characterized pro-angiogenic factor, and an effective blockade of the VEGF pathway has been demonstrated with multiple agents such as neutralizing antibody, receptor tyrosine kinase inhibitors, and ribozyme or AS molecules. Promising preclinical data have documented the potential of these agents for tumor growth inhibition and even tumor regression, yet the translation of novel therapeutics targeting the VEGF pathway to the clinic has proved a substantial challenge in itself. Angiozyme, an anti-VEGF1 RNA developed by Sirna Therapeutics, was the first ribozyme to undergo human trials against several cancers [75], but its use eventually was suspended due to a suboptimal clinical response.

Hepatocyte growth factor/scatter factor (HGF/SF) elicits a number of biological activities, including invasion and migration, through the activation of its tyrosine kinase receptor c-Met, the overexpression of which is implicated in prostate cancer development and progression. Targeting the HGF/SF receptor by way of a hammerhead ribozyme is considered an important therapeutic approach in these cancers [76].

Ribozyme has been particularly useful in targeting oncogenic mRNAs resulting from random chromosomal translocation. For example, translocation between chromosome 9 and 22 fuses the genes of breakpoint cluster region (BCR) and c-ABL (Abelson) to produce BCR-ABL, resulting in chronic myelogenous leukemia (CML). Ribozymes with the guide sequences that target the fused sequence of the transcript effectively destroy the BCR-ABL fusion transcript found in the cancerous cells of leukemia patients [77, 78]. More importantly, the inhibition is highly specific, as the ribozyme cleaves only the targeted chimeric transcript and not the normal cellular counterpart, which makes this a promising approach in cancer therapy. In a variation of this technology, "maxizymes" – dimeric ribozymes capable of targeting two different regions of a chimeric oncogenic mRNA – were designed [79].

In the management of asthma, hammerhead and hairpin ribozymes targeting conserved sequences within IL-4, IL-5, intercellular adhesion molecule (ICAM), and nuclear factor-kappa B (NF-κB) mRNA have been already designed [80]. The results of pharmacokinetics studies in mice revealed that, when administered intratracheally, these ribozymes have improved therapeutic benefits compared to traditional drugs.

The most common gene defect in cystic fibrosis is a deletion of Phe508 from the cystic fibrosis transmembrane receptor (CFTR), which renders the protein dysfunctional and unstable. A partial correction of the phenotype could be achieved, however, by spliceosome-mediated RNA *trans*-splicing [81].

2.3
RNAi against Viruses

The pioneering study on RNAi-mediated inhibition of virus replication in animal cells was reported almost nine years ago, and employed respiratory syncytial virus

(RSV) as a prototype model [7] to open up a new direction in anti-viral drugs. RSV (an RNA virus) is a major of cause respiratory infection in infants and the elderly, and claims several millions of lives worldwide on a annual basis [82]. Although, to date, there is no report of any reliable vaccine or antiviral agent, the use of siRNA against essential RSV transcripts was shown to inhibit virus growth in cultured lung epithelial cells, and also in a mouse model [7, 83]. Silencing of the RdRP subunits, L and P, reduced the progeny viral titer by 10^3 to 10^4-fold [84], while no cytopathic effect was detected in uninfected siRNA-treated cells. Following such success against RSV, a large number of other viruses have been targeted by siRNA both *in vitro* and *in vivo*, with highly encouraging results having been obtained [84, 85]. One company involved in the production of RNAi – Alnylam Pharmaceuticals – has translated this breakthrough into a successful anti-RSV siRNA termed ALN-RSV01, designed to combat RSV N, and has de3monstrated clear benefits in Phase II clinical trials [86]. Currently, ALN-RSV01 remains, from a clinical standpoint, the most advanced antiviral siRNA.

As implied earlier, HIV-1 often escapes its commonly applied therapies because of its high mutation rate and the complexity of the pathogenesis of AIDS [87]. RNAi and ribozymes targeting various HIV sequences that encode the structural proteins Gag and Env, the RT, the regulatory proteins Tat and Rev, and two accessory proteins Nef and Vif, were all shown to be effective – albeit to various degrees – in inhibiting HIV growth [43, 85, 87–93]. The RNAi targeting of nontranslated RNA sequences in the viral long terminal repeats (LTR)s, which contains important regulatory elements required for HIV replication, was also effective. In principle, it might also be possible to inhibit the virus infection by silencing the expression of essential cell factors that are critically involved in the viral life cycle. In fact, an inactivation of the cellular receptor (CD4) and coreceptor (CXCR4, CCR5) by specific siRNA caused a reduction in the HIV infection of T cells [94–97].

Influenza A virus causes widespread infection in the human respiratory tract, yet existing vaccines and drug therapy are of limited value in its treatment. Typically, siRNAs against influenza's nucleocapsid or a component of RdRP, abolish viral transcription and replication in cell lines and embryonated chicken eggs [98–103]. These results should provide a basis for the development of siRNA as preventive treatment and therapeutic drug for influenza infection in humans.

RNAi has also been used against several other viruses [85, 104–108], including Dengue virus, flock house virus (FHV), rhesus rotavirus (RRV), Semliki forest virus (SFV), and poliovirus. All hepatitis viruses, such as hepatitis B virus (HBV), hepatitis C virus (HCV), and hepatitis delta virus (HDV), are major public health hazards, against each of which RNAi appears to be an attractive option [109–114]. siRNA against the core region of HBV, when cotransfected with the full-length HBV DNA into Huh-7 and HepG2 cells, lead to an effective inhibition of viral growth in both cell culture and mice [115, 116]. Essentially similar results were obtained with siRNA against the S gene of HBV. HCV, a Flaviviridae with a small RNA genome, is a uniquely difficult virus to study because of a lack of *in vitro* cell culture models. Recently, however, subgenomic replicon systems for HCV have been developed, with siRNAs against not only viral mRNAs (e.g., NS3 and NS5B) but also the untranslated regions (e.g., 5′-UTR) being shown to effectively inhibit viral replication [113, 117, 118]. In HDV, the results of recent studies have confirmed that the delta antigen mRNA can be successfully targeted by siRNAs in cell culture.

In contrast to viral mRNAs, the genomic and antigenomic RNAs of many viruses are resistant to siRNA action. This is due either to their encapsidated nature (as in negative-strand RNA viruses) or possibly to an extensive secondary structure (as in HDV) [7, 119]. The results of early studies had indeed confirmed that the RNAi machinery is exclusively cytoplasmic, and fails to knock down unexported nuclear pre-mRNA.

2.4 RNAi Targeting Virus-Related Cellular Genes

Recently, cellular functions that are essential for virus replication and related pathology have been explored as potential targets for antiviral therapy. In a typical approach, a high-throughput genome-wide silencing of cellular genes by siRNA is conducted, after which the cells are infected with the virus in question [120]. The reduction of virus growth by a specific siRNA reveals the essentiality of the cellular gene in virus growth [121]. Such strategies have led to the discovery of a variety of cellular functions that are important for RSV [122], Dengue virus [123], HCV [124], influenza [125–127], HIV [128, 129], and West Nile virus [130]. Selected cellular proteins, targeted as antiviral agents, have included HIV coreceptors CXCR4 and CCR5 [95, 131], and cyclin-dependent kinases for adenovirus, papillomavirus, herpesviruses and also HIV [132, 133]. Genome-wide analysis will not only expand the repertoire of such cellular genes but also establish the use of RNA-based drugs against them.

2.5 RNAi as an Antiparasitic Approach

Parasites are lower eukaryotes, and cause diseases of considerable medical and veterinary importance throughout Africa, Asia, and the Americas. The advantages of RNAi have offered new hope, which various siRNA drugs having been shown to cause a substantial inhibition of a number of parasites during the past few years [134]. Among those parasites in which RNAi has been extensively demonstrated is the protozoan parasite, *Trypanosoma brucei* (African sleeping sickness) [28, 134], and *Schistosoma* sp. or flukes that cause liver, kidney, and/or intestinal damage [135].

The Apicomplexan parasite, *Toxoplasma*, may cause blindness, mental retardation, and miscarriage. The genome of *Toxoplasma gondii* also contains orthologs with a significant similarity to traditional RNAi-related genes, such as AGO, Dicer, RdRp, and various RDEs [136]. The expression of dsRNA in *T. gondii* was shown to knock down specific gene expression [137, 138], while more recently deep sequencing also revealed a family of miRNAs [136]. Although promising, the function and parasitic targets of these miRNAs are currently unknown.

Vector-borne infections, including many mosquito-borne viruses and arboviruses in general, pose a tremendous burden to human society, particularly in the developing countries. The results of early studies showed that dsRNA could be used to knock down gene expression both in the adult mosquito and immortalized cell lines [139]. The recent completion of the genome sequence of *Anopheles gambiae*, the main vector for human malaria in

Africa, revealed a family of functional RNAi homologs [140, 141]. Thus, the RNAi mechanism could potentially be used to develop transgenic vectors that would have an innate resistance to the development and growth of arthropod-borne pathogens. Moreover, as with viruses, host mosquito functions can also be targeted by RNAi to inhibit the malaria parasite's growth [142].

2.6
RNAi against Genetic Disorders and Cancer

The fastidious sequence-specificity of siRNA allows the specific targeting of the dominant and codominant mutations that underlie many noninfectious diseases. Recently, siRNAs have proven highly effective against a number of genetic abnormalities, including autoimmune diseases, asthma, age-related macular degeneration (AMD), diabetes and diabetic retinopathy (DR), central nervous system (CNS) disorders, obesity, amyotrophic lateral sclerosis (ALS; also known as Lou Gehrig's disease), and various forms of cancer.

The specific inhibition of the oncogenic K-RAS V12 expression in human tumor cells by siRNA resulted in a loss of anchorage-independent growth and tumorigenicity. As siRNA-mediated gene silencing is highly sequence-specific, siRNA can be designed to silence cancer-derived transcripts that harbor point mutations. The guardian of the genome, p53, is inactivated by point mutation in over 50% of human cancers. A single base difference in siRNA discriminated between mutant and wild-type tumor suppressor p53 in cells expressing both forms, and this resulted in a restoration of the wild-type protein function [143].

Unnatural fusion proteins are relatively common in cancer. The transfection of leukemic cells with siRNAs targeting a BCR–ABL protooncogene fusion transcript induced apoptosis [144] which was comparable to that triggered by the ABL tyrosine kinase inhibitor STI 571 571 (signal transduction inhibitor 571). In Ewing's sarcoma, the transcription factor EWS (named after the sarcoma) is fused to a variety of other transcription factors such as FLI (Feline Leukemia Integration) and Ets (named after the leukemia virus, E26) at various break points; the NPM-ALK (nucleophosmin-anaplastic lymphoma kinase) fusion protein, for example, is found in about 75% of pediatric anaplastic large-cell lymphomas. Specific siRNA drugs designed against such chimeric mRNAs may constitute a potent and specific form of anticancer therapy. Osteosarcoma is the most common highly malignant bone tumor, with a primary appearance during the second and third decades of life. Moreover, the lesion is associated with a high risk of relapse that often results from resistance developed towards chemotherapy agents. Interestingly, the apurinic endonuclease 1 (APE1) is usually overexpressed in human osteosarcoma. By using siRNA against APE1, protein levels were reduced by more than 90% within 24 h, remained low for 72 h, and then had returned to normal levels by 96 h; there was also a clear loss of APE1 endonuclease activity following APE1-siRNA treatment. A decrease in APE1 levels in siRNA-treated human osteogenic sarcoma cells led to an enhanced cell sensitization to DNA damaging and chemotherapeutic agents, thus improving the prognosis.

At least eight human neurodegenerative disorders, including Huntington's disease (HD) and spinobulbar muscular atrophy (SBMA; Kennedy's disease) are caused by an expansion of trinucleotide repeats, the most common among which is a repeat of the CAG codon, coding for glutamine [145]. The transfection of cells expressing the CAG expanded androgen receptor mRNA shows a response to siRNA treatment via a reduction in mutated RNA level and, more importantly, by a rescue of the polyglutamine-induced toxicity. This example represents a proof of principle that siRNA technology can be applied to diseases associated with mutated transcripts arising from one allele (e.g., other neurodegenerative disorders) without affecting the other (healthy) allele.

In the past, anti-VEGF siRNAs have received much attention due to their efficacy in preventing angiogenesis [146–148]. In the mouse model, the targeting of VEGF by injecting siRNA into tumors resulted in growth inhibition, tumor cell killing, and a sensitization of the treated cells to other therapies. Anti-VEGF siRNAs represent a major area of therapy in the treatment of AMD and DR.

2.7 Antisense RNA in Therapy

Therapy with AS has been widely used to specifically and selectively inhibit the expression of selected genes at the mRNA level [149]. In fact, the use of AS-RNA against cancer-associated mRNA may lead to specific protein silencing and death of the cancer cells. Indeed, the anti-apoptotic protein Bcl-2 and telomerase were both successfully silenced in cultured cells, leading to an inhibition of cell growth. The insulin-like growth factor 1 receptor (IGF-1R) is an important signaling molecule in cancer cells, and plays an essential role in the establishment and maintenance of the transformed phenotype. Hence, the inhibition of IGF-1R signaling appears to be a promising strategy to interfere with the growth and survival of cancer cells [150–152]. AS-RNA, designed to inhibit IGF-1R gene expression, caused an efficient reduction in IGF-1-dependent proliferation and survival in a number of human and rodent cancer cell lines. Furthermore, a decrease in tumor size occurred when cells carrying the AS-IGF-1R were injected into syngeneic mice [150] when, in addition to blocking tumor growth, the AS-RNA treatment also inhibited metastasis. Many other genes have been selected as targets for AS-therapy, including HER-2/neu, protein kinase A (PKA), transforming growth factor alpha (TGF-α), TGF-β, EGFR, P12, MDM2, BRCA, Bcl-2, ER, VEGF, MDR, ferritin, transferrin receptor, IRE, C-fos, HSP27, C-myc, C-raf, and metallothionein. In many of these studies, a specific inhibition of tumor cell growth was demonstrated. The combination of AS-RNAs with chemotherapeutic agents may offer important advantages in cancer treatment, with several AS drugs – especially Oblimersen (G3139) – showing promising results in animal experiments, and being entered into clinical trials.

2.8 RNA Aptamers in Therapy

Currently, aptamers are being tested in a variety of disorders and infections, including cardiovascular diseases, neurological disorders, cancer and infections [153–156]. Several highly specific, nucleic acid aptamers capable of targeting select HIV proteins have been described that have effectively blocked viral replication. Aptamers directed against the HIV-1 Rev protein were tested in a surrogate animal model harboring human tissue. For example, in the animal model of human thymopoiesis, which used a humanized SCID (severe combined immunodeficiency) mouse, differentiated thymocytes derived from reconstituted grafts expressed anti-Rev aptamers and showed a significant resistance to HIV-1 infection upon challenge.

Blood clotting, which can trigger heart attacks and strokes, is a significant area of RNA aptamer application. Although anticlotting drugs are currently available, most have serious drawbacks; for example, heparin and its antidote (as used during and after surgery, respectively) can often cause adverse reactions, while the dosage of another common blood-thinning drug, warfarin, is difficult to regulate and also has no antidote. To date, one trillion RNA aptamers have been screened for their ability to block specific protein factors crucial to the blood-clotting process [157]. Subsequently, when a clot-stopping aptamer was selected its antidote was also designed; this was simply another length of RNA with a complementary sequence that would adhere to the first RNA portion, so as to disable it. In the test tube, however, different amounts of antidote either regulated or reversed the aptamer's anticlotting ability. Confirmatory studies in animals are currently under way, however.

The treatment of immunomediated glomerulonephritides is presently based on a limited series of drugs. Recently, the details of several original and innovative approaches to treat inflammatory and proliferative glomerular diseases have been reported, including RNA drugs designed to limit the effect of proinflammatory cytokines and growth factors [158]. The application of peptide aptamers that bind specifically to the IGF-1R represents a novel approach to target IGF-1R signaling in cancer. The integration of peptide aptamers into targeted protein degradation vehicles, and their transduction into cells, will allow a temporary elimination of the receptor protein.

Myasthenia gravis (MG) is a neuromuscular disorder associated with muscular weakness and fatigability. The pathogenesis of MG mainly results from an antibody-mediated autoimmune response to nicotinic acetylcholine receptors (AChRs) located in the postsynaptic muscle cell membrane. Recently, aptamers were successfully used to treat experimental autoimmune myasthenia gravis (EAMG) in rats [159]. In this case, the clinical symptoms of EAMG were efficiently inhibited by a truncated RNA aptamer, but not by a control scrambled RNA. Moreover, the loss of AChR in the animals induced by the antibody was also significantly blocked with the modified RNA aptamer.

Aptamers can also be used for the treatment of parasites [160]. For example, African trypanosomes (which cause sleeping sickness in humans and Nagana in cattle) multiply in the blood and escape the immune response of the infected host by antigenic variation – that is, by the parasite making periodic changes of its surface antigen, known as a variant surface glycoprotein (VSG). Aptamers that bind to VSGs with subnanomolar affinity are capable of recognizing different VSG variants and binding to the surface of live trypanosomes, while aptamers tethered to an antigenic side group are capable of directing antibodies to the surface of the parasite *in vitro*.

2.9
Spiegelmers in Therapy

Spiegelmer that inhibits the action of the migraine-associated target calcitonin gene-related peptide 1 (alpha-CGRP) was identified as a lead compound for *in vivo* studies [161, 162]. Gonadotropin-releasing hormone (GnRH) is a key peptide hormone in the regulation of mammalian reproduction, and is the trigger signal for a cascade of hormones responsible for controlling the production of luteinizing hormone (LH) and follicle-stimulating hormone (FSH). Consequently, both GnRH and its receptor have been identified as therapeutic targets for sex steroid-dependent conditions such as prostate cancer, breast cancer, and endometriosis, as well as in assisted-reproduction techniques. A spiegelmer with a high affinity for GnRH was isolated which acted as an antagonist to GnRH in Chinese hamster ovary (CHO) cells that stably expressed the human GnRH receptor [163]. In a castrated rat model, the spiegelmer further demonstrated a strong GnRH antagonist activity. Taken together, the results of these studies suggest that spiegelmers might be of substantial interest in the development of new pharmaceutical approaches against GnRH and other targets.

2.10 PNA in Therapy

Previously, PNAs have demonstrated significant promise against papillomavirus-induced human cancers, with cervical carcinomas being caused by infections with HPVs in essentially all cases. The expression of the *E6* and *E7* genes from high-risk HPV16 and HPV18 is crucial for the development, immortalization and maintenance of the malignant phenotype of cervical carcinoma, and these constitute important targets for anti-cancer therapies. Different PNAs directed against the HPV18 *E6* and *E7* genes were able to regulate the growth of HeLa-S cervical cancer cells [164]. Telomerase activity, which is below detectable levels in almost all types of diploid cell, is reactivated in most immortal and cancer cells. In recent studies, PNAs directed against the human telomerase reverse transcriptase (hTERT) effectively arrested the growth of prostate cancer cells [165]. In another study, an upregulation of the c-*myc* oncogene in Burkitt's lymphoma cells was inhibited by PNA complementary to a specific unique E mu intronic sequence, and blocked the expression of the c-*myc* oncogene under E mu control [166, 167].

PNAs were also effective when used as antivirals. The RNA genome of HCV contains a well-defined and highly conserved secondary structure that functions as an internal ribosomal entry site (IRES) that is necessary for translation and viral replication. Not only PNA and LNA, but also a combination of PNA and hammerhead ribozymes, can invade critical sequences within the HCV IRES, and thereby inhibit translation [67, 168, 169]. In cells infected with pseudotyped HIV-1 virions, the PNAs exhibited a dramatic reduction in HIV-1 replication.

2.11 Immunotherapy by RNA

With the realization that cancer or a malignant tumor is a disease of defective genetic programming, various attempts have been made to identify tumor-specific proteins and to train the body's immune system against these prospective antigens. The strategy of the immunotherapy of cancer is based on knowledge that non-self (or foreign) proteins are proteolytically degraded inside bone marrow-derived dendritic cells (DCs) to produce short peptides that associate with the major histocompatibility complex (MHC) and are transported to the cell surface. Naïve cytotoxic T cells (CTLs) recognize the displayed peptide–MHC complexes, and undergo an activation process to kill the targets. In the cancer patient, the capture of tumor antigens by DCs, or the stimulation of tumor-specific CTLs, are apparently inefficient. In RNA immunotherapy, the strategy is to transfect the DCs with mRNA for specific tumor antigens or with a total tumor-derived mRNA population, and to introduce these into the patient [170]. In both animals and human volunteers, DCs loaded with tumor mRNAs were indeed shown to stimulate the CTL response. As with other mammalian cells, cationic lipid reagents and electroporation have been used for mRNA transfection into DCs; interestingly, mRNA alone is effective, which reflects the extraordinary sensitivity of the immune system to small amounts of antigen. The success of the mRNA loading obviates the need for difficult and laborious alternatives such as cloning the mRNAs into cDNA or expression and the purification of tumor-specific proteins. In fact, in direct comparison, mRNA-loaded DCs often fare better than those transfected with cDNAs or proteins. Endosomal compartments of DCs also contain specialized TLRs that are activated by single-stranded RNA, which results in the activation of interferon-gamma (IFN-γ); this apparent side effect may actually be beneficial for therapy against viruses and cancer. Although RNA immunotherapy does not require an understanding of the mechanism of tumorigenesis, it does offer a natural and biological anticancer treatment that can be custom-designed and administered against a particular tumor within a matter of hours.

3
The Design, Synthesis, Delivery, and Pharmacokinetics of RNA

3.1
Design and Synthesis of an Effective RNA Drug

A number of algorithms, along with various software, have been developed to aid in the design of ribozymes, siRNA, and aptamers, and to analyze their interactions with prospective targets *in silico*. This is despite many such materials being based on the results of semi-empirical and thermodynamic studies, such that the final selections must be tested experimentally to determine their efficacy. Many such programs and the details of other RNA-related resources are available at various web sites.

The principles of ribozyme design were discussed above (see Sects 1.4.2 and 1.4.3). The siRNAs are 21–23 nt long dsRNAs with 2 nt overhangs, with synthetic siRNAs being generally created to conform to the sequence $NA(N)_{19}TT$, where N can be any nucleotide, although variations of the overhangs have been tried with success. The $(N)_{19}$ core should be perfectly complementary to the target RNA sequence, though some mismatch may be tolerated, especially if they are near the termini of the siRNA. Recently, a set of eight rules has been suggested for the rational design of the $(N)_{19}$ core of the siRNA, based on experimental analysis and thermodynamic annealing parameters [171]; these include (among others): a 30–52% GC content; an absence of internal repeats or hairpins; and most importantly, three or more A/Us at positions 15–19 of the sense strand, so that preferential unwinding occurs from this end of the siRNA duplex resulting in the formation of a RISC with the AS strand (see Fig. 6).

Both, siRNA and degradative ribozymes should be designed away from the 5′ and 3′ termini of the target mRNA, as these regions interact with, and are protected by, translational machinery or factors.

RNA can be synthesized either exogenously (*in vitro*) and then delivered into cells, or transcribed endogenously (*in vivo*) from DNA clones introduced into cells. Exogenous RNA can be prepared either by using synthetic chemistry or by the transcription of DNA clones. A major issue in the therapeutic use of RNA is its stability. The only difference between DNA and RNA (besides the T $\rightarrow$ U change) is the 2′-OH group of the ribose ring in the RNA; this has important consequences, the most prominent of which is a hydrolysis of the phosphodiester bond of the RNA, catalyzed by a nucleophilic attack of the electrons from the extra oxygen atom (this is fundamentally the same reaction that is catalyzed by ribozymes). Thus, much of the effort in synthetic RNA chemistry has been directed towards modifying the 2′-OH group. An ideal modification must provide an improved stability and a better pharmacokinetics, without affecting the base-pairing characteristics or function of the RNA [172]. The most notable modifications include amino, fluoro, methyl, and allyl derivatives of the 2′-OH group, although another common modification is to replace the oxygen atoms of the phosphodiester bonds with sulfur. Although the resultant phosphothioester bond cannot be hydrolyzed, it is now clear that the phosphorothioate RNA exhibits a significantly higher cellular toxicity and a nonspecific binding to proteins. Currently, the trend is to use a limited number of phosphorothioate linkages combined with 2′-modifications. In another approach, an inverted T is added at the 3′ end to form a 3′-3′ phosphodiester linkage, which causes the RNA to become resistant to 3′ exonucleases. In designing a siRNA, deoxythymidines (dTs) are used to substitute for the two T overhangs, with the hope of increasing the nuclease-resistance at the 3′ end. Ribozyme cores are also susceptible to hydrolysis by endonucleases that primarily attack pyrimidine nucleotides [173]. Thus, a 2′-modification of the two U nucleotides of the CUGAUG consensus (see Fig. 3a) produces a more stable hammerhead, without compromising the ribozyme activity. In yet another approach, deoxyribozymes have been developed that exhibit certain advantages over RNA ribozymes, such as a greater stability, an improved catalytic efficiency, and a potentially lower toxicity [174–176].

Clearly, chemical modifications are only possible *in vitro* and not *in vivo*, when the RNA is being produced through transcription inside a cell [177]. On the other hand, the *in vivo* synthesis of RNA from recombinant DNA clones obviates the need for manufacturing and delivering the RNA, and utilizes the natural transcriptional machinery of the target cell itself [178]. The RNA is only produced inside the cell, thus avoiding losses in blood and any degradation by the serum nucleases. A variety of vectors has been used for the *in vivo* expression of RNA, the most common being either plasmids or viral vectors with strong promoters [177–179]. Small RNA molecules such as ribozymes and siRNA are often transcribed from a RNAP III promoter engineered into these vectors, such as the U6, H1, or 7SK promoters of mouse or human origin. A controlled expression of RNA is often achieved through the use of inducible and tissue-specific promoters.

As with small molecules, RNA drugs are highly amenable to automated high-throughput screening (HTS) procedures, based either on a direct binding to targets or a function-based alteration of reporter gene expression. When combined with the chemical modification of synthetic RNA, such assays can be adapted to use almost any type of read-out format, including (but not limited to) fluorescence intensity (FI), fluorescence lifetime (FLT), fluorescence polarization (FP), fluorescence resonance energy transfer (FRET), solid-state (membrane or bead) binding assay, enzyme-linked assay, and radioactivity.

3.2 Delivery and Pharmacokinetics of RNA

The issues of the delivery of an exogenously prepared RNA drug are not unlike those of DNA transfection and gene therapy. In cell culture, Oligofectamine (Life Technologies, Gaithersburg, MD, USA), and TransIT-TKO reagent (Mirus Corp., Madison, WI, USA) have each been used by many laboratories, with much success [7, 180]. Cellular permeation is also improved by conjugation with specific peptides such as helical peptides, Tat protein of HIV, and Antennapedia of *Drosophila* [181, 182]. In live animals, the consistent delivery of sufficient quantities of RNA remains a challenge. In mice, a "hydrodynamic injection" through the tail vein effectively delivers the RNA into the hepatocytes [109], with the optimum amount being 10–15% of the animal's body weight injected within 5–7 s. The injection of a large bolus is believed to result in short-term right-heart failure and the backflow of a large volume into the liver. Unfortunately, a hydrodynamic injection through the tail vein may not transport the RNA to all cells of the body, and is an impossible procedure in human subjects! In the case of respiratory viruses, an intranasal delivery route has been used successfully [102, 183, 184].

Once the RNA has been delivered, the major issues are its pharmacokinetic properties – namely, its stability in the tissues and body fluids, its metabolism and urinary excretion – and the potential toxicity of the large amounts of RNA needed for an intended therapeutic effect [53, 182]. Although generally well tolerated, each RNA must be tested for these parameters, because it may have unique effects on cellular gene expression. This is particularly important when designing second-generation RNA drugs conjugated to novel non-RNA moieties, as this might result in unique conjugates that do not exist in Nature. It must also be remembered that each tissue or organ may have unique interactions with RNA, and that the uptake and distribution of RNA in tumor tissues are typically poor when compared to normal tissues. While working with live animals and human patients, considerations must also be given to the possibility that viral vectors may cause systemic infections and immune reactions.

4 An RNA Drug for Every Disease?

In this chapter, the details have been provided, so far, of a relatively large number of recent and emerging clinical applications of various forms of RNA. Yet, despite the phenomenal prospect of this approach, it must be borne in mind that RNA is a relatively new entrant in the pharmaceutical arena, and considerable investigations must still be conducted before RNA-based drugs can become common items in the family medicine cabinet. The main areas where improvements are desirable are the cost, delivery, stability, and specificity.

On the point of cost, when compared to most small-molecule drugs, RNA is in fact highly specific, primarily because essentially all of its clinical applications are sequence-dependent [185]. dsRNA that are longer than about 35 bp tend to trigger the so-called "interferon response," in which the dsRNA binds to and activates dsRNA-activated protein kinase (PKR). Among the many cellular proteins that are substrates of PKR, one strategically important protein is the translation initiation factor, eIF2α, the phosphorylation of which leads to global translational shut-off and, on occasion, cell death. Primarily by virtue of their shorter length, siRNAs do not activate the IFN response, and this is key to their target-specific effect. However, siRNAs do occasionally affect off-target gene expression, especially when large amounts are applied [186, 187]; this is due in part to their ability to tolerate some degree of mismatch. The mechanism or extent of such nonspecificity, and its potential impact on the clinical applications of RNA, remain an area of active debate [185, 188–192]. Moreover, certain siRNA sequences may also trigger a nonspecific immune reaction that is often undesirable but may sometimes enhances the desired antiviral effect of the siRNA [193, 194]. Appropriate sequence design and chemical modifications of the siRNA may reduce any undesired immunostimulatory effect [195–199]. On a practical note, it should be realized that there is no chemical or medicine that is totally free of side effects, especially when subjected to exquisitely sensitive molecular biological screening such as microarray analysis; rather, the "real-life" issue in medicine is the balance between risk and benefit.

In principle, multiple RNA-based strategies can be applied to a given target; for example, an mRNA can be silenced by AS RNA, ribozyme, or siRNA. Currently, there is no clear *a priori* guideline to choose one siRNA over the others, primarily because very few studies have been conducted to compare them under identical conditions. The general consensus is that siRNAs may perform as well as or better than AS and ribozyme. It can be envisaged that their relative effectiveness would be influenced by a variety of factors, including tissue or cell type, transfection technique, target sequence, and chemical modification.

From an entrepreneurial perspective, the annual worldwide market for RNA-based therapy is estimated to be as high as US$ 150 billion. The excitement in this area is underscored by the growing number of biotechnology companies that have added RNA-based drugs or reagents to their R&D portfolio; some of the major names included in this group are (in alphabetical order): Alnylam Pharmaceuticals (with Cubist and others); Ambion (Life Technologies); Antisense Pharma GmbH; Argos Therapeutics; AVI BioPharma; Dharmacon (Thermo Scientific); EpiGenesis Pharmaceuticals; Gilead Sciences; Imgenex Corporation; Isis Pharmaceuticals; Lorus Therapeutics; Marina Biotech; MDRNA; Merck & Co.; MethylGene; NOXXON Pharma AG; Qiagen NV; Quark Pharmaceuticals; Ribozyme Pharmaceuticals (and Atugen); RXi Pharmaceuticals; Silence Therapeutics; Sirnaomics; SomaGenics; SomaLogic; Tekmira; and Virxsys. It should be noted that, as RNA-based therapeutics is a relatively uncharted territory with various scientific and technical hurdles, the biotechnology industry in this area is rapidly evolving with new ventures, mergers, and partnerships. For example, RxI Pharma is a spin-off of the CytRx Corporation, Atugen AG became

Silence Therapeutics AG in 2005, which then merged with Intradigm in 2010, and Schering-Plough became incorporated into Merck in 2006, among many others.

During the past few years, RNA drugs of virtually every category described here have progressed through different stages of development, including clinical trials, and some have produced encouraging results. Indeed, the breadth and scope of the emerging clinical applications of RNA are matched only by the diversity of the biological tasks assigned to RNA by Mother Nature. It is fair to say, that the full potential of RNA as a pharmaceutical entity has only just begun to be appreciated, and that RNA drugs against a variety of diseases and infections will achieve blockbuster status in the foreseeable future.

Acknowledgments

The authors apologize to readers and colleagues for often referring to comprehensive reviews rather than to original research reports, due to limitations of space. The studies conducted at the authors' laboratory were generously supported by Burroughs Wellcome Foundation, American Heart Association Southeast Affiliate (**AL G970031**), and NIH, USA (**AI045803, EY013826, F32 AI049682, AI37938**). Titus Barik is also acknowledged for assisting with the sequence analysis and other computational projects.

References

1 Li, G.W., Xie, X.S. (2011) Central dogma at the single-molecule level in living cells. *Nature*, **475**, 308–315.

2 Berget, S.M. (1995) Exon recognition in vertebrate splicing. *J. Biol. Chem.*, **270**, 2411–2414.

3 Burge, C.B., Tuschl, T., Sharp, P.A. (1999) Splicing of Precursors to mRNAs by the Spliceosomes, in: Gesteland, R.F., Cech, T.R., Atkins, J.F. (Eds) *The RNA World II*, Cold Spring Harbor Laboratory Press, New York, pp. 525–560.

4 Talerico, M., Berget, S.M. (1994) Intron definition in splicing of small *Drosophila* introns. *Mol. Cell. Biol.*, **14**, 3434–3445.

5 Barik, S. (2004) Control of nonsegmented negative-strand RNA virus replication by siRNA. *Virus Res.*, **102**, 27–35.

6 Ahlquist, P. (2002) RNA-dependent RNA polymerases, viruses, and RNA silencing. *Science*, **296**, 1270–1273.

7 Bitko, V., Barik, S. (2001) Phenotypic silencing of cytoplasmic genes using sequence-specific double-stranded short interfering RNA and its application in the reverse genetics of wild type negative-strand RNA viruses. *BMC Microbiol.*, **1**, 34.

8 Domingo, E., Holland, J.J. (1997) RNA virus mutations and fitness for survival. *Annu. Rev. Microbiol.*, **51**, 151–178.

9 Barik, S., Rud, E.W., Luk, D., Banerjee, A.K., Kang, C.Y. (1990) Nucleotide sequence analysis of the L gene of vesicular stomatitis virus (New Jersey serotype): identification of conserved domains in L proteins of nonsegmented negative-strand RNA viruses. *Virology*, **175**, 332–337.

10 Mathews, D.H., Moss, W.N., Turner, D.H. (2010) Folding and finding RNA secondary structure. *Cold Spring Harbor Perspect. Biol.*, **2**, a003665.

11 Nagaswamy, U., Larios-Sanz, M., Hury, J., Collins, S., Zhang, Z., Zhao, Q., Fox, G.E. (2002) NCIR: a database of non-canonical interactions in known RNA structures. *Nucleic Acids Res.*, **30**, 395–397.

12 Valadkhan, S., Jaladat, Y. (2010) The spliceosomal proteome: at the heart of the largest cellular ribonucleoprotein machine. *Proteomics*, **10**, 4128–4141.

13 Will, C.L., Lührmann, R. (2011) Spliceosome structure and function. *Cold Spring Harbor Perspect. Biol.*, **3**, a003707.

14 Ward, A.J., Cooper, T.A. (2010) The pathobiology of splicing. *J. Pathol.*, **220**, 152–163.

15 Mercatante, D.R., Mohler, J.L., Kole, R. (2002) Cellular response to an antisense-mediated shift of Bcl-x pre-mRNA splicing and antineoplastic agents. *J. Biol. Chem.*, **277**, 49374–49382.

16 Cech, T.R., Uhlenbeck, O.C. (1994) Ribozymes. Hammerhead nailed down. *Nature*, **372**, 39–40.
17 Been, M.D., Barfod, E.T., Burke, J.M., Price, J.V., Tanner, N.K., Zaug, A.J., Cech, T.R. (1987) Structures involved in *Tetrahymena* rRNA self-splicing and RNA enzyme activity. *Cold Spring Harbor Symp. Quant. Biol.*, **52**, 147–157.
18 Kruger, K., Grabowski, P.J., Zaug, A.J., Sands, J., Gottschling, D.E., Cech, T.R. (1982) Self-splicing RNA: autoexcision and autocyclization of the ribosomal RNA intervening sequence of *Tetrahymena*. *Cell*, **31**, 147–157.
19 Murphy, W.J., Watkins, K.P., Agabian, N. (1986) Identification of a novel Y branch structure as an intermediate in trypanosome mRNA processing: evidence for trans splicing. *Cell*, **47**, 517–525.
20 Sutton, R.E., Boothroyd, J.C. (1986) Evidence for *trans* splicing in trypanosomes. *Cell*, **47**, 527–535.
21 Uhlenbeck, O.C. (1987) A small catalytic oligoribonucleotide. *Nature*, **328**, 596–600.
22 McCall, M.J., Hendry, P., Mir, A.A., Conaty, J., Brown, G., Lockett, T.J. (2000) Small, efficient hammerhead ribozymes. *Mol. Biotechnol.*, **14**, 5–17.
23 Long, M.B., Jones, J.P. III, Sullenger, B.A., Byun, J. (2003) Ribozyme-mediated revision of RNA and DNA. *J. Clin. Invest.*, **112**, 312–338.
24 Sullenger, B.A., Cech, T.R. (1994) Ribozyme-mediated repair of defective mRNA by targeted, *trans*-splicing. *Nature*, **371**, 619–622.
25 Puttaraju, M., Jamison, S.F., Mansfield, S.G., Garcia-Blanco, M.A., Mitchell, L.G. (1999) Spliceosome-mediated RNA *trans*-splicing as a tool for gene therapy. *Nat. Biotechnol.*, **17**, 246–252.
26 Mansfield, S.G., Clark, R.H., Puttaraju, M., Kole, J., Cohn, J.A., Mitchell, L.G., Garcia-Blanco, M.A. (2003) 5' exon replacement and repair by spliceosome-mediated RNA *trans*-splicing. *RNA*, **9**, 1290–1297.
27 Fire, A., Xu, S., Montgomery, M.K., Kostas, S.A., Driver, S.E., Mello, C.C. (1998) Potent and specific genetic interference by double-stranded RNA in *Caenorhabditis elegans*. *Nature*, **391**, 806–811.
28 Ullu, E., Djikeng, A., Shi, H., Tschudi, C. (2002) RNA interference: advances and questions. *Philos. Trans. R. Soc. Lond. B, Biol. Sci.*, **29**, 65–70.
29 Couto, L.B., High, K.A. (2010) Viral vector-mediated RNA interference. *Curr. Opin. Pharmacol.*, **10**, 534–542.
30 Mowa, M.B., Crowther, C., Arbuthnot, P. (2010) Therapeutic potential of adenoviral vectors for delivery of expressed RNAi activators. *Expert Opin. Drug Delivery*, **7**, 1373–1385.
31 Kawamata, T., Tomari, Y. (2010) Making RISC. *Trends Biochem. Sci.*, **35**, 368–376.
32 Schwarz, D.S., Hutvágner, G., Du, T., Xu, Z., Aronin, N., Zamore, P.D. (2003) Asymmetry in the assembly of the RNAi enzyme complex. *Cell*, **115**, 199–208.
33 Khvorova, A., Reynolds, A., Jayasena, S.D. (2003) Functional siRNAs and miRNAs exhibit strand bias. *Cell*, **115**, 209–216. [Erratum in: *Cell*, **115**, 505, (2003).]
34 Bartel, D.P. (2004) MicroRNAs: genomics, biogenesis, mechanism, and function. *Cell*, **116**, 281–297.
35 Shruti, K., Shrey, K., Vibha, R. (2011) Micro RNAs: tiny sequences with enormous potential. *Biochem. Biophys. Res. Commun.*, **407**, 445–449.
36 Starega-Roslan, J., Koscianska, E., Kozlowski, P., Krzyzosiak, W.J. (2011) The role of the precursor structure in the biogenesis of microRNA. *Cell. Mol. Life Sci.*, **68**, 2859–2871.
37 Czech, B., Hannon, G.J. (2011) Small RNA sorting: matchmaking for Argonautes. *Nat. Rev. Genet.*, **12**, 19–31.
38 Lagos-Quintana, M., Rauhut, R., Meyer, J., Borkhardt, A., Tuschl, T. (2003) New microRNAs from mouse and human. *RNA*, **9**, 175–179.
39 Tinoco, I. Jr, Bustamante, C. (1999) How RNA folds. *J. Mol. Biol.*, **293**, 271–281.
40 Bartel, D.P., Szostak, J.W. (1993) Isolation of new ribozymes from a large pool of random sequences. *Science*, **261**, 1411–1418.
41 Ellington, A.D., Szostak, J.W. (1990) In vitro selection of RNA molecules that bind specific ligands. *Nature*, **346**, 818–822.
42 Brody, E.N., Gold, L. (2000) Aptamers as therapeutic and diagnostic agents. *J. Biotechnol.*, **74**, 5–13.
43 Zhou, J., Rossi, J.J. (2009) The therapeutic potential of cell-internalizing aptamers. *Curr. Top. Med. Chem.*, **9**, 1144–1157.

44 Tuerk, C., Gold, L. (1990) Systematic evolution of ligands by exponential enrichment: RNA ligands to bacteriophage T4 DNA polymerase. *Science*, **249**, 505–510.
45 Vater, A., Klussmann, S. (2003) Toward third-generation aptamers: Spiegelmers and their therapeutic prospects. *Curr. Opin. Drug Discov. Delivery*, **6**, 253–261.
46 Corradini, R., Sforza, S., Tedeschi, T., Totsingan, F., Manicardi, A., Marchelli, R. (2011) Peptide nucleic acids with a structurally biased backbone. Updated review and emerging challenges. *Curr. Top. Med. Chem.*, **11**, 1535–1554.
47 Nielsen, P.E. (2002) PNA technology. *Methods Mol. Biol.*, **208**, 3–26.
48 Koshkin, A.A., Singh, S.K., Nielsen, P., Rajwanshi, V.K., Kumar, R., Meldgaard, M., Olsen, C.E., Wengel, J. (1998) LNA (Locked Nucleic Acids): synthesis of the adenine, cytosine, guanine, 5-methylcytosine, thymine and uracil bicyclonucleoside monomers, oligomerisation, and unprecedented nucleic acid recognition. *Tetrahedron*, **54**, 3607–3630.
49 Dorn, A., Kippenberger, S. (2008) Clinical application of CpG-, non-CpG-, and antisense oligodeoxynucleotides as immunomodulators. *Curr. Opin. Mol. Ther.*, **10**, 10–20.
50 Lai, J.C., Benimetskaya, L., Santella, R.M., Wang, Q., Miller, P.S., Stein, C.A. (2003) G3139 (oblimersen) may inhibit prostate cancer cell growth in a partially bis-CpG-dependent non-antisense manner. *Mol. Cancer Ther.*, **2**, 1031–1043.
51 Mulhbacher, J., St-Pierre, P., Lafontaine, D.A. (2010) Therapeutic applications of ribozymes and riboswitches. *Curr. Opin. Pharmacol.*, **10**, 551–556.
52 Seyhan, A.A. (2011) RNAi: a potential new class of therapeutic for human genetic disease. *Hum. Genet.*, **130**, 583–605.
53 Castanotto, D., Rossi, J.J. (2009) The promises and pitfalls of RNA-interference-based therapeutics. *Nature*, **457**, 426–433.
54 Chen, C.J., Banerjea, A.C., Harmison, G.G., Haglund, K., Schubert, M. (1992) Multitarget-ribozyme directed to cleave at up to nine highly conserved HIV-1 env RNA regions inhibits HIV-1 replication – potential effectiveness against most presently sequenced HIV-1 isolates. *Nucleic Acids Res.*, **20**, 4581–4589.
55 Hotchkiss, G., Maijgren-Steffensson, C., Ahrlund-Richter, L. (2004) Efficacy and mode of action of hammerhead and hairpin ribozymes against various HIV-1 target sites. *Mol. Ther.*, **10**, 172–180.
56 Scherer, L., Rossi, J.J., Weinberg, M.S. (2007) Progress and prospects: RNA-based therapies for treatment of HIV infection. *Gene Ther.*, **14**, 1057–1064.
57 Wong-Staal, F., Poeschla, E.M., Looney, D.J. (1998) A controlled, phase 1 clinical trial to evaluate the safety and effects in HIV-1 infected humans of autologous lymphocytes transduced with a ribozyme that cleaves HIV-1 RNA. *Hum. Gene Ther.*, **9**, 2407–2425.
58 Zhou, C., Bahner, I.C., Larson, G.P., Zaia, J.A., Rossi, J.J., Kohn, E.B. (1994) Inhibition of HIV-1 in human T-lymphocytes by retrovirally transduced anti-tat and rev hammerhead ribozymes. *Gene*, **149**, 33–39.
59 Sun, L.Q., Wang, L., Gerlach, W.L., Symonds, G. (1995) Target sequence-specific inhibition of HIV-1 replication by ribozymes directed to tat RNA. *Nucleic Acids Res.*, **23**, 2909–2913.
60 Macpherson, J.L., Boyd, M.P., Arndt, A.J., Todd, A.V., Fanning, G.C., Ely, J.A., Elliott, F., Knop, A., Raponi, M., Murray, J., Gerlach, W., Sun, L.Q., Penny, R., Symonds, G.P., Carr, A., Cooper, D.A. (2005) Long-term survival and concomitant gene expression of ribozyme-transduced CD4+ T-lymphocytes in HIV-infected patients. *J. Gene Med.*, **7**, 552–564.
61 Cooper, D., Penny, R., Symonds, G., Carr, A., Gerlach, W., Sun, L.Q., Ely, J. (1999) A marker study of therapeutically transduced CD4+ peripheral blood lymphocytes in HIV discordant identical twins. *Hum. Gene Ther.*, **10**, 1401–1421.
62 Bauer, G., Valdez, P., Kearns, K., Bahner, I., Wen, S.F., Zaia, J.A., Kohn, D.B. (1997) Inhibition of human immunodeficiency virus-1 (HIV-1) replication after transduction of granulocyte colony-stimulating factor-mobilized CD34+ cells from HIV-1-infected donors using retroviral vectors containing anti-HIV-1 genes. *Blood*, **89**, 2259–2267.
63 Trang, P., Lee, J., Kilani, A.F., Kim, J., Liu, F. (2001) Effective inhibition of herpes simplex virus 1 gene expression and growth

by engineered RNase P ribozyme. *Nucleic Acids Res.*, **29**, 5071–5078.

64 Trang, P., Kilani, A., Lee, J., Hsu, A., Liou, K., Kim, J., Nassi, A., Kim, K., Liu, F. (2002) RNase P ribozymes for the studies and treatment of human cytomegalovirus infections. *J. Clin. Virol.*, **25** (Suppl. 2), S63–S74.

65 Zou, H., Lee, J., Umamoto, S., Kilani, A.F., Kim, J., Trang, P., Zhou, T., Liu, F. (2003) Engineered RNase P ribozymes are efficient in cleaving a human cytomegalovirus mRNA *in vitro* and are effective in inhibiting viral gene expression and growth in human cells. *J. Biol. Chem.*, **278**, 37265–37274.

66 Jia, Q., Sun, R. (2003) Inhibition of gammaherpesvirus replication by RNA interference. *J. Virol.*, **77**, 3301–3306.

67 Romero-López, C., Díaz-González, R., Barroso-del Jesus, A., Berzal-Herranz, A. (2009) Inhibition of hepatitis C virus replication and internal ribosome entry site-dependent translation by an RNA molecule. *J. Gen. Virol.*, **90**, 1659–1669.

68 Welch, P.J., Yei, S., Barber, J.R. (1998) Ribozyme gene therapy for hepatitis C virus infection. *Clin. Diagn. Virol.*, **10**, 163–171.

69 Pan, W.H., Xin, P., Morrey, J.D., Clawson, G.A. (2004) A self-processing ribozyme cassette: utility against human papillomavirus 11 E6/E7 mRNA and hepatitis B virus. *Mol. Ther.*, **9**, 596–606.

70 Kijima, H., Yamazaki, H., Nakamura, M., Scanlon, K.J., Osamura, R.Y., Ueyama, Y. (2004) Ribozyme against mutant K-ras mRNA suppresses tumor growth of pancreatic cancer. *Int. J. Oncol.*, **24**, 559–564.

71 Tsuchida, T., Kijima, H., Hori, S., Oshika, Y., Tokunaga, T., Kawai, K., Yamazaki, H., Ueyama, Y., Scanlon, K.J., Tamaoki, N., Nakamura, M. (2000) Adenovirus-mediated anti- K-ras ribozyme induces apoptosis and growth suppression of human pancreatic carcinoma. *Cancer Gene Ther.*, **7**, 373–383.

72 Byun, J., Lan, N., Long, M., Sullenger, B.A. (2003) Efficient and specific repair of sickle beta-globin RNA by *trans*-splicing ribozymes. *RNA*, **9**, 1254–1263.

73 Rogers, C.S., Vanoye, C.G., Sullenger, B.A., George, A.L. Jr (2002) Functional repair of a mutant chloride channel using a *trans*-splicing ribozyme. *J. Clin. Invest.*, **110**, 1783–1179.

74 Suyama, E., Wadhwa, R., Kaur, K., Miyagishi, M., Kaul, S.C., Kawasaki, H., Taira, K. (2004) Identification of metastasis-related genes in a mouse model using a library of randomized ribozymes. *J. Biol. Chem.*, **279**, 38083–38086.

75 Weng, D.E., Usman, N. (2001) Angiozyme: a novel angiogenesis inhibitor. *Curr. Oncol. Rep.*, **3**, 141–146.

76 Davies, G., Watkins, G., Mason, M.D., Jiang, W.G. (2004) Targeting the HGF/SF receptor c-met using a hammerhead ribozyme transgene reduces *in vitro* invasion and migration in prostate cancer cells. *Prostate*, **60**, 317–324.

77 Cobaleda, C., Sanchez-Garcia, I. (2000) *In vivo* inhibition by a site-specific catalytic RNA subunit of RNase P designed against the BCR-ABL oncogenic products: a novel approach for cancer treatment. *Blood*, **95**, 731–737.

78 Kato, Y., Kuwabara, T., Toda, H., Warashina, M., Taira, K. (2000) Suppression of BCR- ABL mRNA by various ribozymes in HeLa cells. *Nucleic Acids Symp. Ser.*, **44**, 283–284.

79 Oshima, K., Kawasaki, H., Soda, Y., Tani, K., Asano, S., Taira, K. (2003) Maxizymes and small hairpin-type RNAs that are driven by a tRNA promoter specifically cleave a chimeric gene associated with leukemia *in vitro* and *in vivo*. *Cancer Res.*, **63**, 6809–6814.

80 Popescu, F.D. (2005) Antisense- and RNA interference-based therapeutic strategies in allergy. *J. Cell. Mol. Med.*, **9**, 840–853.

81 Liu, X., Jiang, Q., Mansfield, S.G., Puttaraju, M., Zhang, Y., Zhou, W., Cohn, J.A., Garcia- Blanco, M.A., Mitchell, L.G., Engelhardt, J.F. (2002) Partial correction of endogenous ΔF508 CFTR in human cystic fibrosis airway epithelia by spliceosome-mediated RNA trans-splicing. *Nat. Biotechnol.*, **20**, 47–52.

82 Maggon, K., Barik, S. (2004) New drugs and treatment for respiratory syncytial virus. *Rev. Med. Virol.*, **14**, 149–168.

83 Bitko, V., Musiyenko, A., Shulyayeva, O., Barik, S. (2005) Inhibition of respiratory viruses by nasally administered siRNA. *Nat. Med.*, **11**, 50–55.

84 Barik, S. (2010) siRNA for influenza therapy. *Virus Res.*, **2**, 1448–1457.
85 Haasnoot, J., Westerhout, E.M., Berkhout, B. (2007) RNA interference against viruses: strike and counterstrike. *Nat. Biotechnol.*, **25**, 1435–1443.
86 DeVincenzo, J., Lambkin-Williams, R., Wilkinson, T., Cehelsky, J., Nochur, S., Walsh, E., Meyers, R., Gollob, J., Vaishnaw, A. (2010) A randomized, double-blind, placebo-controlled study of an RNAi-based therapy directed against respiratory syncytial virus. *Proc. Natl Acad. Sci. USA*, **107**, 8800–8805.
87 Rossi, J.J., June, C.H., Kohn, D.B. (2007) Genetic therapies against HIV. *Nat. Biotechnol.*, **25**, 1444–1454.
88 Berkhout, B. (2009) Toward a durable anti-HIV gene therapy based on RNA interference. *Ann. N.Y. Acad. Sci.*, **1175**, 3–14.
89 Jacque, J.M., Triques, K., Stevenson, M. (2002) Modulation of HIV-1 replication by RNA interference. *Nature*, **418**, 435–438.
90 Novina, C.D., Murray, M.F., Dykxhoorn, D.M., Beresford, P.J., Riess, J., Lee, S.K., Collman, R.G., Lieberman, J., Shankar, P., Sharp, P.A. (2002) siRNA-directed inhibition of HIV-1 infection. *Nat. Med.*, **8**, 681–686.
91 Coburn, G.A., Cullen, B.R. (2002) Potent and specific inhibition of human immunodeficiency virus type 1 replication by RNA interference. *J. Virol.*, **76**, 9225–9231.
92 Lee, N.S., Dohjima, T., Bauer, G., Li, H., Li, M.J., Ehsani, A., Salvaterra, P., Rossi, J. (2002) Expression of small interfering RNAs targeted against HIV-1 rev transcripts in human cells. *Nat. Biotechnol.*, **20**, 500–505.
93 Sarver, N., Cantin, E.M., Chang, P.S., Zaia, J.A., Ladne, P.A., Stephens, D.A., Rossi, J.J. (1990) Ribozymes as potential anti-HIV-1 therapeutic agents. *Science*, **247**, 1222–1225.
94 Anderson, J., Banerjea, A., Planelles, V., Akkina, R. (2003) Potent suppression of HIV type 1 infection by a short hairpin anti-CXCR4 siRNA. *AIDS Res. Hum. Retroviruses*, **19**, 699–706.
95 Qin, X.F., An, D.S., Chen, I.S., Baltimore, D. (2003) Inhibiting HIV-1 infection in human T cells by lentiviral-mediated delivery of small interfering RNA against CCR5. *Proc. Natl Acad. Sci. USA*, **100**, 183–188.
96 Nevot, M., Martrus, G., Clotet, B., Martínez, M.A. (2011) RNA interference as a tool for exploring HIV-1 robustness. *J. Mol. Biol.*, **413**, 84–96.
97 Lee, M.T., Coburn, G.A., McClure, M.O., Cullen, B.R. (2003) Inhibition of human immunodeficiency virus type 1 replication in primary macrophages by using Tat- or CCR5-specific small interfering RNAs expressed from a lentivirus vector. *J. Virol.*, **77**, 11964–11972.
98 Ge, Q., Filip, L., Bai, A., Nguyen, T., Eisen, H.N., Chen, J. (2004) Inhibition of influenza virus production in virus-infected mice by RNA interference. *Proc. Natl Acad. Sci. USA*, **101**, 8676–8681.
99 Tompkins, S.M., Lo, C.Y., Tumpey, T.M., Epstein, S.L. (2004) Protection against lethal influenza virus challenge by RNA interference *in vivo*. *Proc. Natl Acad. Sci. USA*, **101**, 8682–8686.
100 Barik, S. (2010) siRNA for influenza therapy. *Viruses*, **2**, 1448–1457.
101 Barik, S. (2009) Treating respiratory viral diseases with chemically modified, second generation intranasal siRNAs. *Methods Mol. Biol.*, **487**, 331–341.
102 Barik, S. (2011) Intranasal delivery of antiviral siRNA. *Methods Mol. Biol.*, **721**, 333–338.
103 Bitko, V., Barik, S. (2007) Intranasal antisense therapy: preclinical models with a clinical future? *Curr. Opin. Mol. Ther.*, **9**, 119–125.
104 Burnett, J.C., Rossi, J.J., Tiemann, K. (2011) Current progress of siRNA/shRNA therapeutics in clinical trials. *Biotechnol. J.*, **6**, 1130–1146.
105 Adelman, Z.N., Blair, C.D., Carlson, J.O., Beaty, B.J., Olson, K.E. (2001) Sindbis virus-induced silencing of dengue viruses in mosquitoes. *Insect Mol. Biol.*, **10**, 265–273.
106 Adelman, Z.N., Sanchez-Vargas, I., Travanty, E.A., Carlson, J.O., Beaty, B.J., Blair, C.D., Olson, K.E. (2002) RNA silencing of dengue virus type 2 replication in transformed C6/36 mosquito cells transcribing an inverted-repeat RNA derived from the virus genome. *J. Virol.*, **76**, 12925–12933.

107 Caplen, N.J., Zheng, Z., Falgout, B., Morgan, R.A. (2002) Inhibition of viral gene expression and replication in mosquito cells by siRNA-triggered RNA interference. *Mol. Ther.*, **6**, 243–251.

108 Dector, M.A., Romero, P., Lopez, S., Arias, C.F. (2002) Rotavirus gene silencing by small interfering RNAs. *EMBO Rep.*, **3**, 1175–1180.

109 McCaffrey, A.P., Nakai, H., Pandey, K., Huang, Z., Salazar, F.H., Xu, H., Wieland, S.F., Marion, P.L., Kay, M.A. (2003) Inhibition of hepatitis B virus in mice by RNA interference. *Nat. Biotechnol.*, **21**, 639–644.

110 Sen, A., Steele, R., Ghosh, A.K., Basu, A., Ray, R., Ray, R.B. (2003) Inhibition of hepatitis C virus protein expression by RNA interference. *Virus Res.*, **96**, 27–35.

111 Seo, M.Y., Abrignani, S., Houghton, M., Han, J.H. (2003) Small interfering RNA-mediated inhibition of hepatitis C virus replication in the human hepatoma cell line Huh-7. *J. Virol.*, **77**, 810–812.

112 Wilson, J.A., Jayasena, S., Khvorova, A., Sabatinos, S., Rodrigue-Gervais, I.G., Arya, S., Sarangi, F., Harris-Brandts, M., Beaulieu, S., Richardson, C.D. (2003) RNA interference blocks gene expression and RNA synthesis from hepatitis C replicons propagated in human liver cells. *Proc. Natl Acad. Sci. USA*, **100**, 2783–2788.

113 Takigawa, Y., Nagano-Fujii, M., Deng, L., Hidajat, R., Tanaka, M., Mizuta, H., Hotta, H. (2004) Suppression of hepatitis C virus replicon by RNA interference directed against the NS3 and NS5B regions of the viral genome. *Microbiol. Immunol.*, **48**, 591–598.

114 Konishi, M., Wu, C.H., Wu, G.Y. (2003) Inhibition of HBV replication by siRNA in a stable HBV-producing cell line. *Hepatology*, **38**, 842–850.

115 Giladi, H., Ketzinel-Gilad, M., Rivkin, L., Felig, Y., Nussbaum, O., Galun, E. (2003) Small interfering RNA inhibits hepatitis B virus replication in mice. *Mol. Ther.*, **8**, 769–776.

116 Hamasaki, K., Nakao, K., Matsumoto, K., Ichikawa, T., Ishikawa, H., Eguchi, K. (2003) Short interfering RNA-directed inhibition of hepatitis B virus replication. *FEBS Lett.*, **543**, 51–54.

117 Kapadia, S.B., Brideau-Andersen, A., Chisari, F.V. (2003) Interference of hepatitis C virus RNA replication by short interfering RNAs. *Proc. Natl Acad. Sci. USA*, **100**, 2014–2018.

118 Yokota, T., Sakamoto, N., Enomoto, N., Tanabe, Y., Miyagishi, M., Maekawa, S., Yi, L., Kurosaki, M., Taira, K., Watanabe, M., Mizusawa, H. (2003) Inhibition of intracellular hepatitis C virus replication by synthetic and vector-derived small interfering RNAs. *EMBO Rep.*, **4**, 1–7.

119 Chang, J., Taylor, J.M. (2003) Susceptibility of human hepatitis delta virus RNAs to small interfering RNA action. *J. Virol.*, **77**, 9728–9731.

120 Hong-Geller, E., Micheva-Viteva, S.N. (2010) Functional gene discovery using RNA interference-based genomic screens to combat pathogen infection. *Curr. Drug Discov. Technol.*, **7**, 86–94.

121 Surabhi, R.M., Gaynor, R.B. (2002) RNA interference directed against viral and cellular targets inhibits human immunodeficiency virus type 1 replication. *J. Virol.*, **76**, 12963–12973.

122 Bitko, V., Oldenburg, A., Garmon, N.E., Barik, S. (2003) Profilin is required for viral morphogenesis, syncytium formation, and cell-specific stress fiber induction by respiratory syncytial virus. *BMC Microbiol.*, **3**, 9.

123 Sessions, O.M., Barrows, N.J., Souza-Neto, J.A., Robinson, T.J., Hershey, C.L., Rodgers, M.A., Ramirez, J.L., Dimopoulos, G., Yang, P.L., Pearson, J.L., Garcia-Blanco, M.A. (2009) Discovery of insect and human dengue virus host factors. *Nature*, **458**, 1047–1050.

124 Tai, A.W., Benita, Y., Peng, L.F., Kim, S.S., Sakamoto, N., Xavier, R.J., Chung, R.T. (2009) A functional genomic screen identifies cellular cofactors of hepatitis C virus replication. *Cell Host Microbe*, **5**, 298–307.

125 Hao, L., Sakurai, A., Watanabe, T., Sorensen, E., Nidom, C.A., Newton, M.A., Ashlquist, P., Kawaoka, Y. (2008) *Drosophila* RNAi screen identifies host genes important for influenza virus replication. *Nature*, **454**, 890–893.

126 König, R., Stertz, S., Zhou, Y., Inoue, A., Hoffmann, H.H., Bhattacharyya, S., Alamares, J.G., Tscherne, D.M., Ortigoza, M.B., Liang, Y., Gao, Q., Andrews, S.E., Bandyopadhyay, S., De Jesus, P., Tu, B.P.,

Pache, L., Shih, C., Orth, A., Bonamy, G., Miraglia, L., Ideker, T., García-Sastre, A., Young, J.A., Palese, P., Shaw, M.L., Chanda, S.K. (2010) Human host factors required for influenza virus replication. *Nature*, **463**, 813–817.

127 Karlas, A., Machuy, N., Shin, Y., Pleissner, K.P., Artarini, A., Heuer, D., Becker, D., Khalil, H., Ogilvie, L.A., Hess, S., Mäurer, A.P., Müller, E., Wolff, T., Rudel, T., Meyer, T.F. (2010) Genome-wide RNAi screen identifies human host factors crucial for influenza virus replication. *Nature*, **463**, 818–822.

128 Brass, A.L., Dykxhoorn, D.M., Benita, Y., Yan, N., Engelman, A., Xavier, R.J., Lieberman, J., Elledge, S.J. (2008) Identification of host proteins required for HIV infection through a functional genomic screen. *Science*, **319**, 921–926.

129 König, R., Zhou, Y., Elleder, D., Diamond, T.L., Bonamy, G.M., Irelan, J.T., Chiang, C.Y., Tu, B.P., De Jesus, P.D., Lilley, C.E., Seidel, S., Opaluch, A.M., Caldwell, J.S., Weitzman, M.D., Kuhen, K.L., Bandyopadhyay, S., Ideker, T., Orth, A.P., Miraglia, L.J., Bushman, F.D., Young, J.A., Chanda, S.K. (2008) Global analysis of host-pathogen interactions that regulate early-stage HIV-1 replication. *Cell*, **135**, 49–60.

130 Krishnan, M.N., Ng, A., Sukumaran, B., Gilfoy, F.D., Uchil, P.D., Sultana, H., Brass, A.L., Adametz, R., Tsui, M., Qian, F., Montgomery, R.R., Lev, S., Mason, P.W., Koski, R.A., Elledge, S.J., Xavier, R.J., Agaisse, H., Fikrig, E. (2008) RNA interference screen for human genes associated with West Nile virus infection. *Nature*, **455**, 242–245.

131 Parra, J., Portilla, J., Pulido, F., Sánchez-de la Rosa, R., Alonso-Villaverde, C., Berenguer, J., Blanco, J.L., Domingo, P., Dronda, F., Galera, C., Gutiérrez, F., Kindelán, J.M., Knobel, H., Leal, M., López-Aldeguer, J., Mariño, A., Miralles, C., Moltó, J., Ortega, E., Oteo, J.A. (2011) Clinical utility of maraviroc. *Clin. Drug Invest.*, **31**, 527–542.

132 Guendel, I., Agbottah, E.T., Kehn-Hall, K., Kashanchi, F. (2010) Inhibition of human immunodeficiency virus type-1 by cdk inhibitors. *AIDS Res. Ther.*, **7**, 7.

133 Schang, L.M. (2005) Advances on cyclin-dependent kinases (CDKs) as novel targets for antiviral drugs. *Curr. Drug Targets Infect. Disord.*, **5**, 29–37.

134 Kolev, N.G., Tschudi, C., Ullu, E. (2011) RNA interference in protozoan parasites: achievements and challenges. *Eukaryot. Cell*, **10**, 1156–1163.

135 Boyle, J.P., Wu, X.J., Shoemaker, C.B., Yoshino, T.P. (2003) Using RNA interference to manipulate endogenous gene expression in *Schistosoma mansoni* sporocysts. *Mol. Biochem. Parasitol.*, **128**, 205–215.

136 Braun, L., Cannella, D., Ortet, P., Barakat, M., Sautel, C.F., Kieffer, S., Garin, J., Bastien, O., Voinnet, O., Hakimi, M.A. (2010) A complex small RNA repertoire is generated by a plant/fungal-like machinery and effected by a metazoan-like Argonaute in the single-cell human parasite *Toxoplasma gondii*. *PLoS Pathog.*, **6**, e1000920.

137 Adams, B., Musiyenko, A., Kumar, R., Barik, S. (2005) A novel class of dual-family immunophilins. *J. Biol. Chem.*, **280**, 24308–24314.

138 Al-Anouti, F., Quach, T., Ananvoranich, S. (2003) Double-stranded RNA can mediate the suppression of uracil phosphoribosyltransferase expression in *Toxoplasma gondii*. *Biochem. Biophys. Res. Commun.*, **302**, 316–323.

139 Levashina, E.A., Moita, L.F., Blandin, S., Vriend, G., Lagueux, M., Kafatos, F.C. (2001) Conserved role of a complement-like protein in phagocytosis revealed by dsRNA knockout in cultured cells of the mosquito, *Anopheles gambiae*. *Cell*, **104**, 709–718.

140 Campbell, C.L., Black, W.C. IV, Hess, A.M., Foy, B.D. (2008) Comparative genomics of small RNA regulatory pathway components in vector mosquitoes. *BMC Genomics*, **9**, 425.

141 Hoa, N.T., Keene, K.M., Olson, K.E., Zheng, L. (2003) Characterization of RNA interference in an *Anopheles gambiae* cell line. *Insect Biochem. Mol. Biol.*, **33**, 949–957.

142 Prudêncio, M., Rodrigues, C.D., Hannus, M., Martin, C., Real, E., Gonçalves, L.A., Carret, C., Dorkin, R., Röhl, I., Jahn-Hoffmann, K., Luty, A.J., Sauerwein, R., Echeverri, C.J., Mota, M.M. (2008)

Kinome-wide RNAi screen implicates at least 5 host hepatocyte kinases in *Plasmodium* sporozoite infection. *PLoS Pathog.*, **4**, e1000201.
143 Martinez, L.A., Naguibneva, I., Lehrmann, H., Vervisch, A., Tchenio, T., Lozano, G., Harel-Bellan, A. (2002) Synthetic small inhibiting RNAs: efficient tools to inactivate oncogenic mutations and restore p53 pathways. *Proc. Natl Acad. Sci. USA*, **99**, 14849–14854.
144 Scherr, M., Battmer, K., Winkler, T., Heidenreich, O., Ganser, A., Eder, M. (2003) Specific inhibition of bcr-abl gene expression by small interfering RNA. *Blood*, **101**, 1566–1569.
145 McMurray, C.T. (2010) Mechanisms of trinucleotide repeat instability during human development. *Nat. Rev. Genet.*, **11**, 786–799.
146 Filleur, S., Courtin, A., Ait-Si-Ali, S., Guglielmi, J., Merle, C., Harel-Bellan, A., Clezardin, P., Cabon, F. (2003) SiRNA-mediated inhibition of vascular endothelial growth factor severely limits tumor resistance to antiangiogenic thrombospondin-1 and slows tumor vascularization and growth. *Cancer Res.*, **63**, 3919–3922.
147 Campa, C., Harding, S.P. (2011) Anti-VEGF compounds in the treatment of neovascular age related macular degeneration. *Curr. Drug Targets*, **12**, 173–181.
148 Gatto, B., Cavalli, M. (2006) From proteins to nucleic acid-based drugs: the role of biotech in anti-VEGF therapy. *Anticancer Agents Med. Chem.*, **6**, 287–301.
149 Goodchild, J. (2011) Therapeutic oligonucleotides. *Methods Mol. Biol.*, **764**, 1–15.
150 Chernicky, C.L., Yi, L., Tan, H., Gan, S.U., Ilan, J. (2000) Treatment of human breast cancer cells with antisense RNA to the type I insulin-like growth factor receptor inhibits cell growth, suppresses tumorigenesis, alters the metastatic potential and prolongs survival *in vivo*. *Cancer Gene Ther.*, **7**, 384–395.
151 Heidegger, I., Pircher, A., Klocker, H., Massoner, P. (2011) Targeting the insulin-like growth factor network in cancer therapy. *Cancer Biol. Ther.*, **11**, 701–707.
152 Scotlandi, K., Maini, C., Manara, M.C., Benini, S., Serra, M., Cerisano, V., Strammiello, R., Baldini, N., Lollini, P.-L., Nanni, P., Nicoletti, G., Picci, P. (2002) Effectiveness of insulin-like growth factor I receptor antisense strategy against Ewing's sarcoma cells. *Cancer Gene Ther.*, **9**, 296–307.
153 Kanwar, J.R., Roy, K., Kanwar, R.K. (2011) Chimeric aptamers in cancer cell-targeted drug delivery. *Crit. Rev. Biochem. Mol. Biol.*, **46**, 459–477.
154 Ni, X., Castanares, M., Mukherjee, A., Lupold, S.E. (2011) Nucleic Acid aptamers: clinical applications and promising new horizons. *Curr. Med. Chem.*, **18**, 4206–4214.
155 Wang, P., Yang, Y., Hong, H., Zhang, Y., Cai, W., Fang, D. (2011) Aptamers as therapeutics in cardiovascular diseases. *Curr. Med. Chem.*, **18**, 4169–4174.
156 Yang, Y., Ren, X., Schluesener, H.J., Zhang, Z. (2011) Aptamers: selection, modification and application to nervous system diseases. *Curr. Med. Chem.*, **18**, 4159–4168.
157 Becker, R.C., Oney, S., Becker, K.C., Sullenger, B. (2009) Antidote-controlled antithrombotic therapy targeting factor IXa and von Willebrand factor. *Ann. N. Y. Acad. Sci.*, **1175**, 61–70.
158 Floege, J., Ostendorf, T., Janssen, U., Burg, M., Radeke, H.H., Vargeese, C., Gill, S.C., Green, L.S., Janjiæ, N. (1999) Novel approach to specific growth factor inhibition in vivo: antagonism of platelet-derived growth factor in glomerulonephritis by aptamers. *Am. J. Pathol.*, **154**, 169–179.
159 Hwang, B., Han, K., Lee, S.W. (2003) Prevention of passively transferred experimental autoimmune myasthenia gravis by an *in vitro* selected RNA aptamers. *FEBS Lett.*, **31**, 85–89.
160 Lorger, M., Engstle,r, M., Homann, M., Goringer, H.U. (2003) Targeting the variable surface of African trypanosomes with variant surface glycoprotein-specific, serum-stable RNA aptamers. *Eukaryot. Cell*, **2**, 84–94.
161 Edvinsson, L., Nilsson, E., Jansen-Olesen, I. (2007) Inhibitory effect of BIBN4096BS, CGRP(8-37), a CGRP antibody and an RNA-Spiegelmer on CGRP induced vasodilatation in the perfused and non-perfused rat middle cerebral artery. *Br. J. Pharmacol.*, **150**, 633–640.
162 Vater, A., Jarosch, F., Buchner, K., Klussmann, S. (2003) Short bioactive Spiegelmers to migraine-associated

calcitonin gene-related peptide rapidly identified by a novel approach: tailored-SELEX. *Nucleic Acids Res.*, **31**, e130.

163 Wlotzka, B., Leva, S., Eschgfaller, B., Burmeister, J., Kleinjung, F., Kaduk, C., Muhn, P., Hess-Stumpp, H., Klussmann, S. (2002) *In vivo* properties of an anti-GnRH Spiegelmer: an example of an oligonucleotide-based therapeutic substance class. *Proc Natl Acad. Sci. USA*, **99**, 8898–9902.

164 Braun, K., Ehemann, V., Waldeck, W., Pipkorn, R., Corban-Wilhelm, H., Jenne, J., Gissmann, L., Debus, J. (2004) HPV18 E6 and E7 genes affect cell cycle, pRB and p53 of cervical tumor cells and represent prominent candidates for intervention by use peptide nucleic acids (PNAs). *Cancer Lett.*, **209**, 37–49.

165 Folini, M., Bandiera, R., Millo, E., Gandellini, P., Sozzi, G., Gasparini, P., Longoni, N., Binda, M., Daidone, M.G., Berg, K., Zaffaroni, N. (2007) Photochemically enhanced delivery of a cell-penetrating peptide nucleic acid conjugate targeting human telomerase reverse transcriptase: effects on telomere status and proliferative potential of human prostate cancer cells. *Cell Prolif.*, **40**, 905–920.

166 Cutrona, G., Carpaneto, E.M., Ponzanelli, A., Ulivi, M., Millo, E., Scarfì, S., Roncella, S., Benatti, U., Boffa, L.C., Ferrarini, M. (2003) Inhibition of the translocated c-myc in Burkitt's lymphoma by a PNA complementary to the E mu enhancer. *Cancer Res.*, **63**, 6144–6148.

167 Matis, S., Mariani, M.R., Cutrona, G., Cilli, M., Piccardi, F., Daga, A., Damonte, G., Millo, E., Moroni, M., Roncella, S., Fedeli, F., Boffa, L.C., Ferrarini, M. (2009) PNAEmu can significantly reduce Burkitt's lymphoma tumor burden in a SCID mice model: cells dissemination similar to the human disease. *Cancer Gene Ther.*, **16**, 786–793.

168 Alotte, C., Martin, A., Caldarelli, S.A., Di Giorgio, A., Condom, R., Zoulim, F., Durantel, D., Hantz, O. (2008) Short peptide nucleic acids (PNA) inhibit hepatitis C virus internal ribosome entry site (IRES) dependent translation *in vitro*. *Antiviral Res.*, **80**, 280–287.

169 Nulf, C.J., Corey, D. (2004) Intracellular inhibition of hepatitis C virus (HCV) internal ribosomal entry site (IRES)-dependent translation by peptide nucleic acids (PNAs) and locked nucleic acids (LNAs). *Nucleic Acids Res.*, **32**, 3792–3798.

170 Boudreau, J.E., Bonehill, A., Thielemans, K., Wan, Y. (2011) Engineering dendritic cells to enhance cancer immunotherapy. *Mol. Ther.*, **19**, 841–853.

171 Reynolds, A., Leake, D., Boese, Q., Scaringe, S., Marshall, W.S., Khvorova, A. (2004) Rational siRNA design for RNA interference. *Nat. Biotechnol.*, **22**, 326–330.

172 Grunweller, A., Wyszko, E., Bieber, B., Jahnel, R., Erdmann, V.A., Kurreck, J. (2003) Comparison of different antisense strategies in mammalian cells using locked nucleic acids, 2'-O-methyl RNA, phosphorothioates and small interfering RNA. *Nucleic Acids Res.*, **31**, 3185–3193.

173 Usman, N., Blatt, L.M. (2000) Nuclease-resistant synthetic ribozymes: developing a new class of therapeutics. *J. Clin. Invest.*, **106**, 1197–1202.

174 Breaker, R.R., Joyce, G.F. (1994) A DNA enzyme that cleaves RNA. *Chem. Biol.*, **1**, 223–229.

175 Emilsson, G.M., Breaker, R.R. (2002) Deoxyribozymes: new activities and new applications. *Cell Mol. Life. Sci.*, **59**, 596–607.

176 Santoro, S.W., Joyce, G.F., Sakthivel, K., Gramatikova, S., Barbas, C.F. III (2000) RNA cleavage by a DNA enzyme with extended chemical functionality. *J. Am. Chem. Soc.*, **122**, 2433–2439.

177 Manjunath, N., Dykxhoorn, D.M. (2010) Advances in synthetic siRNA delivery. *Discov. Med.*, **9**, 418–430.

178 Miyagishi, M., Taira, K. (2004) RNAi expression vectors in mammalian cells. *Methods Mol. Biol.*, **252**, 483–491.

179 Rubinson, D.A., Dillon, C.P., Kwiatkowski, A.V., Sievers, C., Yang, L., Kopinja, J., Rooney, D.L., Ihrig, M.M., McManus, M.T., Gertler, F.B., Scott, M.L., Van Parijs, L. (2003) A lentivirus-based system to functionally silence genes in primary mammalian cells, stem cells and transgenic mice by RNA interference. *Nat. Genet.*, **33**, 401–406.

180 Elbashir, S.M., Harborth, J., Lendeckel, W., Yalcin, A., Weber, K., Tuschl, T. (2001)

Duplexes of 21-nucleotide RNAs mediate RNA interference in cultured mammalian cells. *Nature*, **411**, 494–498.

181 Dorsett, Y., Tuschl, T. (2004) siRNAs: applications in functional genomics and potential as therapeutics. *Nat. Rev. Drug. Discov.*, **3**, 318–329.

182 Behlke, M.A. (2006) Progress towards in vivo use of siRNAs. *Mol. Ther.*, **13**, 644–670.

183 Bitko, V., Barik, S. (2007) Respiratory viral diseases: access to RNA interference therapy. *Drug Discov. Today Ther. Strateg.*, **4**, 273–276.

184 Bitko, V., Barik, S. (2008) Nasal delivery of siRNA. *Methods Mol. Biol.*, **442**, 75–82.

185 Chi, J.T., Chang, H.Y., Wang, N.N., Chang, D.S., Dunphy, N., Brown, P.O. (2003) Genomewide view of gene silencing by small interfering RNAs. *Proc. Natl Acad. Sci. USA*, **100**, 6343–6346.

186 Jackson, A.L., Bartz, S.R., Schelter, J., Kobayashi, S.V., Burchard, J., Mao, M., Li, B., Cavet, G., Linsley, P.S. (2003) Expression profiling reveals off-target gene regulation by RNAi. *Nat. Biotechnol.*, **21**, 635–637.

187 Robbins, M., Judge, A., Ambegia, E., Choi, C., Yaworski, E., Palmer, L., McClintock, K., MacLachlan, I. (2008) Misinterpreting the therapeutic effects of small interfering RNA caused by immune stimulation. *Hum. Gene Ther.*, **19**, 991–999.

188 Bridge, A.J., Pebernard, S., Ducraux, A., Nicoulaz, A.L., Iggo, R. (2003) Induction of an interferon response by RNAi vectors in mammalian cells. *Nat. Genet.*, **34**, 263–264.

189 Demidov, V.V., Frank-Kamenetskii, M.D. (2004) Two sides of the coin: affinity and specificity of nucleic acid interactions. *Trends Biochem. Sci.*, **29**, 62–71.

190 Persengiev, S.P., Zhu, X., Green, M.R. (2004) Nonspecific, concentration-dependent stimulation and repression of mammalian gene expression by small interfering RNAs (siRNAs). *RNA*, **10**, 12–18.

191 Semizarov, D., Frost, L., Sarthy, A., Kroeger, P., Halbert, D.N., Fesik, S.W. (2003) Specificity of short interfering RNA determined through gene expression signatures. *Proc. Natl Acad. Sci. USA*, **100**, 6347–6352.

192 Sledz, C.A., Holko, M., de Veer, M.J., Silverman, R.H., Williams, B.R. (2003) Activation of the interferon system by short-interfering RNAs. *Nat. Cell. Biol.*, **5**, 834–839.

193 Jurk, M., Chikh, G., Schulte, B., Kritzler, A., Richardt-Pargmann, D., Lampron, C., Luu, R., Krieg, A.M., Vicari, A.P., Vollmer, J. (2011) Immunostimulatory potential of silencing RNAs can be mediated by a non-uridine-rich toll-like receptor 7 motif. *Nucleic Acid Ther.*, **21**, 201–214.

194 Stewart, C.R., Karpala, A.J., Lowther, S., Lowenthal, J.W., Bean, A.G. (2011) Immunostimulatory motifs enhance antiviral siRNAs targeting highly pathogenic avian influenza H5N1. *PLoS ONE*, **6**, e21552.

195 Reynolds, A., Leake, D., Boese, Q., Scaringe, S., Marshall, W.S., Khvorova, A. (2004) Rational siRNA design for RNA interference. *Nat. Biotechnol.*, **22**, 326–330.

196 Gaglione, M., Messere, A. (2010) Recent progress in chemically modified siRNAs. *Mini Rev. Med. Chem.*, **10**, 578–595.

197 Gantier, M.P., Tong, S., Behlke, M.A., Irving, A.T., Lappas, M., Nilsson, U.W., Latz, E., McMillan, N.A., Williams, B.R. (2010) Rational design of immunostimulatory siRNAs. *Mol. Ther.*, **18**, 785–795.

198 Hamm, S., Latz, E., Hangel, D., Müller, T., Yu, P., Golenbock, D., Sparwasser, T., Wagner, H., Bauer, S. (2010) Alternating 2'-O-ribose methylation is a universal approach for generating non-stimulatory siRNA by acting as TLR7 antagonist. *Immunobiology*, **215**, 559–569.

199 Judge, A., MacLachlan, I. (2008) Overcoming the innate immune response to small interfering RNA. *Hum. Gene Ther.*, **19**, 111–124.

26
Epigenetics of the Immune System

Rena Levin-Klein and Yehudit Bergman
The Hebrew University Medical School, Institute for Medical Research, Israel-Canada, Department of Developmental Biology and Cancer Research, Jerusalem, Israel

Epigenetic Regulation and Epigenomics: Advances in Molecular Biology and Medicine, First Edition. Edited by Robert A. Meyers.

Keywords

Chromatin modification
Alterations in the structure of the chromatin, including DNA methylation, histone modifications such as methylation, acetylation, phosphorylation and ubiquitylation, nucleosome repositioning, and long-range chromatin interactions. These modifications regulate DNA accessibility, without changing the DNA sequence, and may be inherited by daughter cells.

Hematopoiesis
The stepwise, hierarchic differentiation of the different cell types that constitute the blood, including erythrocytes, platelets, and cells from the immune system, from the hematopoietic stem cells.

Genomic editing
Targeted modification of the DNA sequence at the immunoglobulin genes in activated B cells. This includes somatic hypermutations, which add to immunoglobulin diversity, and class switch recombination, which directs the character of the immune response.

Tissue-specific transcription factors
Transcriptional regulators which control the expression patterns of specific lineages, such as Pax5 in B cells and GATA3 in $T_{H}2$ cells. These transcription factors are only expressed in their restricted lineage, but may be primed for expression at earlier stages of development.

V(D)J recombination
A somatic rearrangement of the variable (V), diversity (D), and joining (J) regions of the antigen receptor genes, leading to repertoire diversity of both B-cell and T-cell receptors.

The various cells of the immune system all originate from the hematopoietic stem cell, yet each serves a distinct function in the immune response. The differentiation process of the immune system is multistaged and, in the adaptive immune system, includes defined steps of targeted mutations in the genome, such as V(D)J recombination of antigen receptors in B and T cells, and somatic hypermutations at the variable region of the immunoglobulin receptors. Epigenetic marks, such as DNA methylation, histone modifications, chromatin topology, subnuclear localization and replication timing, regulate the accessibility and the stable expression or repression of genomic loci. In this chapter, in which attention is focused on the adaptive immune system, the role of epigenetic marks is discussed in the regulation of the various stages of immune cell development. This discussion sequence includes the potentiation of various cell lineages in hematopoietic stem cells, the stepwise activation and repression of key loci during the differentiation process, the targeting of somatic mutations and, finally, the stable commitment of cellular expression programs in fully differentiated cells.

1
The Immune System: An Introduction

The immune system consists of a large number of cell types which are dedicated to protecting the body of the host from various pathogenic and intrinsic dangers. The immune system itself consists of the myeloid and lymphoid lineages, which are generally responsible for the innate and adaptive immune responses, respectively. The innate immune cells respond rapidly to general danger signals, and are the first line of defense against pathogens. While they themselves do not confer long-lasting protection – as they do not generate a "memory" of previous challenges – they contribute to activating the adaptive immune system. The cells of the adaptive immune system respond more slowly to danger signals, but their response is more specific. Each lymphocyte produces a receptor which binds an exclusive target, known as an *antigen*. Lymphocytes which are challenged with their specific antigen become activated, undergo clonal expansion, and respond strongly to neutralize the perceived danger. In addition to the specificity of the response, once the threat has been dealt with the lymphocytes generate a population of "memory cells" which can respond more rapidly and efficiently to repeated challenges [1].

All cells of the immune system originate from the same adult cell type, the hematopoietic stem cell (HSC), which also gives rise to the other cellular components of the blood, namely the red blood cells (erythroid lineage) and the platelets (megakaryocytic lineage). The HSCs reside in the bone marrow, where differentiation to most of the various hematopoietic lineages also takes place. The differentiation occurs in a stepwise, hierarchical manner (Fig. 1), with the multipotent, self-renewing HSCs differentiating into multipotent progenitors (MPPs), which have lost the ability to self renew. The MPPs can further differentiate into progenitors with a more limited potential, into either common lymphoid progenitors (CLPs), which can give rise to all cells from the lymphoid lineage (but not to myeloid cells), or common myeloid progenitors (CMPs), which no longer have lymphoid potential. The CMPs can further differentiate into megakaryocyte/erythrocyte progenitors (MEPs) or granulocyte/macrophage progenitors (GMPs), while the CLPs differentiate into progenitor cells committed to one of the three main lymphoid lineages: B, T, and natural killer (NK) cells. The GMPs give rise to granulocytes, which consist of neutrophils, basophils, and eosinophils, and also to macrophages and dendritic cells. Different external signals from cytokines and growth factors contribute to the differentiation fate of each cell [2] (Fig. 1).

The cells from the hematopoietic system have been studied extensively. It is possible to isolate cells at various stages of differentiation, and this has led to a deeper understanding of the mechanisms underlying cell fate upon differentiation. This – and the fact that the immune system is auxiliary to the body – has led to the immune system becoming a useful model for understanding the mechanisms of differentiation and cell commitment. In this chapter, the role of epigenetics is discussed in determining and maintaining cell fate. Attention will be focused on cells from the adaptive immune system, the aim being to demonstrate how epigenetic markings contribute to immune cell function at various stages of development.

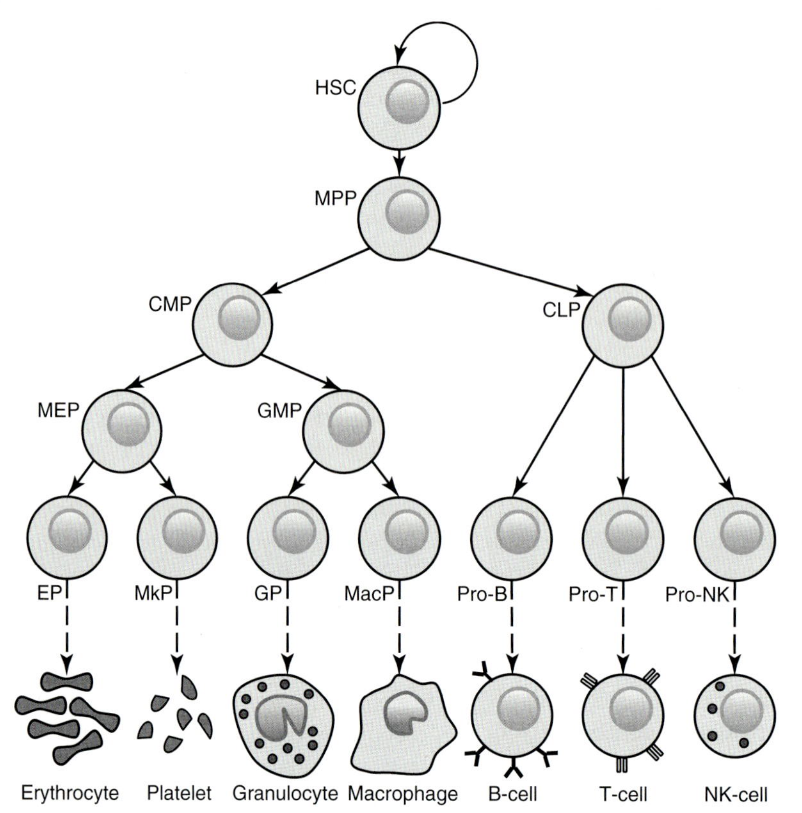

Fig. 1 Hierarchy of differentiation in the hematopoietic system. The HSC differentiates in a stepwise manner. Over the course of the differentiation process, progenitor cells choose a specific branch of the hematopoietic system and lose the potential to differentiate into other lineages, until the cells reach a stage of committed progenitors, which can give rise to only one lineage. The differentiation of a committed progenitor to a mature cell is a multistage process, represented here as a dashed arrow. HSC, hematopoietic stem cell; MPP, multipotent progenitor; CMP, common myeloid progenitor; CLP, common lymphoid progenitor; MEP, megakaryocyte/erythrocyte progenitor; GMP, granulocyte/macrophage progenitor; EP, erythrocyte progenitor; MkP, megakaryocyte progenitor; GP, granulocyte progenitor; MacP, macrophage progenitor; Pro, progenitor; NK, natural killer.

2 Epigenetic Markings

Epigenetic markings are characteristics of the overlying structure of the nuclear genomic DNA in a cell, which affects the accessibility or activation of genomic loci, but does not change the underlying nucleotide sequence. Epigenetic markings are inherited through cell divisions and, as such, allow the stable maintenance of cellular expression programs. While all cells in the body share the same genomic sequence, their expression phenotypes are highly divergent. The different cell identities are defined and maintained by epigenetic markings on the chromatin. The epigenetic markings themselves are far more plastic and changeable than the underlying genomic sequence; thus, under certain circumstances, they will allow changes in a cell's expression program during differentiation or cellular reprogramming.

One such epigenetic marking is methylation of the cytosine base at its fifth carbon [3]. This covalent modification appears in mammalian genomes almost exclusively at cytosine bases which are followed by guanine (a CpG dinucleotide). The CpG sequence is palindromic and usually symmetrically methylated on both DNA strands. This provides the cell with an instructive mechanism for maintaining the mark upon cell cycle progression through the S phase, as a methylated CpG on the older strand serves as a template for methylation of the corresponding CpG on the newly synthesized strand. Proteins from the DNA methyltransferase (DNMT) family are responsible for placing methylation marks on the DNA. DNMT3a and DNMT3b function as *de novo* DNMTs, which are capable of methylating completely unmethylated substrates [4], while DNMT1 functions primarily on hemimethylated CpGs and thus serves as the "maintenance enzyme" during replication [5]. DNA methylation appears to serve as a stable repressive mark at promoters and enhancers. In some cases, this repressive function appears to derive from its ability to block the binding of specific transcription factors, while in other cases repression is brought about by the binding of methyl-binding proteins, such as MBD2, which attach themselves specifically to methylated CpG sites and recruit other repressive factors, such as HP1, to the vicinity of the promoter/enhancer, yielding a heterochromatic structure [3]. Surprisingly, as promoter methylation has been firmly established to serve a silencing function, CpG methylation within the gene body correlates with high transcription [6], though the significance of these marks has yet to be fully understood. Immediately following fertilization, the genome undergoes a massive wave of demethylation which erases most of the parental methylation patterns. Around the time of implantation, new methylation patterns are established by a wave of *de novo* methylation which marks most of the CpG sites in the genome [7]. Specific CpG-rich sequences, known as *CpG islands*, appear to be protected from this wave of methylation [8–10], though some may gain methylation marks during differentiation in a cell type-specific manner [11]. During development, many methylated locations undergo targeted demethylation, either through a passive process of not maintaining methylation during cell division, or through an active process, which is not dependent on cell cycle progression. Although the mechanism of active DNA demethylation is not yet fully understood, there is evidence that it is dependent on DNA repair pathways [12].

A different level of epigenetic marks is present on the nucleosomes which package the DNA. Various covalent modifications on the tails of the histones are interpreted by the cellular machinery as either activating or repressive marks [13, 14]. As such, the acetylation of lysine residues on histones H3 and H4 is seen to make the chromatin structure more permissive and accessible, while methylation of these histones appears to have varying effects, depending on the specific residues modified as well as on the level of methylation that a specific lysine or arginine residue accumulates. H3K27me3 – a mark placed and maintained by the polycomb repressive complex 2 (PRC2) – has a repressive effect on the chromatin, while H3K4me3 has an activating effect and is seen at the promoters of transcribed genes. These two marks are sometimes seen together at the same location, leaving what is termed a *bivalent mark*, which maintains a silent but poised domain in the chromatin [15]. H3K4me1 is seen at many active enhancers [16], and the methylation of H3K36 and H3K79 are seen in correlation with RNA polymerase elongation [17]. H3K9 methylation is seen to have a repressive effect, with H3K9me3 recruiting heterochromatin proteins to domains marked by it [18, 19]. The histone modifications appear to be translated in the cell by proteins which can specifically bind them, such as proteins with chromo-, bromo-, and plant homeodomain (PHD) finger-domains [20]. The acetylation modifications are placed on the histones by histone acetyl transferases (HATs), and removed by histone deacetylases (HDACs). These two enzymes are often found together at the same genomic locations, and seem to function by maintaining a balance in the level of acetylation and accessibility at either an active or poised state [21]. The methylation of histones is carried out by histone methyltransferases (HMTs), such as EZH2 and EZH1 for H3K27 methylation, while demethylation is performed by histone demethylases such as lysine (K)-specific demethylase 1 (LSD1), which can demethylate both H3K4 and H3K9 [14]. Other histone modifications, such as phosphorylation and ubiquitylation, also exist and contribute to the "histone code" [22]. While many of these marks are seen to be faithfully inherited following DNA replication, the mechanism of transmitting this information to the daughter cells remains the subject of intense research [23].

Aside from the modifications on the histones, the actual positioning of the nucleosomes on the DNA can confer either a permissive or repressive state to the chromatin. Many active promoters appear to have a nucleosome-free region. The nucleosomes can be repositioned, or even evicted, from the DNA by nucleosome-remodeling complexes, such as the Swi/Snf complex or the NuRD complex. Each of these remodelers can either activate or repress the areas they act upon, in a context-specific manner [24].

The higher-order topology and localization of the chromatin fibers represents an additional level of epigenetic control. In specially repressed domains, the nucleosomes are densely packed into heterochromatin, which is also segregated to the periphery of the nucleus. Active domains are often more diffused and more centrally located. Large loci which are activated are also often seen to be looped and contracted in a way such that distant enhancers are brought into close proximity with the genes they activate. These long-range interactions can be mediated by CTCF (CCCTC-binding factor) and cohesin complexes [25]. CTCF also has an additional function as an insulator which demarks the boundaries between some active and silent loci [26].

During recent years, noncoding RNA has been recognized as a potent player in the regulation of cellular functions [27]. In many cases, it appears to play a role in marking or regulating genomic accessibility in cis, playing either activating or repressive roles. In some cases, the transcript recruits chromatin modifiers to the DNA, such as the Xist RNA, which recruits PRC2 to inactivate the X chromosome in females [28]. In other cases, these sterile transcripts – which sometimes are referred to as *germline transcripts* – premark an area of the genome that may become active at a later developmental stage [27]. The transcript may do this by remodeling the chromatin with the RNA polymerase, which moves and marks nucleosomes during its passage along the DNA [29].

A final marking of the chromatin to be discussed here is the timing of DNA replication during the S phase of the cell cycle. There is a strong correlation between the timing of replication and the level of activity of a genomic region, with actively transcribed regions replicating earlier in the S phase than repressed regions. Some regions which are expressed in a tissue-specific manner replicate late in developmental stages, where the region is repressed and early in the specific tissue where the region is active. A small portion of the genome replicates in an asynchronous manner, with one allele replicating earlier in the S phase than the allele on the homologous chromosome. Many monoallelically expressed genes – such as imprinted genes, the X chromosome or the olfactory genes –fall into such asynchronous regions. Asynchronous replication may serve as an early marker for future monoallelic expression, as these regions replicate in such a way at developmental stages when there is no expression difference between the alleles. However, the precise mechanistic relationship between replication timing and gene expression is still not well understood [30].

3 HSCs: Epigenetic Basis for Stem Cell Characteristics

The HSCs are found in the bone marrow, and have the capability to give rise to all of the various cell types of the immune system, as well as erythrocytes and platelets. They are additionally characterized by their ability to self renew and replenish the cells in the blood. The HSCs must be capable of a high level of plasticity to allow the differentiation into multiple cell lineages, each with its own unique transcription program that is silent in cells from different lineages [2].

The results of previous studies have shown that proteins which shape the chromatin landscape and write epigenetic marks are important for the maintenance of HSC identity, both for self renewal and for allowing differentiation into the full scope of cells found in the blood. The lack of both *de novo* DNMTs, DNMT3a and DNMT3b, causes the HSCs to lose their ability to self renew, but not their multipotency [31], which indicates that *de novo* methylation is not necessary for the normal hematopoietic differentiation process. Either low levels [32] or a complete lack [32, 33] of DNMT1 results in severe hypomethylation of the genome, the loss of self renewal, and a restriction of the lineages into which the cells can differentiate. This confirms that DNA methylation plays a functional role in suppressing stem cell differentiation, as well as directing differentiation to various lineages.

Similar to DNMT proteins, the Polycomb group (PcG) proteins bring about the repression of genomic loci. Their mechanism of action is through the enzymatic activity of PRC2, which marks the chromatin with H3K27me3 and recruits the heterochromatinization complex, PRC1. The perturbation of PcG proteins was seen to have various effects on HSCs; for example, the depletion of BMI1 (a component of PRC1) in HSCs ablates self renewal, while the overexpression of BMI1 augments the HSC repopulation capacity [34]. An overexpression of EZH2 likewise prevents bone marrow exhaustion, thus augmenting self renewal [35]. However, the deletion of EZH2 does not appear to affect HSC integrity, possibly due to compensation by EZH1 [36]. The partial depletion of two other constitutive

PRC2 components – embryonic ectoderm development (EED) [37, 38] and suppressor of zeste 12 homolog (SUZ12) [39] – surprisingly causes hematopoietic expansion and higher self-renewal capabilities, as seen by the sequential adoptive transfer of the stem cell pool. Taken together, the data concerning PRC2 components may indicate that a homeostatic level of PRC2 function is necessary for the normal balance between self renewal and differentiation, while PRC1 plays a more straightforward role in maintaining the self-renewing stem cell pool.

Mi-2β, a chromatin remodeler which is part of the NuRD complex, has also been shown to be essential for self renewal [40]. Indeed, its absence causes a loss of HSC quiescence and induces proliferation and differentiation, leading to HSC depletion in the bone marrow. The differentiation is limited to the erythroid lineage, with no lymphoid or myeloid cells, which demonstrates that Mi-2β activity is also essential for HSC multipotency. While the key genes regulated by these epigenetic modifiers are still largely unknown, these data confirm that the HSC identity is very closely linked to the epigenetic state of the cell.

The HSCs have the incredible capability to differentiate into many different cell types, with all of the various lineages of the blood being primed in the stem cells, via several different mechanisms. One method of priming appears to be a promiscuous transcription in the HSCs of both hematopoietic lineage genes, as well as genes characteristic of nonhematopoietic tissues [41, 42]. A single stem cell can transcribe low-level genes that are typical of different lineages where they are active; examples are the β-globin gene from the erythroid lineage and myeloperoxidase from the myeloid lineage [43]. During the differentiation process, a "narrowing" of transcriptional activity is observed. At the MPP stage, where the cells are multipotent but can no longer self renew, the non-hematopoietic transcription is lost while the multilineage hematopoietic transcription continues. As the MPPs differentiate into more restricted progenitors (such as lymphoid progenitors), the transcripts of other lineages are no longer manufactured in the cells, whereas lymphoid transcription is increased, with higher levels than in the earlier, unrestricted stages [41]. The genes that are transcribed at low levels in HSCs are those which are poised and can be either activated or repressed, depending on the direction taken during differentiation. The epigenetic markings on the chromatin reflect and maintain this poised state. The HSCs employ a number of different epigenetic mechanisms for poising genes for expression.

One such method is bivalent marking of the chromatin (Fig. 2a), which is observed in CpG island-containing promoters that are marked with both repressive H3K27me3 and active H3K4me3 modifications [44–46]. Bivalent genes are silent or expressed at an extremely low level in HSCs. However, over the course of differentiation most of these genes "choose" to be either active, losing the H3K27me3 mark, or repressed, losing the H3K4me3 mark. EBF1 from the B-cell lineage and GATA3 from the T-cell lineage are examples of lineage-specific genes marked with bivalent modifications in HSCs [45]. Depending on the decisions during differentiation, the genes are resolved in a different manner; some will be resolved early, in the progenitor stages, while others are resolved only much later, when the cells are close to their terminal differentiation [45]. There may be a histone code which primes the stage and lineage in which bivalent domains are resolved. For example, it seems that in human HSCs, bivalent genes which become activated in erythrocytes are also marked with higher levels of H3K4me1, H3K9me1, H4K20me1, histone variant H2A.Z, and RNA polymerase II than are those genes that are repressed [44]. Similarly, bivalent domains in HSCs which are resolved to an active state in progenitor cells or T cells are marked with known activating modifications (H3K79me2 and H3ac). Many bivalent domains which are resolved to an active state late in the differentiation process (such as GATA3 at the T-cell stage) are additionally marked with H3K9me3, as opposed to domains which are activated at the earlier progenitor stage of MPPs or CMPs. This additional repressive mark may be necessary for a prolonged maintenance of the bivalent domain [45]. It is still not entirely clear how bivalent domains are created and maintained up to the point where they are resolved. Recently acquired data have shown that the PRC1 protein BMI1 plays a role in maintaining the repressive half of the bivalent mark on B-cell lineage-specific genes *EBF1* and *PAX5* [46]. In addition, studies have been conducted to show that the PRC2 complex can bind the H3K27me3 mark that it deposits [47, 48], which suggests the existence of a mechanism for H3K27me3 propagation (Fig. 2).

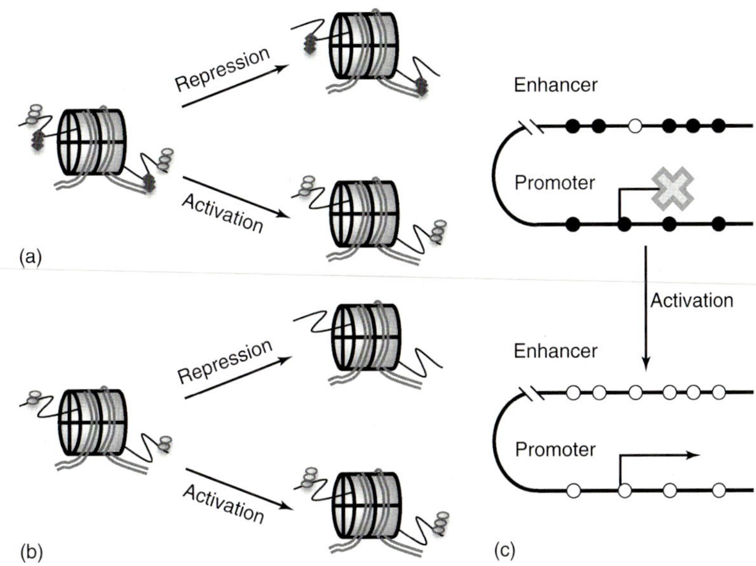

Fig. 2 Epigenetic mechanisms of poising lineage-specific genes in HSCs. (a) Bivalent promoters which are marked with both H3K4me3 (ovals) and H3K27me3 (diamonds) may be either activated or repressed during differentiation. When repressed, only the H3K27me3 mark remains on the nucleosome whereas, when the gene is activated, only the H3K4me3 remains on the histone tails; (b) Genes marked with H3K4me2 can be either repressed during differentiation by losing the H3K4 methylation, or activated by adding a third methyl group; (c) Promoters which are methylated at CpG sequences (filled circles) are not transcribed in HSCs. If the gene is activated during differentiation, the promoter becomes demethylated (empty circles) and transcription is initiated. Some of these genes have enhancers which are partially unmethylated in HSCs, and this poises the promoter for activation.

The HSCs employ an additional epigenetic method to poise or pre-prime critical genes that may be expressed later in development by marking them with H3K4me2, but not with H3K4me3 [49–51] (Fig. 2b). These genes are also hyperacetylated, but are not expressed at high levels [50]. Upon differentiation, the genes either lose H3K4 methylation and remain silent, or gain a third methyl group on H3K4, pushing them into an active state [49]. This situation is true for genes of the erythroid, myeloid, and lymphoid lineages such as *GATA1*, *c-fms*, and *RAG2* [49, 51]. As opposed to bivalent domains, the H3K4me2 mark is not specifically centered at the promoter, and may be present over larger portions of the genes [49]. The mark is also found specifically at genes which do not contain CpG islands. This method of poising genes appears to be a feature of mature tissues, as no similar marks are found in embryonic stem cells (ESCs) [49].

Many lineage-specific genes that are expressed only later in the differentiation process appear to be maintained in a silent state by DNA methylation [52, 53] (Fig. 2c). The promoters of these genes become demethylated during the differentiation process, which may explain why a depletion of DNMT1 forces differentiation [33]. Such demethylation can be seen at the promoters of Lck, POU2af1, and CXCR2, which are expressed and hypomethylated in the T-cell, B-cell, and granulocyte lineages, respectively, and are all methylated in HSCs [52, 53]. Some genes appear to first become poised for activation specifically at their enhancer regions, despite their promoters being in a silent conformation. For example, the CD19 enhancer contains an unmethylated CpG site in multipotent cells, at a time when the promoter is heavily methylated and prior to the expression of the gene itself in pro-B cells [54].

How far back in development does the preparation for hematopoietic lineage expression go? A few examples exist that show seeds of epigenetic priming already in the early stages of embryonic development, prior to the specification of HSCs. The *Ptcrα* and *IL12* genes are hypomethylated at specific CpG sites at their enhancers in ESCs, which correspond to the preimplantation embryo [55, 56]. This demethylated window is essential for poising the respective promoters for activation during T-cell commitment. Similarly, the enhancer of the VpreB1-λ5 locus, which is expressed in pre-B cells, is also already marked in the early embryo with activating histone modifications, which are lost in lineages such as the liver which can never express these genes [57]. Overall, epigenetic markings appear to maintain the delicate balance which provides a framework for HSC identity, both as a multipotent and self-renewing stem cell.

4 B Cells

The B cells are responsible for the humoral adaptive immune response. Each mature B cell expresses a B-cell receptor (BCR) which can bind specific antigens. The B cells undergo a strictly controlled differentiation process from the CLP stage until they are ready to leave the bone marrow. The cells pass through the pro-B, then large and small pre-B-cell stages, during which they rearrange their immunoglobulin (Ig) loci to produce the BCRs and activate the B-cell expression programs. During the immature B-cell stage, they continue activating the B-cell programs and, following maturation, leave the bone marrow to colonize the lymph nodes and the spleen as mature, naïve B cells [58]. Upon activation, the B cells differentiate into a number different types of effector cell, including plasma cells which secrete the Ig proteins in their soluble form as antibodies. The B cells undergo further genomic editing to refine the affinity of the BCR for the antigen, and to provide the antibodies with the correct isotype for the immune response [59]. In this section, the way in which epigenetic regulation shapes the B cells throughout their development will be discussed.

4.1 B-Cell Specification and Development

The B-cell lineage differentiation program from HSCs to mature B cells has been extensively studied. Previously, cells have been isolated from the various stages and the transcription factors involved in the lineage commitment and stepwise differentiation of B cells have been studied [60]. Consequently, an intricate and delicate interplay was identified between transcription factor expression, binding to the DNA, and epigenetic changes in the chromatin, with each process reinforcing and driving the other [61]. An understanding the epigenetic mechanics of B-cell differentiation may serve as a general model for somatic cell differentiation and the acquisition of cell identity.

A few transcription factors have been identified as master regulators of lymphoid and B-cell identity. Some of these are already expressed in the HSCs, thus priming the multipotent cells for lymphoid (and later B-cell-specific) differentiation. One such regulator is Ikaros, which is expressed throughout lymphoid cell development and is seen to regulate lymphoid genes at various stages of their development (see Sects 4.2, 4.3, and 5.1) [62, 63]. Ikaros recruits various chromatin-remodeling complexes to the DNA sequences to which it binds, and in this way changes them epigenetically [62]. Ikaros has been shown to activate many lymphoid-specific genes in the HSCs, while repressing genes from other programs [42]. It also plays an important role in maintaining and balancing the multipotency and self-renewal capacity of HSCs [42].

Another transcription factor that is expressed in HSCs and regulates lymphoid development is E2A. This has two isoforms, E12 and E47, both of which play roles in B-cell differentiation, although E47 appears to have a role at an earlier stage in driving B-cell development [64]. In a recent study, E47 binding was mapped in pre-pro-B and pro-B cells [65]. The E47 binding was associated with H3K4me1 at enhancers and H3K4me3 at promoters of genes, with the number of E47-bound sites rising by almost 200% from pre-pro-B cells to pro-B cells. This indicated an increase in the number of genes regulated, most likely due to cooperative effects with other transcription factors. E2A has been associated in multipotent cells with the demethylation of DNA at enhancers of genes that are expressed only later in B-cell development, thus priming the early developmental stage for the potential of the B-cell lineage [54].

After B-cell specification, an essential transcription factor which is activated is EBF1. This is bivalently marked in HSCs [45, 46], becomes active in the CLP stage, and is necessary for transition to the pro-B stage. EBF1 binding has a strong effect on the chromatin state, and can either activate or repress its various target genes, marking them with H3K4me3 and H3K27me3, respectively [66]. In addition,

EBF1 can also poise the chromatin on some genes, such as *Egr3* and *CD40*, for future activation in later developmental stages by marking them with H3K4me2 [66]. The precise mechanism of how the same transcription factor can lead to different – even opposing – responses from target genes is not yet known. However, it is possible to speculate that additional transcription factors which bind the target promoters contribute to the decision to activate, repress, or poise the chromatin. This possibility is supported by the fact that, when EBF1 is ectopically expressed in T cells, genes which are usually activated by EBF1 are merely poised, hinting that additional factors lead to the activation in the B-cell lineage. EBF1 is known to operate in concert with E2A and, later, Pax5, as well as other transcription factors, thereby creating a complex network [65].

Finally, Pax5 is a transcription factor that is activated in the pro-B stage. Pax5 contains both a transactivation and a repression domain, which are capable of recruiting HATs and HDACs to target genes; this allows Pax5 to function both as an activator and as a repressor [67]. Pax5 is essential for the maintenance of B-cell identity. Indeed, the conditional loss of Pax5 in mature B cells causes dedifferentiation to the lymphoid progenitor stage, and subsequent differentiation to the T-cell lineage *in vivo*, and can even be diverted to the myeloid lineage in cell culture [68]. Pax5 transcription is epigenetically regulated in a stage-specific manner. The Pax5 promoter is bivalently marked in HSCs, an epigenetic mark which is maintained by BMI1 [46]. EBF1 binding helps to resolve the bivalent mark to an active H3K4me3 mark in pro-B cells [69]. In addition to the promoter, Pax5 contains an intronic enhancer which is methylated on its CpG residues in early embryonic development. Such methylation is lost, however, in multipotent hematopoietic cells, which enables an activation of the gene once the promoter is derepressed [69].

Pax5 is also important for the epigenetic regulation of B-cell effector genes which act downstream of it. Pax5 activates at least 170 different genes and changes their chromatin landscape [70]; this epigenetic remodeling is often in cooperation with other upstream B-cell regulators. Two such genes which have been studied are *MB-1* [71, 72] (also called *CD79a*) which encodes Igα (a signaling component of the BCR), and the *CD19* gene [54]. The promoters of both genes are methylated in the HSCs, and must undergo demethylation in order for the genes to be expressed [54, 72]. The CpG methylation at the *MB-1* promoter appears to be maintained in part by the chromatin remodeler Mi-2β [71], with the *MB-1* undergoing demethylation in a stepwise manner. At the CLP stage, EBF1 and E2A bind and recruit the Swi/Snf complexes to the 5′ portion of the promoter and induce CpG demethylation [71]. This sets the stage for Pax5 to bind the 3′ end of the promoter in pro-B cells, bringing about a complete demethylation and expression of the *MB-1* gene [71].

Similar to Pax5, the *CD19* gene has an enhancer which is methylated early in development and undergoes demethylation in multipotent cells, most likely in an E2A-dependent manner [54]. Although the enhancer remains demethylated in all hematopoietic lineages, this is not sufficient for *CD19* expression, as the *CD19* promoter is methylated and remains in this state until the pro-B cell stage, when Pax5 binds and induces DNA demethylation [54]. Aberrant Pax5 expression in the myeloid lineage – but not in nonhematopoietic lineages – results in *CD19* expression [73]. This emphasizes the importance of epigenetic priming for the activation of lineage-specific programs.

Pax5 is also responsible for the epigenetic silencing of genes from opposing lineages. During B-cell development, *c-fms* (a myeloid-specific gene) is silenced by histone modifications and CpG methylation of an intronic enhancer in a Pax5-dependent manner [74, 75]. These epigenetic marks are lost on Pax5 depletion [75]. This highlights the importance of Pax5 in the creation, as well as the maintenance, of the chromatin landscape of B cells. Altogether, it appears that, during B-cell development, transcription factors help shape the epigenetic landscape of the chromatin, while the epigenetic marks ensure stage-specific activation of the transcription factors. These two cellular pathways reinforce each other to ensure correct differentiation.

4.2 Rearrangement of Immunoglobulin Genes

The *Ig* genes are the functional units of the B cells that provide the cells with their clonal specificity in the adaptive immune response. Each B cell has its own unique BCR which is composed of a heavy and a light chain. In order for the cell to produce a functional Ig protein, the Ig loci must be edited in a process known as *"V(D)J rearrangement,"* which brings the various parts of the Ig receptor into a single gene segment. In the germline conformation, the *Ig* loci have multiple variable (V) genes, and a smaller number of diversity (D) (on the heavy chain), and joining (J) (on both the heavy and light chains) regions; these are separated from one another by large expanses of intergenic DNA (Fig. 3a). During the rearrangement process, the *RAG1* and *RAG2* proteins cleave the DNA at a specific sequence known as the recombination signal sequence (RSS), and the DNA is then repaired by the nonhomologous end-joining (NHEJ) pathway so that a *V*, *D*, and *J* gene are now in one continuous exon. The choice of the *V*, *D*, and *J* is what differs between B cells and what is responsible for the highly diverse antibody repertoire.

The rearrangement process is highly regulated and occurs at specific developmental stages (Fig. 3b). The *IgH* locus is rearranged in two steps at the pro-B stage; first, a D is rearranged to a J region, then a single V is attached to the previously rearranged DJ. In the pre-B stage, the light chain loci are rearranged. There are two light chain loci in the genome, the *Igκ* and λ loci. In mice, the *Igκ* locus is the preferred target for rearrangement, and if rearrangement fails to produce a productive *κ* light chain then the *Igλ* locus will undergo rearrangement. The RAG proteins are expressed at all cell stages where rearrangement takes place, and recognizes the same RSS sequences at all rearrangement target loci. Despite this, rearrangement of specific loci only takes place at the "correct" developmental stage. This indicates that an epigenetic mechanism regulates locus accessibility for rearrangement (Fig. 3).

Indeed, *in vitro*, RAG proteins can cleave naked DNA, but not DNA which is packed in nucleosomal form [76]. The chromatin of the antigen receptor loci only become accessible to *in vitro* cleavage by the RAG machinery when extracted from the correct cell type, while the other loci are seen to be refractory to rearrangement [77]. There appear to be a number of epigenetic marks which correlate with Ig accessibility. Prior to rearrangement, the *Ig* loci are marked with repressive histone marks such as

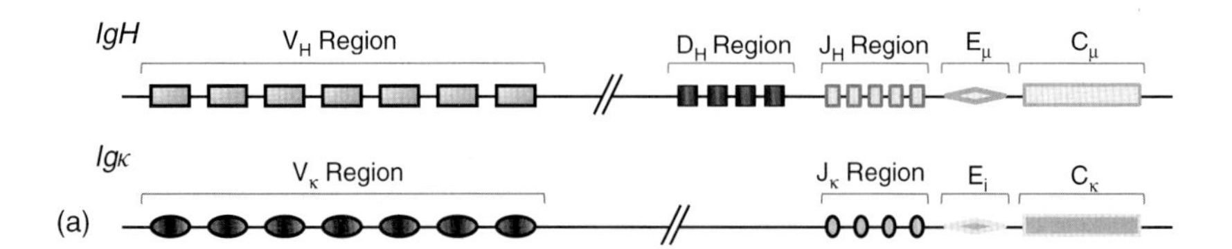

IgH
V_H Region
D_H Region
J_H Region
E_μ
C_μ
Igκ
V_κ Region
J_κ Region
E_i
C_κ
(a)

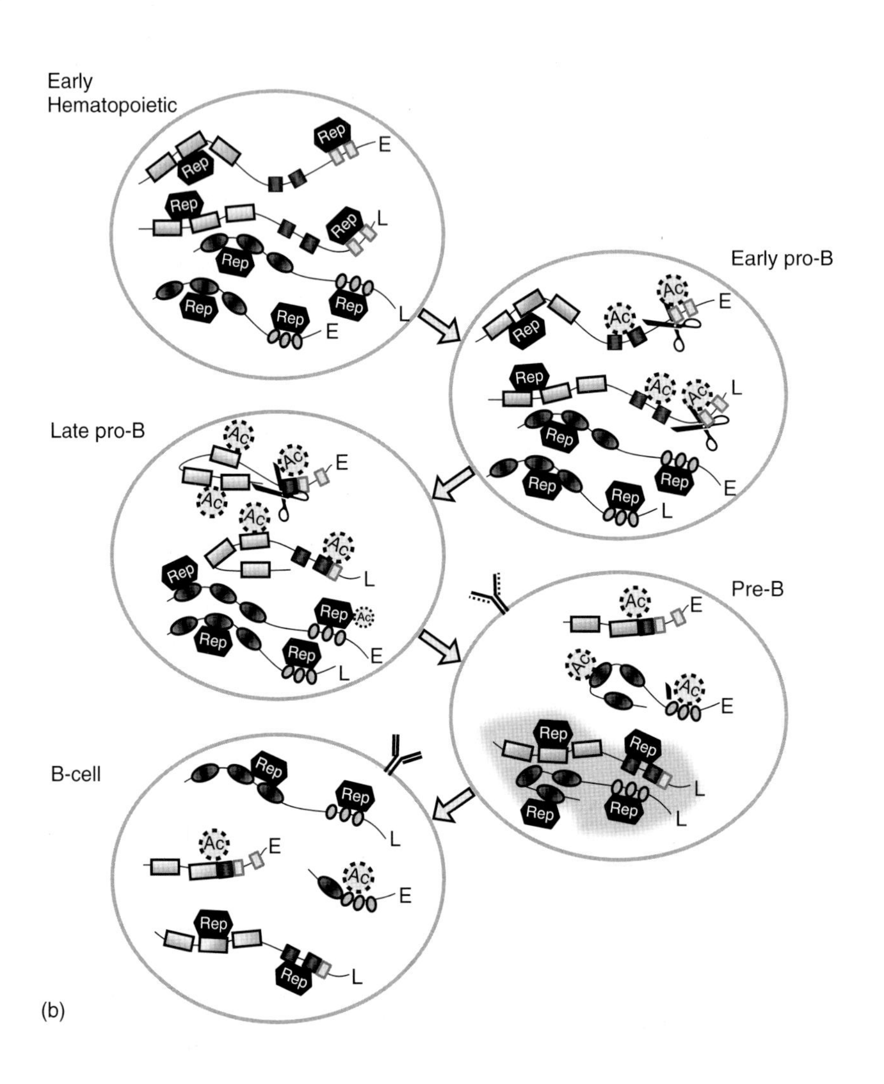

Early Hematopoietic
Early pro-B
Late pro-B
Pre-B
B-cell
Rep
Ac
E
L
(b)

H3K9me2 [78, 79]; this mark is removed from the V_H region in a Pax5-dependent manner [79]. The *Ig* genes are also methylated on their DNA, probably from the early embryonic stages [53], a mark which is faithfully maintained throughout development [80]. However, the rearranged *Ig* genes undergo demethylation immediately before rearrangement, a stage which appears crucial for the rearrangement process [80–82].

In preparation for rearrangement, the *Ig* loci gain activating histone marks, germline transcription in both the sense and antisense directions, and their chromatin is remodeled [78,83–88]. Acetylation seems to mark the accessibility of the chromatin. The loci are acetylated in a stepwise, stage-specific manner, with the D_H and J_H regions becoming acetylated before the V_H regions [85–89]. Higher levels of acetylation seem to correlate with the frequency at which a D or V region is chosen for rearrangement, in both the heavy and light chains [83, 90, 91].

Germline transcription is a prevalent mark in the *Ig* loci. The V, D, and J regions are all seen to be transcribed prior to rearrangement [78, 83, 92–96]; the V_H and D_H regions are also transcribed in the antisense direction [84, 91, 97–99]. In addition to signifying accessible chromatin, these transcripts may play a role in recruiting chromatin-modifying factors to the *Ig* loci, possibly even repressive modifiers in some cases [91].

H3K4me3, which is a mark of active promoters, is a third active epigenetic mark that appears on the J and D regions of *Ig* genes at the stage immediately prior to rearrangement [78, 82, 83, 100]. Aside from making the chromatin accessible, this mark appears to play an important role in RAG protein targeting and anchoring to the chromatin. *RAG2* contains a PHD finger domain which recognizes and binds H3K4me3 [101, 102], an interaction

Fig. 3 *Ig* loci and model of Ig rearrangement during B-cell development. (a) Schematic drawing of the *IgH* and *Igκ* loci. The loci are not drawn to scale and the amounts of V, D, and J segments do not represent actual number of times that these segments appear in the genome; (b) In early hematopoietic cells, both the *IgH* (rectangles) and *Igκ* (ovals) are decorated with repressive epigenetic marks ("Rep" in black hexagon) such as CpG methylation and H3K9me2. The two alleles of each loci replicate asynchronously in the cell cycle with one allele replicating early (E) in the S phase and the other replicating late (L). In the early pro-B stage, the D_H and J_H regions lose their repressive marks on both alleles and gain activating marks (Ac in gray circle) such as H3Ac, H3K4me3, and germline transcription. In addition the RAG proteins (scissors) bind these regions. Once the DJ rearrangement is complete, in the late pro-B stage, the V_H region becomes activated and contracted (depicted as a folding-over of the V region), probably on both alleles. Only the early-replicating allele is rearranged. At this stage, the J_κ region begins to gain activating marks on the early-replicating allele, but does not lose its repressive marks. In the pre-B cell stage, the cells express a cell surface pre-BCR from the rearranged heavy allele. The unrearranged IgH allele is recruited to the pericentromeric heterochromatin (amorphic gray cloud), along with the late-replicating *Igκ* allele. The early-replicating *Igκ* locus gains activating marks and loses repressive marks throughout the locus. The V_κ region contracts into close proximity with the J_κ region. The RAG proteins cleave and rearrange this allele, leading the cells to differentiate into B cells, which express a complete BCR. The late-replicating *Ig* alleles are no longer in the heterochromatic compartment, but retain repressive epigenetic marks, whereas the early-replicating, rearranged alleles retain active marks.

that stimulates cleavage activity of the RAG proteins [103]. The mapping of *RAG2* to the genome shows that it binds globally to regions of the DNA which are marked with H3K4me3. In the *Ig* loci, *RAG2* mapped specifically to the J and D regions in a stage-specific manner [100]. Mutations in the PHD domain, or a reduction of H3K4 methylation, lead to a decrease in recombination [101, 102]. The V regions are not marked with H3K4me3 but, instead, with H3K4me2 [78, 83]; this may poise the *V* genes for rearrangement, similar to the role of this mark in the HSCs (see Sect. 3).

The regions encompassing all the V segments are very large; consequently, the *V* genes must be brought into physical proximity of the locus enhancer as well as the D or J regions, in order to allow recombination. This is particularly true for the distal Vs, which are separated from their recombination partners by over two megabases of DNA. Interestingly, the entire V_H locus is seen to contract in pro-B cells in preparation for the V to DJ rearrangement, bringing all of the V segments close to their rearrangement partners [104–107]. This is also the case for the V_κ region in the later pre-B-cell stage [82]. This contraction is probably mediated by Rad21 (one of the cohesin subunits), which binds CTCF proteins. CTCF recognizes specific CTCF-binding sequences in the V regions [108, 109]. While the CTCF binding is constitutive, cohesin binds these regions in a developmental stage-specific manner. Of note, CTCF and cohesin are infrequently seen in the D and J regions of the *Ig* loci, which are much smaller and already in close proximity to the *Ig* intronic enhancers [108]. While the precise mechanics that regulate the contraction process are still unclear, it appears that the transcription factors Pax5, YY1, and Ikaros are involved [105, 110, 111].

The $\mathrm{V_H}$ region appears to use an additional mechanism in order to provide the distal *V* genes with an equal chance of recombination as the more proximally located genes. This is achieved by the $\mathrm{V_H}$ region marking the proximal genes with a repressive H3K27me3 mark [36, 112], the removal of which by depletion of EZH2 leads to a stark preference of recombination with the proximal V_H genes. Interestingly, during early embryonic development, this H3K27me3 is naturally absent from the proximal $\mathrm{V_H}$ region and, indeed, the repertoire of V_H genes in these cells is strongly biased toward the proximal V_H genes [36].

Each mature B cell expresses only one functional IgH and IgL chain, from a single parental allele. This is physiologically important in order to maintain the clonal specificity of the BCRs during the immune response. The rearrangement of a single *Ig* allele appears to be epigenetically programmed into the developing B cell in a clonal manner; this has been shown extensively on the *Igκ* locus in particular [80–83, 106,113–115]. The *Ig* loci are already marked in the early developing embryo to be treated in a monoallelic manner. Similar to other monoallelically expressed genes, such as the imprinted genes and the X chromosome in females, the *Ig* loci undergo asynchronous replication, with the DNA of one allele replicating earlier in the S phase of the cell cycle than the other allele [116]. The decision as to which allele is early is not clonally maintained in stem cells [117], but becomes stabilized in differentiated cells [116]. The early-replicating allele appears to be the preferred substrate for the rearrangement process [116]. In the pre-B stage of development, the late-replicating allele is preferentially recruited to the pericentromeric heterochromatin compartment in an Ikaros-dependent manner, while the early-replicating allele becomes hyperacetylated [82]. A single allele is also preferentially demethylated on the DNA, and is chosen for an initial rearrangement reaction [80, 82, 114].

In some cases, the initial rearrangement is unsuccessful and the product is out of frame or self-reactive. In such cases, the cell undergoes secondary rearrangements, which can use downstream J segments on the same allele, or move on to the second allele. However, in the case of a successful rearrangement, further *RAG* activity at the locus must be inhibited. Following *IgH* rearrangement, signals from the pre-BCR lead to a decontraction of the *IgH* V region, deacetylation of the histones, and a repositioning of the unrearranged *IgH* locus to heterochromatic regions of the nucleus [118]. This ensures that the *IgH* does not undergo additional rearrangements in the pre-B-cell stage. Similarly, following *Igκ* rearrangement, signaling from a full BCR leads to a silencing of the *RAG* genes and a complete cessation of the rearrangement reaction, while the lack of a productive *κ* rearrangement in mice leads to accessibility and rearrangement on the *Igλ* locus [78]. As with many other processes in the immune system, the rearrangement of the *Ig* genes is tightly bound to epigenetic regulation, before, during, and after the actual rearrangement reaction (Fig. 3b).

4.3 Somatic Hypermutation (SHM) and Class Switch Recombination (CSR)

In addition to Ig rearrangement, the B cells can undergo two further processes of genomic editing which actively change the DNA sequence in the cell relative to other cells in the organism. These processes occur at the *Ig* loci after B-cell activation through the BCR, and are both mediated by the activation-induced cytosine deaminase (AID) enzyme [119].

The first process is somatic hypermutation (SHM), in which the V regions of the rearranged Ig heavy and light chains undergo cytosine deamination. The resulting uracils are recognized as a mismatch by the DNA repair machinery, and the DNA sequence is repaired by an error-prone DNA polymerase, thus introducing mutations to the V region that recognizes the antigen. Some of these mutations enhance antibody affinity, such that the activated B cell is made more effective for both current and potential future challenges [119].

The second process is class switch recombination (CSR), in which the constant heavy region isotype is edited [120]. The heavy region DNA is cleaved at discrete "switch" regions that are immediately upstream of the isotype class cassettes. In this case, the most upstream region is a donor S region, and the downstream regions are acceptor switch regions. When the DNA is repaired, through the NHEJ pathway, a new isotype cassette is put in place, whilst all of the genetic information between the new and old cassettes is discarded from the genome. The isotype helps to define the nature of the immune response that the body will mobilize in order to fight off a perceived danger.

As both of these processes involve irreversible changes of the genome sequence, the process is tightly regulated. An incorrect targeting of the CSR process can cause certain lymphoid malignancies, whilst an incorrect SHM can contribute to genome instability in other cancers [121].

Part of the mechanism of correct targeting is epigenetic. SHM, which is found near the transcription initiation site of the *Ig* genes, mostly on the V region [122,

123], is transcription-dependent [124] and is blocked by the presence of nucleosomes in untranscribed regions [125]. However, transcription of the *Ig* locus is insufficient to allow SHM. In a mouse model with two prerearranged *Igκ* genes (that are present in all B cells), both are transcribed in the mature B cells, though only one gene underwent DNA demethylation [126]. The fact that only the demethylated allele undergoes SHM suggests that the level of DNA methylation might play a role in SHM [126]. Indeed, AID-mediated deamination of the cytosine base appears to be blocked by methylated cytosine *in vitro* [127]. However, this may be a general mechanism to protect the gene bodies of most transcribed genes in the genome since, as opposed to the *Ig* gene – where the active allele is demethylated in the gene body [80, 81, 126] – most active genes are hypermethylated in the gene body [6]. The areas that undergo SHM are associated with H3ac, which may be a consequence of the active transcription and is in place prior to the activation of SHM [128]. The constant region is less acetylated and, under normal circumstances, does not undergo SHM [122]. However, treatment with trichostatin A (TSA; an inhibitor of HDAC) in an SHM-inducible cell line causes a higher acetylation of the constant region, which also begins to gain hypermutations [122]. The two other histone modifications specifically associated with mutation hotspots during SHM are phosphorylation at H2BS14 [128], and monoubiquitylation of H2A and H2B [129]. The phosphorylation mark is AID-dependent, and so is probably not involved in AID targeting; rather, it may be involved in recruiting the error-prone repair machinery [128]. As yet, it is unclear as to whether the ubiquitylation mark is AID-dependent and plays a role in the recruitment of AID, or whether it has a role in the repair mechanism [129].

CSR is also targeted to specific loci, known as the *"switch regions"*. The heavy chain changes its isotype in response to external cues, so that the BCRs and antibodies produced from daughter plasma cells can be used in the correct type of immune response. Consequently, different switch regions will be activated in response to distinct types of stimulation; for example, lipopolysaccharide (LPS) together with transforming growth factor β (TGFβ) induces switching to IgA, whereas LPS in combination with interleukin-4 (IL4) stimulates IgG1 and IgE switching [120]. The switch regions which become activated for recombination are marked epigenetically in a number of different ways. Sterile transcripts are seen at the specific acceptor S regions, which are activated for recombination [130–132], as well as a low level of antisense transcripts [133]. The RNA transcript itself – and not just the act of transcription – is seen to play an important role in the recombination reaction [130]. In addition, there appears to be a marked hyperacetylation of histones H3 and H4 at these acceptor regions [132,134–136]; the H4 acetylation mark appears to be AID-dependent, while the H3 acetylation is in place before AID acts on the locus [135]. Another activating mark present is H3K4me3, which is placed there by the MLL3–MLL4 complex, and is seen to be functionally important for CSR [137]. Interestingly, the deletion of a component of the MLL complex (PAX transcription activation domain interacting protein; PTIP) specifically lowers H3K4me3 levels at the switch regions, and results in defective class switching. Another result of PTIP deletion in B cells is higher genomic mutation rates, possibly through inefficient double-strand break repair [137]. One mark that is found at both donor and acceptor switch regions on CSR stimulation is, surprisingly, H3K9me3, which is normally considered to be a repressive histone mark [132, 134]. This mark is placed together with the acetylation and heightened transcription. The donor S region, which at the time of CSR activation is already transcribed as part of the heavy chain, is differentially marked from the upstream and downstream IgH regions by stalled RNA Pol II and a higher level of H3ac and H3K4me3 [138]. Levels of H4K20me1 (a repressive histone mark) are lowered at the donor switch region by signals which induce CSR [138]. In this way, both the donor and acceptor switch regions are made accessible and can be distinguished from the surrounding chromatin.

Specific S regions which are not supposed to be recombined under certain conditions do not gain these epigenetic marks [131, 134, 136, 137]. One mechanism which facilitates the correct selection of S regions is by a transcriptional repression of the S region [139]. Ikaros has been shown to bind certain switch promoters to repress germline transcription and activate histone modifications. When only low levels of Ikaros are present in the cell, isotypes that are normally repressed are chosen for recombination, regardless of the direction dictated by outside stimulation [139].

From these data, it appears that germline transcription and active histone marks play an important role in AID targeting for both SHM and CSR. However, many of these marks are present at active loci throughout the genome. Indeed, a recent study where AID-binding sites were mapped in IL4- and LPS-stimulated B cells showed that AID binds many hundreds of sites in addition to the *Ig* loci [140]. These sites are marked with active histone modifications and RNA polymerase transcription, along with RNA polymerase stalling, which may facilitate AID binding. However, high levels of mutation are still targeted specifically to the V and switch regions. These are the only regions which appear to bind phosphorylated (i.e., the active form) AID, and also bind replication protein A (RPA), which is a cofactor of AID activity. The mechanisms that specifically target the active form of AID, together with all of its cofactors, to the correct locations are, as yet, unresolved.

5
T Cells

The T cells play an extremely central role in the adaptive immune response. Each T cell expresses a T-cell receptor (TCR) which can bind an antigen peptide presented by a neighboring cell's major histocompatibility complex (MHC). The T cells can be divided into two main classes:

- The $CD4^+$ T cells, which interact with MHC class II complexes and serve as coordinators of the immune response, but do not themselves neutralize pathogenic elements.
- The $CD8^+$ T cells (also called *cytotoxic T lymphocytes*), which interact with MHC class I complexes and directly cause the death of cells that carry an antigen to which their TCRs can bind.

As opposed to other immune cells, most stages of T-cell development take place in the thymus, and not the bone marrow. Following activation, the T cells proliferate rapidly and differentiate into various effector cell types. Once the danger is cleared, however, these effector cells are no longer necessary and the number of T cells contracts, leaving a quiescent population of memory cells that respond more rapidly to challenges by their respective antigens than do the naïve cells from which they originated. At this point, the discussion will center on the epigenetic regulation of T-cell development, and cell identity from the thymus to the memory cell.

5.1
T-Cell Receptor Rearrangement

During their development in the thymus, T cells undergo many stages before they become mature, functioning T cells. At the first stage – the double-negative (DN) stage – the developing T cells (known as thymocytes) express neither CD4 nor CD8, which are functional markers of the helper and cytotoxic T cells, respectively. It is during this stage that the thymocytes begin to rearrange their TCR loci. The rearrangement concludes during the next developmental stage – known as the double-positive (DP) stage – when both CD4 and CD8 are expressed and the TCRs are tested for functionality (Fig. 4).

The mammalian genome contains three different TCR loci which can give rise to four types of TCR subunit chain (Fig. 4a). TCRβ and TCRγ each have a distinct locus, while TCRs α and δ are located in the same chromosomal area, and even share some of the same variable genes. TCRs β and δ are analogous to the IgH chain, in the fact that they undergo two stages of rearrangement – D to J and V to DJ – whereas TCRs α and γ are analogous to the Ig light chains, which have only a V to J rearrangement step. The TCRα protein makes heterodimers with TCRβ, and TCRγ with TCRδ, to give rise to functional TCRs. During T-cell development in the thymus, the TCR loci undergo rearrangement in a stage-specific manner. Typically, TCRs β, γ, and δ undergo a rearrangement during the DN stages, while TCRα is rearranged only in the DP stage, after a productive TCRβ rearrangement has been produced and tested with a preTCRα molecule. As the same *RAG* machinery is responsible for the rearrangement process in all of these loci, but they proceed at discrete time points (as with the *Ig* locus in B cells), it can be deduced that there must be a level of regulation beyond the sequence of the DNA which draws the RAG proteins to the antigen receptor loci at the correct time. In the *TCRα* locus, the problem is even more complex as it shares a locus with *TCRδ*, which can undergo rearrangement with some of the same *V* genes at an early stage, during which the *TCRα* rearrangement is somehow blocked.

As with the *Ig* loci during B-cell development, the *TCR* loci are seen to become accessible in a developmental, stage-specific manner. Such accessibility appears to be regulated mostly by the enhancers and promoters in the C and J regions. For instance, at the *TCRβ* locus (Fig. 4b) the Eβ enhancer interacts with the PD promoters located upstream of the two D regions, and brings about acetylation of the DJ region [141]. The stimulation of acetylation appears to be a key role of the enhancer, since treatment with an HDAC inhibitor can relieve a bock in *TCRβ* rearrangement caused by the deletion of Eβ [142]. An additional level of accessibility mediated by the enhancer is chromatin remodeling, brought about by the Swi/Snf complex, which is recruited by the enhancer and PD promoters, and is essential for *TCRβ* rearrangement [143]. Both, the enhancer [142] and promoters [144] are also mediators of CpG demethylation of the *TCRβ* locus (Fig. 4b) which, taken together, show that these regions are master regulators of TCR accessibility.

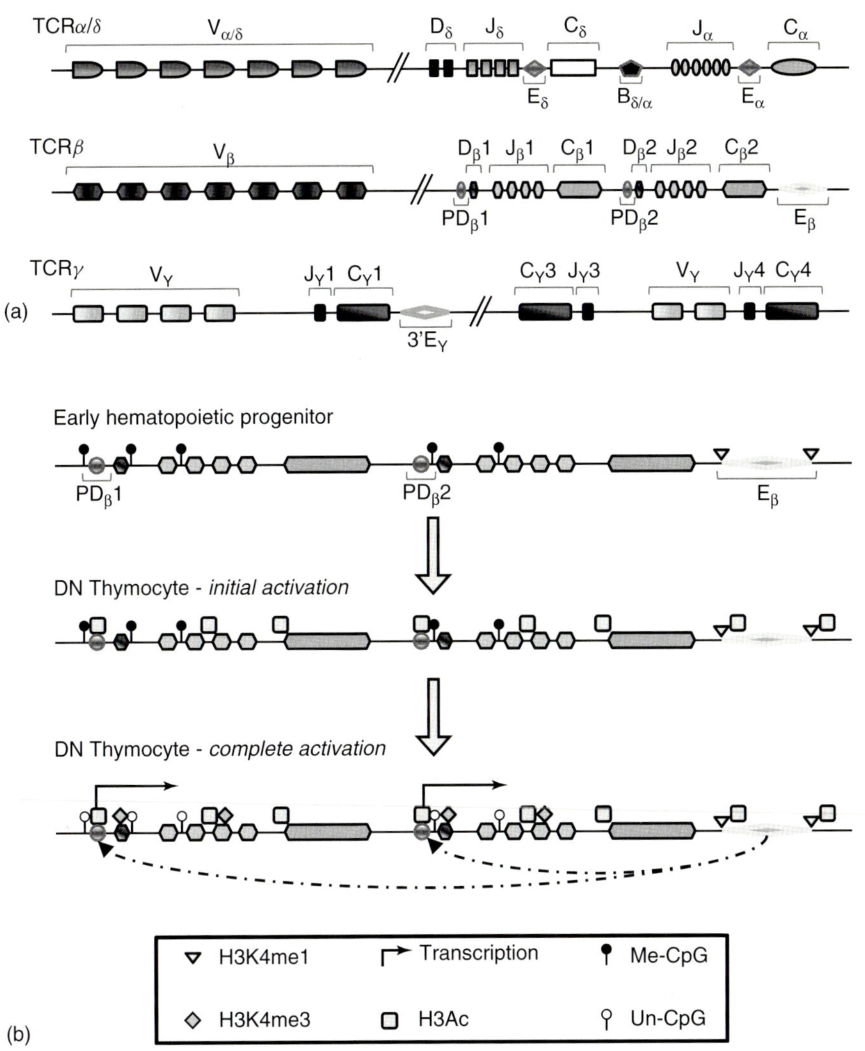

Fig. 4 *TCR* loci and model of chromatin accessibility at the *TCRβ* locus. (a) Schematic drawings of the murine *TCR* loci. The loci are not drawn to scale, and the amounts of V, D, and J segments do not represent actual number of times that these segments appear in the genome. E, enhancer element; B, enhancer-blocking element (homologous to human element, inactive in mice); PD, promoter of D region; (b) In early hematopoietic progenitors, the TCRβ *D-J-*C_β locus is in a closed, inaccessible conformation and the DNA is methylated. The intronic enhancer is marked with H3K4me1 prior to locus activation. In the DN thymocyte stage, the enhancer generates acetylated chromatin throughout the locus. This leads to activation of the PD1 and PD2, which interact physically with the enhancer (dashed lines), resulting in H3K4me3, germline transcription and demethylation of CpGs throughout the locus. Altogether, this sets the ground for rearrangement of the *TCRβ* locus.

The V region of the *TCRβ* locus appears to be regulated independently of the DJ regions [142], in keeping with the fact that it is rearranged at a different stage. E47, which is an isoform of E2A, recruits HATs to the V region and mediates V_β accessibility [145]. Following rearrangement, E47 binding is lost from the V_β region and the acetylation levels are decreased, which probably prevents further *TCRβ* rearrangements [145].

RAG-mediated restriction of the DNA has been seen to be blocked by nucleosomes *in vitro* [76]. The nucleosomes at the TCR J_α and J_β loci are highly positioned in thymocytes at stages where rearrangement does not occur, thus blocking the accessibility of the RAG machinery [29]. A significant movement and eviction of nucleosomes occurs at the stages when the *TCR* loci are rearranged. The remodeling at the *TCRα* locus is dependent on the elongation of a long germline transcript which originates from a promoter upstream of the J region [29]. Another important role of germline transcription at the *TCRα* locus is to coordinate the order of J recombination. In the *TCRα* locus, the J_α segments are chosen for recombination in a $5' \rightarrow 3'$ manner, whereas in the V region, proximal (closer to the J_α region) Vs are chosen over distal Vs. Germline transcription from the upstream J promoters activates these areas for rearrangement, while at the same time keeping the downstream promoters inaccessible [146, 147]. In this way, the order of rearrangement is maintained. If the initial *TCRα* rearrangement is nonproductive, then the downstream promoter will be activated, as the upstream transcript is no longer being produced, due to the rearrangement-dependent excision of the upstream promoter. In keeping with the ordered rearrangement of the J_α segments, the upstream J_α segments are marked with higher H3K4me3, and bind RAG2 more strongly than do the downstream segments [100]. On the flip side, the proximal V_α segments are marked with higher levels of acetylation and germline transcription than the distal segments, which correlates with these segments being preferentially chosen for rearrangement [148]. This once again shows how chromatin structure reflects the locus' defined order in the rearrangement process.

Although the *TCRδ* region shares the same locus as the *TCRα*, the rearrangement to the J_δ or J_α regions takes place at two distinct stages. An enhancer-blocking element was found between the C_δ and J_α region in human cells, which may be responsible for preventing the δ enhancer from making the α region accessible during the DN stage [149]. Rearrangements of the δ region regularly make use of upstream V segments which are usually not seen in α rearrangements. This is reflected by the level of acetylation on these distal V segments, which is high in DN cells but low in DP cells [148]. It is also reflected by the level of contraction of the $V_{\alpha/\delta}$ region, which provides a means for bringing the V segments into close proximity with the J segments [150, 151]. In the DN stage, the entire V region is contracted, making both distal and proximal V segments available for *TCRδ* rearrangement. In contrast, in the DP stage, only the proximal V region appears to be in a contracted conformation, while the distal V segments are decontracted, supporting the preference of proximal V segments in initial *TCRα* rearrangements [150]. The contraction of the V region appears to be a general mechanism of organizing the vast V region to be spatially available for rearrangement, as it is also seen at the *TCRβ* locus in the DN stage [151]. The contraction is reversible in all cases, and is alleviated in the stage following rearrangement [151], possibly as part of the feedback inhibition mechanism which prevents additional rearrangements once a functional rearrangement takes place.

Similarly to the *Ig* genes, the *TCR* genes are monoallelically expressed, with the loci being replicated asynchronously during the S phase [116]. In the DN stage of development – which is the stage when *TCRβ* undergoes rearrangement – either one [151] or both [152] of the *TCRβ* alleles are packaged in pericentromeric chromatin per cell. Only the allele that escapes the heterochromatin appears to provide a substrate for the RAG machinery [152], thus ensuring that no more than a single allele is rearranged and expressed in the mature T cell. Taken as a whole, the rearrangement of the TCR loci is subject to complex epigenetic regulation, which ensures the correct timing and location of the rearrangement process during T-cell development.

5.2 Developmental Regulation of *CD4* and *CD8* Expression

Following TCR rearrangement, the thymocytes in the DP stage undergo positive selection, whereas cells with TCRs that interact with the MHCs – and which may, therefore, be potentially functional for the immune response – are given "life" signals, and the remainder of the cells are allowed to die. Cells with TCRs that recognize MHC class I keep expressing *CD8* and then begin to repress *CD4*, whereas cells that recognize MHC class II repress *CD8* and maintain *CD4* expression. As soon as only one of the two cell-surface markers remains, the thymocytes enter the single positive (SP) stage. At this point, the cell will continue to express only one of the T-cell surface markers for the rest of its life, while the choice of *CD4* or *CD8* will be inherited by all of its daughter cells. During and after transition to the SP state, the cells undergo negative selection, where thymocytes that recognize self antigens are disposed of. Now, the cells are ready to exit the thymus as mature T cells.

The *CD4* and *CD8* genes are tightly regulated at different developmental stages. Typically, their expression pattern is transiently silenced in the DN stage, transiently activated in the DP stage, and finally permanently silenced or activated in the SP stage. This makes such genes particularly interesting from an epigenetic point of view. Indeed, it has been theorized that a permanent activation or repression is sustained in a primarily epigenetic manner, whereas any "temporary" silencing and/or activation is achieved in a manner that is more dependent on the direct action of transcription factors [153]. So far, research into the developmental regulation of these genes has supported this theory. For example, while the *CD8* locus has a large number of enhancers and silencers which effect transcription in a cell stage-specific manner [154], the *CD4* locus appears simpler, with one silencing element and two enhancers [155]. Deletion of the *CD4*-silencing element in the DP or $CD8^+$ SP thymocyte stages abolishes *CD4* repression [156]. However, deletion in mature $CD8^+$ T lymphocytes does not affect *CD4* silencing, which shows that the silencer is necessary for the establishment, but not for the propagation of the repressed *CD4* gene [156]. Similarly, deletion of the *CD4* enhancer in mature $CD4^+$ T cells did not lower the levels of *CD4* expression, which remained high even after many cell divisions, whereas deletion prior to the SP stage significantly lowered

CD4 expression [157]. This demonstrated, yet again, an epigenetic mode of maintaining the state of expression of the *CD4* gene once the expression pattern has been established. The *CD4* locus is marked with H3ac and H3K4me3 in DP and SP *CD4*-expressing cells [158, 159]. The acetylation marks are placed there by HATs, which are recruited by Mi-2β, a component of the NuRD complex [158]. Mi-2β is already in place at the DN stage, poising the gene for expression, but is antagonized by Ikaros [160] and Runx family transcription factors [159, 161], which repress *CD4* expression. Upon differentiation to the DP stage, Ikaros is evicted from the *CD4* locus [160], allowing *CD4* expression, but the Runx protein remains bound to the silencer, possibly poising the locus for the potential of silencing *CD4* expression should the cell differentiate to a $CD8^+$ SP cell [159]. Runx binding is lost in $CD4^+$ SP cells which are fully committed to *CD4* expression, whereas when the $CD8^+$ lineage is chosen, Runx remains bound to the silencer and the promoter gains repressive H3K9me2 and H3K27me3 histone marks. These repressive marks are absent in the DN stage, where the repression is temporary [159].

Although the many enhancers in the *CD8* locus make it more difficult to study, sufficient data exist regarding the epigenetic regulation of the locus. The Swi/Snf-like BAF (Brahma-related gene (BRG1)/Brahma (BRM)-associated factor) complex plays a role in creating $CD8^+$ SP cells, remodeling the chromatin so that the *CD8* locus is in an open and active conformation [162]. The Brg subunit of the BAF complex is specifically important for *CD8* activation. In addition, the Baf57 subunit of the BAF complex helps mediate silencing of the *CD4* locus, making the BAF complex into a key epigenetic regulator of the transition to $CD8^+$ SP cells [162]. A number of transcription factors have also been identified that activate *CD8* while silencing *CD4*, and which may play a role in recruiting epigenetic modifiers. For example, Runx proteins are activators of the *CD8* locus and have a clear role in repressing the *CD4* locus [159, 161, 163]. An additional transcription factor involved in this choice is the Myc-associated zinc-finger protein-related (MAZR) [164, 165]. This protein is involved in complex, stage-specific regulation of the *CD8* locus. At the DN stage, MAZR represses *CD8* by recruiting the NCoR complex to *CD8* enhancers [164]. However, at the DP stage, MAZR actually encourages *CD8* expression over *CD4* by repressing Th-pok, an activator of *CD4*, thus overriding the repression of *CD8* [165].

Both, the *CD4* and *CD8* genes alter their nuclear localization in cells where they are repressed, in comparison to cells where they are expressed [166, 167]. In DN cells, both the *CD4* and *CD8* loci reside in the cell periphery, in the pericentromeric heterochromatin [166]. However, on transition to the DP stage – where both loci are expressed – the *CD4* and *CD8* loci relocate to a more central location in the nucleus [166]. When a single coreceptor has been selected, and the cells differentiate into the SP stage, the coreceptor that is repressed moves back into the pericentromeric heterochromatin domain [166, 167]. An additional higher order dynamic of the coreceptor loci is the movement relative to its subchromosomal territory [168]. In cells where CD8 or CD4 are silent, the genes are found compacted with other genes from the same region on the chromosome. However, in cells where the coreceptors are expressed, the loci are seen to loop out from their chromosomal territories [168]. The regulatory elements of the *CD8* locus interact with each other, and appear to form a hub only in cells which are $CD8^+$, both DP and SP [168]. Overall, the expression patterns of CD4 and CD8 are regulated by epigenetic marks throughout T-cell development, with the histone marks becoming less plastic and more stable during the maturation process.

5.3 $CD4^+$ Cell Fates and Lineage Plasticity

The cells of the $CD4^+$ T-cell lineage can be seen as the master regulators of the adaptive immune response. Upon antigen recognition, these T cells do not directly neutralize the threat of a foreign body but, instead, secrete cytokines which recruit and coordinate the active response of innate and other portions of the adaptive immune system (cytotoxic T cells and B cells). Depending on the type of T-cell activation, naïve $CD4^+$ T cells can differentiate into a variety of $CD4^+$ subsets, leading to different modes of response from the body. There are four main subsets of $CD4^+$ T cells, each of which is characterized by a different set of cytokines and transcription factors that drive their expression [169]:

- T_H1 cells, which promote the cellular immune response mediated by $CD8^+$ T cells, are characterized by expression of the cytokine interferon (IFN)-γ, as well as the prominent transcription factor, T-bet.
- T_H2 cells, which promote a more antibody central response, are especially useful for protection against helminth infections. The response is characterized by expression of the cytokines IL4, IL5, and IL13, all of which are located in the same genomic locus, as well as the transcription factor GATA3.
- T_H17 cells, which were characterized more recently than the T_H1 and T_H2 subsets. These seem to contribute to the defense against extracellular pathogens, and are characterized by the IL17 family of cytokines and the transcription factor RORC (RAR-related orphan receptor gamma).
- T_{Reg} cells, which are involved in immunosuppression and tolerance of the antigens they recognize. They are central for averting autoimmune diseases, and are characterized by the transcription factor FOXP3 (forkhead box P3). T_{Reg} cells can be further divided into natural T_{Reg} (nT_{Reg}) cells, which are isolated directly from the thymus, and induced T_{Reg} (iT_{Reg}) cells, which are induced in the periphery or in cell culture from naïve cells. iT_{Reg} cells appear to have a less stable identity than nT_{Reg} cells, and undergo transdifferentation relatively easily.

Both, T_H1 and T_H2 cells seem to be highly polarized and are mutually exclusive. The cytokines that they secrete form a positive autocrine feedback loop, which induces an activation of the subtype's transcription factors, and transcription of the cytokines themselves. The transcription factors of T_H1 and T_H2 cells negatively regulate the cytokines of the opposite subtype (i.e., GATA3 represses IFN-γ, while T-bet represses the *IL4* locus), thus helping to preserve the differentiated identity [170–172]. However, while these transcription factors are important for initiating lineage commitment to a specific T_H subset, it has been shown that the inactivation of either T-bet in fully committed T_H1 cells, or of GATA3 in T_H2 cells, does not strongly impair silencing of the opposing

T_H program, though it does significantly lower the expression of the specific T_H subset genes [173, 174].

A large number of studies have concentrated on understanding the regulation of the *IFN-γ* and *IL4* loci in the T-cell subsets. These appear to be regulated on many epigenetic levels, starting with DNA methylation and histone modifications, and culminating with reorganization of the overlying chromatin architecture. Both, the *IFN-γ* [175] and *IL4* [176] loci have a complex pattern of CpG methylation in naïve $CD4^+$ T cells, with certain regulatory regions being methylated while others are demethylated. Upon T_H1 specification, the *IFN-γ* locus undergoes demethylation while *de novo* methylation occurs at the *IL4* locus [176]. When the naïve cells are directed toward T_H2 differentiation, the reverse process occurs, with the *IFN-γ* locus gaining methylation [177] and the *IL4* locus becoming globally hypomethylated [176, 178] (Fig. 5). This methylation plays a functional role in silencing the cytokine loci in the respective subsets. For example, naïve $CD4^+$ T cells lacking DNMT3a cannot methylate these regions upon differentiation, and are unable to maintain subset identity under polarizing conditions [179]. Similarly, $CD4^+$ T cells of the T_H1 and T_H2 subsets lacking DNMT1 [180, 181] or MBD2 [182, 183], which maintain and mediate cellular response to methylation, respectively, incorrectly express IL4 and IFN-γ (Fig. 5).

In T_{Reg} cells the *Foxp3* gene is also highly regulated by DNA methylation [184]. In naïve T cells, the *Foxp3* promoter and enhancers are methylated, and undergo demethylation during differentiation to the T_{Reg} subset (Fig. 5). These regions are completely demethylated in nT_{Reg} cells, whereas they maintain partial methylation in iT_{Reg} cells; this may explain part of the lineage stability of nT_{Reg} cells as opposed to iT_{Reg} cells. Recently, it has been shown that the maintenance of DNA methylation in naïve cells is at least partially achieved by the E3 ligase PIAS1, which binds the *FOXP3* promoter and recruits DNMTs and heterochromatin proteins to the locus. PIAS1 expression is lower in T_{Reg} cells, thus allowing the promoter to become demethylated [185].

In addition to DNA methylation, the modifications on the histones of T_H subset genes change substantially between the various subsets (Fig. 5). Upon differentiation from the naïve to the T_H subset, the *IFN-γ*, *IL4*, and *IL17* loci undergo extensive histone acetylation and H3K4 trimethylation in T_H1, T_H2, and T_H17 cells, respectively [186–190]. HDACs maintain the opposing cytokine locus in a hypoacetylated state, and treatment with HDAC inhibitors has been shown to activate *IL4* and *IFN-γ* expression in T_H1 and T_H2 cells, respectively [191]. Some mechanistic aspects of the deposition of these epigenetic marks have been revealed. For example, during T_H1 specification, T-bet removes Sin3a/HDAC complexes from the *IFN-γ* locus [192]. GATA3 has been shown to recruit the MLL H3K4 methyltransferase complex to the *IL4* locus [193], and this complex was shown to be essential for T_H2 effector function and memory [194].

The repressive histone mark H3K27me3 appears to play a key role in silencing the expression of genes from opposing subsets, and limiting transdifferentiation between the various $CD4^+$ lineages (Fig. 5). The *IL4* and *IFN-γ* loci are marked with H3K27me3 in most subsets where they are not expressed, such as naïve cells [190]. However, the H3K27me3 mark is missing from both cytokine loci in T_{Reg} cells, which may explain why T_{Reg}s can be converted into T_H1 [190] or T_H2 [195] cells under

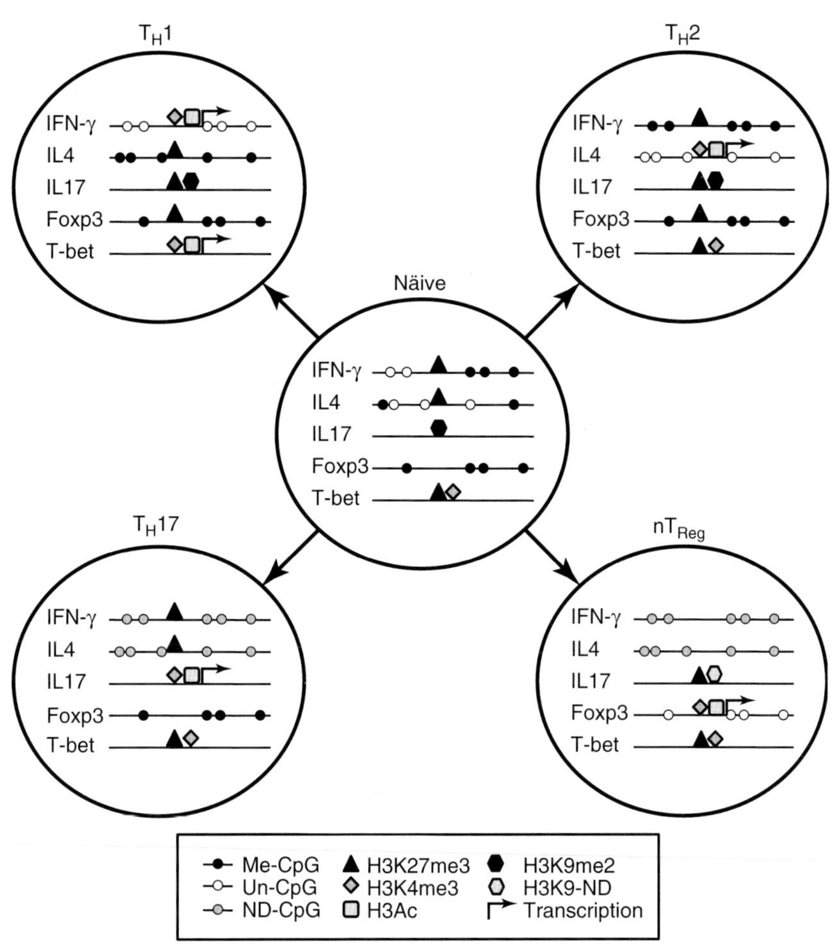

Fig. 5 Epigenetic markings on key CD4$^+$ T-cell genes in various CD4$^+$ subsets. Key cytokine and transcription factor loci are marked with different epigenetic marks in cells where the genes are transcribed or repressed. T-bet is representative of method of regulation in T$_H$ transcription factor loci. Sequences where CpG methylation has been reported to change between the different subsets are marked with black (methylated), white (unmethylated), or gray (unknown status) circles. Histone modifications are marked with various shapes. ND, not determined.

polarizing conditions. The transcription factors that regulate the fate of T$_H$1 and T$_H$2, such as T-bet and GATA3, are marked with bivalent chromatin in naïve cells, as well as in all of the CD4$^+$ T-cell lineages, except for those in which they are specifically expressed, where they retain only the H3K4me3 mark [190]. This suggests that the cells preserve a certain level of plasticity, possibly via their chromatin structure, which may allow them to change to an alternative fate under certain circumstances.

The T$_H$17 lineage appears to be an exception to this rule. The *IL17* locus is strongly marked with H3K27me3 in all CD4$^+$ lineages, aside from naïve and T$_H$17 cells (Fig. 5). RORC, a master transcriptional regulator of the T_H17 cell program is not bivalent, but instead is marked solely with H3K27me3 in all *ex vivo* subsets, aside from T_H17. This indicates that there is a greater epigenetic barrier for differentiating toward the T_H17 lineage than T_H1 or T_H2 from a specified $CD4^+$ subset [190]. The *IL17* locus is also marked with an additional repressive mark, H3K9me2, in non-T_H17 subsets. This epigenetic mark is put in place by the HMT G9a [196], an important developmental regulator [197, 198]. The absence of G9a drives incorrect *IL17* expression under conditions that usually would produce T_H1 or T_H2 cells [196].

An additional level of epigenetic regulation is seen in the higher-order architecture of the chromatin at the cytokine loci. In naïve $CD4^+$ T cells, prior to specification, the *IFN-γ* and *IL4* loci interact with each other, despite their being on different chromosomes [199]. This may facilitate the choosing of one and silencing of the other in T_H1 versus T_H2 differentiation. Upon subset specification, the *IFN-γ* and *IL4* loci are seen to contract and to form long-range looping in T_H1 [200] and T_H2 [201] cells, respectively. The looping is mediated by CTCF and cohesin, which link linearly distant regions close together [200, 202, 203]. CTCF and cohesin binding at the *IFN-γ* locus enhances transcription, and the tissue specificity of this process is mediated by T-bet [203]. CTCF binding is essential for correct transcription at the *IL4* locus, and the depletion of CTCF abrogates *IL4* locus expression, despite the fact that levels of GATA3, the activator of *IL4*, are not affected [202]. The looping is probably very central for gene activation, as the *IL4* locus is quite large, with three cytokine genes regulated in tandem by a single locus control region (LCR) [204]. The contraction of the locus brings the LCR into close proximity with the genes that it regulates [201]. The three-dimensional reorganization process is also assisted by the protein SATB1 (special AT-rich sequence binding protein 1), which appears to help coordinate the long-range interactions and recruit histone modifiers to the *IL4* locus [205]. Taken as a whole, it is possible to see that epigenetic regulation is an important factor in maintaining $CD4^+$ T-cell identity.

5.4 Epigenetic Basis for Memory in $CD8^+$ T Cells

One of the hallmarks of the adaptive immune system is the ability of its cells to remember previous challenges, and to react more swiftly and strongly to a repeated exposure. It does this by creating "memory" cells, which are quiescent until they come in contact with the specific antigen to which their receptors can bind. Upon antigen binding, these cells become quickly activated and differentiate into effector cells. Whilst the same process occurs in naïve cells, the response in memory cells is both more rapid and more potent. Until challenged with an antigen, naïve and memory cells may seem in many ways externally similar, as both are resting cells which do not express the effector genes for the immune response. There are, however, clear epigenetic differences between naïve and memory cells which may explain the differences in the quality and kinetics of the immune response. $CD8^+$ T cells make an interesting platform for studying these differences since, among the lymphoid lineages, they share the most outward similarities to their naïve ancestors, whereas $CD4^+$ T cells undergo further differentiation after activation (see Sect. 5.3) and B cells edit their genomic sequence in a manner which enhances the immune response (see Sect. 4.3).

One striking difference between naïve and memory $CD8^+$ T cells is that the general level of histone acetylation is higher in memory cells, which indicates a higher percentage of accessible genes in the memory cells [206]. This acetylation appears to be significant for the memory function of these cells. Cytokines from $CD4^+$ Th cells are necessary for marking effector genes such as *IFN-γ* with acetylation, which brings about the maturation of memory cells [207, 208]. The signals from the helper T cells can be replaced with treatment with an HDAC inhibitor [208]. In contrast, when acetylation is chemically inhibited, memory $CD8^+$ T cells are unable to begin the rapid transcription of effector genes, such as granzyme B [209].

DNA methylation is also a key player in memory cell formation and identity. $CD8^+$ T cells lacking DNMT1 are globally hypomethylated, and have impaired memory capabilities [210]. In addition, the knockout of MBD2 – which "reads" the DNA methylation and translates the epigenetic information onwards – causes a delay in memory cell formation and also impairs the ability of memory cells to overcome repeated infections [211]. While the presence of global DNA methylation is central for memory cells, the DNA hypomethylation at specific promoters seems to be important for memory cell activation. During the initial activation of naïve $CD8^+$ T cells, the *IL2* and *IFN-γ* promoters undergo demethylation and remain so in memory $CD8^+$ cells, which may contribute to the more rapid activation of these genes [207].

The global mapping of histone H3K4me3 and H3K27me3 modifications in memory, as opposed to naïve $CD8^+$, T cells demonstrates extensive epigenetic differences between these cell types [212]. Many genes are transcribed at a similarly low level in both naïve and resting memory cells, while their chromatin landscapes are completely different; these are genes that are activated in effector $CD8^+$ T cells. In naïve $CD8^+$ T cells these genes have repressive H3K27me3 histone marks at their promoters, whereas in memory cells they are marked with both H3K27me3 and H3K4me3, poising them for activation [212]. In this way, the bivalent genes can be more rapidly activated upon TCR stimulation and cell activation. Taken together, all of these epigenetic states contribute to the inheritable robust response of memory $CD8^+$ T cells.

6 Conclusions

The cells of the immune system are tightly regulated throughout their development, thus ensuring a correct immune response that can protect the host body without causing it harm. As discussed in this chapter, epigenetics plays a central role in the regulation of the immune system at all stages of development, thus ensuring correct differentiation and activity of the immune cells. It has been seen that the epigenetic landscape can confer cellular plasticity in progenitor cells on the one hand, and stable cell identity in fully differentiated cells on the other hand. Unfortunately, the mechanisms that direct these epigenetic identities are still not entirely clear, and remain the subject of intense research. Over the next few years, it is hoped that a better understanding of the mechanisms underlying epigenetic inheritance can be realized, and that this will further the present comprehension of the complex network of cells that constitute the immune system, perhaps enabling a better treatment of situations in which the system is misregulated.

References

1 Abbas, A.K., Janeway, C.A. Jr (2000) Immunology: improving on nature in the twenty-first century. *Cell*, **100**, 129–138.

2 Seita, J., Weissman, I.L. (2010) Hematopoietic stem cell: self-renewal versus differentiation. *Wiley Interdiscip. Rev. Syst. Biol. Med.*, **2**, 640–653.

3 Siegfried, Z., Cedar, H. (1997) DNA methylation: a molecular lock. *Curr. Biol.*, **7**, R305–R307.

4 Okano, M., Bell, D.W., Haber, D.A., Li, E. (1999) DNA methyltransferases Dnmt3a and Dnmt3b are essential for *de novo* methylation and mammalian development. *Cell*, **99**, 247–257.

5 Hermann, A., Goyal, R., Jeltsch, A. (2004) The Dnmt1 DNA-(cytosine-C5)-methyltransferase methylates DNA processively with high preference for hemimethylated target sites. *J. Biol. Chem.*, **279**, 48350–48359.

6 Laurent, L., Wong, E., Li, G., Huynh, T., Tsirigos, A., Ong, C.T., Low, H.M., Kin Sung, K.W., Rigoutsos, I., Loring, J., Wei, C.L. (2010) Dynamic changes in the human methylome during differentiation. *Genome Res.*, **20**, 320–331.

7 Kafri, T., Ariel, M., Brandeis, M., Shemer, R., Urven, L., McCarrey, J., Cedar, H., Razin, A. (1992) Developmental pattern of gene-specific DNA methylation in the mouse embryo and germ line. *Genes Dev.*, **6**, 705–714.

8 Bird, A., Taggart, M., Frommer, M., Miller, O.J., Macleod, D. (1985) A fraction of the mouse genome that is derived from islands of nonmethylated, CpG-rich DNA. *Cell*, **40**, 91–99.

9 Brandeis, M., Frank, D., Keshet, I., Siegfried, Z., Mendelsohn, M., Nemes, A., Temper, V., Razin, A., Cedar, H. (1994) Sp1 elements protect a CpG island from de novo methylation. *Nature*, **371**, 435–438.

10 Macleod, D., Charlton, J., Mullins, J., Bird, A.P. (1994) Sp1 sites in the mouse aprt gene promoter are required to prevent methylation of the CpG island. *Genes Dev.*, **8**, 2282–2292.

11 Meissner, A., Mikkelsen, T.S., Gu, H., Wernig, M., Hanna, J., Sivachenko, A., Zhang, X., Bernstein, B.E., Nusbaum, C., Jaffe, D.B., Gnirke, A., Jaenisch, R., Lander, E.S. (2008) Genome-scale DNA methylation maps of pluripotent and differentiated cells. *Nature*, **454**, 766–770.

12 Wu, S.C., Zhang, Y. (2010) Active DNA demethylation: many roads lead to Rome. *Nat. Rev. Mol. Cell Biol.*, **11**, 607–620.

13 Peterson, C.L., Laniel, M.A. (2004) Histones and histone modifications. *Curr. Biol.*, **14**, R546–R551.

14 Kouzarides, T. (2007) Chromatin modifications and their function. *Cell*, **128**, 693–705.

15 Bernstein, B.E., Mikkelsen, T.S., Xie, X., Kamal, M., Huebert, D.J., Cuff, J., Fry, B., Meissner, A., Wernig, M., Plath, K., Jaenisch, R., Wagschal, A., Feil, R., Schreiber, S.L., Lander, E.S. (2006) A bivalent chromatin structure marks key developmental genes in embryonic stem cells. *Cell*, **125**, 315–326.

16 Heintzman, N.D., Stuart, R.K., Hon, G., Fu, Y., Ching, C.W., Hawkins, R.D., Barrera, L.O., Van Calcar, S., Qu, C., Ching, K.A., Wang, W., Weng, Z., Green, R.D., Crawford, G.E., Ren, B. (2007) Distinct and predictive chromatin signatures of transcriptional promoters and enhancers in the human genome. *Nat. Genet.*, **39**, 311–318.

17 Barski, A., Cuddapah, S., Cui, K., Roh, T.Y., Schones, D.E., Wang, Z., Wei, G., Chepelev, I., Zhao, K. (2007) High-resolution profiling of histone methylations in the human genome. *Cell*, **129**, 823–837.

18 Bannister, A.J., Zegerman, P., Partridge, J.F., Miska, E.A., Thomas, J.O., Allshire, R.C., Kouzarides, T. (2001) Selective recognition of methylated lysine 9 on histone H3 by the HP1 chromo domain. *Nature*, **410**, 120–124.

19 Lachner, M., O'Carroll, D., Rea, S., Mechtler, K., Jenuwein, T. (2001) Methylation of histone H3 lysine 9 creates a binding site for HP1 proteins. *Nature*, **410**, 116–120.

20 Taverna, S.D., Li, H., Ruthenburg, A.J., Allis, C.D., Patel, D.J. (2007) How chromatin-binding modules interpret histone modifications: lessons from professional pocket pickers. *Nat. Struct. Mol. Biol.*, **14**, 1025–1040.

21 Wang, Z., Zang, C., Cui, K., Schones, D.E., Barski, A., Peng, W., Zhao, K. (2009) Genome-wide mapping of HATs

and HDACs reveals distinct functions in active and inactive genes. *Cell*, **138**, 1019–1031.

22 Jenuwein, T., Allis, C.D. (2001) Translating the histone code. *Science*, **293**, 1074–1080.

23 Probst, A.V., Dunleavy, E., Almouzni, G. (2009) Epigenetic inheritance during the cell cycle. *Nat. Rev. Mol. Cell Biol.*, **10**, 192–206.

24 Clapier, C.R., Cairns, B.R. (2009) The biology of chromatin remodeling complexes. *Annu. Rev. Biochem.*, **78**, 273–304.

25 Jhunjhunwala, S., van Zelm, M.C., Peak, M.M., Murre, C. (2009) Chromatin architecture and the generation of antigen receptor diversity. *Cell*, **138**, 435–448.

26 Filippova, G.N. (2008) Genetics and epigenetics of the multifunctional protein CTCF. *Curr. Top. Dev. Biol.*, **80**, 337–360.

27 Corcoran, A.E. (2010) The epigenetic role of non-coding RNA transcription and nuclear organization in immunoglobulin repertoire generation. *Semin. Immunol.*, **22**, 353–361.

28 Zhao, J., Sun, B.K., Erwin, J.A., Song, J.J., Lee, J.T. (2008) Polycomb proteins targeted by a short repeat RNA to the mouse X chromosome. *Science*, **322**, 750–756.

29 Kondilis-Mangum, H.D., Cobb, R.M., Osipovich, O., Srivatsan, S., Oltz, E.M., Krangel, M.S. (2010) Transcription-dependent mobilization of nucleosomes at accessible TCR gene segments *in vivo*. *J. Immunol.*, **184**, 6970–6977.

30 Goren, A., Cedar, H. (2003) Replicating by the clock. *Nat. Rev. Mol. Cell Biol.*, **4**, 25–32.

31 Tadokoro, Y., Ema, H., Okano, M., Li, E., Nakauchi, H. (2007) De novo DNA methyltransferase is essential for self-renewal, but not for differentiation, in hematopoietic stem cells. *J. Exp. Med.*, **204**, 715–722.

32 Broske, A.M., Vockentanz, L., Kharazi, S., Huska, M.R., Mancini, E., Scheller, M., Kuhl, C., Enns, A., Prinz, M., Jaenisch, R., Nerlov, C., Leutz, A., Andrade-Navarro, M.A., Jacobsen, S.E., Rosenbauer, F. (2009) DNA methylation protects hematopoietic stem cell multipotency from myeloerythroid restriction. *Nat. Genet.*, **41**, 1207–1215.

33 Trowbridge, J.J., Snow, J.W., Kim, J., Orkin, S.H. (2009) DNA methyltransferase 1 is essential for and uniquely regulates hematopoietic stem and progenitor cells. *Cell Stem Cell*, **5**, 442–449.

34 Iwama, A., Oguro, H., Negishi, M., Kato, Y., Morita, Y., Tsukui, H., Ema, H., Kamijo, T., Katoh-Fukui, Y., Koseki, H., van Lohuizen, M., Nakauchi, H. (2004) Enhanced self-renewal of hematopoietic stem cells mediated by the polycomb gene product Bmi-1. *Immunity*, **21**, 843–851.

35 Kamminga, L.M., Bystrykh, L.V., de Boer, A., Houwer, S., Douma, J., Weersing, E., Dontje, B., de Haan, G. (2006) The Polycomb group gene Ezh2 prevents hematopoietic stem cell exhaustion. *Blood*, **107**, 2170–2179.

36 Su, I.H., Basavaraj, A., Krutchinsky, A.N., Hobert, O., Ullrich, A., Chait, B.T., Tarakhovsky, A. (2003) Ezh2 controls B cell development through histone H3 methylation and Igh rearrangement. *Nat. Immunol.*, **4**, 124–131.

37 Majewski, I.J., Ritchie, M.E., Phipson, B., Corbin, J., Pakusch, M., Ebert, A., Busslinger, M., Koseki, H., Hu, Y., Smyth, G.K., Alexander, W.S., Hilton, D.J., Blewitt, M.E. (2010) Opposing roles of polycomb repressive complexes in hematopoietic stem and progenitor cells. *Blood*, **116**, 731–739.

38 Lessard, J., Schumacher, A., Thorsteinsdottir, U., van Lohuizen, M., Magnuson, T., Sauvageau, G. (1999) Functional antagonism of the Polycomb-Group genes eed and Bmi1 in hemopoietic cell proliferation. *Genes Dev.*, **13**, 2691–2703.

39 Majewski, I.J., Blewitt, M.E., de Graaf, C.A., McManus, E.J., Bahlo, M., Hilton, A.A., Hyland, C.D., Smyth, G.K., Corbin, J.E., Metcalf, D., Alexander, W.S., Hilton, D.J. (2008) Polycomb repressive complex 2 (PRC2) restricts hematopoietic stem cell activity. *PLoS Biol.*, **6**, e93.

40 Yoshida, T., Hazan, I., Zhang, J., Ng, S.Y., Naito, T., Snippert, H.J., Heller, E.J., Qi, X., Lawton, L.N., Williams, C.J., Georgopoulos, K. (2008) The role of the chromatin remodeler Mi-2beta in hematopoietic stem cell self-renewal and multilineage differentiation. *Genes Dev.*, **22**, 1174–1189.

41 Akashi, K., He, X., Chen, J., Iwasaki, H., Niu, C., Steenhard, B., Zhang, J., Haug, J., Li, L. (2003) Transcriptional accessibility for

genes of multiple tissues and hematopoietic lineages is hierarchically controlled during early hematopoiesis. *Blood*, **101**, 383–389.

42 Ng, S.Y., Yoshida, T., Zhang, J., Georgopoulos, K. (2009) Genome-wide lineage-specific transcriptional networks underscore Ikaros-dependent lymphoid priming in hematopoietic stem cells. *Immunity*, **30**, 493–507.

43 Hu, M., Krause, D., Greaves, M., Sharkis, S., Dexter, M., Heyworth, C., Enver, T. (1997) Multilineage gene expression precedes commitment in the hemopoietic system. *Genes Dev.*, **11**, 774–785.

44 Cui, K., Zang, C., Roh, T.Y., Schones, D.E., Childs, R.W., Peng, W., Zhao, K. (2009) Chromatin signatures in multipotent human hematopoietic stem cells indicate the fate of bivalent genes during differentiation. *Cell Stem Cell*, **4**, 80–93.

45 Weishaupt, H., Sigvardsson, M., Attema, J.L. (2010) Epigenetic chromatin states uniquely define the developmental plasticity of murine hematopoietic stem cells. *Blood*, **115**, 247–256.

46 Oguro, H., Yuan, J., Ichikawa, H., Ikawa, T., Yamazaki, S., Kawamoto, H., Nakauchi, H., Iwama, A. (2010) Poised lineage specification in multipotential hematopoietic stem and progenitor cells by the polycomb protein Bmi1. *Cell Stem Cell*, **6**, 279–286.

47 Hansen, K.H., Bracken, A.P., Pasini, D., Dietrich, N., Gehani, S.S., Monrad, A., Rappsilber, J., Lerdrup, M., Helin, K. (2008) A model for transmission of the H3K27me3 epigenetic mark. *Nat. Cell Biol.*, **10**, 1291–1300.

48 Margueron, R., Justin, N., Ohno, K., Sharpe, M.L., Son, J., Drury, W.J., III, Voigt, P., Martin, S.R., Taylor, W.R., De Marco, V., Pirrotta, V., Reinberg, D., Gamblin, S.J. (2009) Role of the polycomb protein EED in the propagation of repressive histone marks. *Nature*, **461**, 762–767.

49 Orford, K., Kharchenko, P., Lai, W., Dao, M.C., Worhunsky, D.J., Ferro, A., Janzen, V., Park, P.J., Scadden, D.T. (2008) Differential H3K4 methylation identifies developmentally poised hematopoietic genes. *Dev. Cell*, **14**, 798–809.

50 Maes, J., Maleszewska, M., Guillemin, C., Pflumio, F., Six, E., Andre-Schmutz, I., Cavazzana-Calvo, M., Charron, D., Francastel, C., Goodhardt, M. (2008) Lymphoid-affiliated genes are associated with active histone modifications in human hematopoietic stem cells. *Blood*, **112**, 2722–2729.

51 Attema, J.L., Papathanasiou, P., Forsberg, E.C., Xu, J., Smale, S.T., Weissman, I.L. (2007) Epigenetic characterization of hematopoietic stem cell differentiation using miniChIP and bisulfite sequencing analysis. *Proc. Natl Acad. Sci. USA*, **104**, 12371–12376.

52 Ji, H., Ehrlich, L.I., Seita, J., Murakami, P., Doi, A., Lindau, P., Lee, H., Aryee, M.J., Irizarry, R.A., Kim, K., Rossi, D.J., Inlay, M.A., Serwold, T., Karsunky, H., Ho, L., Daley, G.Q., Weissman, I.L., Feinberg, A.P. (2010) Comprehensive methylome map of lineage commitment from haematopoietic progenitors. *Nature*, **467**, 338–342.

53 Borgel, J., Guibert, S., Li, Y., Chiba, H., Schubeler, D., Sasaki, H., Forne, T., Weber, M. (2010) Targets and dynamics of promoter DNA methylation during early mouse development. *Nat. Genet.*, **42**, 1093–1100.

54 Walter, K., Bonifer, C., Tagoh, H. (2008) Stem cell-specific epigenetic priming and B cell-specific transcriptional activation at the mouse Cd19 locus. *Blood*, **112**, 1673–1682.

55 Xu, J., Pope, S.D., Jazirehi, A.R., Attema, J.L., Papathanasiou, P., Watts, J.A., Zaret, K.S., Weissman, I.L., Smale, S.T. (2007) Pioneer factor interactions and unmethylated CpG dinucleotides mark silent tissue-specific enhancers in embryonic stem cells. *Proc. Natl Acad. Sci. USA*, **104**, 12377–12382.

56 Xu, J., Watts, J.A., Pope, S.D., Gadue, P., Kamps, M., Plath, K., Zaret, K.S., Smale, S.T. (2009) Transcriptional competence and the active marking of tissue-specific enhancers by defined transcription factors in embryonic and induced pluripotent stem cells. *Genes Dev.*, **23**, 2824–2838.

57 Szutorisz, H., Canzonetta, C., Georgiou, A., Chow, C.M., Tora, L., Dillon, N. (2005) Formation of an active tissue-specific chromatin domain initiated by epigenetic marking at the embryonic stem cell stage. *Mol. Cell. Biol.*, **25**, 1804–1820.

58 Hardy, R.R., Hayakawa, K. (2001) B cell development pathways. *Annu. Rev. Immunol.*, **19**, 595–621.

59 McHeyzer-Williams, L.J., McHeyzer-Williams, M.G. (2005) Antigen-specific memory B cell development. *Annu. Rev. Immunol.*, **23**, 487–513.

60 Bryder, D., Sigvardsson, M. (2010) Shaping up a lineage--lessons from B lymphopoiesis. *Curr. Opin. Immunol.*, **22**, 148–153.

61 Ramirez, J., Lukin, K., Hagman, J. (2010) From hematopoietic progenitors to B cells: mechanisms of lineage restriction and commitment. *Curr. Opin. Immunol.*, **22**, 177–184.

62 Ng, S.Y., Yoshida, T., Georgopoulos, K. (2007) Ikaros and chromatin regulation in early hematopoiesis. *Curr. Opin. Immunol.*, **19**, 116–122.

63 Thompson, E.C., Cobb, B.S., Sabbattini, P., Meixlsperger, S., Parelho, V., Liberg, D., Taylor, B., Dillon, N., Georgopoulos, K., Jumaa, H., Smale, S.T., Fisher, A.G., Merkenschlager, M. (2007) Ikaros DNA-binding proteins as integral components of B cell developmental-stage-specific regulatory circuits. *Immunity*, **26**, 335–344.

64 Beck, K., Peak, M.M., Ota, T., Nemazee, D., Murre, C. (2009) Distinct roles for E12 and E47 in B cell specification and the sequential rearrangement of immunoglobulin light chain loci. *J. Exp. Med.*, **206**, 2271–2284.

65 Lin, Y.C., Jhunjhunwala, S., Benner, C., Heinz, S., Welinder, E., Mansson, R., Sigvardsson, M., Hagman, J., Espinoza, C.A., Dutkowski, J., Ideker, T., Glass, C.K., Murre, C. (2010) A global network of transcription factors, involving E2A, EBF1 and Foxo1, that orchestrates B cell fate. *Nat. Immunol.*, **11**, 635–643.

66 Treiber, T., Mandel, E.M., Pott, S., Gyory, I., Firner, S., Liu, E.T., Grosschedl, R. (2010) Early B cell factor 1 regulates B cell gene networks by activation, repression, and transcription- independent poising of chromatin. *Immunity*, **32**, 714–725.

67 Cobaleda, C., Schebesta, A., Delogu, A., Busslinger, M. (2007) Pax5: the guardian of B cell identity and function. *Nat. Immunol.*, **8**, 463–470.

68 Mikkola, I., Heavey, B., Horcher, M., Busslinger, M. (2002) Reversion of B cell commitment upon loss of Pax5 expression. *Science*, **297**, 110–113.

69 Decker, T., Pasca di Magliano, M., McManus, S., Sun, Q., Bonifer, C., Tagoh, H., Busslinger, M. (2009) Stepwise activation of enhancer and promoter regions of the B cell commitment gene Pax5 in early lymphopoiesis. *Immunity*, **30**, 508–520.

70 Schebesta, A., McManus, S., Salvagiotto, G., Delogu, A., Busslinger, G.A., Busslinger, M. (2007) Transcription factor Pax5 activates the chromatin of key genes involved in B cell signaling, adhesion, migration, and immune function. *Immunity*, **27**, 49–63.

71 Gao, H., Lukin, K., Ramirez, J., Fields, S., Lopez, D., Hagman, J. (2009) Opposing effects of SWI/SNF and Mi-2/NuRD chromatin remodeling complexes on epigenetic reprogramming by EBF and Pax5. *Proc. Natl Acad. Sci. USA*, **106**, 11258–11263.

72 Maier, H., Ostraat, R., Gao, H., Fields, S., Shinton, S.A., Medina, K.L., Ikawa, T., Murre, C., Singh, H., Hardy, R.R., Hagman, J. (2004) Early B cell factor cooperates with Runx1 and mediates epigenetic changes associated with mb-1 transcription. *Nat. Immunol.*, **5**, 1069–1077.

73 Walter, K., Cockerill, P.N., Barlow, R., Clarke, D., Hoogenkamp, M., Follows, G.A., Richards, S.J., Cullen, M.J., Bonifer, C., Tagoh, H. (2010) Aberrant expression of CD19 in AML with t(8;21) involves a poised chromatin structure and PAX5. *Oncogene*, **29**, 2927–2937.

74 Tagoh, H., Ingram, R., Wilson, N., Salvagiotto, G., Warren, A.J., Clarke, D., Busslinger, M., Bonifer, C. (2006) The mechanism of repression of the myeloid-specific c-fms gene by Pax5 during B lineage restriction. *EMBO J.*, **25**, 1070–1080.

75 Tagoh, H., Schebesta, A., Lefevre, P., Wilson, N., Hume, D., Busslinger, M., Bonifer, C. (2004) Epigenetic silencing of the c-fms locus during B-lymphopoiesis occurs in discrete steps and is reversible. *EMBO J.*, **23**, 4275–4285.

76 Golding, A., Chandler, S., Ballestar, E., Wolffe, A.P., Schlissel, M.S. (1999) Nucleosome structure completely inhibits in vitro cleavage by the V(D)J recombinase. *EMBO J.*, **18**, 3712–3723.

77 Stanhope-Baker, P., Hudson, K.M., Shaffer, A.L., Constantinescu, A., Schlissel, M.S. (1996) Cell type-specific chromatin structure determines the targeting of V(D)J recombinase activity in vitro. *Cell*, **85**, 887–897.

78 Xu, C.R., Feeney, A.J. (2009) The epigenetic profile of Ig genes is dynamically regulated during B cell differentiation and is modulated by pre-B cell receptor signaling. *J. Immunol.*, **182**, 1362–1369.

79 Johnson, K., Pflugh, D.L., Yu, D., Hesslein, D.G., Lin, K.I., Bothwell, A.L., Thomas-Tikhonenko, A., Schatz, D.G., Calame, K. (2004) B cell-specific loss of histone 3 lysine 9 methylation in the V(H) locus depends on Pax5. *Nat. Immunol.*, **5**, 853–861.

80 Mostoslavsky, R., Singh, N., Kirillov, A., Pelanda, R., Cedar, H., Chess, A., Bergman, Y. (1998) Kappa chain monoallelic demethylation and the establishment of allelic exclusion. *Genes. Dev.*, **12**, 1801–1811.

81 Mostoslavsky, R., Kirillov, A., Ji, Y.H., Goldmit, M., Holzmann, M., Wirth, T., Cedar, H., Bergman, Y. (1999) Demethylation and the establishment of kappa allelic exclusion. *Cold Spring Harbor Symp. Quant. Biol.*, **64**, 197–206.

82 Goldmit, M., Ji, Y., Skok, J., Roldan, E., Jung, S., Cedar, H., Bergman, Y. (2005) Epigenetic ontogeny of the Igk locus during B cell development. *Nat. Immunol.*, **6**, 198–203.

83 Fitzsimmons, S.P., Bernstein, R.M., Max, E.E., Skok, J.A., Shapiro, M.A. (2007) Dynamic changes in accessibility, nuclear positioning, recombination, and transcription at the Ig kappa locus. *J. Immunol.*, **179**, 5264–5273.

84 Osipovich, O.A., Subrahmanyam, R., Pierce, S., Sen, R., Oltz, E.M. (2009) Cutting edge: SWI/SNF mediates antisense Igh transcription and locus-wide accessibility in B cell precursors. *J. Immunol.*, **183**, 1509–1513.

85 Chowdhury, D., Sen, R. (2001) Stepwise activation of the immunoglobulin mu heavy chain gene locus. *EMBO J.*, **20**, 6394–6403.

86 Maes, J., O'Neill, L.P., Cavelier, P., Turner, B.M., Rougeon, F., Goodhardt, M. (2001) Chromatin remodeling at the Ig loci prior to V(D)J recombination. *J. Immunol.*, **167**, 866–874.

87 Morshead, K.B., Ciccone, D.N., Taverna, S.D., Allis, C.D., Oettinger, M.A. (2003) Antigen receptor loci poised for V(D)J rearrangement are broadly associated with BRG1 and flanked by peaks of histone H3 dimethylated at lysine 4. *Proc. Natl Acad. Sci. USA*, **100**, 11577–11582.

88 Maes, J., Chappaz, S., Cavelier, P., O'Neill, L., Turner, B., Rougeon, F., Goodhardt, M. (2006) Activation of V(D)J recombination at the IgH chain J_H locus occurs within a 6-kilobase chromatin domain and is associated with nucleosomal remodeling. *J. Immunol.*, **176**, 5409–5417.

89 Hesslein, D.G., Pflugh, D.L., Chowdhury, D., Bothwell, A.L., Sen, R., Schatz, D.G. (2003) Pax5 is required for recombination of transcribed, acetylated, 5' IgH V gene segments. *Genes Dev.*, **17**, 37–42.

90 Espinoza, C.R., Feeney, A.J. (2005) The extent of histone acetylation correlates with the differential rearrangement frequency of individual V_H genes in pro-B cells. *J. Immunol.*, **175**, 6668–6675.

91 Chakraborty, T., Chowdhury, D., Keyes, A., Jani, A., Subrahmanyam, R., Ivanova, I., Sen, R. (2007) Repeat organization and epigenetic regulation of the D_H-Cmu domain of the immunoglobulin heavy-chain gene locus. *Mol. Cell*, **27**, 842–850.

92 Yancopoulos, G.D., Alt, F.W. (1985) Developmentally controlled and tissue-specific expression of unrearranged V_H gene segments. *Cell*, **40**, 271–281.

93 Thompson, A., Timmers, E., Schuurman, R.K., Hendriks, R.W. (1995) Immunoglobulin heavy chain germ-line J_H-C mu transcription in human precursor B lymphocytes initiates in a unique region upstream of DQ52. *Eur. J. Immunol.*, **25**, 257–261.

94 Corcoran, A.E., Riddell, A., Krooshoop, D., Venkitaraman, A.R. (1998) Impaired immunoglobulin gene rearrangement in mice lacking the IL7 receptor. *Nature*, **391**, 904–907.

95 Martin, D.J., van Ness, B.G. (1990) Initiation and processing of two kappa immunoglobulin germ line transcripts in mouse B cells. *Mol. Cell. Biol.*, **10**, 1950–1958.

96 Singh, N., Bergman, Y., Cedar, H., Chess, A. (2003) Biallelic germline transcription at

the kappa immunoglobulin locus. *J. Exp. Med.*, **197**, 743–750.

97 Bolland, D.J., Wood, A.L., Johnston, C.M., Bunting, S.F., Morgan, G., Chakalova, L., Fraser, P.J., Corcoran, A.E. (2004) Antisense intergenic transcription in V(D)J recombination. *Nat. Immunol.*, **5**, 630–637.

98 Bolland, D.J., Wood, A.L., Afshar, R., Featherstone, K., Oltz, E.M., Corcoran, A.E. (2007) Antisense intergenic transcription precedes Igh D-to-J recombination and is controlled by the intronic enhancer Emu. *Mol. Cell. Biol.*, **27**, 5523–5533.

99 Featherstone, K., Wood, A.L., Bowen, A.J., Corcoran, A.E. (2010) The mouse immunoglobulin heavy chain V-D intergenic sequence contains insulators that may regulate ordered V(D)J recombination. *J. Biol. Chem.*, **285**, 9327–9338.

100 Ji, Y., Resch, W., Corbett, E., Yamane, A., Casellas, R., Schatz, D.G. (2010) The *in vivo* pattern of binding of RAG1 and RAG2 to antigen receptor loci. *Cell*, **141**, 419–431.

101 Liu, Y., Subrahmanyam, R., Chakraborty, T., Sen, R., Desiderio, S. (2007) A plant homeodomain in RAG-2 that binds Hypermethylated lysine 4 of histone H3 is necessary for efficient antigen-receptor-gene rearrangement. *Immunity*, **27**, 561–571.

102 Matthews, A.G., Kuo, A.J., Ramon- Maiques, S., Han, S., Champagne, K.S., Ivanov, D., Gallardo, M., Carney, D., Cheung, P., Ciccone, D.N., Walter, K.L., Utz, P.J., Shi, Y., Kutateladze, T.G., Yang, W., Gozani, O., Oettinger, M.A. (2007) RAG2 PHD finger couples histone H3 lysine 4 trimethylation with V(D)J recombination. *Nature*, **450**, 1106–1110.

103 Shimazaki, N., Tsai, A.G., Lieber, M.R. (2009) H3K4me3 stimulates the V(D)J RAG complex for both nicking and hairpinning in trans in addition to tethering in cis: implications for translocations. *Mol. Cell*, **34**, 535–544.

104 Kosak, S.T., Skok, J.A., Medina, K.L., Riblet, R., Le Beau, M.M., Fisher, A.G., Singh, H. (2002) Subnuclear compartmentalization of immunoglobulin loci during lymphocyte development. *Science*, **296**, 158–162.

105 Fuxa, M., Skok, J., Souabni, A., Salvagiotto, G., Roldan, E., Busslinger, M. (2004) Pax5 induces V-to-DJ rearrangements and locus contraction of the immunoglobulin heavy-chain gene. *Genes Dev.*, **18**, 411–422.

106 Sayegh, C.E., Jhunjhunwala, S., Riblet, R., Murre, C. (2005) Visualization of looping involving the immunoglobulin heavy-chain locus in developing B cells. *Genes Dev.*, **19**, 322–327.

107 Jhunjhunwala, S., van Zelm, M.C., Peak, M.M., Cutchin, S., Riblet, R., van Dongen, J.J., Grosveld, F.G., Knoch, T.A., Murre, C. (2008) The 3D structure of the immunoglobulin heavy-chain locus: implications for long-range genomic interactions. *Cell*, **133**, 265–279.

108 Degner, S.C., Wong, T.P., Jankevicius, G., Feeney, A.J. (2009) Cutting edge: developmental stage-specific recruitment of cohesin to CTCF sites throughout immunoglobulin loci during B lymphocyte development. *J. Immunol.*, **182**, 44–48.

109 Garrett, F.E., Emelyanov, A.V., Sepulveda, M.A., Flanagan, P., Volpi, S., Li, F., Loukinov, D., Eckhardt, L.A., Lobanenkov, V.V., Birshtein, B.K. (2005) Chromatin architecture near a potential 3' end of the igh locus involves modular regulation of histone modifications during B-Cell development and in vivo occupancy at CTCF sites. *Mol. Cell. Biol.*, **25**, 1511–1525.

110 Liu, H., Schmidt-Supprian, M., Shi, Y., Hobeika, E., Barteneva, N., Jumaa, H., Pelanda, R., Reth, M., Skok, J., Rajewsky, K., Shi, Y. (2007) Yin Yang 1 is a critical regulator of B-cell development. *Genes Dev.*, **21**, 1179–1189.

111 Reynaud, D., Demarco, I.A., Reddy, K.L., Schjerven, H., Bertolino, E., Chen, Z., Smale, S.T., Winandy, S., Singh, H. (2008) Regulation of B cell fate commitment and immunoglobulin heavy-chain gene rearrangements by Ikaros. *Nat. Immunol.*, **9**, 927–936.

112 Xu, C.R., Schaffer, L., Head, S.R., Feeney, A.J. (2008) Reciprocal patterns of methylation of H3K36 and H3K27 on proximal vs. distal IgV_H genes are modulated by IL7 and Pax5. *Proc. Natl Acad. Sci. USA*, **105**, 8685–8690.

113 Skok, J.A., Brown, K.E., Azuara, V., Caparros, M.L., Baxter, J., Takacs, K., Dillon, N., Gray, D., Perry, R.P., Merkenschlager, M., Fisher, A.G. (2001) Nonequivalent nuclear location of immunoglobulin alleles in B lymphocytes. *Nat. Immunol.*, **2**, 848–854.

114 Goldmit, M., Schlissel, M., Cedar, H., Bergman, Y. (2002) Differential accessibility at the kappa chain locus plays a role in allelic exclusion. *EMBO J.*, **21**, 5255–5261.
115 Hewitt, S.L., Farmer, D., Marszalek, K., Cadera, E., Liang, H.E., Xu, Y., Schlissel, M.S., Skok, J.A. (2008) Association between the Igk and Igh immunoglobulin loci mediated by the 3' Igk enhancer induces 'decontraction' of the Igh locus in pre-B cells. *Nat. Immunol.*, **9**, 396–404.
116 Mostoslavsky, R., Singh, N., Tenzen, T., Goldmit, M., Gabay, C., Elizur, S., Qi, P., Reubinoff, B.E., Chess, A., Cedar, H., Bergman, Y. (2001) Asynchronous replication and allelic exclusion in the immune system. *Nature*, **414**, 221–225.
117 Dutta, D., Ensminger, A.W., Zucker, J.P., Chess, A. (2009) Asynchronous replication and autosome-pair non-equivalence in human embryonic stem cells. *PLoS ONE*, **4**, e4970.
118 Roldan, E., Fuxa, M., Chong, W., Martinez, D., Novatchkova, M., Busslinger, M., Skok, J.A. (2005) Locus 'decontraction' and centromeric recruitment contribute to allelic exclusion of the immunoglobulin heavy-chain gene. *Nat. Immunol.*, **6**, 31–41.
119 Li, Z., Luo, Z., Ronai, D., Kuang, F.L., Peled, J.U., Iglesias-Ussel, M.D., Scharff, M.D. (2007) Targeting AID to the Ig genes. *Adv. Exp. Med. Biol.*, **596**, 93–109.
120 Stavnezer, J. (1996) Immunoglobulin class switching. *Curr. Opin. Immunol.*, **8**, 199–205.
121 Perez-Duran, P., de Yebenes, V.G., Ramiro, A.R. (2007) Oncogenic events triggered by AID, the adverse effect of antibody diversification. *Carcinogenesis*, **28**, 2427–2433.
122 Woo, C.J., Martin, A., Scharff, M.D. (2003) Induction of somatic hypermutation is associated with modifications in immunoglobulin variable region chromatin. *Immunity*, **19**, 479–489.
123 Peters, A., Storb, U. (1996) Somatic hypermutation of immunoglobulin genes is linked to transcription initiation. *Immunity*, **4**, 57–65.
124 Fukita, Y., Jacobs, H., Rajewsky, K. (1998) Somatic hypermutation in the heavy chain locus correlates with transcription. *Immunity*, **9**, 105–114.
125 Shen, H.M., Poirier, M.G., Allen, M.J., North, J., Lal, R., Widom, J., Storb, U. (2009) The activation-induced cytidine deaminase (AID) efficiently targets DNA in nucleosomes but only during transcription. *J. Exp. Med.*, **206**, 1057–1071.
126 Fraenkel, S., Mostoslavsky, R., Novobrantseva, T.I., Pelanda, R., Chaudhuri, J., Esposito, G., Jung, S., Alt, F.W., Rajewsky, K., Cedar, H., Bergman, Y. (2007) Allelic 'choice' governs somatic hypermutation in vivo at the immunoglobulin kappa-chain locus. *Nat. Immunol.*, **8**, 715–722.
127 Larijani, M., Frieder, D., Sonbuchner, T.M., Bransteitter, R., Goodman, M.F., Bouhassira, E.E., Scharff, M.D., Martin, A. (2005) Methylation protects cytidines from AID-mediated deamination. *Mol. Immunol.*, **42**, 599–604.
128 Odegard, V.H., Kim, S.T., Anderson, S.M., Shlomchik, M.J., Schatz, D.G. (2005) Histone modifications associated with somatic hypermutation. *Immunity*, **23**, 101–110.
129 Borchert, G.M., Holton, N.W., Edwards, K.A., Vogel, L.A., Larson, E.D. (2010) Histone H2A and H2B are monoubiquitinated at AID-targeted loci. *PLoS ONE*, **5**, e11641.
130 Lorenz, M., Jung, S., Radbruch, A. (1995) Switch transcripts in immunoglobulin class switching. *Science*, **267**, 1825–1828.
131 Nambu, Y., Sugai, M., Gonda, H., Lee, C.G., Katakai, T., Agata, Y., Yokota, Y., Shimizu, A. (2003) Transcription-coupled events associating with immunoglobulin switch region chromatin. *Science*, **302**, 2137–2140.
132 Chowdhury, M., Forouhi, O., Dayal, S., McCloskey, N., Gould, H.J., Felsenfeld, G., Fear, D.J. (2008) Analysis of intergenic transcription and histone modification across the human immunoglobulin heavy-chain locus. *Proc. Natl Acad. Sci. USA*, **105**, 15872–15877.
133 Perlot, T., Li, G., Alt, F.W. (2008) Antisense transcripts from immunoglobulin heavy-chain locus V(D)J and switch regions. *Proc. Natl Acad. Sci. USA*, **105**, 3843–3848.
134 Kuang, F.L., Luo, Z., Scharff, M.D. (2009) H3 trimethyl K9 and H3 acetyl K9 chromatin modifications are associated with class switch recombination. *Proc. Natl Acad. Sci. USA*, **106**, 5288–5293.

135 Wang, L., Whang, N., Wuerffel, R., Kenter, A.L. (2006) AID-dependent histone acetylation is detected in immunoglobulin S regions. *J. Exp. Med.*, **203**, 215–226.

136 Li, Z., Luo, Z., Scharff, M.D. (2004) Differential regulation of histone acetylation and generation of mutations in switch regions is associated with Ig class switching. *Proc. Natl Acad. Sci. USA*, **101**, 15428–15433.

137 Daniel, J.A., Santos, M.A., Wang, Z., Zang, C., Schwab, K.R., Jankovic, M., Filsuf, D., Chen, H.T., Gazumyan, A., Yamane, A., Cho, Y.W., Sun, H.W., Ge, K., Peng, W., Nussenzweig, M.C., Casellas, R., Dressler, G.R., Zhao, K., Nussenzweig, A. (2010) PTIP promotes chromatin changes critical for immunoglobulin class switch recombination. *Science*, **329**, 917–923.

138 Wang, L., Wuerffel, R., Feldman, S., Khamlichi, A.A., Kenter, A.L. (2009) S region sequence, RNA polymerase II, and histone modifications create chromatin accessibility during class switch recombination. *J. Exp. Med.*, **206**, 1817–1830.

139 Sellars, M., Reina-San-Martin, B., Kastner, P., Chan, S. (2009) Ikaros controls isotype selection during immunoglobulin class switch recombination. *J. Exp. Med.*, **206**, 1073–1087.

140 Yamane, A., Resch, W., Kuo, N., Kuchen, S., Li, Z., Sun, H.W., Robbiani, D.F., McBride, K., Nussenzweig, M.C., Casellas, R. (2010) Deep-sequencing identification of the genomic targets of the cytidine deaminase AID and its cofactor RPA in B lymphocytes. *Nat. Immunol.*, **12**, 62–69.

141 Oestreich, K.J., Cobb, R.M., Pierce, S., Chen, J., Ferrier, P., Oltz, E.M. (2006) Regulation of TCRbeta gene assembly by a promoter/enhancer holocomplex. *Immunity*, **24**, 381–391.

142 Mathieu, N., Hempel, W.M., Spicuglia, S., Verthuy, C., Ferrier, P. (2000) Chromatin remodeling by the T cell receptor (TCR)-beta gene enhancer during early T cell development: implications for the control of TCR-beta locus recombination. *J. Exp. Med.*, **192**, 625–636.

143 Osipovich, O., Cobb, R.M., Oestreich, K.J., Pierce, S., Ferrier, P., Oltz, E.M. (2007) Essential function for SWI-SNF chromatin-remodeling complexes in the promoter-directed assembly of Tcrb genes. *Nat. Immunol.*, **8**, 809–816.

144 Whitehurst, C.E., Schlissel, M.S., Chen, J. (2000) Deletion of germline promoter PD beta 1 from the TCR beta locus causes hypermethylation that impairs D beta 1 recombination by multiple mechanisms. *Immunity*, **13**, 703–714.

145 Agata, Y., Tamaki, N., Sakamoto, S., Ikawa, T., Masuda, K., Kawamoto, H., Murre, C. (2007) Regulation of T cell receptor beta gene rearrangements and allelic exclusion by the helix-loop-helix protein, E47. *Immunity*, **27**, 871–884.

146 Abarrategui, I., Krangel, M.S. (2007) Noncoding transcription controls downstream promoters to regulate T-cell receptor alpha recombination. *EMBO J.*, **26**, 4380–4390.

147 Abarrategui, I., Krangel, M.S. (2006) Regulation of T cell receptor-alpha gene recombination by transcription. *Nat. Immunol.*, **7**, 1109–1115.

148 Hawwari, A., Krangel, M.S. (2005) Regulation of TCR delta and alpha repertoires by local and long-distance control of variable gene segment chromatin structure. *J. Exp. Med.*, **202**, 467–472.

149 Zhong, X.P., Krangel, M.S. (1997) An enhancer-blocking element between alpha and delta gene segments within the human T cell receptor alpha/delta locus. *Proc. Natl Acad. Sci. USA*, **94**, 5219–5224.

150 Shih, H.Y., Krangel, M.S. (2010) Distinct contracted conformations of the Tcra/Tcrd locus during Tcra and Tcrd recombination. *J. Exp. Med.*, **207**, 1835–1841.

151 Skok, J.A., Gisler, R., Novatchkova, M., Farmer, D., de Laat, W., Busslinger, M. (2007) Reversible contraction by looping of the Tcra and Tcrb loci in rearranging thymocytes. *Nat. Immunol.*, **8**, 378–387.

152 Schlimgen, R.J., Reddy, K.L., Singh, H., Krangel, M.S. (2008) Initiation of allelic exclusion by stochastic interaction of Tcrb alleles with repressive nuclear compartments. *Nat. Immunol.*, **9**, 802–809.

153 Taniuchi, I., Ellmeier, W., Littman, D.R. (2004) The CD4/CD8 lineage choice: new insights into epigenetic regulation during T cell development. *Adv. Immunol.*, **83**, 55–89.

154 Ellmeier, W., Sunshine, M.J., Losos, K., Littman, D.R. (1998) Multiple developmental stage-specific enhancers regulate CD8 expression in developing thymocytes and

in thymus-independent T cells. *Immunity*, **9**, 485–496.

155 Adlam, M., Siu, G. (2003) Hierarchical interactions control CD4 gene expression during thymocyte development. *Immunity*, **18**, 173–184.

156 Zou, Y.R., Sunshine, M.J., Taniuchi, I., Hatam, F., Killeen, N., Littman, D.R. (2001) Epigenetic silencing of CD4in T cells committed to the cytotoxic lineage. *Nat. Genet.*, **29**, 332–336.

157 Chong, M.M., Simpson, N., Ciofani, M., Chen, G., Collins, A., Littman, D.R. (2010) Epigenetic propagation of CD4 expression is established by the Cd4 proximal enhancer in helper T cells. *Genes Dev.*, **24**, 659–669.

158 Williams, C.J., Naito, T., Arco, P.G., Seavitt, J.R., Cashman, S.M., De Souza, B., Qi, X., Keables, P., Von Andrian, U.H., Georgopoulos, K. (2004) The chromatin remodeler Mi-2beta is required for CD4 expression and T cell development. *Immunity*, **20**, 719–733.

159 Yu, M., Wan, M., Zhang, J., Wu, J., Khatri, R., Chi, T. (2008) Nucleoprotein structure of the CD4 locus: Implications for the mechanisms underlying CD4 regulation during T cell development. *Proc. Natl Acad. Sci. USA*, **105**, 3873–3878.

160 Naito, T., Gomez-Del Arco, P., Williams, C.J., Georgopoulos, K. (2007) Antagonistic interactions between Ikaros and the chromatin remodeler Mi-2beta determine silencer activity and Cd4 gene expression. *Immunity*, **27**, 723–734.

161 Taniuchi, I., Osato, M., Egawa, T., Sunshine, M.J., Bae, S.C., Komori, T., Ito, Y., Littman, D.R. (2002) Differential requirements for Runx proteins in CD4 repression and epigenetic silencing during T lymphocyte development. *Cell*, **111**, 621–633.

162 Chi, T.H., Wan, M., Zhao, K., Taniuchi, I., Chen, L., Littman, D.R., Crabtree, G.R. (2002) Reciprocal regulation of CD4/CD8 expression by SWI/SNF-like BAF complexes. *Nature*, **418**, 195–199.

163 Sato, T., Ohno, S., Hayashi, T., Sato, C., Kohu, K., Satake, M., Habu, S. (2005) Dual functions of Runx proteins for reactivating CD8 and silencing CD4 at the commitment process into CD8 thymocytes. *Immunity*, **22**, 317–328.

164 Bilic, I., Koesters, C., Unger, B., Sekimata, M., Hertweck, A., Maschek, R., Wilson, C.B., Ellmeier, W. (2006) Negative regulation of CD8 expression via Cd8 enhancer-mediated recruitment of the zinc finger protein MAZR. *Nat. Immunol.*, **7**, 392–400.

165 Sakaguchi, S., Hombauer, M., Bilic, I., Naoe, Y., Schebesta, A., Taniuchi, I., Ellmeier, W. (2010) The zinc-finger protein MAZR is part of the transcription factor network that controls the CD4 versus CD8 lineage fate of double-positive thymocytes. *Nat. Immunol.*, **11**, 442–448.

166 Delaire, S., Huang, Y.H., Chan, S.W., Robey, E.A. (2004) Dynamic repositioning of CD4 and CD8 genes during T cell development. *J. Exp. Med.*, **200**, 1427–1435.

167 Merkenschlager, M., Amoils, S., Roldan, E., Rahemtulla, A., O'Connor, E., Fisher, A.G., Brown, K.E. (2004) Centromeric repositioning of coreceptor loci predicts their stable silencing and the CD4/CD8 lineage choice. *J. Exp. Med.*, **200**, 1437–1444.

168 Ktistaki, E., Garefalaki, A., Williams, A., Andrews, S.R., Bell, D.M., Foster, K.E., Spilianakis, C.G., Flavell, R.A., Kosyakova, N., Trifonov, V., Liehr, T., Kioussis, D. (2010) CD8 locus nuclear dynamics during thymocyte development. *J. Immunol.*, **184**, 5686–5695.

169 Zhou, L., Chong, M.M., Littman, D.R. (2009) Plasticity of CD4+ T cell lineage differentiation. *Immunity*, **30**, 646–655.

170 Chang, S., Aune, T.M. (2007) Dynamic changes in histone-methylation 'marks' across the locus encoding interferon-gamma during the differentiation of T helper type 2 cells. *Nat. Immunol.*, **8**, 723–731.

171 Djuretic, I.M., Levanon, D., Negreanu, V., Groner, Y., Rao, A., Ansel, K.M. (2007) Transcription factors T-bet and Runx3 cooperate to activate Ifng and silence Il4 in T helper type 1 cells. *Nat. Immunol.*, **8**, 145–153.

172 Yagi, R., Junttila, I.S., Wei, G., Urban, J.F., Zhao, K., Paul, W.E., Zhu, J. Jr (2010) The transcription factor GATA3 actively represses RUNX3 protein-regulated production of interferon-gamma. *Immunity*, **32**, 507–517.

173 Zhu, J., Min, B., Hu-Li, J., Watson, C.J., Grinberg, A., Wang, Q., Killeen, N., Urban,

J.F., Guo, L., Paul, W.E. Jr (2004) Conditional deletion of Gata3 shows its essential function in T(H)1-T(H)2 responses. *Nat. Immunol.*, **5**, 1157–1165.

174 Mullen, A.C., Hutchins, A.S., High, F.A., Lee, H.W., Sykes, K.J., Chodosh, L.A., Reiner, S.L. (2002) Hlx is induced by and genetically interacts with T-bet to promote heritable T(H)1 gene induction. *Nat. Immunol.*, **3**, 652–658.

175 Schoenborn, J.R., Dorschner, M.O., Sekimata, M., Santer, D.M., Shnyreva, M., Fitzpatrick, D.R., Stamatoyannopoulos, J.A., Wilson, C.B. (2007) Comprehensive epigenetic profiling identifies multiple distal regulatory elements directing transcription of the gene encoding interferon-gamma. *Nat. Immunol.*, **8**, 732–742.

176 Lee, D.U., Agarwal, S., Rao, A. (2002) Th2 lineage commitment and efficient IL4 production involves extended demethylation of the IL4 gene. *Immunity*, **16**, 649–660.

177 Winders, B.R., Schwartz, R.H., Bruniquel, D. (2004) A distinct region of the murine IFN-gamma promoter is hypomethylated from early T cell development through mature naive and Th1 cell differentiation, but is hypermethylated in Th2 cells. *J. Immunol.*, **173**, 7377–7384.

178 Kim, S.T., Fields, P.E., Flavell, R.A. (2007) Demethylation of a specific hypersensitive site in the Th2 locus control region. *Proc. Natl Acad. Sci. USA*, **104**, 17052–17057.

179 Gamper, C.J., Agoston, A.T., Nelson, W.G., Powell, J.D. (2009) Identification of DNA methyltransferase 3a as a T cell receptor-induced regulator of Th1 and Th2 differentiation. *J. Immunol.*, **183**, 2267–2276.

180 Lee, P.P., Fitzpatrick, D.R., Beard, C., Jessup, H.K., Lehar, S., Makar, K.W., Perez-Melgosa, M., Sweetser, M.T., Schlissel, M.S., Nguyen, S., Cherry, S.R., Tsai, J.H., Tucker, S.M., Weaver, W.M., Kelso, A., Jaenisch, R., Wilson, C.B. (2001) A critical role for Dnmt1 and DNA methylation in T cell development, function, and survival. *Immunity*, **15**, 763–774.

181 Makar, K.W., Perez-Melgosa, M., Shnyreva, M., Weaver, W.M., Fitzpatrick, D.R., Wilson, C.B. (2003) Active recruitment of DNA methyltransferases regulates interleukin 4 in thymocytes and T cells. *Nat. Immunol.*, **4**, 1183–1190.

182 Hutchins, A.S., Artis, D., Hendrich, B.D., Bird, A.P., Scott, P., Reiner, S.L. (2005) Cutting edge: a critical role for gene silencing in preventing excessive type 1 immunity. *J. Immunol.*, **175**, 5606–5610.

183 Hutchins, A.S., Mullen, A.C., Lee, H.W., Sykes, K.J., High, F.A., Hendrich, B.D., Bird, A.P., Reiner, S.L. (2002) Gene silencing quantitatively controls the function of a developmental trans-activator. *Mol. Cell*, **10**, 81–91.

184 Floess, S., Freyer, J., Siewert, C., Baron, U., Olek, S., Polansky, J., Schlawe, K., Chang, H.D., Bopp, T., Schmitt, E., Klein-Hessling, S., Serfling, E., Hamann, A., Huehn, J. (2007) Epigenetic control of the foxp3 locus in regulatory T cells. *PLoS Biol.*, **5**, e38.

185 Liu, B., Tahk, S., Yee, K.M., Fan, G., Shuai, K. (2010) The ligase PIAS1 restricts natural regulatory T cell differentiation by epigenetic repression. *Science*, **330**, 521–525.

186 Fields, P.E., Kim, S.T., Flavell, R.A. (2002) Cutting edge: changes in histone acetylation at the IL4 and IFN-gamma loci accompany Th1/Th2 differentiation. *J. Immunol.*, **169**, 647–650.

187 Kaneko, T., Hosokawa, H., Yamashita, M., Wang, C.R., Hasegawa, A., Kimura, M.Y., Kitajiama, M., Kimura, F., Miyazaki, M., Nakayama, T. (2007) Chromatin remodeling at the Th2 cytokine gene loci in human type 2 helper T cells. *Mol. Immunol.*, **44**, 2249–2256.

188 Avni, O., Lee, D., Macian, F., Szabo, S.J., Glimcher, L.H., Rao, A. (2002) T(H) cell differentiation is accompanied by dynamic changes in histone acetylation of cytokine genes. *Nat. Immunol.*, **3**, 643–651.

189 Akimzhanov, A.M., Yang, X.O., Dong, C. (2007) Chromatin remodeling of interleukin-17 (IL17)-IL17F cytokine gene locus during inflammatory helper T cell differentiation. *J. Biol. Chem.*, **282**, 5969–5972.

190 Wei, G., Wei, L., Zhu, J., Zang, C., Hu-Li, J., Yao, Z., Cui, K., Kanno, Y., Roh, T.Y., Watford, W.T., Schones, D.E., Peng, W., Sun, H.W., Paul, W.E., O'Shea, J.J., Zhao, K. (2009) Global mapping of H3K4me3 and H3K27me3 reveals specificity and plasticity in lineage fate determination of

differentiating CD4+ T cells. *Immunity*, **30**, 155–167.
191 Bird, J.J., Brown, D.R., Mullen, A.C., Moskowitz, N.H., Mahowald, M.A., Sider, J.R., Gajewski, T.F., Wang, C.R., Reiner, S.L. (1998) Helper T cell differentiation is controlled by the cell cycle. *Immunity*, **9**, 229–237.
192 Chang, S., Collins, P.L., Aune, T.M. (2008) T-bet dependent removal of Sin3A-histone deacetylase complexes at the Ifng locus drives Th1 differentiation. *J. Immunol.*, **181**, 8372–8381.
193 Nakata, Y., Brignier, A.C., Jin, S., Shen, Y., Rudnick, S.I., Sugita, M., Gewirtz, A.M. (2010) c-Myb, Menin, GATA-3, and MLL form a dynamic transcription complex that plays a pivotal role in human T helper type 2 cell development. *Blood*, **116**, 1280–1290.
194 Yamashita, M., Hirahara, K., Shinnakasu, R., Hosokawa, H., Norikane, S., Kimura, M.Y., Hasegawa, A., Nakayama, T. (2006) Crucial role of MLL for the maintenance of memory T helper type 2 cell responses. *Immunity*, **24**, 611–622.
195 Wan, Y.Y., Flavell, R.A. (2007) Regulatory T-cell functions are subverted and converted owing to attenuated Foxp3 expression. *Nature*, **445**, 766–770.
196 Lehnertz, B., Northrop, J.P., Antignano, F., Burrows, K., Hadidi, S., Mullaly, S.C., Rossi, F.M., Zaph, C. (2010) Activating and inhibitory functions for the histone lysine methyltransferase G9a in T helper cell differentiation and function. *J. Exp. Med.*, **207**, 915–922.
197 Epsztejn-Litman, S., Feldman, N., Abu-Remaileh, M., Shufaro, Y., Gerson, A., Ueda, J., Deplus, R., Fuks, F., Shinkai, Y., Cedar, H., Bergman, Y. (2008) De novo DNA methylation promoted by G9a prevents reprogramming of embryonically silenced genes. *Nat. Struct. Mol. Biol.*, **15**, 1176–1183.
198 Feldman, N., Gerson, A., Fang, J., Li, E., Zhang, Y., Shinkai, Y., Cedar, H., Bergman, Y. (2006) G9a-mediated irreversible epigenetic inactivation of Oct-3/4 during early embryogenesis. *Nat. Cell Biol.*, **8**, 188–194.
199 Spilianakis, C.G., Lalioti, M.D., Town, T., Lee, G.R., Flavell, R.A. (2005) Interchromosomal associations between alternatively expressed loci. *Nature*, **435**, 637–645.
200 Hadjur, S., Williams, L.M., Ryan, N.K., Cobb, B.S., Sexton, T., Fraser, P., Fisher, A.G., Merkenschlager, M. (2009) Cohesins form chromosomal cis-interactions at the developmentally regulated IFNG locus. *Nature*, **460**, 410–413.
201 Spilianakis, C.G., Flavell, R.A. (2004) Long-range intrachromosomal interactions in the T helper type 2 cytokine locus. *Nat. Immunol.*, **5**, 1017–1027.
202 Ribeiro de Almeida, C., Heath, H., Krpic, S., Dingjan, G.M., van Hamburg, J.P., Bergen, I., van de Nobelen, S., Sleutels, F., Grosveld, F., Galjart, N., Hendriks, R.W. (2009) Critical role for the transcription regulator CCCTC-binding factor in the control of Th2 cytokine expression. *J. Immunol.*, **182**, 999–1010.
203 Sekimata, M., Perez-Melgosa, M., Miller, S.A., Weinmann, A.S., Sabo, P.J., Sandstrom, R., Dorschner, M.O., Stamatoyannopoulos, J.A., Wilson, C.B. (2009) CCCTC-binding factor and the transcription factor T-bet orchestrate T helper 1 cell-specific structure and function at the interferon-gamma locus. *Immunity*, **31**, 551–564.
204 Lee, G.R., Fields, P.E., Griffin, T.J., Flavell, R.A. (2003) Regulation of the Th2 cytokine locus by a locus control region. *Immunity*, **19**, 145–153.
205 Cai, S., Lee, C.C., Kohwi-Shigematsu, T. (2006) SATB1 packages densely looped, transcriptionally active chromatin for coordinated expression of cytokine genes. *Nat. Genet.*, **38**, 1278–1288.
206 Dispirito, J.R., Shen, H. (2010) Histone acetylation at the single-cell level: a marker of memory CD8+ T cell differentiation and functionality. *J. Immunol.*, **184**, 4631–4636.
207 Northrop, J.K., Thomas, R.M., Wells, A.D., Shen, H. (2006) Epigenetic remodeling of the IL2 and IFN-gamma loci in memory CD8 T cells is influenced by CD4 T cells. *J. Immunol.*, **177**, 1062–1069.
208 Northrop, J.K., Wells, A.D., Shen, H. (2008) Cutting edge: chromatin remodeling as a molecular basis for the enhanced functionality of memory CD8 T cells. *J. Immunol.*, **181**, 865–868.
209 Araki, Y., Fann, M., Wersto, R., Weng, N.P. (2008) Histone acetylation facilitates rapid and robust memory CD8 T cell response through differential expression of effector

molecules (eomesodermin and its targets: perforin and granzyme B). *J. Immunol.*, **180**, 8102–8108.

210 Chappell, C., Beard, C., Altman, J., Jaenisch, R., Jacob, J. (2006) DNA methylation by DNA methyltransferase 1 is critical for effector CD8 T cell expansion. *J. Immunol.*, **176**, 4562–4572.

211 Kersh, E.N. (2006) Impaired memory CD8 T cell development in the absence of methyl-CpG-binding domain protein 2. *J. Immunol.*, **177**, 3821–3826.

212 Araki, Y., Wang, Z., Zang, C., Wood, W.H., III, Schones, D., Cui, K., Roh, T.Y., Lhotsky, B., Wersto, R.P., Peng, W., Becker, K.G., Zhao, K., Weng, N.P. (2009) Genome-wide analysis of histone methylation reveals chromatin state-based regulation of gene transcription and function of memory CD8 + T cells. *Immunity*, **30**, 912–925.

27 Epigenetic Medicine

Randy Jirtle, Autumn Bernal, and David Skaar
Duke University, Radiation Oncology, 139 Environmental Safety DUMC, Durham, NC 27710, USA

Epigenetic Regulation and Epigenomics: Advances in Molecular Biology and Medicine, First Edition. Edited by Robert A. Meyers.

Keywords

Agouti viable yellow allele
A mouse *Agouti* allele carrying a retrotransposable insertion element that is controlled by epigenetic modifications and utilized for examining epigenetic changes during early development.

DNA methyltransferase (DNMT)
Any of a class of enzymes that transfer methyl groups to the adenine or cytosine bases of DNA, using *S*-adenosyl methionine (SAM) as the methyl donor.

Epigenetic programming
The process by which DNA methylation, histone modifications, and small regulatory RNAs work to control gene expression in tissue, gender, and developmental-specific manners.

Histone deacetylase
Any of a class of enzymes that remove acetyl groups from an ε-*N*-acetyl lysine amino acid on a histone tail.

Imprinted genes
Genes expressed in a parent of origin manner that are controlled by epigenetic mechanisms established in early development.

Although the field of epigenetic medicine is relatively new, it continues to make great advances due to an increased understanding of imprinted genes and the origins of epigenetic markings, the environmental effects on the epigenome, and how these elements relate to both health and disease. Following its establishment in gametes and during embryonic development, the epigenome is recognized as a critical regulator that employs DNA methylation and histone modifications to control cellular differentiation and also the expression of imprinted genes that are critical to development. The epigenome is an epigenetic "memory" that converts environmental exposures to phenotypes, resulting in lifelong – and even trans-generational – effects on health. As a consequence, the susceptibility of humankind has been increased not only to metabolic and psychiatric disorders but also to cancer. An understanding of the epigenome, combined with knowledge of its origins and plasticity, has led to the creation of new methods for diagnosis and treatment. These include histone deacetylase inhibitors and methylation inhibitors that are capable of specifically targeting dysregulated genes, and consequently affecting abnormal cells more effectively than by applying "traditional" therapies.

1
Introduction to the Epigenome

The epigenome controls gene expression during development and throughout life, its role being to guide and control correct tissue differentiation and function. Unlike the static DNA of cells, the epigenome is malleable and consists of marks that vary between tissues, developmental stages, and maternally and paternally inherited alleles. The engagement of these marks during development is a precise orchestration of events referred to as *epigenetic programming*, the disruption of which results in aberrant gene expression. Ultimately, this has been shown to lead to a litany of disease phenotypes such as cancer, developmental abnormalities, psychosis, and metabolic disorders.

As the epigenome's role in disease is quite clear, its potential as a biomarker and therapeutic target has begun to unfold. Studies conducted in both mouse and human have indicated that epigenetic marks are not only susceptible to environmental modifications, but that alterations are also persistent through mitosis and meiosis and correlate with several adult onset diseases. Due to their persistence across tissues and their ease of analysis, these marks can be used as biomarkers for exposures and warning signs for disease development. Yet, their use in such a role has only recently begun to gain momentum.

Whilst the epigenome clearly plays a major role in the development of complex disorders, such as metabolic syndrome and psychiatric diseases, epigenetic therapies have not yet achieved maturity in these situations. However, with epigenetic medicine having undergone major developments and achieving success in the form of cancer therapies, the consideration is that the treatment of other diseases thought to result from epigenetic aberrations might also benefit. In this chapter, following a brief introduction to epigenetic modifications and programming, the role of the epigenome in medicine will be outlined – first as a potential biomarker for exposures and disease diagnostics, and second as a therapeutic target for complex diseases.

1.1
Epigenetic Marks: From DNA to Chromatin Structure

The etiologies of many diseases are associated with persistent aberrant gene expression changes that often cannot be explained or associated with genetic mutations, or with the genetic inheritance of DNA polymorphisms. Yet, gene expression changes can be inherited both mitotically and meiotically through epigenetic mechanisms, without changing the DNA sequence. As its name implies, the epigenome lies "above the genome," and comprises DNA methylation, histone modifications, and small regulatory RNAs that help to control nucleosomal packaging, chromatin conformation and, ultimately, gene expression.

DNA methylation is controlled by a group of enzymes, the DNA methyltransferases (DNMTs), that catalyze the covalent addition of a methyl group (CH_3) from *S*-adenosylmethionine (SAM) to the 5′ position of a cytosine adjacent to a guanine (CpG dinucleotide). This reaction occurs through the one-carbon metabolism pathway – a network of biochemical reactions that transfer methyl groups among various donors and acceptors for use in cellular methylation reactions, including that of DNA [1]. The cycle begins when folate donates a methyl group with the help of 5,10-methylenetetrahydrofolate reductase (MTHFR), converting it to 5-methyltetrahydrofolate (5me-THF). The methyl group is then transferred from 5me-THF through a series of biochemical steps to methionine, and finally to *S*-adenosylhomocysteine (SAH), where it is then reassigned to a cytosine via a DNMT [2]. Metabolites involved in this pathway, which include folate, methionine, choline, vitamin B_{12}, and betaine, can be modulated through dietary intake, thus influencing the equilibrium of the pathway and the supply of methyl groups for these reactions [3, 4].

Currently, three enzymes have been identified with DNA methyltransferase activity, namely DNMT1, DNMT3a, and DNMT3b [5]. DNMT3a and 3b catalyze *de novo* methylation during gametogenesis, early development, and carcinogenesis [6]. DNMT1, with its high preference for hemimethylated substrates, primarily maintains these methylation patterns during DNA replication [7]. Although, in mammals, 60–90% of CpGs are methylated [8], CpG islands with a GC content >55% are typically hypomethylated compared to their CpG-poor counterparts within intergenic and intronic regions [9]. Typically, DNA methylation represses gene expression by lying within the DNA major groove and blocking transcription factors from binding and promoting gene expression.

As the DNA wraps around histones to form the nucleosomes, modifications to the histone tails also function to control gene expression, and may even mark DNA sites for methylation [10–12]. The histone variants H2A, H2B, H3, and H4 contain various histone modifications such as methylation, acetylation, phosphorylation, sumoylation, and ubiquitylation [13]. Histone methylation and acetylation have been most investigated for their role in controlling epigenetic programming. For example, methylated DNA is associated with chromatin enriched for the methylation of lysines at all histone variants, and is devoid of histone H3 and H4 acetylation [14]. This histone modification pattern promotes a closed chromatin conformation and suppresses gene transcription. Conversely, histone acetylation opens up chromatin to enhance gene transcription. Histone lysine 4 (H3K4) methylation also prevents DNA methylation, and is often identified in active promoter regions [10] (Fig. 1). Histone methylases (HMTs), histone acetylases (HATs), and histone deacetylases (HDACs) organize these marks in varying combinations with DNA methylation levels. Although the diversity of the combinations are still unknown, further definition of the relationship and interactions between these elements will help to determine how they dictate nucleosome and chromatin structures to control gene expression in normal and diseased states.

Small regulatory RNAs must also be acknowledged for their role in epigenetic

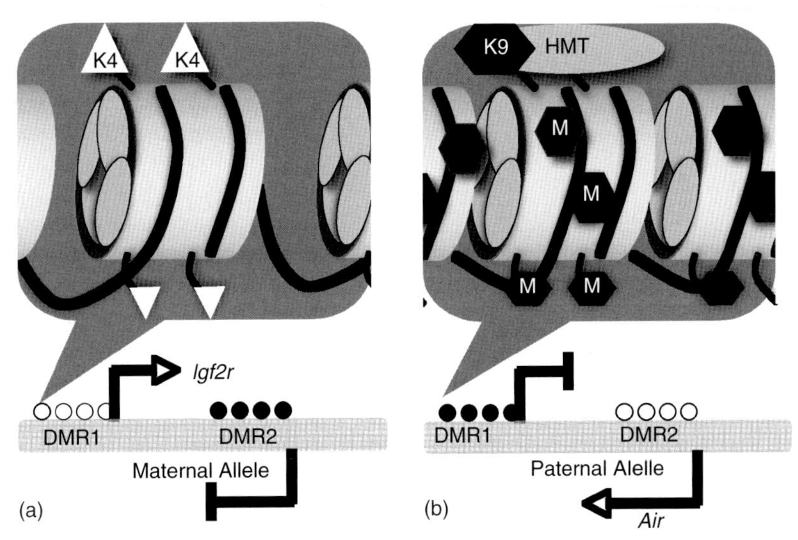

Fig. 1 Imprinting regulation of *Igf2r*. (a) Maternal expression of *Igf2r* involves two differentially methylated regions, DMR1 and DMR2, and the paternal expression of the Air ncRNA antisense transcript. The methylated DMR2 on the maternal allele is inherited through the germline (Primary Imprint) and suppresses Air transcription. On this allele, DMR1 hypomethylation and histone trimethylation at Lys4 (H3K4me3) are activating marks that lead to *Igf2r* expression; (b) The paternal allele is unmethylated at DMR2, resulting in expression of Air, recruitment of histone methyltransferases, histone trimethylation at Lys9 (H3K9me3), and paternal methylation of DMR1 during early development in peripheral tissues (Secondary Imprint). These repressive marks inhibit *Igf2r* transcription. Panel (a) reproduced from Dindot, S., Person, R., Strivens, M., Garcia, R., Beaudet, A. (2009) Epigenetic profiling at mouse imprinted gene clusters reveals novel epigenetic and genetic features at differentially methylated regions. *Genome Res.*, **19**, 1374–1383. Panel (b) reproduced from Nagano, T., Mitchell, J., Sanz, L., Pauler, F., Ferguson-Smith, A., Feil, R., Fraser, P. (2008) The Air noncoding RNA epigenetically silences transcription by targeting G9a to chromatin. *Science*, **322**, 1717–1720.

regulation. Previously, noncoding RNAs (ncRNAs) have been recognized for their part in maintaining silenced gene expression patterns [15]. The production of ncRNAs generally occurs from an unmethylated genomic region in order to silence gene expression at promoter regions in *cis*, and in some cases with the help of repressive histone modifications [14, 16] (Fig. 1). The RNA-dependent production of DNA methylation remains controversial; several reports have indicated that ncRNAs induce methylation, while others have described no impact on methylation, or even demethylation [17–19]; ncRNAs may, instead, interact with histone marks to modify gene expression [20]. Other small RNAs, known as *Piwi-interacting RNAs*, may also influence DNA methylation of the genome, on the basis of Piwi loss-of-function studies which have demonstrated a reduced DNA methylation in transposons [21]. The involvement of ncRNAs in DNA and histone epigenetic modifications indicates the complexity of these processes, and offers additional targets for modifying the epigenome in disease and therapy.

1.2 Imprinted Genes and Epigenetic Programming

The function of the complicated network of regulatory elements described above is to control imprinted genes and epigenetic programming. Imprinting is a unique regulatory process in which genes are expressed in a monoallelic, parent-of-origin-dependent manner [22, 23]. This distinctive expression pattern is regulated by epigenetic marks, such as DNA methylation and histone modifications. The contrast in DNA methylation marks between the maternal and paternal alleles is termed *differential methylation*, and the corresponding allelic region is termed a differentially methylated region (DMR). DMRs often lie in imprint control regions (ICRs), which are allelic locations that are essential to imprinted gene expression and repression [24]. Between the maternal and paternal alleles, ICRs can vary in DNA methylation levels, histone methylation and acetylation levels, and regulatory RNA expression (Fig. 1). It has been predicted that 154 human genes are to be imprinted [25]; many of these regulate growth and proliferation, playing key roles in these processes during early development.

Correct epigenetic programming ensures that imprinted gene expression is accurately controlled during early development and throughout life, and is an essential process for establishing and orchestrating tissue-specific gene expression. Prior to fertilization, the gametes carry sex-specific epigenetic marks, including primary imprint marks on imprinted genes that were developed during gametogenesis. Following fertilization, global demethylation occurs across the paternal and maternal genomes in order to create totipotency during embryogenesis [26]. Throughout this demethylation process, DNMT1 functions to maintain primary methylation marks at imprinted genes. DNMT3a and DNMT3b then catalyze *de novo* methylation across the remainder of the genome to establish tissue-specific methylation marks and promote cell differentiation [27]. During *de novo* methylation, secondary (i.e., established after fertilization) imprint marks are established in somatic tissues and gametes [28] (Fig. 2). These imprints are maintained in somatic and placental tissues via histone modifications, but are reprogrammed in the gametes during gametogenesis. Gametic reprogramming follows a loss of several epigenetic marks, including DNA methylation, H3 lysine 9 trimethylation (H3K9me3), H3K27me3, and H3 lysine 9 acetylation (H3K9ac) [29]. Methylation is then re-established in the gametes according to the sex of the individual, who will carry the imprint marks of either their mother or father [28, 30]. On fertilization of the next generation, these gamete-specific DNA methylation profiles will be transmitted to the zygote, and the epigenetic programming cycle will repeat itself [27] (see Fig. 2).

Epigenetic processes are critical for appropriate prenatal and postnatal development. The maintenance of these methylation marks throughout life is carried out primarily through the action of DNMT1 during each mitotic cell division [31, 32]; however, additional studies have also suggested that the DNMT3a and 3b interact with DNMT1 and function to maintain methylation in the genome, despite their preference for unmethylated DNA [11, 33]. Histone modifications, such as acetylation and methylation, are also thought to be transmitted during cell division in somatic cells, permanently affecting lifelong gene expression [34].

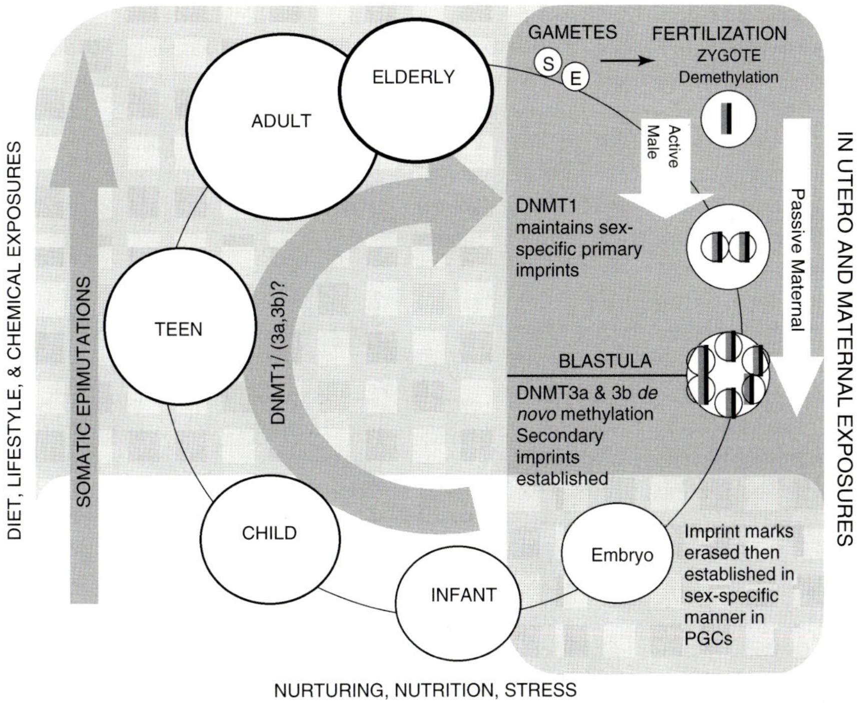

Fig. 2 Epigenetic programming is susceptible to developmental exposures. Prior to fertilization, the male and female gametes carry genomic methylation patterns and sex-specific imprints that were established during gametogenesis. Following fertilization, active demethylation of the paternal genome occurs and passive demethylation of the maternal genome initiates to create totipotentcy. Throughout this demethylation process, DNMT1 maintains primary imprints. Around implantation, DNMT3a and 3b methylate the genome to aid in tissue differentiation. Secondary marks at imprinted genes are established at this time. In the developing embryo, the epigenetic marks of imprinted genes are cleared in the primordial germ cells (PGCs) and established in a sex-specific manner prior to birth in the males and shortly following birth in female oocytes. These epigenetic changes during embryonic development are affected by chemical, nutritional, and behavioral environmental exposures. As an individual ages, DNMT1 maintains methylation marks during cellular replication. DNMT3a and 3b quite possibly are active in this process. During this period, diet, lifestyle, and chemical exposures can alter the epigenome and lead to somatic epimutations. Germline epimutations can occur early in development when marks are established in PGCs or throughout life during the male.

1.3
Vulnerable Epigenetic Processes

The labile processes of imprinting and epigenetic reprogramming during periods of rapid growth and development are particularly sensitive to environmental perturbations such as diet [35], nurturing [36], and toxicant exposure [37, 38]. Due to the haploid expression of imprinted genes, fewer insults are necessary to completely deregulate imprinted gene expression, according to Knudson's two-hit hypothesis for carcinogenesis [39]. Many imprinting disorders have been identified and are apparent at birth, such as transient neonatal diabetes mellitus, Beckwith–Weidman, Silver–Russell, Angelman, and Prader–Willi syndromes [40]. Apart from these germline disorders, imprint deregulation is also thought to play a role in autism, metabolic syndrome, psychiatric disorders, and cancer. The activation of imprinted oncogenes or the silencing of imprinted tumor suppressor genes is termed loss of imprinting (LOI), and this has been described in many types of cancer, including those of the bladder, brain, breast, cervix, colon, lung, prostate gland, and ovary [41–48]. Imprinted genes are thought to be most susceptible to epigenetic aberrations during gametogenesis when primary sex-specific marks are laid down. However, the establishment of secondary imprint marks after fertilization, and the maintenance of these marks in the gametes and in other tissues, could also be deregulated throughout life (Fig. 2).

Just as imprinted genes can be deregulated, the remainder of the genome is also susceptible to epigenetic modifications during both *de novo* and maintenance methylation. The establishment and maintenance of DNA methylation marks changes throughout life. In fact, an individual's global and CpG island DNA methylation levels change over time [49], and these changes are related to both age and differences in exposure [50, 51]. These patterns are even seen to diverge between monozygotic twins [52]. This phenomenon is most likely due to differing environmental exposures, and may also explain the discordant phenotypes often seen among monozygotic twins [53]. Epigenetic changes in gametic DNA also could be affected throughout life, with the potential to persist through generations. Although the female gametes are established shortly after birth, the male germline epigenome is established perinatally, but must then be maintained during pubertal and adult spermatogenesis. Consequently, exposures that occur during every life stage can alter the epigenome at imprinted genes and globally, with varying consequences that have yet to be fully determined [54].

2
The Epigenome: A Biomarker for Exposure

Although the molecular mechanisms linking environmental exposures to observed epigenetic changes are still unclear, the results of various investigations support the roles that diet and the one-carbon metabolism cycle play in supporting healthy, balanced methylation patterns [2, 55]. Beyond that, the number of molecular pathways leading to DNA methylation and demethylation are unknown, and they may also vary between tissue type and disease state [56]. Pathways determined thus far to play a role in epigenetic changes are DNA repair and damage pathways [57], stress pathways [58–60], and cell-signaling pathways, such as ERK (extracellular signal-regulated kinase), JNK (c-Jun N-terminal kinase), and Ras [61–63]. Although the specifics are still unclear, it is known that when these pathways are disrupted in disease or through environmental exposures, they can all lead to the same phenotype: persistent aberrant alterations in the epigenome. Because of the epigenome's susceptibility to modifications due to environmental exposures, and the persistence of these changes, it can serve as a biomarker for early life events [64, 65]. Several studies in animals and humans have lent momentum to the thought that epigenetic marks can provide a history of developmental exposures. With further investigations, epigenetic biomarkers may also help to predict the risk for disease onset and progression resulting from epigenetic predispositions that lead to aberrant gene expression [66].

2.1
The Agouti Viable Yellow Mouse Model

The Agouti Viable Yellow Mouse (A^{vy}) is one of the best mammalian models for determining that early exposures can lead to persistent epigenetic changes. The *Agouti* gene encodes for a paracrine signaling molecule that is normally expressed from a hair-cycle promoter during a specific developmental stage, leading to a yellow subapical pigment band on the black hair shaft and brown coat color [67]. In the A^{vy} mouse, this normal expression is usurped due to the spontaneous insertion of an intracisternal a particle (IAP) retrotransposable element upstream of the normal promoter site in the *Agouti* gene (see Fig. 3). The insertion leads to ectopic expression of agouti in all cells – not just hair follicles – and throughout the animal's lifetime rather than during a specific stage in development. This overexpression of agouti causes a yellow coat coloration, obesity, and an

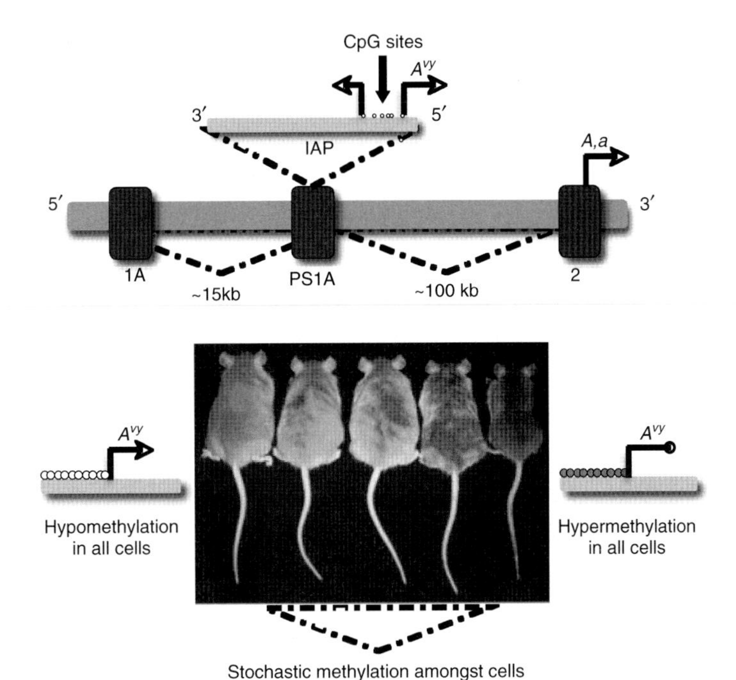

Fig. 3 A^{vy} epigenetic regulation. The A^{vy} metastable epiallele results from the insertion of an intracisternal a particle (IAP) retrotransposable element upstream of the normal Agouti transcription start site. Stochastic methylation of CpG sites upstream of the IAP cryptic promoter lead to ectopic Agouti expression (hypomethylation of IAP) or Agouti expression from the normal promoter (hypermethylation of IAP). The ectopic expression of Agouti in all cells leads to completely yellow, obese animals (left), while suppression of the cryptic promoter leads to brown, thin, and healthy animals (right). These coat colors are a direct sensor of methylation state in the animal.

increased incidence of diabetes and cancer [68]. Phenotypic variation, however, occurs due to stochastic and environmentally influenced methylation of upstream CpG sites that affect *Agouti* transcription from the IAP cryptic promoter. Thus, genetically identical animals can have a normal amount of variability in *Agouti* expression, along with the corresponding range of coat colors and disease phenotypes [68] (Fig. 3). Because of the A^{vy} allele's stochastic nature and vulnerability to environmentally induced changes *in utero*, it is considered a metastable epiallele [69, 70].

The analyses of this metastable epiallele in mice show that environmental exposures *in utero* contribute to epigenetic modifications that could lead to the manifestation of cancers, diabetes, obesity, and other developmental disorders later in life [68, 70]. For example, *in utero* supplementation with folic acid (an important methyl group donor) increased methylation and induced phenotypic changes in A^{vy} offspring [35]. Alcohol exposure, *in utero*, also increased methylation at the allele and induced phenotypic effects consistent with fetal alcohol syndrome [71]. In another mouse study, alcohol exposure *in utero* led to aberrant methylation across the entire genome, consistent with the A^{vy} studies [72]. Studies conducted in humans, in which the effect of ethanol on the epigenome was examined, showed that alcohol consumption decreases the methylation levels of *H19* and *IGF2* DMRs [73]. The culturing of A^{vy} zygotes to the blastocyst stage prior to implantation also leads to hypomethylation of the metastable epiallele [74]. Human *in vitro* fertilization (IVF) results are consistent with those obtained with mice, with changes in genome-wide methylation [75], hypomethylation at imprinted genes, and increased rates of cancer and imprinting disorders all seen to result from IVF [76]. The endocrine disruptor, bisphenol A, also causes hypomethylation of the A^{vy} metastable epiallele [38]. In the same study, dietary genistein, a soy isoflavone, was able to counteract the bisphenol A-induced hypomethylation, demonstrating the lability of the epigenome as well as the potential for preventing epigenotoxicity with dietary adjustments [38]. In several other mammalian studies, bisphenol A has shown to alter the epigenome [77–79], further supporting the use of the A^{vy} model as an accurate indicator of compounds that can alter the epigenome (see Table 1).

Tab. 1 Mouse and human prenatal exposures lead to lasting epigenetic modifications.

Species	*Exposure*	*Methylation change*
Mouse (A^{vy})	Folate	A^{vy} hypermethylation
Mouse (A^{vy})	Ethanol	A^{vy} hypermethylation
Mouse (A^{vy})	*In-vitro* culture	A^{vy} hypomethylation
Mouse (A^{vy})	Bisphenol A	A^{vy} hypomethylation
Mouse (A^{vy})	Genistein	A^{vy} hypermethylation
Human	Famine	*IGF2* and *INSIGF* hypomethylation; *IL10, LEP, ABCA1, GNASAS, MEG3* hypermethylation
Human	*In-vitro* fertilization	Hypomethylation and hypermethylation changes in imprinted and nonimprinted genes

Thus, the A^{vy} mouse model has served as a biosensor for early exposures that can alter the epigenome and persist into the adult life. Future studies should continue to examine other chemical, nutritional, and stress exposures that may play a role in disease development later in life. These mice could also be used to analyze genome-wide and imprint-specific methylation changes that result from early exposures. The tissue specificity of alterations, the persistence of modifications, and the heritability of these changes should also be considered. Animal biosensors, such as A^{vy}, could eventually play an even larger role in epigenetic medicine if they were to become tools for the examination of epigenetic therapies such as pharmaceuticals, nutritional interventions, and behavioral treatment.

2.2 Human Biomarkers of Exposure

Other metastable epialleles similar to A^{vy} have been identified and characterized in mice, although comparable alleles have yet to be fully described in the human genome, despite retrotransposable elements being present in both species [80–83]. Nonetheless, unique human exposures are providing insight into the impact that early developmental exposures might have on an individual's epigenetic predisposition for disease. These exposures are also providing clues for potential biomarkers in humans, which could be used to identify susceptible populations and provide insight into new therapies.

One such exposure was the Dutch Hunger Winter, which occurred during World War II due to the disruption of infrastructure and extreme food rationing to tens of thousands of people in the Netherlands. A study of 2414 people born around the time of the Dutch Hunger Winter showed that gestational exposure to famine led to a litany of phenotypes typical of metabolic syndrome in adults, including glucose intolerance, coronary heart disease, increased lipid profile, altered blood coagulation, increased stress responsiveness, increased microalbuminuria, obstructive airway diseases, and obesity [84]. When the persistence of these phenotypes was examined in the following generation, it was found that the offspring (F2) of exposed parents (F1) had an increased neonatal adiposity and poor health almost twice as frequently in later life due to miscellaneous causes; however, the birth weight and cardiovascular disease were not affected in the F2 generation (mean age 32 years) [85]. The results of other studies in which the effects of low birth weight and poor nutrition on adult disease were examined concurred with these findings. An analysis of the Helsinki Birth Cohort, born between 1924 and 1944, showed that low birth weight, low weight at one year of age, and a high body mass index (BMI) at 11 years of age correlated with higher rates of coronary heart disease, type II diabetes, and impaired glucose tolerance [86].

Other complex diseases may also be affected by poor prenatal nutrition. Women exposed to famine during the Dutch Hunger Winter are at an increased risk of breast cancer [84]. Complex psychological disorders, such as schizophrenia and addiction were also seen at significantly higher levels in the Dutch Famine offspring [87–89]. These findings support the developmental origins of disease hypothesis proposed by David Barker, which states simply that early environmental exposures can alter developmental programming and affect diseases later in life.

If these early exposures, such as famine, are leading to disease later in life, might they leave an imprint in individuals that could indicate disease susceptibility, and even provide information on how to prevent disease onset? Increasing evidence suggests that epigenetic changes comprise this imprint, and that they play a large role in the developmental origins of these complex diseases. Epigenetic changes aid developmental plasticity, which allows an organism to adapt to its environment, whether fetal or postnatal. A divergence between the early environment and the current environment is hypothesized to increase the risk of complex diseases, because epigenetic changes established in the fetal environment lead to gene expression changes that are then maladaptive later in life.

Persistent epigenetic changes observed following exposure to famine support this hypothesis. In the Dutch Famine cohort, epigenetic changes were seen in several imprinted genes. First, *IGF2* hypomethylation was observed several decades after exposure [90]. Later, six additional imprinted loci were also determined to be differentially methylated, namely *IL10*, *GNASAS*, *INSIGF*, *LEP*, *ABCA1*, and *MEG3* [91]. Interestingly, all of these genes are implicated in growth, metabolic, or cardiovascular functions. The alterations of these imprinted genes indicates the stability and persistence of epigenetic changes following environmental perturbations, although the persistence of these alterations to future generations has not yet been examined. In order to further catalog epigenomic regions that can serve as biosensors for early exposure, it is clear that epigenome-wide studies are required [65]. Locating these regions will also help to evaluate the effect of developmental influences on complex human diseases [64].

3 Targeting the Epigenome in Complex Disease

In addition to the epigenome's potential as a biomarker for early exposures that may lead to complex diseases, the results of various research studies have indicated that the epigenome could also become a target for the prevention and therapy of these disorders. Whilst metabolic syndrome, psychiatric disorders, and cancer are all thought to be largely controlled by the epigenetic programming of genes, these diseases lack any concrete preventive or therapeutic strategies due to their complex etiologies and phenotypes. The labile nature of the epigenome means that it is not only susceptible to early changes that can lead to disease, but that it may also be amenable to therapeutic drug targeting. Consequently, as this field continues to grow, the epigenome might also become a biomarker for tracking the efficacy and toxicity of therapies.

3.1 Metabolic Syndrome

Metabolic syndrome consists of risk determinants for type 2 diabetes, in addition to cardiovascular-related disorders such as obesity, insulin resistance, and hypertension. The development of metabolic syndrome is thought to be largely controlled by the perinatal and postnatal periods, which are important stages for establishing adipocyte number, type, and growth. The pancreas, liver, cardiovascular system, muscles, and the hypothalamic pituitary axis (HPA) are all important targets of maternal nutritional programming during this period, and are greatly influenced by lipid metabolism and storage. It is hypothesized that, if the maternal–fetal environment is limited nutritionally, then programming (including epigenetic programming) will occur in favor of lipid storage, which then alters the organogenesis of key organs and predisposes individuals to metabolic syndrome [92]. As seen in the Dutch Famine cohort, changes in human imprinted loci correlate with increased incidences of metabolic disease. However, non-imprinted loci are also important for correct metabolism during development, and epigenetic changes at these loci can increase the susceptibility to metabolic disease [93–95].

The results of several mammalian studies have been consistent with those of the Dutch Famine studies, and have indicated that epigenetic changes occur in offspring due to the maternal nutritional status. A dietary protein restriction in rats was shown to induce phenotypes typical of altered metabolic programming, such as hypertension, dyslipidemia and impaired glucose metabolism [96]. This model shows that several of the genes involved in metabolic programming are epigenetically altered in response to prenatal nutrition. For instance, protein-restricted rats bear offspring with decreased methylation of the glucocorticoid receptor (GR) and the peroxisome proliferator-activated receptor α (PPARα), two key genes involved in metabolic balance [96, 97]. These methylation changes can persist to the F2 generation [98]. Folic acid supplementation *in utero* and during the pubertal period was also found to reverse hepatic methylation and phenotypic changes [99]. Finally, protein restriction was shown to cause the differential methylation of over 200 promoter regions within mouse fetal liver genes, including the *Lxra* (liver-X-receptor alpha) gene, a nuclear receptor involved in controlling the metabolism of cholesterol and fatty acids [100].

Methylation changes due to protein restriction affect other tissues in addition to the liver. Protein restriction *in utero* induces the hypomethylation of angiotensin receptor promoter (*Agtr1b*) and also increased protein expression in the adrenal glands, which has been shown to increase hypertension later in life [101]. This epigenetic effect was determined to be modulated by maternal glucocorticoid levels of the HPA axis; this effect was confirmed when the treatment of dams with a glucocorticoid inhibitor, metyrapone, was seen to prevent the epigenetic and gene expression changes [102].

Until now, the number of studies conducted in humans to examine growth restriction and epigenetic changes has been limited. In human low-birthweight and normal-birthweight subjects, methylation was analyzed in response to high-fat overfeeding. The low-birthweight subjects showed insulin resistance and reduced PPARγ coactivator 1α (*PGC-1α*) expression, which correlated with an increased methlyation that occurred with both control and high-fat diets. In contrast, the normal-birthweight individuals showed an increased *PGC-1α* activity when they received the high-fat diet, but this returned to normal when they were fed the control diet [103]. *PGC-1α* methylation is significantly increased in human diabetic islet cells compared to non-diabetic islet cells [104]. Thus, maternal and fetal nutrition might impact the onset of diabetes through epigenetic mechanisms.

Overfeeding can also induce obesity, diabetes, and epigenetic changes in the offspring, which indicates that a nutritional imbalance – in either direction – will disturb normal epigenetic programming. Recently, it has been hypothesized that high-fat diets strain metabolic plasticity, leading to overt developmental disruption; consequently, even a high-fat diet matched to the early environment will worsen the already maladapted phenotype [105]. In rodents, neonatal overfeeding increases methylation at the hypothalamic insulin receptor promoter and increases blood glucose levels [106]. Early overfeeding also leads to hypermethylation of binding sites at the neurohormone, proopiomelanocortin (*Pomc*), which governs the effects of leptin and insulin in the hypothalamus [107]. In humans, maternal obesity also alters offspring methylation. Maternal obesity has been shown to be associated with the DNA methylation of *PGC-1α* in cord blood from the newborn offspring [108].

Interestingly, adipose tissue carries unique epigenetic marks, depending on location and type (which are presumed indicators of developmental lineage) as well as rates of lipogenesis and lipolysis. Two of the developmental genes identified in a mouse study, *Hoxa5* and *Hoxc9*, showed varying levels of expression between different fat deposits, and were also predicted to be imprinted in humans [109]. The disruption of these imprinted genes would render adipose programming particularly susceptible to persistent changes throughout life and in future generations (Table 2).

While less extensively studied, histone modifications are also susceptible to rearrangement due to restricted uterine growth and occur in genes relating to diabetes [98, 110]. Both HATs and HDACs have been found to regulate the PPAR family, which consists of PPARα, PPARγ, and PPARδ. PGC-1α has also been shown to associate with HATs upon binding to PPARγ, and this association enhances PPARγ transcriptional activity [111]. HDACs can repress transcription at the PGC-1α promoter, and this repression has been correlated with diabetes [112].

Consequently, HDACs appear as likely candidates for treating metabolic

Tab. 2 Epigenetic modifications result from maternal nutritional imbalances.

Species	*Exposure*	*Methylation change*	*Tissue(s)*
Rat	Dietary protein restriction	*GR* and *PPARα receptor* hypomethylation	Liver
Rat	Dietary protein restriction followed by folate supplementation	Reversal of methylation changes	Liver
Rat	Dietary protein restriction	*Agtr1b* hypomethylation	Adrenal glands
Mouse	Dietary protein restriction	Global methylation changes	Liver
Rat	Intrauterine growth-restriction	Histone code modifications repress glucose transporter 4 expression	Muscle
Human	Low birthweight	*PGC-1α* hypermethylation	Skeletal muscle
Rat	Overfeeding	Hypermethylation at insulin receptor	Hypothalamus
Rat	Overfeeding	*Pomc* hypermethylation	Hypothalamus
Human	High-fat diet	*PGC-1α* hypermethylation	Cord blood

syndrome and diabetes. Sodium butyrate, an HDAC inhibitor, can increase insulin gene expression in rat islet cells [113], while valproic acid (another widely used HDAC inhibitor) is able to block adipogenesis [114]. The use of HDAC inhibitors for the treatment of metabolic syndrome has not yet been extensively examined, however, and their specificity is questionable for targeting this complex disease. Other therapies that target signalers of epigenetic changes, such as leptin, may be more specific and result in a lower potential for adverse side effects [115].

Pharmaceuticals derived from botanicals, which have similar labile epigenomes to mammals, also represent potential therapeutics. In fact, many botanicals have been found to target epigenetic regulatory genes, and some of these have initially been identified as candidates for antidiabetic drugs. For example, compound NZ-01 (from *Ligustrum lucidum* L.), which was identified in a random screening of botanicals and has antidiabetic activities, led to exposure-related *Dnmt1* and *-3b* expression changes [116]. Thus, as the genes and epigenetic changes involved in metabolic programming are further defined, targets can be established for a myriad of epigenetic therapies and their efficacy in treating metabolic syndrome can be determined.

In order to further define potential epigenetic targets, epigenomic mapping must be completed to determine which genes and tissues are altered, and at which developmental stages in response to nutritional insults and in diseased individuals. Global networks must then be created to predict the effects that exposures have on specific pathways involved in metabolic programming, and with this aim the field of nutrigenomics will continue to attract much attention. Nutrigenomics seeks also to determine how nutrition can influence metabolic pathways, and the role(s) that epigenetics plays in these changes. Although, at present, the field is moving slowly, the aim of many ongoing studies is to determine the signature profiles of exposures, which would include target genes, pathways, and biomarkers. Ultimately, nutrient sensors – such as transcription factors and nuclear receptors such as PPARα – should be identified and monitored in response to both nutritional deprivation and supplementation [117]. Epigenetic changes at these sensors might serve as sensitive endpoints to examine the effects of maternal nutrition. However, until more is known regarding the etiology of metabolic disorders, preventative strategies – such as nutritional supplements and counseling during fetal development – may represent the best methods for attenuating the epigenetic changes associated with a poor maternal diet.

3.2 Psychological Disorders

While epigenetic therapies for metabolic disorders are advancing only slowly, the use of epigenetic therapy for psychological disorders is expanding more quickly. Psychiatric diseases share many features that are consistent with epigenetic deregulation: discordance between monozygotic twins, late age of onset, parent-of-origin and sex effects, fluctuating and episodic disease course, and a relationship to environmental factors, such as stress [118–120]. The findings of the Dutch Famine studies also indicated that early nutrition may influence the onset of schizophrenia and addiction disorders [89]. A closer examination of another famine, the Chinese Great Leap Forward

Famine, showed that conception and birth during a famine increased the risk of developing schizophrenia [121]. These phenomena are thought to result from the major effects that maternal nutrition and stress have on epigenetic programming in the fetal brain.

Schizophrenic patients carry unique epigenetic signatures. In particular, reelin (*RELN*) and glutamate decarboxylase (*GAD*$_{67}$) promoter hypermethylations are seen in the brains of schizophrenic individuals [122, 123]. Reelin is a protein that aids neuronal migration and positioning during brain development. Interestingly, methylation of the neocortical *RELN* promoter is increased significantly after puberty in both schizophrenic and autistic individuals, coinciding with the onset and worsening of the diseases [124]. Methylation changes have also been observed in other genes and brain regions in schizophrenic patients. Telencephalic gamma-aminobutyric acid (GABA)ergic neurons of schizophrenic individuals have increased expressions of *DNMT1* and *DNMT3a* that are also detected in peripheral blood lymphocytes [125], and this may also help to distinguish them from bipolar patients [126]. Further identification of such methylation patterns that can be detected noninvasively will become essential in the diagnosis, treatment and monitoring of schizophrenic patients.

Similarly to schizophrenia, bipolar disorder presents with genome-wide epigenetic abnormalities [127] and discordance between monozygotic twins. Monozygotic twins discordant for bipolar disorder contrast in their methylation levels of several genes [128], some of which lay on the X-chromosome, suggesting inadequate X-inactivation [129]. Hypomethylation of the *membrane-bound catechol-O-methyltransferase* (*MB-COMT*) promoter is shared between schizophrenic and bipolar individuals, and could be a risk factor for developing either disease [130]. Other genes of the extended dopaminergic system also display differential methylation patterns, and their analysis suggests that *MB-COMT* promoter hypomethylation influences the promoter methylation of *RELN* and dopamine receptor genes [131]. It is hoped that continued investigations of these systems will lead to the discovery of targets for improved therapies.

Aberrant epigenetic programming may also play a role in autistic spectrum disorders (ASDs) [132, 133]. One of the most well-studied links between epigenetic mechanisms and autism is *methyl CpG binding protein 2* (*MECP2*) gain of function from duplication [134, 135]. Others have hypothesized that autistic disorders result from a developmental deregulation of the locus coeruleus-noradrenergic (LC-NA) system. LC-NA developmental genes are under an exquisite degree of epigenetic control and thus, the deregulation of multiple epigenetic regulatory processes could result in ASD [133]. Specific imprinted genes are also involved in LC-NA deployment, and in mediating behavioral responses to novel environmental conditions [136]. The maternal methylation balance might also play a role in dictating autism in children. In one study of autistic children, a functional polymorphism in the mother's reduced folate carrier (*RFC1*) gene was significantly increased [137]. These mothers showed DNA hypomethylation and significantly elevated levels of plasma homocysteine, adenosine, and SAH, indicating the importance of the one-carbon metabolism cycle in controlling methylation levels. Due to the spectrum in severity of phenotypes, it is likely that several pathways exist that may lead to ASD and several different treatment options.

Epigenetic changes have also been implicated in depression and suicidal behavior. Maternal behavior largely impacts offspring methylation patterns that could lead to psychological disorders later in life. In particular, the results of one study showed that early maternal behavior can alter the HPA in the mouse brain through epigenetic mechanisms [36]. In this case, pups that had been less nurtured by their mothers had different methylation patterns at the *GR* gene, *Nr3c1*, than those that were more heavily nurtured. However, the methylation changes at *GR* that influenced the HPA axis were reversed when the pups were later placed with heavily grooming mothers. Prenatal exposure to maternal depression and anxiety also increases the methylation of *Nr3c1* and salivary cortisol stress levels [138]. These findings should have major repercussions if they can be applied to cases of childhood neglect or abuse, as these situations have also been shown to alter HPA stress responses and increase the risk of suicide. More specifically, human suicide abused victims have higher levels of methylation of the promoter for the hippocampal *GR* [139]. Additionally, maternal care has been associated with methylation of the *estrogen receptor-alpha 1b* promoter and *estrogen receptor-alpha 1b* expression in the medial preoptic area of female offspring [140]. The changes in this gene may be responsible for the transmission of maternal behavior across generations. Consequently, if disorders can be detected and treated at an early stage, then the persistence of abnormal behavior might also be prevented (Table 3).

The results of the above studies confirm the ability of the epigenome not

Tab. 3 Epigenetic changes in psychiatric disorders and potential therapy angles.

Species	*Psychiatric disease*	*Methylation change*	*Tissue(s)*	*Potential therapy?*
Human	Schizophrenia	*RELN* and GAD_{67} promoter hypermethylation	Brain	HDAC inhibitors; valproate, clozapine, sulpiride, MS-275
Human	Schizophrenia	Increased *DNMT1* and *DNMT3a*	Telencephalic GABAergic neurons, PBMCs	Methyl transferase inhibitors
Human	Bipolar disorder and schizophrenia	*MB-COMT* hypomethylation	Frontal lobe brain	–
Human	Autistic spectrum disorder	*MECP2* gain of function	Liver	–
Human	Depression/suicidal behavior	Hypermethylation of hippocampal *GR*	Brain	Fluoxetine increases HDAC expression; imipramine inhibits HDACs

PBMC, peripheral blood mononuclear cell.

only to change in response to the early environment, but also to be reversible or persistent later in life and in future generations. Since epigenetic abnormalities have been recognized in psychiatric disorders, schizophrenia, depression, and drug addiction, HDAC inhibitors are of major interest as they have the potential to modulate multiple pathways. Valproate (or valproic acid), which previously has been used for the treatment of acute mania, bipolar disorder, and schizophrenia, also inhibits HDAC activity [141]. Valproate is often coadministered with other antipsychotics, such as clozapine and sulpiride, and this combination has been shown to enhance an open-state chromatin conformation. In schizophrenic and bipolar patients, the ability to increase the accessibility to chromatin might counteract the hypermethylation and the reduced *RELN* and GAD_{67} expression that is thought to contribute to the disease etiology [142]. Both clozapine and sulpiride have also been shown to reduce *RELN* and GAD_{67} promoter methylation and to increase histone acetylation. At these genes, it was shown recently that HDAC inhibitors are able to disassociate the repressor complex of DNMTs, MeCP2, and HDAC1 from the promoter regions [143].

Epigenetic therapies for psychiatric disorders might also be found among cancer therapies. For example, the benzamide HDAC inhibitor MS-275, which currently is undergoing clinical trials for cancer, can also inhibit HDAC activity in the brain, thus increasing histone acetylation in multiple areas of the brain and *RELN* [144]. Interestingly, MS275 might also have antidepressant properties [145], with an increased selectivity and greater efficacy than valproate in increasing histone acetylation [146]. Three other schizophrenia/bipolar medications – fluoxetine, imipramine, and haloperidol – are known to induce epigenetic changes in the brain [120, 147, 148]. Fluoxetine (a selective serotonin reuptake inhibitor antidepressant) increases HDAC expression, while imipramine (a tricyclic antidepressant) downregulates HDACs [149].

The way in which both of these drugs target opposing mechanisms, yet still function effectively in the treatment of depression, has been of great interest to those investigators seeking to increase the specificity of epigenetic therapies in medicine. Currently, there is great unease regarding the long-term effects of epigenetic therapies on the entire genome, since epigenetic changes that are beneficial for some genes might be disadvantageous for others, leading to unwanted expressions or suppressions. The long-term effects on peripheral tissues must also be considered, as well as to the brain, as the pleiotrophic effects of these drugs are seen in different cell types. Additionally, if HDAC inhibitors are to be used for treating psychological disorders, the interactions between histone marks and methylation marks will be very important to determine, as DNA methylation signatures within the human brain vary significantly between regions [150]. The testing of potential HDAC inhibitors in humans will also need to be streamlined, and to that end it has been proposed that cell-based assays be performed in living patients such that the endpoints of GAD_{67} levels, histone acetylation levels, and chromatin immunoprecipitation signatures can be determined [151]. Ultimately, further information concerning complex psychological disorders needs to be acquired based on genome-scale approaches, so that biomarkers can be established not only to target novel pharmaceuticals but also to monitor the efficacy of drugs.

4
Cancer as an Epigenetic Disease

Epigenetics determine a cell's potency to differentiate and to set its behavior as a terminally differentiated cell type. In a terminally differentiated cell, it is the epigenome that silences genes that would result in abnormal proliferation, and activates the genes that are specific to that cell type. On this basis, cancer is essentially a epigenetic disease, as it is such gene expression that sets the cell's shape, size, membrane receptors, secreted signaling molecules, response to physical contact, decision on when to divide, decision on when to die, and every other factor that defines cancerous behavior.

The extent to which oncogenesis is the result of genetics or epigenetics, and whether cancer cells are stem cells that have become unregulated or mature cells that dedifferentiate to a stem-cell like phenotype, remain the subjects of much debate. What is known, however, is that the disruption of a limited set of master regulator pathways is an obligate event in carcinogenesis [152, 153], and results in a revised epigenetic signature where the key element is the silencing of tumor suppressor genes [154, 155]. Whatever the root cause that disposes a cell to becoming cancerous, when the result is an epigenetic signature, it is that very signature – and how it controls gene expression – that becomes the main target for treatment. Today, experimental drugs that are able to target the altered epigenetic regulation of tumor promotion and gene suppression are undergoing clinical trials. Clearly, it is hoped that the epigenetic approach may, in time, become the most effective method for the treatment of certain cancers.

4.1
History of Cancer Epigenetics

The concept that the cancer phenotype is epigenetic, and that it is reversible, began with early studies using mouse teratomas. Teratocarcinomas are highly malignant tumors that incorporate a heterogeneous assortment of cell types, including differentiated cells representative of the three primary germ layers (endo-, meso-, and ectoderm), as well as embryonic carcinoma cells. The first report that embryonic carcinoma cells were in fact multipotent stem cells capable of spontaneous differentiation into the other cells types seen in teratocarcinomas, was made in 1959 [156]. Shortly afterwards, in 1960, these discoveries were followed by claims that the differentiated daughter cells of the embryonic cells were benign [157, 158], while the first methods to modulate the differentiation of these cells *in vitro* were developed in 1961.

Further studies conducted during the 1960s led to additional examples of the role of differentiation in modulating carcinogenesis. In 1965, leukemic stem cells grown *in vitro* were seen to differentiate into macrophages and granulocytes [159, 160]. In an embryonal carcinoma of the testis, the cell of origin was shown to be the primordial germ cell [161] while, from a structural standpoint, the primordial germ cell and the carcinoma cells that developed from it were seen to be equally undifferentiated [162]. These findings implied that the embryonal carcinomas were not the result of dedifferentiation, but rather that they were due to the aberrant behavior of the stem cells.

This information inspired the conduct of further experiments into the capability of the blastocyst to reverse a variety of embryonic cancers, using assays of tumor-forming potential of blastocyst/carcinoma cells fusions [163–165]. The results of these experiments confirmed that, within the blastocyst environment, injected tumor cells would be differentiated along with the zygotic cells, and in doing so would lose their tumorigenic potential [165, 166]. This "resetting" was dependent, however, on the origin of the embryonic carcinoma corresponding to a cell type that was already present in the blastocyst; while the leukemia and sarcoma cells were still able to form tumors, the neuroblastoma was only marginally regulated [167].

In 1974, the idea of an epigenetic resetting of cancer cells was supported by the results of experiments with chimeric mice created from cancer cells. In this case, the injection of embryonic carcinoma cells into blastocysts led to the production of viable, teratoma-free, offspring, with markers from the carcinoma cells being detected in multiple cell types from all three germ layers [168–170]. Male chimeric mice were shown to produce functional sperm derived from the carcinoma cells, and thereby to produce offspring carrying markers from the original carcinoma line [169].

These results established that the development of an embryonic carcinoma is an epigenetic mechanism, as evidenced by the resetting of embryonic carcinoma cells by an environmental stimulus. However, there followed several years of controversial discussions related to the differences between teratomas and adult cancers arising in mature tissues. The main issue of contention was whether the nonembryonic cancers had arisen from cells that dedifferentiated and reverted to a corrupted stem cell-like state, or from stem cells that were present in mature tissues for the purposes of regeneration, but which become deregulated, much like the embryonic teratoma examples.

Experiments parallel to those with chimeric mice created from embryonic tumor cells helped to highlight the difference between these embryonic cells and adult tumor cells. Blastocyst transplant experiments have been performed using leukemia, lymphoma, and breast cancer. Cells derived by nuclear transfer from the tumor cells and then implanted into blastocysts showed that this environment could regulate the tumor nuclei, as the created cells develop normally, without abnormal proliferation [171, 172]. However, the chimeras generated from these cells, unlike those from embryonic cancer cells, showed an increased occurrence and severity of cancer after doxycycline-induced expression of a *RAS* transgene, as well as the occurrence of other cancer types. These findings indicated that, in these tumors, there is either a genetic basis for tumorigenesis or the epigenome is not truly reset.

The discovery of cancer stem cells in the early 2000s renewed the concept of undifferentiated pluripotent cells being the originators of cancers [173–177], similar to the undifferentiated cells of embryonic teratomas. Additionally, the search for adult tissue-specific stem cells highlighted similarities between cancer cells and adult stem cells. Cancerous cells were seen to have no functional gap junction intercellular communication (GJIC) [178, 179]; these observations helped lead to the search for organ-specific adult stem cells, via the isolation of normal, contact-insensitive adult cells that behave like GJIC-deficient cancer cells [180]. Finally, it transpired that when the "immortalizing virus" SV40 was used to transform normal adult breast cells, breast stem cells were effectively immortalized and blocked from differentiation, but normal breast epithelial cells were not [181].

While these results cannot completely exclude the dedifferentiation of cells in tumorigenesis, they are all points of correspondence between stem cells, either embryonic or adult tissue-specific, and cancer cells. The similarities found between cancer cells and stem cells indicate that the phenotype of cancer cells may have a basis in the same epigenetic mechanisms controlling stem cell growth and differentiation. Therefore, an understanding of the epigenome of normal cells – either stem or terminally differentiated – and how epigenomic changes relate to proliferation, differentiation, senescence and apoptosis, is guiding new approaches in cancer treatment. It has been proposed by G.B. Pierce, who conducted the initial investigations into early teratoma differentiation, that a promising alternative to cytotoxic treatments would be "... direction of differentiation of malignant to benign cells" [182]. The recent development of technologies to map the epigenome, combined with the ever-increasing capabilities of epigenome analysis, may provide the ability to achieve just that.

4.2 Epigenetic Markers in Cancerous Cells

While there are genetic mutations associated with cancer risk, the ability to measure epigenetic markers was accompanied by the determination of epigenotypes characteristic of cancer cells. While these epigenetic alterations may result from gene mutations in the master regulators of the epigenome, from errors in epigenome maintenance accumulated over time, or from inborn epigenetic abnormalities set at gametogenesis or post-fertilization, it is the regulation of gene expression by the epigenome that makes a cancer cell just that.

4.2.1 Cytosine Methylation

Of particular relevance to cancer is hypermethylation-induced gene silencing at promoters of DNA repair genes (Fig. 4). One particular example is the methylation-mediated silencing of the mismatch repair gene *MLH1*, which has been seen to give rise to colorectal cancers [183]. Likewise, a loss of expression of the DNA repair gene *MGMT* due to methylation makes cells sensitive to cancer induction by environmental exposure to alkylating carcinogens [184]. One connection between hypomethylation and cancer, which is independent of any specific gene regulation, is the genomic instability of unmethylated chromosomes, particularly in the pericentromeric regions [185, 186].

Beyond these specific effects, an examination of global methylation shows that cancer cells are characterized by the widespread hypomethylation of genes downstream from the promoters, but are hypermethylated at CpG islands that are usually unmethylated in normal cells (Fig. 4). Hypomethylation may allow the upregulation of the *MDR1* (multidrug resistance 1) gene [187], and a loss of normal patterns of silencing and expression of imprinted genes may be tumorigenic [188]. A fundamentally important epigenetic effect in carcinogenesis is the silencing of tumor suppressor genes by hypermethylation of promoter CpGs [189]. In fact, this is the most commonly observed mechanism of tumor suppressor inactivation.

Loss of tumor suppressor activity has been shown to disrupt almost all of the key pathways that are activated in cancer cells [152], including insensitivity to growth restriction signals, resistance to apoptosis, metastasis, angiogenesis, and

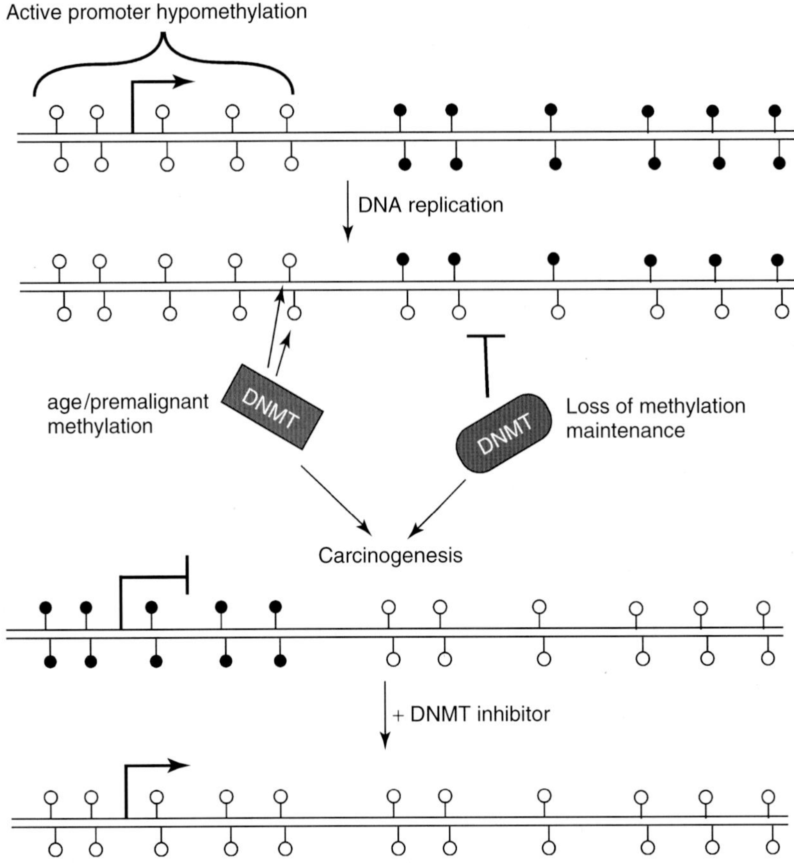

Fig. 4 Dysregulation of DNA methylation control in carcinogenesis. Maintenance of CpG methylation is necessary for correct gene expression. Active genes have hypomethylated promoters (open circles), and typically are methylated at downstream coding sequences (filled circles). DNA replication results in hemimethylated DNA, which must be recognized by maintenance methyltransferases to replicate methylation on the newly synthesized strand. Aging and carcinogenesis result in methylation spreading into the promoter, while maintenance of methylation downstream is lost. This results in silenced genes with hypermethylated promoters, which can be reversed over multiple cell cycles by methyltransferase inhibitors.

others (Table 4) [190]. With regards to the known tumor suppressor genes that are linked to cancer susceptibility when mutated in the germline, the epigenetic silencing of these genes has been observed in tumors in individuals without any of the tumorigenic mutations. Examples of these include *RB1, VHL, p16(INK4A), MLH1, BRCA1,* and *APC*. The hypermethlyation of *RB1, VHL,* and *BRCA1* is only seen in the same tumor types where germline mutations are observed (retinoblastoma, renal, breast, and ovarian cancer, respectively) [191–194]. Alternatively, *p16* has known germline mutations in melanoma [195] and pancreatic cancer [196], but is also inactivated by promoter hypermethylation in a variety of cancers,

Tab. 4 Methylated genes in key pathways for cancer development and growth.

Pathway	*Example gene*	*Cancer type(s)*
Growth signal autonomy	*RASSF1A, SOCS1*	Lung, bladder, ovarian, breast, lymphoma, MDS, gastric
Insensitivity to anti-growth	$P15^{INK4B}$	AML, MDS
signals	$P16^{INK4A}$	Melanoma, lymphoma, bladder
Apoptosis resistance	*DAPK*	Lymphoma
Invasion and metastasis	*CDH1, TIMP3*	GI, esophageal
Angiogenesis	*THBS1*	T-cell lymphoma, neuroblastoma, endometrial
Genetic instability	*MGMT*	Lymphoma, colon
	LMNA	Lymphoma
	MLH1	Colon
	CHFR	Gastric

MDS, myelodysplastic syndrome; AML, acute myeloid leukemia; GI, gastrointestinal.

including gastrointestinal, respiratory tract, gynecological, and hematopoietic [194, 197–199].

In other cancers, there is an apparent systematic promoter hypermethylation for multiple tumor suppressor genes; this was first recognized in a set of colorectal cancers, and labeled the CpG island methylator phenotype (CIMP) [200]. This phenotype has been associated in colon tumors with mutations in the *BRAF* oncogene and promoter methylation of the *MLH1* mismatch repair gene [201]. CIMP has also been seen in many other types of cancer, including glioblastoma, gastric, liver, pancreatic, esophageal, ovarian, and acute lymphoblastic and acute myeloid leukemias [201, 202].

Notably, there are distinctions between $CIMP^{+}$ and $CIMP^{-}$ tumors, which have characteristic differences in clinical and molecular features. It has been suggested that the existence of CIMP indicates an underlying molecular defect responsible for hypermethylation and epigenetic instability in cancer cells [202].

4.2.2 Methylation as a Regulator of Micro-RNAs

One class of noncoding transcripts, the micro-RNAs (miRNAs), has become very prominent in recent years for their functions in regulation of expression; it transpires that one function of miRNAs is as oncogenes and tumor suppressor genes [203]. miRNAs function at a post-transcriptional level by interacting with mRNAs via the RNA-induced silencing complex (RISC), blocking the message from progressing to translation. A comparison of the expression profiles of normal and cancerous cells showed that most normally expressed miRNAs are not expressed in cancer [204], with epigenetic silencing being a common mechanism for this shifted expression.

In one case, the miRNA miR-127 is a specific inhibitor of the oncogene *BCL-6*, while miR-127 is often underexpressed in cancer cells [205]. miR-127 is embedded in a CpG island, and treatment with methylation inhibitors and HDAC inhibitors can reactivate its expression, resulting in the downregulation of *BCL-6*

[206]. However, epigenetic therapies have shown no effect on miRNAs in lung cancer, which indicates that there are most likely different mechanisms responsible for miRNA downregulation in different cancer types [207].

4.2.3 Cancer Cell Clustering by Methylation Profile

The profiling of cancer cell lines has produced strong correlations between methylation patterns and cancer cell type [208], using over 400 genes chosen for cancer-relevance, including a majority of described cancer consensus genes [209]. In normal cells, the majority of regions (ca. 76%) analyzed had a mean methylation of under 15%, while almost all of the remaining regions (19%) had methylation in the range of 15% to 85%. In cancer cell lines, however, the average methylation shifted towards the medium range, with 49% of regions methylated less than 15%, and 49% in the range of 15–85%. However, as observed previously in other studies, for transcription start sites there is an approximately 1 kb core region with reduced DNA methylation [210, 211]. The maintenance of this unmethylated open window was observed in the cancer cell lines, which indicated that, despite the general promoter methylation increase in cancer, hypomethylation in these areas remained.

Possibly the most interesting result of this multigene methylation analysis is the product of cluster analysis of the results, grouping cell types by methylation similarities. When cluster analysis was performed on the methylation data from both normal cells and cancer cell lines combined, the normal cells all clustered together, with very little variability in methylation, despite the different cell types included. The cancer cell results all clustered separately from the normal, with greater variability in methylation level and methylated regions. It was notable that the cancer cells formed clusters based on cell type of origin, with colon cancer, central nervous system, and melanoma forming the strongest clusters. Non-small-cell lung cancer, renal carcinoma, and ovarian cancer formed weaker groups, while breast cancer samples showed no strong similarities to each other. These results were comparable to the findings from earlier studies that used gene expression or copy number variation [212, 213].

4.2.4 Age and Environmental Cancer Risks through Methylation

While methylation patterns are, for the most part, laid down in gametogenesis and early embryonic development, changes are observed over the lifespan of individuals, with implications for cancer. As described previously (see Sect. 1.3), methylation changes with age, consistent with the correlation of increased risk of cancer with age [214]. The implications are that the occurrence of diseases with an age-correlated risk – particularly cancer – may be related to these epigenetic changes.

With age, global methylation tends to decrease in somatic cells, while promoter methylation increases (Fig. 4); this pattern is similar to that described as the "average" cancer cell methylation profile, and suggests that abnormal methylation in cancer may be either sudden and tumor-specific, or a natural accumulation that reaches a tipping point. Methylation seems to spread from downstream intragenic sites towards the promoter, with gene silencing as a possible result [215]. There are, however, other specific links between methylation in aging cells and cancer cells. The tumor suppressor gene *RASSF1A* accumulates methylation with age in breast tissue, with methylation correlated to the risk of cancer

[216]. In colon cancer, estrogen receptors function as tumor suppressors, with aging normal colon epithelial cells showing an increased methylation of the estrogen receptor (ER) gene, and colon cancers showing a dense methylation of this gene [217]. Similarly, in normal bladder tissue adjacent to transitional cell carcinoma, there is an increased methylation of *DBC1* (deleted in bladder cancer precursor 1) with age, and *DBC1* is completely methylated in a large fraction of bladder tumors [218].

As previously described, environmental influences affect methylation patterns, and there are many correlated risk factors for cancer development. While the traditional test for carcinogenesis of a substance is its potential as a mutagen (the Ames test being the original "gold standard"), epigenetic alterations by toxins and environmental factors may be even more significant. For example, the association of tobacco smoke and cancer is well known, but such smoke is not a strong mutagen. What has been determined, however, is that lung cancers tied to smoking show a high methylation of several tumor suppressor genes [219, 220], including genes encoding cell adhesion proteins, apoptosis accelerators, and mitosis inhibitors. Additionally, synuclein-gamma, which is not normally expressed in lung, has been shown to be activated by cigarette smoke, via demethylation apparently due to the downregulation of DNMT3B, and this activation promotes spreading of the tumor [221].

4.2.5 **Histone Modifications**

The regulation of gene expression by covalent modifications of histones has been previously described (see Sect. 1.1), and a few alterations in histone modification are characteristic of human cancers, with loss of acetylation at Lys16 and trimethylation at Lys20 of histone H4 commonly seen [222] in association with the hypomethylation of repetitive DNA sequences (Fig. 5). Recently, it has been the deacetylation of histones that has become the most effective target in epigenetic cancer therapies. This typical deacetylation has been seen as an early event in mouse models, indicating that it is crucial in cancer development [222], and studies in gastrointestinal tumors have found it to be involved in invasive growth and metastasis [223]. Similar to the changes in DNA methylation observed due to age and environment, there are also changes in histone acetylation with time, as shown particularly well by the differences between monozygotic twins that occur with age [52].

The mechanisms by which HDACs and histone hypoacetylation are involved in carcinogenesis are not well understood; indeed, the indications are that more than one mechanism is involved. In the specific case of acute promyelocytic leukemia, chromosomal rearrangements produce fusion proteins that inappropriately recruit HDACs and repress the genes that regulate the normal differentiation and proliferation of myeloid cells [224]. As a more general mechanism, the expression of different HDACs is elevated in gastric, prostate, colon, breast, and cervical cancers. These results suggest that an aberrant recruitment of HDACs to promoter regions of tumor suppressor genes may be a common phenomenon in tumor development and progression. One such example is p21, an inhibitor of cell-cycle progression that has a reduced expression in a number of tumor types. In tumors as diverse as neuroblastoma, multiple myeloma, or alveolar rhabdomyosarcoma, treatment with HDAC inhibitors restores p21 expression while inhibiting growth or inducing apoptosis [225–227].

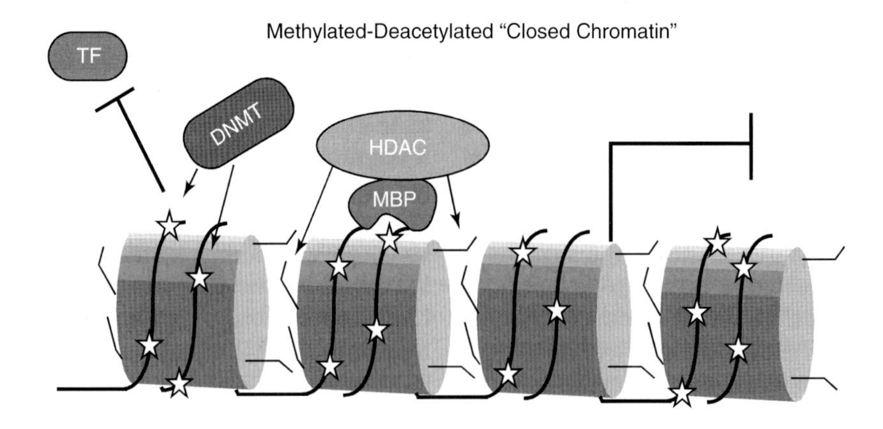

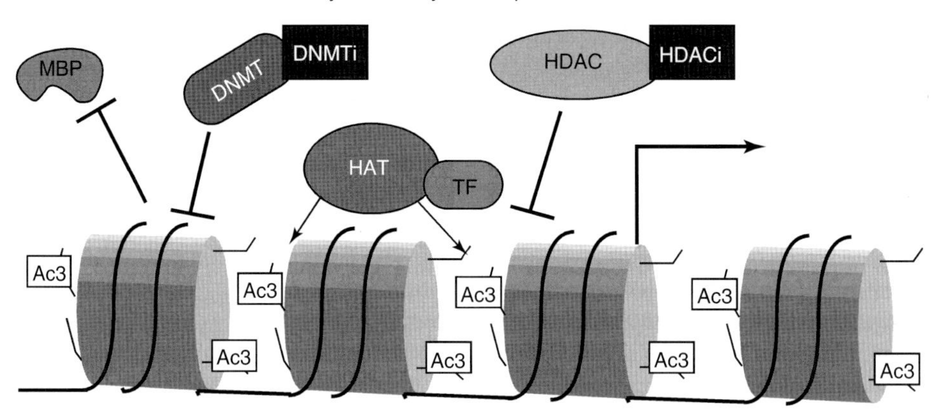

Fig. 5 Relative histone modifications of inactive/closed (upper panel) and active/open (lower panel) chromatin. Closed chromatin is characterized by DNA methylation (stars) generated by a DNA methyltransferase (DNMT), which can recruit a methyl-binding protein (MBP) complexed with a histone deacetylase (HDAC), resulting in deacetylated histone tails. This chromatin is inaccessible to a transcription factor (TF), and silenced. Open chromatin can be generated by the reduction of DNA methylation by a DNMT inhibitor (DNMTi), allowing histone acetyltransferase (HAT) to modify the histone tails, in this example, generating the triacetylated lysine characteristic of active chromatin (Ac3). Loss of methylation reduces HDAC recruitment and, in combination with blocking of deacetylation by an HDAC inhibitor (HDACi), allows transcription factors access to the promoter, activating the gene.

4.3 Epigenetic Drug Targets in Cancer

Both DNA methylation and histone acetylation are currently of great interest as targets for cancer therapy. As described previously, this approach has its basis in investigations into the role of cell-type identity and differentiation in carcinogenesis. Whether cancer cells are stem cells that have lost control over replication and differentiation, or are cells that have dedifferentiated, using epigenetic means to force differentiation and quiescence, apoptosis, or growth reduction, may represent an effective treatment approach. The advantage of epigenetic treatment rather than chemotherapy is the reduced toxicity towards normal cells. As will be described, the mechanism of their action is to reduce the replication of epigenetic marks after DNA replication, so that only dividing cells are affected, and only after multiple cell divisions. Epigenetic drugs may also be more effective in combination with each other, or with other therapies, as HDAC inhibitors may increase sensitivity to chemotherapy, or vice versa. The noncytotoxicity and specificity to actively dividing cells of epigenetic treatments, as well as the potential to reduce the doses of cytotoxic drugs, provides the potential for cancer treatment with reduced effects on healthy cells.

4.3.1 Demethylating Agents

Multiple ways of reducing promoter hypermethylation, and the presumed downregulation of tumor suppressor genes, have been developed; all of these target DNMT, either by blocking its function, or by depleting it from the nucleus. A number of effective inhibitors of DNMT exist as two types – nucleoside analogs and non-nucleoside analogs – which can reverse aberrant promoter hypermethylation. The more frequently used type are the nucleoside analogs, which are variants of cytidine with a modified cytosine ring, such as 5-azacytidine (5-aza-CR), 5-aza-2′-deoxycytidine (5-aza-CdR), 5-fluoro-2′-deoxycytidine, 5,6-dihydro-5-azacytidine, and zebularine [228]. In the cell, these analogs are converted by kinases to nucleotides that are incorporated into DNA during its replication in the S-phase of the cell cycle [229]. These bases act as substrates for DNMT, but the modifications form covalent bonds between the enzyme and the cytosine ring, thus inactivating the enzyme and depleting the nucleus of active DNMT for the post-replicative reproduction of methylation. After multiple cell cycles, this blocking of methylation and depletion of DNMT results in hypomethylated DNA (Fig. 4) [229, 230]. Both 5-aza-CR and 5-aza-CdR were originally designed as chemotherapeutic agents, but showed poor efficacy against solid tumors and severe bone marrow toxicity. However, their efficacy against acute myeloid leukemia (AML) and myelodysplastic syndrome (MDS) led to a reduction of dose levels to a range that favored hypomethylation over cytotoxicity. This required lower doses over a longer period of time, and proved to be successful in the treatment of hematopoietic cancers [231].

The non-nucleotide inhibitors of DNMT are not incorporated into DNA, and are believed to function by blocking the interaction of DNMT with cytosine, as for procaine [232], or by blocking the active site of DNMT1, as for the polyphenol compound of green tea (epigallocatechin-3-gallate) [233], hydralazine [234], and the synthetic compound RG108 [235]. Another approach is by employing RNA interference (RNAi), using an antisense oligonucleotide (MG98) to interfere with translation and initiate the degradation of DNMT mRNA [236].

4.3.2 Histone Deacetylase Inhibitors

Currently, there are 18 recognized HDACs, that have been assigned to four classes. HDAC class I members (HDACs 1, 2, 3, 8) are nuclear, and have histones and other nuclear proteins as targets. The class II HDACs (HDACs 4, 5, 6, 7, 9, 10) shuttle between the nucleus and cytoplasm and have tissue-specific expression; the IIa subclass members (HDACs 4, 5, 7, 9) interact with transcription factors, and their subcellular localization depends on a number of external factors and stimuli. Class IV has only one member (HDAC11), which is nuclear and has no clear function, while class III members (sirtuins 1–7) are involved in transcriptional regulation in response to cellular stress, including free radical levels [237].

Given the observed hypoacetylation of histones in silenced tumor suppressor genes, possibly instigated by promoter hypomethylation and the overexpression of HDACs, the reversal of this deacetylation represents a promising approach to cancer treatment (Fig. 5). The targeting of HDACs may be even more effective, as these regulate a smaller number of genes involved in cell growth, differentiation, and survival, with effects on angiogenesis and immunogenicity [238] (Table 5).

Among at least 15 HDAC inhibitors that have undergone preclinical and clinical trials, only one – vorinostat/SAHA (suberoylanilide hydroxamic acid; Zolinza; Merck, Whitehouse Station, NJ, USA) – has been approved by the US Food and Drug Administration [240]. Vorinostat has been approved for treatment of cutaneous T-cell lymphoma, and is currently being investigated for the treatment of several types of lymphoma and leukemia, as well as solid tumors.

Different HDAC inhibitors have different specificities, with some highly specific to one HDAC class, and others able to act on two or three of the classes (I and IIa, such as valproic acid and phenyl butyrate, or I, II, and IV, such as vorinostat and trichostatin A), while class III is affected only by a specific group of agents. Distinctions in the functions of the classes of drugs are also made by their abilities to reactivate genes with hypermethylated promoters, either in monotherapy or in combination with demethylating agents. For instance, HDAC I/IIa inhibitors must be used in combination with DNMT inhibitors to upregulate methylated genes *in vitro*; with this combination there is a strong synergy in reversing the silencing [241]. There is also evidence that this combination therapy is effective *in vivo* in AML [242], and that DNA methylation is decreased while histone acetylation is

Tab. 5 Classes of genes affected by HDAC inhibitors [239].

Function	***Upregulated***		***Downregulated***	
	mRNA	***Protein***	***mRNA***	***Protein***
Apoptosis	–	Fas, DR5, TRAIL, FasL, Bim, Bmf, Bik, Noxa, Bak	XIAP	Bcl-xL, Bcl-2, Mcl-1, XIAP
Cell-cycle arrest	p21	p21, p53	Cyclin B1	Cyclin B1, cyclin D1, cyclin D2, cyclin E
Angiogenesis	p53, VHL, TSP1, neurofibromin 2	p53, VHL, TSP1	HIF-1α, VEGF, FGF, VEGFR1, VEGFR2, CXCR4	HIF-1α, VEGF, FGF, CXCR4
Metastasis	KAI1, RECK, TIMP1	RhoB, RECK, TIMP1	ITGA5	–

re-established in tumor cells from responders [243]. HDAC III inhibitors, when administered singly, are able to upregulate silenced hypermethylated genes, although as yet there is no indication whether this effect will hold *in vivo*.

HDAC inhibitors are able to induce cell cycle arrest at G_0/G_1 or G_2/M checkpoints [244–248]. p21 is upregulated by HDAC inhibitors [244–247, 249–251], while multiple cyclins that regulate cell cycle transitions (B1, D1, D2, and E) are downregulated by HDAC inhibitors [246, 250, 252, 253].

With regards to angiogenesis, HDAC inhibitors have been seen to downregulate genes coding for angiogenesis-promoting proteins, such as hypoxia-inducible factor-1α and its target, vascular endothelial growth factor (VEGF), VEGF receptors 1 and 2, and CXC chemokine receptor 4 [254, 255]. Simultaneously, HDAC inhibitors upregulate the genes coding for angiogenesis suppressors, such as p53, von Hippel–Lindau, thrombospondin-1, and neurofibromin 2, in epithelial cells and in a number of cancers [256–258].

Possibly the most effective mechanism of HDAC inhibitors is the induction of apoptosis, which achieves the same goal as cytotoxic chemotherapy but with fewer collateral effects. The mechanisms for the induction of apoptosis can be either through extrinsic (death-receptor) and/or intrinsic (mitochondrial) pathways. HDAC inhibitors can activate the transcription of death receptors such as Fas and DR5, as well as their ligands, such as tumor necrosis factor-related apoptosis-inducing ligand (TRAIL) and Fas ligand (FasL) [259]. This transcriptional upregulation results in caspase-8 and -10 activation, and an initiation of the extrinsic apoptosis pathway. Activation of the intrinsic apoptotic pathway occurs by the inactivation of anti-apoptotic proteins, and activation of the pro-apoptotic Bcl-2 family of proteins [238, 260].

Several *in vitro* and *in vivo* studies have also demonstrated effective combinations of HDAC inhibitors with traditional chemotherapy agents, such as doxorubicine, etoposide, gemcitabine, and docetaxel [261–265]. The initial treatment of solid tumors with chemotherapeutics can debulk the tumors, allowing the induction of differentiation by HDAC/DNMT inhibitor treatment. Conversely, an initial treatment with epigenetically acting inhibitors may reactivate the tumor suppressor proteins, which can subsequently respond to conventional chemotherapy [266]. In acute promyelocytic leukemia, a combination of HDAC inhibitors with the differentiating agent all-*trans*-retinoic acid has proven effective [267, 268]. A summary of the possible combinations of epigenetic and traditional treatments, and their effects on cancer cells, is presented in Fig. 6.

HDAC inhibitors can also increase sensitivity to radiotherapy, and act synergistically with other targeted therapies such as proteasome inhibitors, kinase inhibitors, and death receptor agonists [265, 269–271]. One example of a combination therapy of this type is in cancer cell lines, for which HDAC inhibitors have limited efficacy. This may be due to the induction of genes facilitating tumor growth, such as NF-κB [272, 273], an anti-apoptotic factor, and Mcl-1 [274]. However, NF-κB expression can be downregulated by the proteasome inhibitor MC 132 [273], protein kinase inhibitor UCN 01 [275], or the specific inhibitor parthenolide [275], thereby increasing the efficacy of HDAC inhibitors.

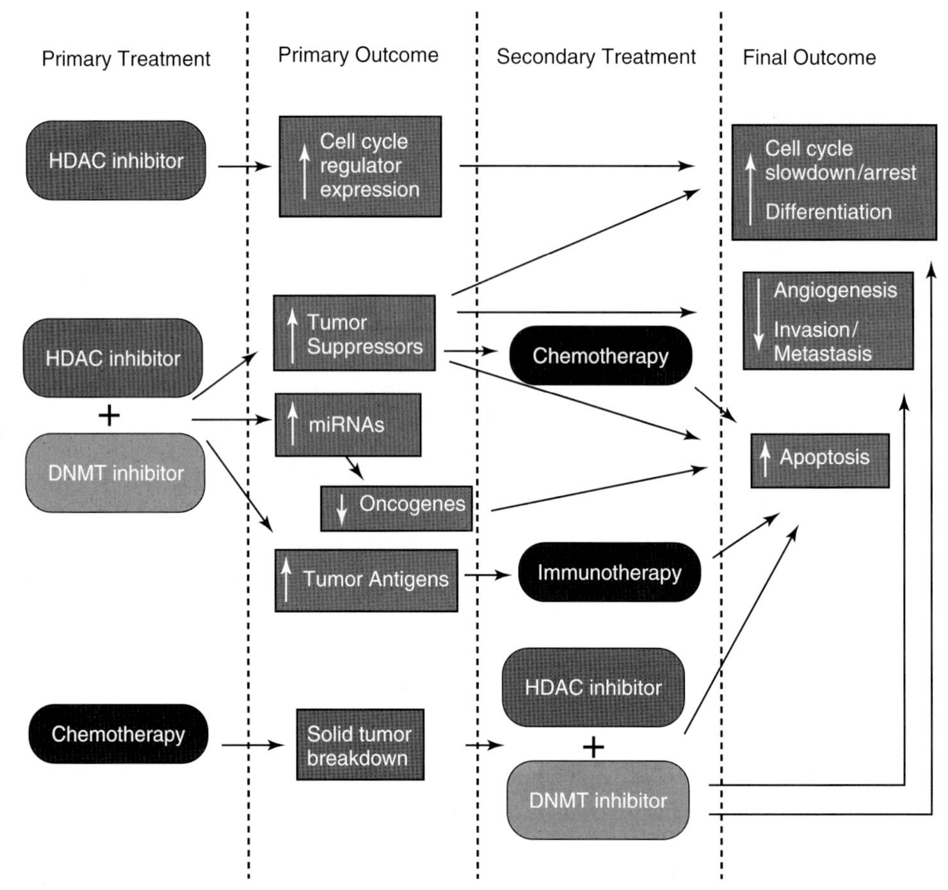

Fig. 6 Regulatory effects of cancer drugs targeting epigenetic modifications. Histone deacetylase (HDAC) inhibitors and DNA methyltransferase (DNMT) inhibitors function as anticancer drugs by upregulating genes that reverse cancer phenotypes. The end results are the restoration of a "normal" phenotype by differentiation or cell cycle arrest of cancer cells, induction of apoptosis, or reduction of tumor growth by blocking angiogenesis and metastasis-promoting genes. The effectiveness of these inhibitors can be increased by their combination with existing chemotherapeutic agents.

5 Summary

Today, the ability to understand epigenetics and its function in growth, development, and cellular identity is changing the way in which complex diseases and the influence of the environment on organisms are considered. Clearly, epigenetic effects on health can occur through changes in the epigenome during the life of an organism, or even across generations. The rapid response of the epigenome to external forces, through changes in DNA methylation and histone modifications, means that many diseases can arise from epigenetic changes, possibly in combination with underlying genetics. Autism, metabolic syndrome, psychiatric disorders, and cancer all have epigenetic components, and many of the known environmental risk factors for these diseases have epigenetic effects. To date, many advances have been made in understanding and treating a large majority of these diseases, particularly cancer. Further elucidation of the epigenomes of healthy organisms and disease states will undoubtedly advance the power of epigenetic medicine.

References

1 Laanpere, M., Altmae, S., Stavreus-Evers, A., Nilsson, T.K., Yngve, A., Salumets, A. (2010) Folate-mediated one-carbon metabolism and its effect on female fertility and pregnancy viability. *Nutr. Rev.*, **68**, 99–113.

2 Ulrich, C.M., Reed, M.C., Nijhout, H.F. (2008) Modeling folate, one-carbon metabolism, and DNA methylation. *Nutr. Rev.*, **66** (Suppl. 1), S27–S30.

3 Dobrovic, A., Kristensen, L.S. (2009) DNA methylation, epimutations and cancer predisposition. *Int. J. Biochem. Cell Biol.*, **41**, 34–39.

4 Waterland, R.A. (2006) Assessing the effects of high methionine intake on DNA methylation. *J. Nutr.*, **136**, 1706S–1710S.

5 Miranda, T.B., Jones, P.A. (2007) DNA methylation: the nuts and bolts of repression. *J. Cell Physiol.*, **213**, 384–390.

6 Ooi, S.K., O'Donnell, A.H., Bestor, T.H. (2009) Mammalian cytosine methylation at a glance. *J. Cell Sci.*, **122**, 2787–2791.

7 Bestor, T.H. (2000) The DNA methyltransferases of mammals. *Hum. Mol. Genet.*, **9**, 2395–2402.

8 Bird, A.P. (1986) CpG-rich islands and the function of DNA methylation. *Nature*, **321**, 209–213.

9 Takai, D., Jones, P.A. (2002) Comprehensive analysis of CpG islands in human chromosomes 21 and 22. *Proc. Natl Acad. Sci. USA*, **99**, 3740–3745.

10 Ikegami, K., Ohgane, J., Tanaka, S., Yagi, S., Shiota, K. (2009) Interplay between DNA methylation, histone modification and chromatin remodeling in stem cells and during development. *Int. J. Dev. Biol.*, **53**, 203–214.

11 Kim, G.D., Ni, J., Kelesoglu, N., Roberts, R.J., Pradhan, S. (2002) Co-operation and communication between the human maintenance and de novo DNA (cytosine-5) methyltransferases. *EMBO J.*, **21**, 4183–4195.

12 Xin, Z., Tachibana, M., Guggiari, M., Heard, E., Shinkai, Y., Wagstaff, J. (2003) Role of histone methyltransferase G9a in CpG methylation of the Prader-Willi syndrome imprinting center. *J. Biol. Chem.*, **278**, 14996–15000.

13 Talbert, P.B., Henikoff, S. (2010) Histone variants – ancient wrap artists of the epigenome. *Nat. Rev. Mol. Cell. Biol.*, **11**, 264–275.

14 Kacem, S., Feil, R. (2009) Chromatin mechanisms in genomic imprinting. *Mamm. Genome*, **20**, 544–556.

15 Royo, H., Cavaille, J. (2008) Non-coding RNAs in imprinted gene clusters. *Biol. Cell*, **100**, 149–166.

16 Sleutels, F., Zwart, R., Barlow, D.P. (2002) The non-coding Air RNA is required for silencing autosomal imprinted genes. *Nature*, **415**, 810–813.

17 Imamura, T., Yamamoto, S., Ohgane, J., Hattori, N., Tanaka, S., Shiota, K. (2004) Non-coding RNA directed DNA demethylation of Sphk1 CpG island. *Biochem. Biophys. Res. Commun.*, **322**, 593–600.

18 Ting, A.H., Schuebel, K.E., Herman, J.G., Baylin, S.B. (2005) Short double-stranded RNA induces transcriptional gene silencing in human cancer cells in the absence of DNA methylation. *Nat. Genet.*, **37**, 906–910.

19 Zaratiegui, M., Irvine, D.V., Martienssen, R.A. (2007) Noncoding RNAs and gene silencing. *Cell*, **128**, 763–776.

20 Pandey, R.R., Mondal, T., Mohammad, F., Enroth, S., Redrup, L., Komorowski, J., Nagano, T., Mancini-Dinardo, D., Kanduri, C. (2008) Kcnq1ot1 antisense noncoding RNA mediates lineage-specific transcriptional silencing through chromatin-level regulation. *Mol. Cell*, **32**, 232–246.

21 Aravin, A.A., Sachidanandam, R., Bourc'his, D., Schaefer, C., Pezic, D., Toth, K.F., Bestor, T., Hannon, G.J. (2008) A piRNA pathway primed by individual transposons is linked to *de novo* DNA methylation in mice. *Mol. Cell*, **31**, 785–799.

22 Falls, J.G., Pulford, D.J., Wylie, A.A., Jirtle, R.L. (1999) Genomic imprinting: implications for human disease. *Am. J. Pathol.*, **154**, 635–647.

23 Murphy, S.K., Jirtle, R.L. (2003) Imprinting evolution and the price of silence. *BioEssays*, **25**, 577–588.

24 Das, R., Hampton, D.D., Jirtle, R.L. (2009) Imprinting evolution and human health. *Mamm. Genome*, **20**, 563–572.

25 Luedi, P.P., Dietrich, F.S., Weidman, J.R., Bosko, J.M., Jirtle, R.L., Hartemink, A.J. (2007) Computational and experimental identification of novel human imprinted genes. *Genome Res.*, **17**, 1723–1730.

26 Shi, L., Wu, J. (2009) Epigenetic regulation in mammalian preimplantation embryo development. *Reprod. Biol. Endocrinol.*, **7**, 59.

27 Okano, M., Bell, D.W., Haber, D.A., Li, E. (1999) DNA methyltransferases Dnmt3a and Dnmt3b are essential for *de novo* methylation and mammalian development. *Cell*, **99**, 247–257.

28 Jirtle, R.L., Skinner, M.K. (2007) Environmental epigenomics and disease susceptibility. *Nat. Rev. Genet.*, **8**, 253–262.

29 Weaver, J.R., Susiarjo, M., Bartolomei, M.S. (2009) Imprinting and epigenetic changes in the early embryo. *Mamm. Genome*, **20**, 532–543.

30 Reik, W., Dean, W., Walter, J. (2001) Epigenetic reprogramming in mammalian development. *Science*, **293**, 1089–1093.

31 Gopalakrishnan, S., Van Emburgh, B.O., Robertson, K.D. (2008) DNA methylation in development and human disease. *Mutat. Res.*, **647**, 30–38.

32 Leonhardt, H., Page, A.W., Weier, H.U., Bestor, T.H. (1992) A targeting sequence directs DNA methyltransferase to sites of DNA replication in mammalian nuclei. *Cell*, **71**, 865–873.

33 Chen, T., Ueda, Y., Dodge, J.E., Wang, Z., Li, E. (2003) Establishment and maintenance of genomic methylation patterns in mouse embryonic stem cells by Dnmt3a and Dnmt3b. *Mol. Cell. Biol.*, **23**, 5594–5605.

34 Margueron, R., Reinberg, D. (2010) Chromatin structure and the inheritance of epigenetic information. *Nat. Rev. Genet.*, **11**, 285–296.

35 Waterland, R.A., Dolinoy, D.C., Lin, J.R., Smith, C.A., Shi, X., Tahiliani, K.G. (2006) Maternal methyl supplements increase offspring DNA methylation at Axin Fused. *Genesis*, **44**, 401–406.

36 Weaver, I.C., Cervoni, N., Champagne, F.A., D'Alessio, A.C., Sharma, S., Seckl, J.R., Dymov, S., Szyf, M., Meaney, M.J. (2004) Epigenetic programming by maternal behavior. *Nat. Neurosci.*, **7**, 847–854.

37 Anway, M.D., Cupp, A.S., Uzumcu, M., Skinner, M.K. (2005) Epigenetic transgenerational actions of endocrine disruptors and male fertility. *Science*, **308**, 1466–1469.

38 Dolinoy, D.C., Huang, D., Jirtle, R.L. (2007) Maternal nutrient supplementation counteracts bisphenol A-induced DNA hypomethylation in early development. *Proc. Natl Acad. Sci. USA*, **104**, 13056–13061.

39 Knudson, A.G. Jr (1986) Genetics of human cancer. *Annu. Rev. Genet.*, **20**, 231–251.

40 Lim, D., Maher, E. (2009) Human imprinting syndromes. *Epigenomics*, **1**, 347–369.

41 Douc-Rasy, S., Barrois, M., Fogel, S., Ahomadegbe, J.C., Stehelin, D., Coll, J., Riou, G. (1996) High incidence of loss of heterozygosity and abnormal imprinting of H19 and IGF2 genes in invasive cervical carcinomas. Uncoupling of H19 and IGF2 expression and biallelic hypomethylation of H19. *Oncogene*, **12**, 423–430.

42 Dowdy, S.C., Gostout, B.S., Shridhar, V., Wu, X., Smith, D.I., Podratz, K.C., Jiang, S.W. (2005) Biallelic methylation and silencing of paternally expressed gene 3 (PEG3) in gynecologic cancer cell lines. *Gynecol. Oncol.*, **99**, 126–134.

43 Hartmann, W., Koch, A., Brune, H., Waha, A., Schuller, U., Dani, I., Denkhaus, D., Langmann, W., Bode, U., Wiestler, O.D., Schilling, K., Pietsch, T. (2005) Insulin-like growth factor II is involved in the proliferation control of medulloblastoma and its cerebellar precursor cells. *Am. J. Pathol.*, **166**, 1153–1162.

44 Jarrard, D.F., Bussemakers, M.J., Bova, G.S., Isaacs, W.B. (1995) Regional loss of imprinting of the insulin-like growth factor II gene occurs in human prostate tissues. *Clin. Cancer Res.*, **1**, 1471–1478.

45 Kim, M.S., Lebron, C., Nagpal, J.K., Chae, Y.K., Chang, X., Huang, Y., Chuang, T., Yamashita, K., Trink, B., Ratovitski, E.A., Califano, J.A., Sidransky, D. (2008)

Methylation of the DFNA5 increases risk of lymph node metastasis in human breast cancer. *Biochem. Biophys. Res. Commun.*, **370**, 38–43.

46 Kim, S.J., Park, S.E., Lee, C., Lee, S.Y., Jo, J.H., Kim, J.M., Oh, Y.K. (2002) Alterations in promoter usage and expression levels of insulin-like growth factor-II and H19 genes in cervical carcinoma exhibiting biallelic expression of IGF-II. *Biochim. Biophys. Acta*, **1586**, 307–315.

47 Kohda, M., Hoshiya, H., Katoh, M., Tanaka, I., Masuda, R., Takemura, T., Fujiwara, M., Oshimura, M. (2001) Frequent loss of imprinting of IGF2 and MEST in lung adenocarcinoma. *Mol. Carcinogen.*, **31**, 184–191.

48 Murphy, S.K., Huang, Z., Wen, Y., Spillman, M.A., Whitaker, R.S., Simel, L.R., Nichols, T.D., Marks, J.R., Berchuck, A. (2006) Frequent IGF2/H19 domain epigenetic alterations and elevated IGF2 expression in epithelial ovarian cancer. *Mol. Cancer Res.*, **4**, 283–292.

49 Bjornsson, H.T., Sigurdsson, M.I., Fallin, M.D., Irizarry, R.A., Aspelund, T., Cui, H., Yu, W., Rongione, M.A., Ekstrom, T.J., Harris, T.B., Launer, L.J., Eiriksdottir, G., Leppert, M.F., Sapienza, C., Gudnason, V., Feinberg, A.P. (2008) Intra-individual change over time in DNA methylation with familial clustering. *J. Am. Med. Assoc.*, **299**, 2877–2883.

50 Christensen, B.C., Houseman, E.A., Marsit, C.J., Zheng, S., Wrensch, M.R., Wiemels, J.L., Nelson, H.H., Karagas, M.R., Padbury, J.F., Bueno, R., Sugarbaker, D.J., Yeh, R.F., Wiencke, J.K., Kelsey, K.T. (2009) Aging and environmental exposures alter tissue-specific DNA methylation dependent upon CpG island context. *PLoS Genet.*, **5**, e1000602.

51 Ronn, T., Poulsen, P., Hansson, O., Holmkvist, J., Almgren, P., Nilsson, P., Tuomi, T., Isomaa, B., Groop, L., Vaag, A., Ling, C. (2008) Age influences DNA methylation and gene expression of COX7A1 in human skeletal muscle. *Diabetologia*, **51**, 1159–1168.

52 Fraga, M.F., Ballestar, E., Paz, M.F., Ropero, S., Setien, F., Ballestar, M.L., Heine-Suner, D., Cigudosa, J.C., Urioste, M., Benitez, J., Boix-Chornet, M., Sanchez-Aguilera, A., Ling, C., Carlsson, E., Poulsen, P., Vaag, A., Stephan, Z., Spector, T.D., Wu, Y.Z., Plass, C., Esteller, M. (2005) Epigenetic differences arise during the lifetime of monozygotic twins. *Proc. Natl Acad. Sci. USA*, **102**, 10604–10609.

53 Javierre, B.M., Fernandez, A.F., Richter, J., Al-Shahrour, F., Martin-Subero, J.I., Rodriguez-Ubreva, J., Berdasco, M., Fraga, M.F., O'Hanlon, T.P., Rider, L.G., Jacinto, F.V., Lopez-Longo, F.J., Dopazo, J., Forn, M., Peinado, M.A., Carreno, L., Sawalha, A.H., Harley, J.B., Siebert, R., Esteller, M., Miller, F.W., Ballestar, E. (2010) Changes in the pattern of DNA methylation associate with twin discordance in systemic lupus erythematosus. *Genome Res.*, **20**, 170–179.

54 Szyf, M. (2009) The early life environment and the epigenome. *Biochim. Biophys. Acta*, **1790**, 878–885.

55 Pogribny, I.P., Basnakian, A.G., Miller, B.J., Lopatina, N.G., Poirier, L.A., James, S.J. (1995) Breaks in genomic DNA and within the p53 gene are associated with hypomethylation in livers of folate/methyl-deficient rats. *Cancer Res.*, **55**, 1894–1901.

56 Szyf, M. (1996) The DNA methylation machinery as a target for anticancer therapy. *Pharmacol. Ther.*, **70**, 1–37.

57 Ma, D.K., Guo, J.U., Ming, G.L., Song, H. (2009) DNA excision repair proteins and Gadd45 as molecular players for active DNA demethylation. *Cell Cycle*, **8**, 1526–1531.

58 Drake, A.J., Tang, J.I., Nyirenda, M.J. (2007) Mechanisms underlying the role of glucocorticoids in the early life programming of adult disease. *Clin. Sci. (Lond.)*, **113**, 219–232.

59 Murgatroyd, C., Wu, Y., Bockmuhl, Y., Spengler, D. (2010) Genes learn from stress: how infantile trauma programs us for depression. *Epigenetics*, **5**, 194–199.

60 Wu, H., Sun, Y.E. (2009) Reversing DNA methylation: new insights from neuronal activity-induced Gadd45b in adult neurogenesis. *Sci. Signal.*, **2**, pe17.

61 Bonilla-Henao, V., Martinez, R., Sobrino, F., Pintado, E. (2005) Different signaling pathways inhibit DNA methylation activity and up-regulate IFN-gamma in human lymphocytes. *J. Leukoc. Biol.*, **78**, 1339–1346.

62 MacLeod, A.R., Rouleau, J., Szyf, M. (1995) Regulation of DNA methylation by the Ras signaling pathway. *J. Biol. Chem.*, **270**, 11327–11337.
63 Richardson, B.C. (2002) Role of DNA methylation in the regulation of cell function: autoimmunity, aging and cancer. *J. Nutr.*, **132**, 2401S–2405S.
64 Heijmans, B.T., Tobi, E.W., Lumey, L.H., Slagboom, P.E. (2009) The epigenome: archive of the prenatal environment. *Epigenetics*, **4**, 526–531.
65 Hoyo, C., Schildkraut, J.M., Murphy, S.K., Chow, W.H., Vaughan, T.L., Risch, H., Marks, J.R., Jirtle, R.L., Calingaert, B., Mayne, S., Fraumeni, J., Jr, Gammon, M.D. (2009) IGF2R polymorphisms and risk of esophageal and gastric adenocarcinomas. *Int. J. Cancer*, **125**, 2673–2678.
66 Weidman, J.R., Dolinoy, D.C., Murphy, S.K., Jirtle, R.L. (2007) Cancer susceptibility: epigenetic manifestation of environmental exposures. *Cancer J.*, **13**, 9–16.
67 Dickies, M.M. (1962) A new viable yellow mutation in the house mouse. *J. Hered.*, **53**, 84–86.
68 Waterland, R.A., Jirtle, R.L. (2003) Transposable elements: targets for early nutritional effects on epigenetic gene regulation. *Mol. Cell. Biol.*, **23**, 5293–5300.
69 Dolinoy, D.C., Jirtle, R.L. (2008) Environmental epigenomics in human health and disease. *Environ. Mol. Mutagen.*, **49**, 4–8.
70 Waterland, R.A., Jirtle, R.L. (2004) Early nutrition, epigenetic changes at transposons and imprinted genes, and enhanced susceptibility to adult chronic diseases. *Nutrition*, **20**, 63–68.
71 Kaminen-Ahola, N., Ahola, A., Maga, M., Mallitt, K.A., Fahey, P., Cox, T.C., Whitelaw, E., Chong, S. (2010) Maternal ethanol consumption alters the epigenotype and the phenotype of offspring in a mouse model. *PLoS Genet.*, **6**, e1000811.
72 Liu, Y., Balaraman, Y., Wang, G., Nephew, K.P., Zhou, F.C. (2009) Alcohol exposure alters DNA methylation profiles in mouse embryos at early neurulation. *Epigenetics*, **4**, 500–511.
73 Ouko, L.A., Shantikumar, K., Knezovich, J., Haycock, P., Schnugh, D.J., Ramsay, M. (2009) Effect of alcohol consumption on CpG methylation in the differentially methylated regions of H19 and IG-DMR in male gametes: implications for fetal alcohol spectrum disorders. *Alcohol. Clin. Exp. Res.*, **33**, 1615–1627.
74 Morgan, H.D., Jin, X.L., Li, A., Whitelaw, E., O'Neill, C. (2008) The culture of zygotes to the blastocyst stage changes the postnatal expression of an epigenetically labile allele, agouti viable yellow, in mice. *Biol. Reprod.*, **79**, 618–623.
75 Katari, S., Turan, N., Bibikova, M., Erinle, O., Chalian, R., Foster, M., Gaughan, J.P., Coutifaris, C., Sapienza, C. (2009) DNA methylation and gene expression differences in children conceived in vitro or in vivo. *Hum. Mol. Genet.*, **18**, 3769–3778.
76 Manipalviratn, S., DeCherney, A., Segars, J. (2009) Imprinting disorders and assisted reproductive technology. *Fertil. Steril.*, **91**, 305–315.
77 Bromer, J.G., Zhou, Y., Taylor, M.B., Doherty, L., Taylor, H.S. (2010) Bisphenol-A exposure in utero leads to epigenetic alterations in the developmental programming of uterine estrogen response. *FASEB J.*, **24**, 2273–2280.
78 Prins, G.S., Tang, W.Y., Belmonte, J., Ho, S.M. (2008) Perinatal exposure to oestradiol and bisphenol A alters the prostate epigenome and increases susceptibility to carcinogenesis. *Basic Clin. Pharmacol. Toxicol.*, **102**, 134–138.
79 Yaoi, T., Itoh, K., Nakamura, K., Ogi, H., Fujiwara, Y., Fushiki, S. (2008) Genome-wide analysis of epigenomic alterations in fetal mouse forebrain after exposure to low doses of bisphenol A. *Biochem. Biophys. Res. Commun.*, **376**, 563–567.
80 Druker, R., Bruxner, T.J., Lehrbach, N.J., Whitelaw, E. (2004) Complex patterns of transcription at the insertion site of a retrotransposon in the mouse. *Nucleic Acids Res.*, **32**, 5800–5808.
81 Dumitrescu, R.G. (2009) Epigenetic targets in cancer epidemiology. *Methods Mol. Biol.*, **471**, 457–467.
82 Kuff, E.L., Lueders, K.K. (1988) The intracisternal A-particle gene family: structure and functional aspects. *Adv. Cancer Res.*, **51**, 183–276.
83 Vasicek, T.J., Zeng, L., Guan, X.J., Zhang, T., Costantini, F., Tilghman, S.M. (1997) Two dominant mutations in the mouse

fused gene are the result of transposon insertions. *Genetics*, **147**, 777–786.

84 Roseboom, T., de Rooij, S., Painter, R. (2006) The Dutch famine and its long-term consequences for adult health. *Early Hum. Dev.*, **82**, 485–491.

85 Painter, R.C., Osmond, C., Gluckman, P., Hanson, M., Phillips, D.I., Roseboom, T.J. (2008) Transgenerational effects of prenatal exposure to the Dutch famine on neonatal adiposity and health in later life. *BJOG*, **115**, 1243–1249.

86 Barker, D.J., Osmond, C., Kajantie, E., Eriksson, J.G. (2009) Growth and chronic disease: findings in the Helsinki Birth Cohort. *Ann. Hum. Biol.*, **36**, 445–458.

87 Franzek, E.J., Sprangers, N., Janssens, A.C., Van Duijn, C.M., Van De Wetering, B.J. (2008) Prenatal exposure to the 1944-45 Dutch 'hunger winter' and addiction later in life. *Addiction*, **103**, 433–438.

88 Hoek, H.W., Brown, A.S., Susser, E. (1998) The Dutch famine and schizophrenia spectrum disorders. *Soc. Psychiatry Psychiatr. Epidemiol.*, **33**, 373–379.

89 Susser, E.S., Lin, S.P. (1992) Schizophrenia after prenatal exposure to the Dutch Hunger Winter of 1944–1945. *Arch. Gen. Psychiatry*, **49**, 983–988.

90 Heijmans, B.T., Tobi, E.W., Stein, A.D., Putter, H., Blauw, G.J., Susser, E.S., Slagboom, P.E., Lumey, L.H. (2008) Persistent epigenetic differences associated with prenatal exposure to famine in humans. *Proc. Natl Acad. Sci. USA*, **105**, 17046–17049.

91 Tobi, E.W., Lumey, L.H., Talens, R.P., Kremer, D., Putter, H., Stein, A.D., Slagboom, P.E., Heijmans, B.T. (2009) DNA methylation differences after exposure to prenatal famine are common and timing- and sex-specific. *Hum. Mol. Genet.*, **18**, 4046–4053.

92 Symonds, M.E., Sebert, S.P., Hyatt, M.A., Budge, H. (2009) Nutritional programming of the metabolic syndrome. *Nat. Rev. Endocrinol.*, **5**, 604–610.

93 Burdge, G.C., Lillycrop, K.A., Jackson, A.A. (2009) Nutrition in early life, and risk of cancer and metabolic disease: alternative endings in an epigenetic tale? *Br. J. Nutr.*, **101**, 619–630.

94 Gluckman, P.D., Hanson, M.A., Buklijas, T., Low, F.M., Beedle, A.S. (2009) Epigenetic mechanisms that underpin metabolic and cardiovascular diseases. *Nat. Rev. Endocrinol.*, **5**, 401–408.

95 Warner, M.J., Ozanne, S.E. (2010) Mechanisms involved in the developmental programming of adulthood disease. *Biochem. J.*, **427**, 333–347.

96 Lillycrop, K.A., Phillips, E.S., Torrens, C., Hanson, M.A., Jackson, A.A., Burdge, G.C. (2008) Feeding pregnant rats a protein-restricted diet persistently alters the methylation of specific cytosines in the hepatic PPAR alpha promoter of the offspring. *Br. J. Nutr.*, **100**, 278–282.

97 Lillycrop, K.A., Phillips, E.S., Jackson, A.A., Hanson, M.A., Burdge, G.C. (2005) Dietary protein restriction of pregnant rats induces and folic acid supplementation prevents epigenetic modification of hepatic gene expression in the offspring. *J. Nutr.*, **135**, 1382–1386.

98 Burdge, G.C., Slater-Jefferies, J., Torrens, C., Phillips, E.S., Hanson, M.A., Lillycrop, K.A. (2007) Dietary protein restriction of pregnant rats in the F0 generation induces altered methylation of hepatic gene promoters in the adult male offspring in the F1 and F2 generations. *Br. J. Nutr.*, **97**, 435–439.

99 Burdge, G.C., Lillycrop, K.A., Phillips, E.S., Slater-Jefferies, J.L., Jackson, A.A., Hanson, M.A. (2009) Folic acid supplementation during the juvenile-pubertal period in rats modifies the phenotype and epigenotype induced by prenatal nutrition. *J. Nutr.*, **139**, 1054–1060.

100 van Straten, E.M., Bloks, V.W., Huijkman, N.C., Baller, J.F., Meer, H., Lutjohann, D., Kuipers, F., Plosch, T. (2010) The liver X-receptor gene promoter is hypermethylated in a mouse model of prenatal protein restriction. *Am. J. Physiol. Regul. Integr. Comp. Physiol.*, **298**, R275–R282.

101 Bogdarina, I., Welham, S., King, P.J., Burns, S.P., Clark, A.J. (2007) Epigenetic modification of the renin-angiotensin system in the fetal programming of hypertension. *Circ. Res.*, **100**, 520–526.

102 Bogdarina, I., Haase, A., Langley-Evans, S., Clark, A.J. (2010) Glucocorticoid effects on the programming of AT1b angiotensin receptor gene methylation and expression in the rat. *PLoS ONE*, **5**, e9237.

103 Brons, C., Jacobsen, S., Nilsson, E., Ronn, T., Jensen, C.B., Storgaard, H., Poulsen, P., Groop, L., Ling, C., Astrup, A., Vaag, A. (2010) Deoxyribonucleic acid methylation and gene expression of PPARGC1A in human muscle is influenced by high-fat overfeeding in a birth-weight-dependent manner. *J. Clin. Endocrinol. Metab.*, **95**, 3048–3056.

104 Ling, C., Del Guerra, S., Lupi, R., Ronn, T., Granhall, C., Luthman, H., Masiello, P., Marchetti, P., Groop, L., Del Prato, S. (2008) Epigenetic regulation of PPARGC1A in human type 2 diabetic islets and effect on insulin secretion. *Diabetologia*, **51**, 615–622.

105 Bruce, K.D., Hanson, M.A. (2010) The developmental origins, mechanisms, and implications of metabolic syndrome. *J. Nutr.*, **140**, 648–652.

106 Plagemann, A., Roepke, K., Harder, T., Brunn, M., Harder, A., Wittrock-Staar, M., Ziska, T., Schellong, K., Rodekamp, E., Melchior, K., Dudenhausen, J.W. (2010) Epigenetic malprogramming of the insulin receptor promoter due to developmental overfeeding. *J. Perinat. Med.*, **38**, 393–400.

107 Plagemann, A., Harder, T., Brunn, M., Harder, A., Roepke, K., Wittrock-Staar, M., Ziska, T., Schellong, K., Rodekamp, E., Melchior, K., Dudenhausen, J.W. (2009) Hypothalamic proopiomelanocortin promoter methylation becomes altered by early overfeeding: an epigenetic model of obesity and the metabolic syndrome. *J. Physiol.*, **587**, 4963–4976.

108 Gemma, C., Sookoian, S., Alvarinas, J., Garcia, S.I., Quintana, L., Kanevsky, D., Gonzalez, C.D., Pirola, C.J. (2009) Maternal pregestational BMI is associated with methylation of the PPARGC1A promoter in newborns. *Obesity (Silver Spring)*, **17**, 1032–1039.

109 Yamamoto, Y., Gesta, S., Lee, K.Y., Tran, T.T., Saadatirad, P., Kahn, C.R. (2010) Adipose depots possess unique developmental gene signatures. *Obesity (Silver Spring)*, **18**, 872–878.

110 Raychaudhuri, N., Raychaudhuri, S., Thamotharan, M., Devaskar, S.U. (2008) Histone code modifications repress glucose transporter 4 expression in the intrauterine growth-restricted offspring. *J. Biol. Chem.*, **283**, 13611–13626.

111 Gray, S.G., De Meyts, P. (2005) Role of histone and transcription factor acetylation in diabetes pathogenesis. *Diabetes Metab. Res. Rev.*, **21**, 416–433.

112 Mootha, V.K., Lindgren, C.M., Eriksson, K.F., Subramanian, A., Sihag, S., Lehar, J., Puigserver, P., Carlsson, E., Ridderstrale, M., Laurila, E., Houstis, N., Daly, M.J., Patterson, N., Mesirov, J.P., Golub, T.R., Tamayo, P., Spiegelman, B., Lander, E.S., Hirschhorn, J.N., Altshuler, D., Groop, L.C. (2003) PGC-1alpha-responsive genes involved in oxidative phosphorylation are coordinately downregulated in human diabetes. *Nat. Genet.*, **34**, 267–273.

113 Powers, A.C., Philippe, J., Hermann, H., Habener, J.F. (1988) Sodium butyrate increases glucagon and insulin gene expression by recruiting immunocytochemically negative cells to produce hormone. *Diabetes*, **37**, 1405–1410.

114 Lagace, D.C., Nachtigal, M.W. (2004) Inhibition of histone deacetylase activity by valproic acid blocks adipogenesis. *J. Biol. Chem.*, **279**, 18851–18860.

115 Vickers, M.H., Gluckman, P.D., Coveny, A.H., Hofman, P.L., Cutfield, W.S., Gertler, A., Breier, B.H., Harris, M. (2008) The effect of neonatal leptin treatment on postnatal weight gain in male rats is dependent on maternal nutritional status during pregnancy. *Endocrinology*, **149**, 1906–1913.

116 Kirk, H., Cefalu, W.T., Ribnicky, D., Liu, Z., Eilertsen, K.J. (2008) Botanicals as epigenetic modulators for mechanisms contributing to development of metabolic syndrome. *Metabolism*, **57**, S16–S23.

117 Afman, L., Muller, M. (2006) Nutrigenomics: from molecular nutrition to prevention of disease. *J. Am. Diet. Assoc.*, **106**, 569–576.

118 Chatkupt, S., Antonowicz, M., Johnson, W.G. (1995) Parents do matter: genomic imprinting and parental sex effects in neurological disorders. *J. Neurol. Sci.*, **130**, 1–10.

119 Ptak, C., Petronis, A. (2010) Epigenetic approaches to psychiatric disorders. *Dialogues Clin. Neurosci.*, **12**, 25–35.

120 Tsankova, N.M., Berton, O., Renthal, W., Kumar, A., Neve, R.L., Nestler, E.J. (2006) Sustained hippocampal chromatin regulation in a mouse model of depression

and antidepressant action. *Nat. Neurosci.*, **9**, 519–525.

121 Song, S., Wang, W., Hu, P. (2009) Famine, death, and madness: schizophrenia in early adulthood after prenatal exposure to the Chinese Great Leap Forward Famine. *Soc. Sci. Med.*, **68**, 1315–1321.

122 Grayson, D.R., Jia, X., Chen, Y., Sharma, R.P., Mitchell, C.P., Guidotti, A., Costa, E. (2005) Reelin promoter hypermethylation in schizophrenia. *Proc. Natl Acad. Sci. USA*, **102**, 9341–9346.

123 Tamura, Y., Kunugi, H., Ohashi, J., Hohjoh, H. (2007) Epigenetic aberration of the human REELIN gene in psychiatric disorders. *Mol. Psychiatry*, **12**, 519, 593–600.

124 Lintas, C., Persico, A.M. (2010) Neocortical RELN promoter methylation increases significantly after puberty. *NeuroReport*, **21**, 114–118.

125 Zhubi, A., Veldic, M., Puri, N.V., Kadriu, B., Caruncho, H., Loza, I., Sershen, H., Lajtha, A., Smith, R.C., Guidotti, A., Davis, J.M., Costa, E. (2009) An upregulation of DNA-methyltransferase 1 and 3a expressed in telencephalic GABAergic neurons of schizophrenia patients is also detected in peripheral blood lymphocytes. *Schizophr. Res.*, **111**, 115–122.

126 Veldic, M., Kadriu, B., Maloku, E., Agis-Balboa, R.C., Guidotti, A., Davis, J.M., Costa, E. (2007) Epigenetic mechanisms expressed in basal ganglia GABAergic neurons differentiate schizophrenia from bipolar disorder. *Schizophr. Res.*, **91**, 51–61.

127 Mill, J., Tang, T., Kaminsky, Z., Khare, T., Yazdanpanah, S., Bouchard, L., Jia, P., Assadzadeh, A., Flanagan, J., Schumacher, A., Wang, S.C., Petronis, A. (2008) Epigenomic profiling reveals DNA-methylation changes associated with major psychosis. *Am. J. Hum. Genet.*, **82**, 696–711.

128 Kuratomi, G., Iwamoto, K., Bundo, M., Kusumi, I., Kato, N., Iwata, N., Ozaki, N., Kato, T. (2008) Aberrant DNA methylation associated with bipolar disorder identified from discordant monozygotic twins. *Mol. Psychiatry*, **13**, 429–441.

129 Rosa, A., Picchioni, M.M., Kalidindi, S., Loat, C.S., Knight, J., Toulopoulou, T., Vonk, R., van der Schot, A.C., Nolen, W., Kahn, R.S., McGuffin, P., Murray, R.M., Craig, I.W. (2008) Differential methylation of the X-chromosome is a possible source of discordance for bipolar disorder female monozygotic twins. *Am. J. Med. Genet. B. Neuropsychiatr. Genet.*, **147B**, 459–462.

130 Abdolmaleky, H.M., Cheng, K.H., Faraone, S.V., Wilcox, M., Glatt, S.J., Gao, F., Smith, C.L., Shafa, R., Aeali, B., Carnevale, J., Pan, H., Papageorgis, P., Ponte, J.F., Sivaraman, V., Tsuang, M.T., Thiagalingam, S. (2006) Hypomethylation of MB-COMT promoter is a major risk factor for schizophrenia and bipolar disorder. *Hum. Mol. Genet.*, **15**, 3132–3145.

131 Abdolmaleky, H.M., Smith, C.L., Zhou, J.R., Thiagalingam, S. (2008) Epigenetic alterations of the dopaminergic system in major psychiatric disorders. *Methods Mol. Biol.*, **448**, 187–212.

132 Badcock, C., Crespi, B. (2006) Imbalanced genomic imprinting in brain development: an evolutionary basis for the aetiology of autism. *J. Evol. Biol.*, **19**, 1007–1032.

133 Mehler, M.F., Purpura, D.P. (2009) Autism, fever, epigenetics and the locus coeruleus. *Brain Res. Rev.*, **59**, 388–392.

134 Lasalle, J.M., Yasui, D.H. (2009) Evolving role of MeCP2 in Rett syndrome and autism. *Epigenomics*, **1**, 119–130.

135 Ramocki, M.B., Peters, S.U., Tavyev, Y.J., Zhang, F., Carvalho, C.M., Schaaf, C.P., Richman, R., Fang, P., Glaze, D.G., Lupski, J.R., Zoghbi, H.Y. (2009) Autism and other neuropsychiatric symptoms are prevalent in individuals with MeCP2 duplication syndrome. *Ann. Neurol.*, **66**, 771–782.

136 Plagge, A., Isles, A.R., Gordon, E., Humby, T., Dean, W., Gritsch, S., Fischer-Colbrie, R., Wilkinson, L.S., Kelsey, G. (2005) Imprinted Nesp55 influences behavioral reactivity to novel environments. *Mol. Cell. Biol.*, **25**, 3019–3026.

137 James, S.J., Melnyk, S., Jernigan, S., Pavliv, O., Trusty, T., Lehman, S., Seidel, L., Gaylor, D.W., Cleves, M.A. (2010) A functional polymorphism in the reduced folate carrier gene and DNA hypomethylation in mothers of children with autism. *Am. J. Med. Genet. B. Neuropsychiatr. Genet.*, **153B**, 1209–1220.

138 Oberlander, T.F., Weinberg, J., Papsdorf, M., Grunau, R., Misri, S., Devlin, A.M. (2008) Prenatal exposure to maternal depression, neonatal methylation of human glucocorticoid receptor gene (NR3C1) and

infant cortisol stress responses. *Epigenetics*, **3**, 97–106.

139 McGowan, P.O., Sasaki, A., D'Alessio, A.C., Dymov, S., Labonte, B., Szyf, M., Turecki, G., Meaney, M.J. (2009) Epigenetic regulation of the glucocorticoid receptor in human brain associates with childhood abuse. *Nat. Neurosci.*, **12**, 342–348.

140 Champagne, F.A., Weaver, I.C., Diorio, J., Dymov, S., Szyf, M., Meaney, M.J. (2006) Maternal care associated with methylation of the estrogen receptor-alpha1b promoter and estrogen receptor-alpha expression in the medial preoptic area of female offspring. *Endocrinology*, **147**, 2909–2915.

141 Haddad, P.M., Das, A., Ashfaq, M., Wieck, A. (2009) A review of valproate in psychiatric practice. *Expert Opin. Drug Metab. Toxicol.*, **5**, 539–551.

142 Guidotti, A., Dong, E., Kundakovic, M., Satta, R., Grayson, D.R., Costa, E. (2009) Characterization of the action of antipsychotic subtypes on valproate-induced chromatin remodeling. *Trends Pharmacol. Sci.*, **30**, 55–60.

143 Kundakovic, M., Chen, Y., Guidotti, A., Grayson, D.R. (2009) The reelin and GAD67 promoters are activated by epigenetic drugs that facilitate the disruption of local repressor complexes. *Mol. Pharmacol.*, **75**, 342–354.

144 Simonini, M.V., Camargo, L.M., Dong, E., Maloku, E., Veldic, M., Costa, E., Guidotti, A. (2006) The benzamide MS-275 is a potent, long-lasting brain region-selective inhibitor of histone deacetylases. *Proc. Natl Acad. Sci. USA*, **103**, 1587–1592.

145 Covington, H.E. III, Maze, I., LaPlant, Q.C., Vialou, V.F., Ohnishi, Y.N., Berton, O., Fass, D.M., Renthal, W., Rush, A.J. III, Wu, E.Y., Ghose, S., Krishnan, V., Russo, S.J., Tamminga, C., Haggarty, S.J., Nestler, E.J. (2009) Antidepressant actions of histone deacetylase inhibitors. *J. Neurosci.*, **29**, 11451–11460.

146 Grayson, D.R., Kundakovic, M., Sharma, R.P. (2010) Is there a future for histone deacetylase inhibitors in the pharmacotherapy of psychiatric disorders? *Mol. Pharmacol.*, **77**, 126–135.

147 Feinberg, A.P. (2010) Genome-scale approaches to the epigenetics of common human disease. *Virchows Arch.*, **456**, 13–21.

148 Shimabukuro, M., Jinno, Y., Fuke, C., Okazaki, Y. (2006) Haloperidol treatment induces tissue- and sex-specific changes in DNA methylation: a control study using rats. *Behav. Brain Funct.*, **2**, 37.

149 Ptak, C., Petronis, A. (2008) Epigenetics and complex disease: from etiology to new therapeutics. *Annu. Rev. Pharmacol. Toxicol.*, **48**, 257–276.

150 Ladd-Acosta, C., Pevsner, J., Sabunciyan, S., Yolken, R.H., Webster, M.J., Dinkins, T., Callinan, P.A., Fan, J.B., Potash, J.B., Feinberg, A.P. (2007) DNA methylation signatures within the human brain. *Am. J. Hum. Genet.*, **81**, 1304–1315.

151 Gavin, D.P., Kartan, S., Chase, K., Jayaraman, S., Sharma, R.P. (2009) Histone deacetylase inhibitors and candidate gene expression: An in vivo and in vitro approach to studying chromatin remodeling in a clinical population. *J. Psychiatr. Res.*, **43**, 870–876.

152 Hanahan, D., Weinberg, R.A. (2000) The hallmarks of cancer. *Cell*, **100**, 57–70.

153 Vogelstein, B., Kinzler, K.W. (2004) Cancer genes and the pathways they control. *Nat. Med.*, **10**, 789–799.

154 Herman, J.G., Baylin, S.B. (2003) Gene silencing in cancer in association with promoter hypermethylation. *N. Engl. J. Med.*, **349**, 2042–2054.

155 Jones, P.A., Baylin, S.B. (2002) The fundamental role of epigenetic events in cancer. *Nat. Rev. Genet.*, **3**, 415–428.

156 Pierce, G.B., Dixon, F.J. Jr (1959) Testicular teratomas. I Demonstration of teratogenesis by metamorphosis of multipotential cells. *Cancer*, **12**, 573–583.

157 Pierce, G.B. Jr, Dixon, F.J. Jr, Verney, E.L. (1960) Teratocarcinogenic and tissue-forming potentials of the cell types comprising neoplastic embryoid bodies. *Lab. Invest.*, **9**, 583–602.

158 Pierce, G.B. Jr, Verney, E.L. (1961) An in vitro and in vivo study of differentiation in teratocarcinomas. *Cancer*, **14**, 1017–1029.

159 Bradley, T.R., Metcalf, D. (1966) The growth of mouse bone marrow cells in vitro. *Aust. J. Exp. Biol. Med. Sci.*, **44**, 287–299.

160 Pluznik, D.H., Sachs, L. (1965) The cloning of normal "mast" cells in tissue culture. *J. Cell Physiol.*, **66**, 319–324.

161 Stevens, L.C. (1967) Origin of testicular teratomas from primordial germ cells in mice. *J. Natl Cancer Inst.*, **38**, 549–552.

162 Pierce, G.B., Stevens, L.C., Nakane, P.K. (1967) Ultrastructural analysis of the early development of teratocarcinomas. *J. Natl Cancer. Inst.*, **39**, 755–773.

163 Gardner, R.L. (1968) Mouse chimeras obtained by the injection of cells into the blastocyst. *Nature*, **220**, 596–597.

164 Markert, C.L., Petters, R.M. (1977) Homozygous mouse embryos produced by microsurgery. *J. Exp. Zool.*, **201**, 295–302.

165 Wells, R.S. (1982) An in vitro assay for growth regulation of embryonal carcinoma by the blastocyst. *Cancer Res.*, **42**, 2736–2741.

166 Pierce, G.B., Lewis, S.H., Miller, G.J., Moritz, E., Miller, P. (1979) Tumorigenicity of embryonal carcinoma as an assay to study control of malignancy by the murine blastocyst. *Proc. Natl Acad. Sci. USA*, **76**, 6649–6651.

167 Pierce, G.B., Pantazis, C.G., Caldwell, J.E., Wells, R.S. (1982) Specificity of the control of tumor formation by the blastocyst. *Cancer Res.*, **42**, 1082–1087.

168 Brinster, R.L. (1974) The effect of cells transferred into the mouse blastocyst on subsequent development. *J. Exp. Med.*, **140**, 1049–1056.

169 Mintz, B., Illmensee, K. (1975) Normal genetically mosaic mice produced from malignant teratocarcinoma cells. *Proc. Natl Acad. Sci. USA*, **72**, 3585–3589.

170 Papaioannou, V.E., McBurney, M.W., Gardner, R.L., Evans, M.J. (1975) Fate of teratocarcinoma cells injected into early mouse embryos. *Nature*, **258**, 70–73.

171 Hochedlinger, K., Blelloch, R., Brennan, C., Yamada, Y., Kim, M., Chin, L., Jaenisch, R. (2004) Reprogramming of a melanoma genome by nuclear transplantation. *Genes Dev.*, **18**, 1875–1885.

172 Hochedlinger, K., Jaenisch, R. (2003) Nuclear transplantation, embryonic stem cells, and the potential for cell therapy. *N. Engl. J. Med.*, **349**, 275–286.

173 Al-Hajj, M., Wicha, M.S., Benito-Hernandez, A., Morrison, S.J., Clarke, M.F. (2003) Prospective identification of tumorigenic breast cancer cells. *Proc. Natl Acad. Sci. USA*, **100**, 3983–3988.

174 Cozzio, A., Passegue, E., Ayton, P.M., Karsunky, H., Cleary, M.L., Weissman, I.L. (2003) Similar MLL-associated leukemias arising from self-renewing stem cells and short-lived myeloid progenitors. *Genes Dev.*, **17**, 3029–3035.

175 Krivtsov, A.V., Twomey, D., Feng, Z., Stubbs, M.C., Wang, Y., Faber, J., Levine, J.E., Wang, J., Hahn, W.C., Gilliland, D.G., Golub, T.R., Armstrong, S.A. (2006) Transformation from committed progenitor to leukaemia stem cell initiated by MLL-AF9. *Nature*, **442**, 818–822.

176 Li, Y., Welm, B., Podsypanina, K., Huang, S., Chamorro, M., Zhang, X., Rowlands, T., Egeblad, M., Cowin, P., Werb, Z., Tan, L.K., Rosen, J.M., Varmus, H.E. (2003) Evidence that transgenes encoding components of the Wnt signaling pathway preferentially induce mammary cancers from progenitor cells. *Proc. Natl Acad. Sci. USA*, **100**, 15853–15858.

177 Singh, S.K., Clarke, I.D., Terasaki, M., Bonn, V.E., Hawkins, C., Squire, J., Dirks, P.B. (2003) Identification of a cancer stem cell in human brain tumors. *Cancer Res.*, **63**, 5821–5828.

178 Loewenstein, W.R. (1966) Permeability of membrane junctions. *Ann. N. Y. Acad. Sci.*, **137**, 441–472.

179 Loewenstein, W.R., Kanno, Y. (1966) Intercellular communication and the control of tissue growth: lack of communication between cancer cells. *Nature*, **209**, 1248–1249.

180 Chang, C.C., Trosko, J.E., el-Fouly, M.H., Gibson-D'Ambrosio, R.E., D'Ambrosio, S.M. (1987) Contact insensitivity of a subpopulation of normal human fetal kidney epithelial cells and of human carcinoma cell lines. *Cancer Res.*, **47**, 1634–1645.

181 Sun, W., Kang, K.S., Morita, I., Trosko, J.E., Chang, C.C. (1999) High susceptibility of a human breast epithelial cell type with stem cell characteristics to telomerase activation and immortalization. *Cancer Res.*, **59**, 6118–6123.

182 Pierce, G.B. (1983) The cancer cell and its control by the embryo. Rous-Whipple Award lecture. *Am. J. Pathol.*, **113**, 117–124.

183 Cunningham, J.M., Christensen, E.R., Tester, D.J., Kim, C.Y., Roche, P.C.,

Burgart, L.J., Thibodeau, S.N. (1998) Hypermethylation of the hMLH1 promoter in colon cancer with microsatellite instability. *Cancer Res.*, **58**, 3455–3460.

184 Esteller, M., Gaidano, G., Goodman, S.N., Zagonel, V., Capello, D., Botto, B., Rossi, D., Gloghini, A., Vitolo, U., Carbone, A., Baylin, S.B., Herman, J.G. (2002) Hypermethylation of the DNA repair gene O(6)-methylguanine DNA methyltransferase and survival of patients with diffuse large B-cell lymphoma. *J. Natl Cancer. Inst.*, **94**, 26–32.

185 Deng, G., Nguyen, A., Tanaka, H., Matsuzaki, K., Bell, I., Mehta, K.R., Terdiman, J.P., Waldman, F.M., Kakar, S., Gum, J., Crawley, S., Sleisenger, M.H., Kim, Y.S. (2006) Regional hypermethylation and global hypomethylation are associated with altered chromatin conformation and histone acetylation in colorectal cancer. *Int. J. Cancer*, **118**, 2999–3005.

186 Widschwendter, M., Jiang, G., Woods, C., Muller, H.M., Fiegl, H., Goebel, G., Marth, C., Muller-Holzner, E., Zeimet, A.G., Laird, P.W., Ehrlich, M. (2004) DNA hypomethylation and ovarian cancer biology. *Cancer Res.*, **64**, 4472–4480.

187 Nakayama, M., Wada, M., Harada, T., Nagayama, J., Kusaba, H., Ohshima, K., Kozuru, M., Komatsu, H., Ueda, R., Kuwano, M. (1998) Hypomethylation status of CpG sites at the promoter region and overexpression of the human MDR1 gene in acute myeloid leukemias. *Blood*, **92**, 4296–4307.

188 Sakatani, T., Kaneda, A., Iacobuzio-Donahue, C.A., Carter, M.G., de Boom Witzel, S., Okano, H., Ko, M.S., Ohlsson, R., Longo, D.L., Feinberg, A.P. (2005) Loss of imprinting of Igf2 alters intestinal maturation and tumorigenesis in mice. *Science*, **307**, 1976–1978.

189 Baylin, S.B. (2005) DNA methylation and gene silencing in cancer. *Nat. Clin. Pract. Oncol.*, **2** (Suppl. 1), S4–11.

190 Widschwendter, M., Jones, P.A. (2002) DNA methylation and breast carcinogenesis. *Oncogene*, **21**, 5462–5482.

191 Dobrovic, A., Simpfendorfer, D. (1997) Methylation of the BRCA1 gene in sporadic breast cancer. *Cancer Res.*, **57**, 3347–3350.

192 Herman, J.G., Latif, F., Weng, Y., Lerman, M.I., Zbar, B., Liu, S., Samid, D., Duan, D.S., Gnarra, J.R., Linehan, W.M., Baylin, S.B. (1994) Silencing of the VHL tumor-suppressor gene by DNA methylation in renal carcinoma. *Proc. Natl Acad. Sci. USA*, **91**, 9700–9704.

193 Stirzaker, C., Millar, D.S., Paul, C.L., Warnecke, P.M., Harrison, J., Vincent, P.C., Frommer, M., Clark, S.J. (1997) Extensive DNA methylation spanning the Rb promoter in retinoblastoma tumors. *Cancer Res.*, **57**, 2229–2237.

194 Yang, H.J., Liu, V.W., Wang, Y., Tsang, P.C., Ngan, H.Y. (2006) Differential DNA methylation profiles in gynecological cancers and correlation with clinico-pathological data. *BMC Cancer*, **6**, 212.

195 Pho, L., Grossman, D., Leachman, S.A. (2006) Melanoma genetics: a review of genetic factors and clinical phenotypes in familial melanoma. *Curr. Opin. Oncol.*, **18**, 173–179.

196 Goldstein, A.M. (2004) Familial melanoma, pancreatic cancer and germline CDKN2A mutations. *Hum. Mutat.*, **23**, 630.

197 Gronbaek, K., de Nully Brown, P., Moller, M.B., Nedergaard, T., Ralfkiaer, E., Moller, P., Zeuthen, J., Guldberg, P. (2000) Concurrent disruption of p16INK4a and the ARF-p53 pathway predicts poor prognosis in aggressive non-Hodgkin's lymphoma. *Leukemia*, **14**, 1727–1735.

198 Luo, D., Zhang, B., Lv, L., Xiang, S., Liu, Y., Ji, J., Deng, D. (2006) Methylation of CpG islands of p16 associated with progression of primary gastric carcinomas. *Lab. Invest.*, **86**, 591–598.

199 Wang, J., Lee, J.J., Wang, L., Liu, D.D., Lu, C., Fan, Y.H., Hong, W.K., Mao, L. (2004) Value of p16INK4a and RASSF1A promoter hypermethylation in prognosis of patients with resectable non-small cell lung cancer. *Clin. Cancer Res.*, **10**, 6119–6125.

200 Toyota, M., Ahuja, N., Ohe-Toyota, M., Herman, J.G., Baylin, S.B., Issa, J.P. (1999) CpG island methylator phenotype in colorectal cancer. *Proc. Natl Acad. Sci. USA*, **96**, 8681–8686.

201 Weisenberger, D.J., Siegmund, K.D., Campan, M., Young, J., Long, T.I., Faasse, M.A., Kang, G.H., Widschwendter, M., Weener, D., Buchanan, D., Koh, H.,

Simms, L., Barker, M., Leggett, B., Levine, J., Kim, M., French, A.J., Thibodeau, S.N., Jass, J., Haile, R., Laird, P.W. (2006) CpG island methylator phenotype underlies sporadic microsatellite instability and is tightly associated with BRAF mutation in colorectal cancer. *Nat. Genet.*, **38**, 787–793.

202 Issa, J.P. (2004) CpG island methylator phenotype in cancer. *Nat. Rev. Cancer*, **4**, 988–993.

203 Hammond, S.M. (2006) MicroRNAs as oncogenes. *Curr. Opin. Genet. Dev.*, **16**, 4–9.

204 Lu, J., Getz, G., Miska, E.A., Alvarez-Saavedra, E., Lamb, J., Peck, D., Sweet-Cordero, A., Ebert, B.L., Mak, R.H., Ferrando, A.A., Downing, J.R., Jacks, T., Horvitz, H.R., Golub, T.R. (2005) MicroRNA expression profiles classify human cancers. *Nature*, **435**, 834–838.

205 Saito, Y., Liang, G., Egger, G., Friedman, J.M., Chuang, J.C., Coetzee, G.A., Jones, P.A. (2006) Specific activation of microRNA-127 with downregulation of the proto-oncogene BCL6 by chromatin-modifying drugs in human cancer cells. *Cancer Cell*, **9**, 435–443.

206 Phan, R.T., Dalla-Favera, R. (2004) The BCL6 proto-oncogene suppresses p53 expression in germinal-centre B cells. *Nature*, **432**, 635–639.

207 Diederichs, S., Haber, D.A. (2006) Sequence variations of microRNAs in human cancer: alterations in predicted secondary structure do not affect processing. *Cancer Res.*, **66**, 6097–6104.

208 Ehrich, M., Turner, J., Gibbs, P., Lipton, L., Giovanneti, M., Cantor, C., van den Boom, D. (2008) Cytosine methylation profiling of cancer cell lines. *Proc. Natl Acad. Sci. USA*, **105**, 4844–4849.

209 Futreal, P.A., Coin, L., Marshall, M., Down, T., Hubbard, T., Wooster, R., Rahman, N., Stratton, M.R. (2004) A census of human cancer genes. *Nat. Rev. Cancer*, **4**, 177–183.

210 Eckhardt, F., Lewin, J., Cortese, R., Rakyan, V.K., Attwood, J., Burger, M., Burton, J., Cox, T.V., Davies, R., Down, T.A., Haefliger, C., Horton, R., Howe, K., Jackson, D.K., Kunde, J., Koenig, C., Liddle, J., Niblett, D., Otto, T., Pettett, R., Seemann, S., Thompson, C., West, T., Rogers, J., Olek, A., Berlin, K., Beck, S. (2006) DNA methylation profiling of human chromosomes 6, 20 and 22. *Nat. Genet.*, **38**, 1378–1385.

211 Mito, Y., Henikoff, J.G., Henikoff, S. (2005) Genome-scale profiling of histone H3.3 replacement patterns. *Nat. Genet.*, **37**, 1090–1097.

212 Garraway, L.A., Widlund, H.R., Rubin, M.A., Getz, G., Berger, A.J., Ramaswamy, S., Beroukhim, R., Milner, D.A., Granter, S.R., Du, J., Lee, C., Wagner, S.N., Li, C., Golub, T.R., Rimm, D.L., Meyerson, M.L., Fisher, D.E., Sellers, W.R. (2005) Integrative genomic analyses identify MITF as a lineage survival oncogene amplified in malignant melanoma. *Nature*, **436**, 117–122.

213 Scherf, U., Ross, D.T., Waltham, M., Smith, L.H., Lee, J.K., Tanabe, L., Kohn, K.W., Reinhold, W.C., Myers, T.G., Andrews, D.T., Scudiero, D.A., Eisen, M.B., Sausville, E.A., Pommier, Y., Botstein, D., Brown, P.O., Weinstein, J.N. (2000) A gene expression database for the molecular pharmacology of cancer. *Nat. Genet.*, **24**, 236–244.

214 Issa, J.P. (1999) Aging, DNA methylation and cancer. *Crit. Rev. Oncol. Hematol.*, **32**, 31–43.

215 Nguyen, C., Liang, G., Nguyen, T.T., Tsao-Wei, D., Groshen, S., Lubbert, M., Zhou, J.H., Benedict, W.F., Jones, P.A. (2001) Susceptibility of nonpromoter CpG islands to *de novo* methylation in normal and neoplastic cells. *J. Natl Cancer Inst.*, **93**, 1465–1472.

216 Euhus, D.M., Bu, D., Milchgrub, S., Xie, X.J., Bian, A., Leitch, A.M., Lewis, C.M. (2008) DNA methylation in benign breast epithelium in relation to age and breast cancer risk. *Cancer Epidemiol. Biomarkers Prev.*, **17**, 1051–1059.

217 Issa, J.P., Ottaviano, Y.L., Celano, P., Hamilton, S.R., Davidson, N.E., Baylin, S.B. (1994) Methylation of the oestrogen receptor CpG island links ageing and neoplasia in human colon. *Nat. Genet.*, **7**, 536–540.

218 Habuchi, T., Takahashi, T., Kakinuma, H., Wang, L., Tsuchiya, N., Satoh, S., Akao, T., Sato, K., Ogawa, O., Knowles, M.A., Kato, T. (2001) Hypermethylation at 9q32-33 tumour suppressor region is age-related in normal urothelium and an early and frequent alteration in bladder cancer. *Oncogene*, **20**, 531–537.

219 Liu, Y., Lan, Q., Siegfried, J.M., Luketich, J.D., Keohavong, P. (2006) Aberrant promoter methylation of p16 and MGMT genes in lung tumors from smoking and never-smoking lung cancer patients. *Neoplasia*, **8**, 46–51.

220 Russo, A.L., Thiagalingam, A., Pan, H., Califano, J., Cheng, K.H., Ponte, J.F., Chinnappan, D., Nemani, P., Sidransky, D., Thiagalingam, S. (2005) Differential DNA hypermethylation of critical genes mediates the stage-specific tobacco smoke-induced neoplastic progression of lung cancer. *Clin. Cancer Res.*, **11**, 2466–2470.

221 Liu, H., Zhou, Y., Boggs, S.E., Belinsky, S.A., Liu, J. (2007) Cigarette smoke induces demethylation of prometastatic oncogene synuclein-gamma in lung cancer cells by downregulation of DNMT3B. *Oncogene*, **26**, 5900–5910.

222 Fraga, M.F., Ballestar, E., Villar-Garea, A., Boix-Chornet, M., Espada, J., Schotta, G., Bonaldi, T., Haydon, C., Ropero, S., Petrie, K., Iyer, N.G., Perez-Rosado, A., Calvo, E., Lopez, J.A., Cano, A., Calasanz, M.J., Colomer, D., Piris, M.A., Ahn, N., Imhof, A., Caldas, C., Jenuwein, T., Esteller, M. (2005) Loss of acetylation at Lys16 and trimethylation at Lys20 of histone H4 is a common hallmark of human cancer. *Nat. Genet.*, **37**, 391–400.

223 Yasui, W., Oue, N., Ono, S., Mitani, Y., Ito, R., Nakayama, H. (2003) Histone acetylation and gastrointestinal carcinogenesis. *Ann. N. Y. Acad. Sci.*, **983**, 220–231.

224 Lin, R.J., Sternsdorf, T., Tini, M., Evans, R.M. (2001) Transcriptional regulation in acute promyelocytic leukemia. *Oncogene*, **20**, 7204–7215.

225 Hecker, R.M., Amstutz, R.A., Wachtel, M., Walter, D., Niggli, F.K., Schafer, B.W. (2010) p21 Downregulation is an important component of PAX3/FKHR oncogenicity and its reactivation by HDAC inhibitors enhances combination treatment. *Oncogene*, **29**, 3942–3952.

226 Mandl-Weber, S., Meinel, F.G., Jankowsky, R., Oduncu, F., Schmidmaier, R., Baumann, P. (2010) The novel inhibitor of histone deacetylase resminostat (RAS2410) inhibits proliferation and induces apoptosis in multiple myeloma (MM) cells. *Br. J. Haematol.*, **149**, 518–528.

227 Panicker, J., Li, Z., McMahon, C., Sizer, C., Steadman, K., Piekarz, R., Bates, S.E., Thiele, C.J. (2010) Romidepsin (FK228/depsipeptide) controls growth and induces apoptosis in neuroblastoma tumor cells. *Cell Cycle*, **9**, 1830–1838.

228 Yoo, C.B., Jones, P.A. (2006) Epigenetic therapy of cancer: past, present and future. *Nat. Rev. Drug Discov.*, **5**, 37–50.

229 Momparler, R.L. (2005) Epigenetic therapy of cancer with 5-aza-2′-deoxycytidine (decitabine). *Semin. Oncol.*, **32**, 443–451.

230 Zhou, L., Cheng, X., Connolly, B.A., Dickman, M.J., Hurd, P.J., Hornby, D.P. (2002) Zebularine: a novel DNA methylation inhibitor that forms a covalent complex with DNA methyltransferases. *J. Mol. Biol.*, **321**, 591–599.

231 Issa, J.P., Garcia-Manero, G., Giles, F.J., Mannari, R., Thomas, D., Faderl, S., Bayar, E., Lyons, J., Rosenfeld, C.S., Cortes, J., Kantarjian, H.M. (2004) Phase 1 study of low-dose prolonged exposure schedules of the hypomethylating agent 5-aza-2′-deoxycytidine (decitabine) in hematopoietic malignancies. *Blood*, **103**, 1635–1640.

232 Villar-Garea, A., Fraga, M.F., Espada, J., Esteller, M. (2003) Procaine is a DNA-demethylating agent with growth-inhibitory effects in human cancer cells. *Cancer Res.*, **63**, 4984–4989.

233 Fang, M.Z., Wang, Y., Ai, N., Hou, Z., Sun, Y., Lu, H., Welsh, W., Yang, C.S. (2003) Tea polyphenol (−)-epigallocatechin-3-gallate inhibits DNA methyltransferase and reactivates methylation-silenced genes in cancer cell lines. *Cancer Res.*, **63**, 7563–7570.

234 Arce, C., Segura-Pacheco, B., Perez-Cardenas, E., Taja-Chayeb, L., Candelaria, M., Duennas-Gonzalez, A. (2006) Hydralazine target: from blood vessels to the epigenome. *J. Transl. Med.*, **4**, 10.

235 Brueckner, B., Garcia Boy, R., Siedlecki, P., Musch, T., Kliem, H.C., Zielenkiewicz, P., Suhai, S., Wiessler, M., Lyko, F. (2005) Epigenetic reactivation of tumor suppressor genes by a novel small-molecule inhibitor of human DNA methyltransferases. *Cancer Res.*, **65**, 6305–6311.

236 Goffin, J., Eisenhauer, E. (2002) DNA methyltransferase inhibitors-state of the art. *Ann. Oncol.*, **13**, 1699–1716.
237 Longo, V.D., Kennedy, B.K. (2006) Sirtuins in aging and age-related disease. *Cell*, **126**, 257–268.
238 Bolden, J.E., Peart, M.J., Johnstone, R.W. (2006) Anticancer activities of histone deacetylase inhibitors. *Nat. Rev. Drug Discov.*, **5**, 769–784.
239 Ma, X., Ezzeldin, H.H., Diasio, R.B. (2009) Histone deacetylase inhibitors: current status and overview of recent clinical trials. *Drugs*, **69**, 1911–1934.
240 Marks, P.A., Breslow, R. (2007) Dimethyl sulfoxide to vorinostat: development of this histone deacetylase inhibitor as an anticancer drug. *Nat. Biotechnol.*, **25**, 84–90.
241 Egger, G., Liang, G., Aparicio, A., Jones, P.A. (2004) Epigenetics in human disease and prospects for epigenetic therapy. *Nature*, **429**, 457–463.
242 Gore, S.D., Baylin, S., Sugar, E., Carraway, H., Miller, C.B., Carducci, M., Grever, M., Galm, O., Dauses, T., Karp, J.E., Rudek, M.A., Zhao, M., Smith, B.D., Manning, J., Jiemjit, A., Dover, G., Mays, A., Zwiebel, J., Murgo, A., Weng, L.J., Herman, J.G. (2006) Combined DNA methyltransferase and histone deacetylase inhibition in the treatment of myeloid neoplasms. *Cancer Res.*, **66**, 6361–6369.
243 Daskalakis, M., Nguyen, T.T., Nguyen, C., Guldberg, P., Kohler, G., Wijermans, P., Jones, P.A., Lubbert, M. (2002) Demethylation of a hypermethylated P15/INK4B gene in patients with myelodysplastic syndrome by 5-Aza-2′-deoxycytidine (decitabine) treatment. *Blood*, **100**, 2957–2964.
244 Cheng, Y.C., Lin, H., Huang, M.J., Chow, J.M., Lin, S., Liu, H.E. (2007) Downregulation of c-Myc is critical for valproic acid-induced growth arrest and myeloid differentiation of acute myeloid leukemia. *Leuk. Res.*, **31**, 1403–1411.
245 Komatsu, N., Kawamata, N., Takeuchi, S., Yin, D., Chien, W., Miller, C.W., Koeffler, H.P. (2006) SAHA, a HDAC inhibitor, has profound anti-growth activity against non-small cell lung cancer cells. *Oncol. Rep.*, **15**, 187–191.
246 Noh, E.J., Lee, J.S. (2003) Functional interplay between modulation of histone deacetylase activity and its regulatory role in G2-M transition. *Biochem. Biophys. Res. Commun.*, **310**, 267–273.
247 Qian, X., Ara, G., Mills, E., LaRochelle, W.J., Lichenstein, H.S., Jeffers, M. (2008) Activity of the histone deacetylase inhibitor belinostat (PXD101) in preclinical models of prostate cancer. *Int. J. Cancer*, **122**, 1400–1410.
248 Rosato, R.R., Maggio, S.C., Almenara, J.A., Payne, S.G., Atadja, P., Spiegel, S., Dent, P., Grant, S. (2006) The histone deacetylase inhibitor LAQ824 induces human leukemia cell death through a process involving XIAP down-regulation, oxidative injury, and the acid sphingomyelinase-dependent generation of ceramide. *Mol. Pharmacol.*, **69**, 216–225.
249 Gui, C.Y., Ngo, L., Xu, W.S., Richon, V.M., Marks, P.A. (2004) Histone deacetylase (HDAC) inhibitor activation of p21WAF1 involves changes in promoter-associated proteins, including HDAC1. *Proc. Natl Acad. Sci. USA*, **101**, 1241–1246.
250 Sakajiri, S., Kumagai, T., Kawamata, N., Saitoh, T., Said, J.W., Koeffler, H.P. (2005) Histone deacetylase inhibitors profoundly decrease proliferation of human lymphoid cancer cell lines. *Exp. Hematol.*, **33**, 53–61.
251 Valentini, A., Gravina, P., Federici, G., Bernardini, S. (2007) Valproic acid induces apoptosis, p16INK4A upregulation and sensitization to chemotherapy in human melanoma cells. *Cancer Biol. Ther.*, **6**, 185–191.
252 Alao, J.P., Stavropoulou, A.V., Lam, E.W., Coombes, R.C., Vigushin, D.M. (2006) Histone deacetylase inhibitor, trichostatin A induces ubiquitin-dependent cyclin D1 degradation in MCF-7 breast cancer cells. *Mol. Cancer*, **5**, 8.
253 Petrella, A., D'Acunto, C.W., Rodriquez, M., Festa, M., Tosco, A., Bruno, I., Terracciano, S., Taddei, M., Paloma, L.G., Parente, L. (2008) Effects of FR235222, a novel HDAC inhibitor, in proliferation and apoptosis of human leukaemia cell lines: role of annexin A1. *Eur. J. Cancer*, **44**, 740–749.
254 Liu, T., Kuljaca, S., Tee, A., Marshall, G.M. (2006) Histone deacetylase inhibitors: multifunctional anticancer agents. *Cancer Treat. Rev.*, **32**, 157–165.
255 Qian, D.Z., Kato, Y., Shabbeer, S., Wei, Y., Verheul, H.M., Salumbides, B., Sanni, T.,

Atadja, P., Pili, R. (2006) Targeting tumor angiogenesis with histone deacetylase inhibitors: the hydroxamic acid derivative LBH589. *Clin. Cancer Res.*, **12**, 634–642.

256 Kang, J.H., Kim, M.J., Chang, S.Y., Sim, S.S., Kim, M.S., Jo, Y.H. (2008) CCAAT box is required for the induction of human thrombospondin-1 gene by trichostatin A. *J. Cell. Biochem.*, **104**, 1192–1203.

257 Kim, M.S., Kwon, H.J., Lee, Y.M., Baek, J.H., Jang, J.E., Lee, S.W., Moon, E.J., Kim, H.S., Lee, S.K., Chung, H.Y., Kim, C.W., Kim, K.W. (2001) Histone deacetylases induce angiogenesis by negative regulation of tumor suppressor genes. *Nat. Med.*, **7**, 437–443.

258 Kwon, H.J., Kim, M.S., Kim, M.J., Nakajima, H., Kim, K.W. (2002) Histone deacetylase inhibitor FK228 inhibits tumor angiogenesis. *Int. J. Cancer*, **97**, 290–296.

259 Insinga, A., Monestiroli, S., Ronzoni, S., Gelmetti, V., Marchesi, F., Viale, A., Altucci, L., Nervi, C., Minucci, S., Pelicci, P.G. (2005) Inhibitors of histone deacetylases induce tumor-selective apoptosis through activation of the death receptor pathway. *Nat. Med.*, **11**, 71–76.

260 Xu, W.S., Parmigiani, R.B., Marks, P.A. (2007) Histone deacetylase inhibitors: molecular mechanisms of action. *Oncogene*, **26**, 5541–5552.

261 Catalano, M.G., Fortunati, N., Pugliese, M., Poli, R., Bosco, O., Mastrocola, R., Aragno, M., Boccuzzi, G. (2006) Valproic acid, a histone deacetylase inhibitor, enhances sensitivity to doxorubicin in anaplastic thyroid cancer cells. *J. Endocrinol.*, **191**, 465–472.

262 Fuino, L., Bali, P., Wittmann, S., Donapaty, S., Guo, F., Yamaguchi, H., Wang, H.G., Atadja, P., Bhalla, K. (2003) Histone deacetylase inhibitor LAQ824 down-regulates Her-2 and sensitizes human breast cancer cells to trastuzumab, taxotere, gemcitabine, and epothilone B. *Mol. Cancer Ther.*, **2**, 971–984.

263 Maiso, P., Carvajal-Vergara, X., Ocio, E.M., Lopez-Perez, R., Mateo, G., Gutierrez, N., Atadja, P., Pandiella, A., San Miguel, J.F. (2006) The histone deacetylase inhibitor LBH589 is a potent antimyeloma agent that overcomes drug resistance. *Cancer Res.*, **66**, 5781–5789.

264 Sanchez-Gonzalez, B., Yang, H., Bueso-Ramos, C., Hoshino, K., Quintas-Cardama, A., Richon, V.M., Garcia-Manero, G. (2006) Antileukemia activity of the combination of an anthracycline with a histone deacetylase inhibitor. *Blood*, **108**, 1174–1182.

265 Sonnemann, J., Kumar, K.S., Heesch, S., Muller, C., Hartwig, C., Maass, M., Bader, P., Beck, J.F. (2006) Histone deacetylase inhibitors induce cell death and enhance the susceptibility to ionizing radiation, etoposide, and TRAIL in medulloblastoma cells. *Int. J. Oncol.*, **28**, 755–766.

266 Gronbaek, K., Hother, C., Jones, P.A. (2007) Epigenetic changes in cancer. *APMIS*, **115**, 1039–1059.

267 Bishton, M., Kenealy, M., Johnstone, R., Rasheed, W., Prince, H.M. (2007) Epigenetic targets in hematological malignancies: combination therapies with HDACis and demethylating agents. *Expert Rev. Anticancer Ther.*, **7**, 1439–1449.

268 Kuendgen, A., Lubbert, M. (2008) Current status of epigenetic treatment in myelodysplastic syndromes. *Ann. Hematol.*, **87**, 601–611.

269 Earel, J.K. Jr, Van Oosten, R.L., Griffith, T.S. (2006) Histone deacetylase inhibitors modulate the sensitivity of tumor necrosis factor-related apoptosis-inducing ligand-resistant bladder tumor cells. *Cancer Res.*, **66**, 499–507.

270 Fiskus, W., Pranpat, M., Balasis, M., Bali, P., Estrella, V., Kumaraswamy, S., Rao, R., Rocha, K., Herger, B., Lee, F., Richon, V., Bhalla, K. (2006) Cotreatment with vorinostat (suberoylanilide hydroxamic acid) enhances activity of dasatinib (BMS-354825) against imatinib mesylate-sensitive or imatinib mesylate-resistant chronic myelogenous leukemia cells. *Clin. Cancer Res.*, **12**, 5869–5878.

271 Hideshima, T., Bradner, J.E., Wong, J., Chauhan, D., Richardson, P., Schreiber, S.L., Anderson, K.C. (2005) Small-molecule inhibition of proteasome and aggresome function induces synergistic antitumor activity in multiple myeloma. *Proc. Natl Acad. Sci. USA*, **102**, 8567–8572.

272 Dai, Y., Rahmani, M., Dent, P., Grant, S. (2005) Blockade of histone deacetylase inhibitor-induced RelA/p65 acetylation and

NF-κB activation potentiates apoptosis in leukemia cells through a process mediated by oxidative damage, XIAP downregulation, and c-Jun N-terminal kinase 1 activation. *Mol. Cell. Biol.*, **25**, 5429–5444.

273 Domingo-Domenech, J., Pippa, R., Tapia, M., Gascon, P., Bachs, O., Bosch, M. (2008) Inactivation of NF-κB by proteasome inhibition contributes to increased apoptosis induced by histone deacetylase inhibitors in human breast cancer cells. *Breast Cancer Res. Treat.*, **112**, 53–62.

274 Inoue, S., Walewska, R., Dyer, M.J., Cohen, G.M. (2008) Downregulation of Mcl-1 potentiates HDACi-mediated apoptosis in leukemic cells. *Leukemia*, **22**, 819–825.

275 Yeow, W.S., Ziauddin, M.F., Maxhimer, J.B., Shamimi-Noori, S., Baras, A., Chua, A., Schrump, D.S., Nguyen, D.M. (2006) Potentiation of the anticancer effect of valproic acid, an antiepileptic agent with histone deacetylase inhibitory activity, by the kinase inhibitor staurosporine or its clinically relevant analogue UCN-01. *Br. J. Cancer*, **94**, 1436–1445.

28
Chromatin Remodeling in Carcinoma Cells

Therese M. Becker
University of Sydney, Westmead Institute for Cancer Research at Westmead Millennium Institute, Westmead Hospital, Westmead, Sydney, New South Wales, Australia

Epigenetic Regulation and Epigenomics: Advances in Molecular Biology and Medicine, First Edition. Edited by Robert A. Meyers.

Keywords

Chromatin

An organized structure of eukaryotic DNA.

Histone modification

Covalent changes to histones.

ATP-dependent chromatin remodeling

Changing the histone–DNA interaction to unwind or pack DNA.

SWI/SNF

A specific chromatin-remodeling complex.

Carcinoma

A cancer derived from epithelial cells.

Tumor suppressor

A protein with a function that opposes tumor development and progression.

Senescence

Secure withdrawal from proliferation.

Carcinoma arises from the malignant transformation of epithelial cells, which involves the deregulation of genetic information leading to excessive proliferation, cell death evasion, and invasive potential. The traditional dogma was that the underlying genetic deregulation emerged through an accumulation of individual mutations that inactivate tumor suppressor genes and activate oncogenes. During recent years, it has become evident that epigenetic changes are potently contributing to tumor formation. Epigenetics describes the modulation of genetic information by mechanisms other than sequence alterations of the underlying DNA. A major part of epigenetic gene regulation is accomplished through chromatin remodeling. In eukaryotic cells, huge amounts of DNA are managed through intricate mechanisms that allow DNA condensation into tightly packable protein–DNA units (chromatin), and chromatin-remodeling processes render DNA either accessible or inaccessible to diverse DNA processes such as replication and transcription. The role of chromatin remodeling in carcinoma cells is reviewed in this chapter.

1 Introduction

The malignant transformation of normal cells into cancer cells involves dynamic changes of the cellular genome. Cancer cells have accumulated a range of genetic alterations that render them unresponsive to normal cellular regulatory mechanisms, and thereby they have acquired the main hallmarks of cancer cells:

- *Excessive proliferation* by becoming independent of external stimuli to activate proliferative signaling pathways and by "ignoring" anti-proliferative signals.
- *Avoidance of cell death* by opposing normal pathways of programmed cell death (apoptosis) and developing immortality mechanisms, which means cancer cells have in principal an unlimited life span.
- *Gain of invasive potential* to invade other tissues and form metastasis (as reviewed by Hanahan and Weinberg [1]).

The chance to accumulate genetic alterations that cause tumor development increases with the number of cell divisions that a single cell undergoes during its life span. Thus, it is not surprising that the majority of tumors arise in tissues with a high proliferation rate. Epithelial cells undergo relatively frequent cell divisions to renew epithelial tissue and to ensure its function. Cancers that arise from epithelial cells are termed *carcinomas*, and comprise over 90% of all cancers [2] (see Table 1).

The genetic changes relating to malignant transformation may be direct mutational changes, which will produce hyperactive oncogenes or inactive tumor suppressor genes [1]. Moreover, proliferation, cell death avoidance, and invasive potential may be achieved by more global changes in gene regulation. Often, transformed cells have reverted their overall gene expression pattern from a growth-arrested, differentiated state in which cells function as part of a specialized tissue to a stem cell-like proliferative state [3]. The stem cell state is normally maintained during embryonic development or during later life stages in long-term progenitor cells, which continue to divide and give rise to cells, which will growth arrest and differentiate. Progenitor cells are characteristic for tissues with a high cell renewal, such as colon epithelial tissue or skin tissue. It has been proposed that, in a 60-year-old individual, the colon epithelial cells – which arise from progenitor cells in the crypts of the colon epithelium – have arisen through about 3000 cell divisions, whereas skin keratinocytes – which originate from progenitors in the basal layer of the skin – have arisen from about 300 divisions by this age [4]. Not surprisingly, in these progenitor cells the proliferation genes are highly expressed, while differentiation gene expression is limited. The switch from one gene expression state to the other can be accomplished by expressing or repressing an entire cassette of genes associated with either proliferation or differentiation. In mammalian cells, these gene cassettes are often organized in adjacent genomic regions that allow coordinated regulation. Such regulation can be achieved by chromatin-remodeling processes that either allow or deny access to the transcriptional machinery or transcriptional repressors [5]. Consequently, tumor cells may achieve a more stem cell-like gene expression pattern by targeting chromatin-remodeling processes during cellular transformation, while alterations of chromatin-remodeling functions will contribute to the development of the carcinoma, and its progression.

Tab. 1 Types of carcinoma.

Carcinoma type	*Commonest tissues*	*Proportion of all cancers (%)*
External epithelia	Skin, large intestine, lung, stomach, cervix	56
Internal epithelia	Breast, prostate, ovary, bladder, pancreas	36

Data derived from Ref. [2].

Some insights into chromatin-remodeling events, and their contribution to carcinoma formation, are provided in this chapter.

2
Chromatin

Eukaryotic cells are characterized by their nuclear compartmentation of DNA, and by the arrangement of DNA molecules into highly organized, dynamic DNA–protein structures referred to as *chromatin*. Within chromatin, the DNA (in units of 146 bp) is wrapped around and covalently bound to histone octamer protein complexes; the resulting structures are termed *nucleosomes*. The nucleosome histone octamers consist of H2A, H2B, H3, and H4 or specialized histone variants [6] (Fig. 1). The nucleosomes can be further folded (condensed) with the aid of the H1 linker histone, and may be organized in several hierarchical steps into tightly condensed chromosomes [7] (Fig. 2). Chromatin is a dynamic structure that is maintained by a series of sophisticated chromatin-remodeling procedures that

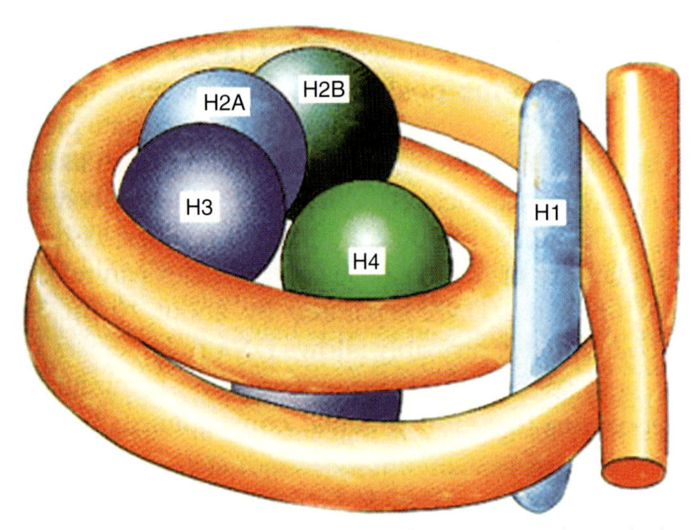

Fig. 1 The nucleosome, the basic unit of chromatin. This schematic shows the winding of the double-stranded DNA molecule around a histone octamer, consisting of H2A, H2B, H3, and H4. The histone linker H1 is also represented. Reproduced with permission from Ref. [7]; © 1996–2011, themededicalbiochemistrypage.org.

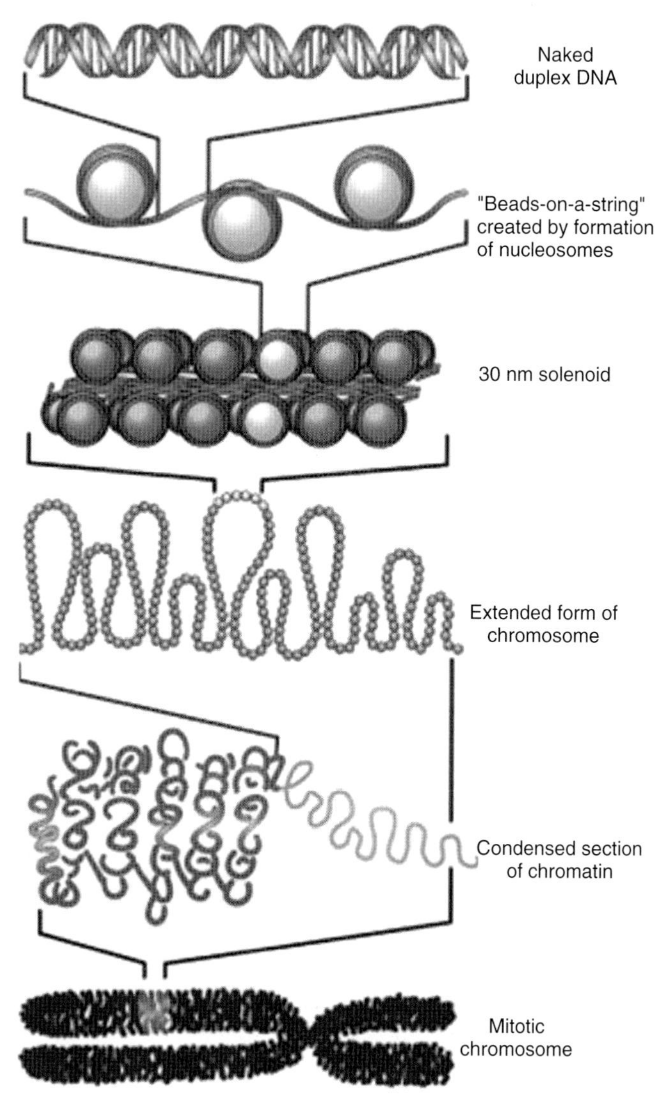

Fig. 2 Chromatin condensation. Schematic demonstrating the formation of higher-order chromatin complexes from the double-stranded DNA molecule over nucleosome formation to condensation into chromosomes. Reproduced with permission from Ref. [7]; © 1996–2011, themededicalbiochemistrypage.org

either allow or deny local and global access to the DNA. Thus, chromatin remodeling regulates all cellular aspects related to DNA maintenance, repair, transcription, and replication (for reviews, see Refs [3, 5]).

3 Chromatin Remodeling

In principle, chromatin remodeling describes a series of coordinated cellular mechanisms that alter the nucleosome configuration such that chromatin and DNA become either accessible or inaccessible to proteins that regulate DNA processes such as transcription, replication, telomere maintenance, and DNA repair.

Chromatin remodeling involves two major steps:

- Histone-modifying enzymes covalently add or remove modifications (e.g., acetylations, methylations, or phosphorylations) at amino acid residues of the N-terminal tail of histones which protrude from the globular nucleosome core. These modifications alter the histones' affinity for binding with the DNA and with specific DNA-processing molecules. For example, acetylated histones, which are found preferentially in actively transcribed and regulatory DNA regions (euchromatin), are highly interactive with bromodomains, which are protein domains found in DNA-interacting enzymes such as the ATP-dependent chromatin-remodeling factors BRG1 and BRM.
- Additionally, chromatin remodeling involves the action of ATP-dependent chromatin-remodeling complexes. These complexes utilize the energy of ATP hydrolyses to affect DNA–histone interactions and, thereby, the repositioning of nucleosomes. These functions alter the DNA–histone interface and either allow or inhibit access of transcriptional repressors or the transcriptional machinery, which includes facilitating molecules such as RNA polymerase II (RNApol-II) and the TATTA box binding protein (TBP) and sequence-specific transcription factors [3].

Both of these processes are tightly cooperative. Histone modifications are often required to form docking stations for ATP-dependent chromatin-remodeling complexes, whereas some histone-modifying proteins are known to function while interacting with ATP- dependent chromatin-remodeling complexes [8].

For more detailed information concerning the physical and biochemical mechanisms of chromatin remodeling around histone octamers, the reader is referred to reviews focusing on this area (e.g., Ref. [9]).

3.1 Histone Modifications

3.1.1 Histone Acetyl Transferases (HATs)

Histone acetyl transferase (HAT) complexes transfer acetyl groups to the N-terminal lysine residues of histones. The acetylations of histones are thought to alter the overall charge of lysine residues, and thereby weaken electrostatic histone–DNA interactions such that acetylated DNA regions become less tightly folded. As a result, acetylation makes DNA more accessible to the transcriptional machinery, and ATP-dependent bromodomain-containing chromatin remodelers usually promote gene expression in acetylated chromatin regions [10]. More open and transcriptionally active chromatin with acetylated histones is also referred to as *euchromatin* [11].

3.1.2 Histone Deacetylases (HDACs)

Histone deacetylase (HDAC) complexes are the direct antagonists of HATs, and remove acetyl groups from lysine residues. Deacetylation promotes a tightening of the DNA–histone interactions which, in conjunction with further modifications, leads to chromatin condensation which prevents transcription from occurring [12] (Fig. 2). The deacetylation function of HDACs is essential for the formation of heterochromatin which, in contrast to the transcriptionally active and highly acetylated euchromatin, does not contain histone acetylations and is characterized by the heterochromatin protein HP1 and increased chromatin condensation, thus not allowing access to transcriptional activators [13].

3.1.3 Histone Methyltransferases

Methylations occur on histones at either lysine or arginine residues. Histone methylation affects four different N-terminal histone tail lysines of H3 (K4, K9, K27, and K36), one H3 core lysine (K79), and the H4 lysine K20. Additionally, methylations have been identified for the N-terminal arginines of H3 (R2, R8, R17, and R26), and for H4 on R3 [14]. Three different degrees of methylation have been identified: monomethylations (me1); dimethylations (me2); and trimethylations (me3). The different orders of methylation state produce a diversity of bio chemical characteristics, and create conditions for a range of specific interactions with effector protein complexes [10, 15]. Histone lysine methylations have been associated with both transcriptional activation or repression: H3K4me2/me3, H3K27me1, and H3K36me2/me3 are found at actively transcribed euchromatic regions (usually in combination with acetylation), while H3K9me2/me3 attracts the heterochromatin protein HP1 and is linked to heterochromatin formation. It is found together with H4K20me2/me3 in inactive heterochromatin regions [10]. Thus, the outcomes of adjacent H3K9me2/me3 and H3K4me/2me3 oppose each other and, indeed, H3 is never found methylated at both K4 and K9. H3K27me3 is linked to facultative heterochromatin in euchromatic regions [15, 16]. For example, EZH2 silencing of the cell cycle inhibitor $p16^{INK4a}$ (see Sect. 4.1.1.3) involves H3K27me3 at $p16^{INK4a}$ regulatory sequences [17]. The effects of histone methylation are listed in Table 2.

3.1.4 Histone Demethylases

Although, originally, histone methylation turnover was thought to be negligible [18], today it has become clear that methylation of histones is a dynamic process, with specialized enzymes that catalyze the removal of methyl groups having been identified; these are termed *histone*

Tab. 2 Effects of histone lysine methylations.

Lysine methylation state	***Effect, chromatin type***
H3K4me2/me3	Euchromatin, Active transcription, Usually combined with acetylation
H3K27me1	Euchromatin, Active transcription, Usually combined with acetylation
H3K36me2/me3	Euchromatin, Active transcription, Usually combined with acetylation
H3K9me2/me3	Heterochromatin No transcription Binds heterochromatin protein HP1
H4K20me2/me3	Heterochromatin No transcription
H3K27me3	Facultative heterochromatin No transcription

demethylases. Currently, the importance of histone demethylases in cancer is emerging, together with their discovery [14].

3.1.5 Histone Modifications in Carcinoma

It has been suggested that combinations of specific histone methylation patterns over an extended genetic region are "read" by DNA-interacting proteins as an "epigenetic code" for gene regulation, and that carcinoma cells have evolved their own epigenetic code to promote proliferation, survival, and invasion [19].

It has been shown that the loss of monoacetylation and trimethylation of histone H4 arises early during the development of carcinoma of the breast, lung, and colon [20], while risk-associated patterns of H3 and H4 acetylation and dimethylation have been reported for prostate cancer[21]. In breast cancer, a low or absent H4K16Ac was identified, while moderate acetylation (H3K9Ac, H3K18Ac, H4K12Ac), lysine methylation (H3K4me2, H4K20me3), and arginine methylation (H4R3me2) were each correlated with aggressive subtypes, and high relative levels of global histone acetylation and methylation were associated with a favorable prognosis [22]. A low expression of H3K18Ac and H3K27me3 also correlated with esophageal squamous cell carcinoma [23], whilst in non-small-cell lung cancer (NSCLC) patients H3K4me2, H3K9Ac were shown to be prognostic markers, together with H2AK5Ac [24]. Low H3K4me2 and H3K18Ac were also significant predictors for overall survival in pancreatic adenocarcinoma, which also displays low levels of H3K9me2 [25]. These data are summarized in Table 3.

3.1.6 Histone-Modifying Enzymes and Carcinoma: Acetylation

The initial evidence for an alteration of HATs in cancer derived from viral oncogenes such as E1A, which target the methyltransferases p300 and CBP. Moreover, mutations of p300 and CBP have been associated with colorectal, breast,

Tab. 3 Histone modifications associated with carcinoma.

Carcinoma type	*Histone modification*	*Reference(s)*
Breast cancer	High: H3R17me2 low/absent: H4K16Ac, H3K9Ac, H3K18Ac, H4K12Ac, H3K4me2, H4K20me3, H4R3me2	[20, 22, 26]
Prostate cancer	Low: H3K18Ac, H4K12Ac	[21]
Hepatocellular carcinoma	Histone hyperacetylation	[27]
Oral carcinoma	Histone hyperacetylation	[28]
Esophageal small-cell cancer	Low: H3K18Ac, H3K27me3	[23]
Non-small-cell lung cancer	High: H3K4me2, H4K8Ac, low: H2AK5Ac, H2BK12Ac	[24]
Pancreatic cancer	Low: H3K4me2, H3K18Ac, H3K9me2	[25]

gastric, pancreatic, and hepatocellular cancers [29, 30]. In the case of HDACs, it appears that overexpression is the most common mode of malignant adaptation, with HDAC1 being overexpressed in prostate [31], gastric [32], colorectal [33, 34], and breast carcinoma [35]. HDAC2 is overexpressed in colon [33, 36], cervical [37], and gastric carcinoma [38], while an increased expression of HDAC3 is seen in colon carcinoma [33] and an overexpression of HDAC6 was identified in breast cancer specimens [39]. The overexpression of HDACs in cancer is proposed to promote facultative heterochromatin, which silences a range of tumor suppressor and anti-apoptotic proteins. Consequently, the specific silencing of HDACs with small interfering RNA (siRNA) molecules reduced tumor cell growth and, most importantly, caused programmed cell death in tumor cells by derepressing the apoptotic pathways. It is, therefore, important to highlight that the use of small molecules targeting and inhibiting HDACs (i.e., HDAC inhibitors) has been suggested as an anti-cancer treatment. Subsequent studies conducted both *in vitro* and *in vivo* have shown much promise, and have linked HDAC inhibitors to cell-cycle arrest, the induction of apoptosis, and the repression of angiogenesis and invasion [40].

3.1.7 **Histone-Modifying Enzymes and Carcinoma: Methylation**

Normally, the histone methyltransferase and polycomb protein EZH2 plays an important role during development and tissue renewal, as it is involved in shutting down the expression of genes that promote growth arrest and differentiation. In proliferating stem cells, EZH2 facilitates the trimethylation of H3K27 and facultative heterochromatin formation. The genes shut down by EZH2 often have a tumor-suppressor function in normal cells, such that an increased EZH2 level leads to excess proliferation and favors tumor formation [41]. The overexpression of EZH2 has been demonstrated in prostate, breast, colorectal, endometrial, hepatocellular, and bladder cancers, as well as in melanoma [42–44], while specific EZH2 silencing using siRNA molecules caused growth arrest in prostate cancer cells [45].

The lysine-specific demethylase 1 (LSD1), which removes methylations from H3K4me2/me3, was found to be downregulated and linked to invasiveness in breast cancer [46], whereas LSD1 was overexpressed and correlated with a poor prognosis in prostate cancer [47]. The histone demethylase JMJD2C, which removes methyl groups from H3K9me3, may destabilize heterochromatin by disallowing HP1 binding; typically, JMJD2C is amplified and thus highly expressed in prostate cancer [48]. JMJD2C is also found amplified in lung sarcomatoid carcinoma, esophageal carcinoma, and squamous cell carcinoma [49, 50]. In contrast, the histone demethylase JMJD3 is specific for the removal of trimethylation from H3K27 in facultative heterochromatin and is, therefore, a pivotal player in the derepression of cell cycle inhibitory genes. JMJD3 has also been linked to keratinocyte growth arrest and differentiation, with one of the derepressed targets being the cell-cycle inhibitor and tumor suppressor $p16^{INK4a}$ (see Sect. 4.1.1.3). It is not surprising, therefore, that JMJD3 inhibition contributes to mouse fibroblast transformation [51, 52]. JARID1B (PLU-1), which specifically demethylates H3K4me3/me2, was found to be overexpressed in both breast and prostate carcinoma [53, 54], while JARID1C was overexpressed in renal cell carcinoma [55]. Of particularly interest is that the expression of JARID1B and JARID1A expression was recently shown to be dynamic within cancer cell populations, and to be associated with drug resistance. For instance, only a small proportion of proliferating melanoma cells expressed JARID1B, and this correlated with a decreased proliferation rate of these cells in culture. However, an elevated expression of JARID1B was associated with an increased resistance against specific BRAF inhibitors targeted against the melanoma-associated, oncogenic $BRAF^{V600E}$ mutation. Consequently, JARID1B-expressing melanoma cells were shown to accumulate in BRAF inhibitor-treated melanoma cell cultures [56]. Similarly, JARID1A, which is also expressed transiently in only a small proportion of NSCLC cells, contributes to resistance to epidermal growth factor receptor (EGFR) tyrosine kinase inhibitors [57]. These data not only highlight a role for histone demethylation in tumor drug resistance, but also emphasize a new challenge related to the dynamic nature of the expression of these demethylases. This, in turn, suggests that a small proportion of tumor cells with adapted chromatin remodeling features may survive drug treatment and have the potential to establish new proliferating subpopulations and, thus, relapse.

Regarding the role of histone methylation and demethylation in cancer, therapeutic targeting is an option. Due to the fact that methylases and demethylases are quite specific with regards to the histone residues that they modify (as opposed to HATs and HDACs, which are quite promiscuous in their acetylation targets), the small molecules required for their inhibition need to be tailored for specific methylases or demethylases (for a review, see Ref. [58]).

3.1.8 Other Histone Modifications

Histones have further been shown to be phosphorylated at serine or tyrosine residues, ubiquitylated or sumoylated at lysine residues, and poly-ADP ribosylated at glutamate residues; they may also be altered by arginine deamination and proline isomerization[10, 59]. Clearly, when these modifications have been more extensively investigated, associations with carcinoma development and progression will emerge. Thus far, histone phosphorylation may have potential links to tumor development, with roles in DNA repair, mitosis, and apoptosis (as reviewed by Wang *et al.* [59]). The H2A.X histone variant is phosphorylated at Ser139, and is linked to DNA double-strand break repair [60], whereas aurora-kinase-B phosphorylates H3 at Ser10 (H3S10), which is crucial for chromosome condensation and appropriate segregation during mitosis [61]. Interestingly, an increased H3S10 phosphorylation has been associated with cell transformation in response to oncogenic RAS signaling [62]. Finally, H2BS14 phosphorylation follows apoptotic stimuli and plays a role in peroxide-induced apoptosis [63, 64].

3.2 ATP-Dependent Chromatin Remodeling

3.2.1 ATP-Dependent Chromatin-Remodeling Complexes

In humans, four families of ATP-dependent chromatin-remodeling complexes have been characterized to date:

- The SWI/SNF (named due to its homology to the yeast "switching defective/ sucrose nonfermenting") complexes are thought to bind promoter regions by being attracted to acetylated histones via the bromodomain of their enzymatic cores BRG1 or BRM, and mainly activate gene expression. More recently, this view was challenged when it was shown that only about 12% of BRG1 complexes bound to active promoters while, quite unexpectedly, a higher BRG1 occupancy was found at distant regulatory enhancer or repressor motives. This distant BRG1–DNA interaction is proposed to confer predominantly gene-repressive functions (as reviewed in Ref. [3]). The underlying mechanisms of distant gene regulation remain to be further defined, but will no doubt reveal important chromatin-remodeling insights and have implications on malignant cell transformation.

- The ISWI ("imitation switching defective") complexes have so-called SANT and SLIDE domains, which mediate binding to unmodified histones.
- The NuRD/Mi-2/CHD ("nucleosome remodeling and acetylation"/ "chromatin helicase DNA-binding") complexes have chromodomains that specifically interact with methylated histones.
- The INO80 ("inositol requiring 80") complexes have SPLIT domains for histone interaction.

All four remodeling complexes have been associated with activation and repression of transcription (for a review, see Ref. [8]).

3.2.2 **SWI/SNF in Carcinoma**

SWI/SNF complexes are the most extensively studied human ATP-dependent chromatin-remodeling complexes, and thus have provided most of the data implicating ATP-dependent chromatin remodeling in malignant transformation and tumor progression. SWI/SNF complexes consist of either BRG1 or BRM as their enzymatic ATPase core. Both core proteins are able to interact via their bromodomain with acetylated histones, and to catalyze ATP hydrolyses to alter DNA–histone interactions. The SWI/SNF are multisubunit complexes that contain, in addition to BRG1 or BRM, seven to eleven subunits termed BRG1-associated factors (BAFs), and are approximately 2 MDa in size. Specific BAF subunit composition is thought to alter enzymatic activity, to facilitate binding to DNA sequence-specific transcription factors, and to target the remodeling complex to DNA and nucleosomes. Protein subunits in human cells include BAF155, BAF170, BAF180, BAF200, BAF270A, BAF270B, BAF53A, BAF53B, BAF57, BAF60a, BAF60b, BAF60c, BRG1, BRM, β-actin, and BAF47 (the BAF-associated number refers to the protein size, in kDa; see Table 4) [65, 66]. The subunit composition of SWI/SNF chromatin-remodeling factors can vary, depending on the biological context. For example, during the development stage, pluripotent, proliferating stem cells express mainly BRG1, which is paired with specific proliferation-associated BAFs, including BAF155 and BAF53A. However, once the cells have differentiated, growth arrest-associated BAFs make up the SWI/SNF complex. In this case, BRG1 is replaced by BRM, while BAF155 is no longer part of the complex which, instead, contains BAF170 and BAF53B rather than BAF53A [3]. This highlights the fact that cells rely on different chromatin-remodeling factor compositions, depending on their proliferative fate and that, during tumorigenesis, cells may revert to a more "stem cell-like" chromatin-remodeling control by switching to a proliferative SWI/SNF composition. Moreover, as the multiple subunits of chromatin-remodeling complexes influence the enzymatic activity

Tab. 4 Human SWI/SNF subunits.

Subunit	Gene	Relevance
BRG1	*SMARCA4*	Proliferating stem cells; context-dependent potential oncogenic and tumor-suppressor activity
BRM	*SMRACA2*	Differentiated cells; potential tumor suppressor
BAF155	*SMARCC1*	Proliferating stem cells, potential oncogene
BAF47 (hSNF5, INI)	*SMARCB1*	Bona fide tumor suppressor
BAF57	*SMARCE1*	Mediates hormone receptor-dependent transcription
BAF53A	*ACTL6A*	Proliferating stem cells, potential oncogene
BAF53B	*ACTL6B*	Differentiated cells
BAF180	*PBRM1*	Potential tumor suppressor
BAF270A (BAF250A)	*ARID1A*	Potential tumor suppressor
BAF270B (BAF250B)	*ARID1B*	–
BAF170	*SMARCC2*	Differentiated cells
BAF200	*ARID2*	–
BAF60a	*SMARCD1*	Potential tumor suppressor
BAF60b	*SMARCD2*	–
BAF60c	*SMARCD3*	–
β-Actin	*ACTB*	–

and specificity of the entire complex, alterations in any of the subunits may be associated with selective advantages for cancer cells.

3.2.2.1 SWI/SNF Subunits Linked to Cancer

BAF47 Germline mutations of the BAF47 subunit of SWI/SNF, which originally was identified as INI1 and is also known as *hSNF5*, predispose carriers to rhabdoid carcinomas [67, 68]. BAF47 is also frequently found to be somatically impaired in rhabdoid carcinomas [69], as well as in a subset of central primitive neuroectodermal tumors (cPNETs) and medulloblastomas [70]. BAF47 fulfills the criteria of a tumor suppressor gene, as mice heterozygous for BAF47 are tumor-prone (a complete BAF47 knockout is developmentally lethal) [71], while the re-expression of BAF47 in rhabdoid carcinoma cells causes cell-cycle arrest and an onset of senescence [72]. Arrest and senescence are associated with expression of the cell-cycle inhibitor $p16^{INK4a}$ [73, 74]. Importantly, the reintroduction of BAF47 does not affect the SWI/SNF-driven regulation of a range of other target genes [75]; this suggests that BAF47 provides specificity for the $p16^{INK4a}$ promoter and, indeed, it has been shown that BAF47 can bind the $p16^{INK4a}$ promoter [76, 77]. With $p16^{INK4a}$ regulation as a predominant target of BAF47, it is no surprise that BAF47 loss is associated with melanoma progression and chemoresistance [78], since it is known that $p16^{INK4a}$ plays a prominent role in melanoma predisposition and development (for reviews, see Refs [79, 80]) as well as chemoresistance [81]. Interestingly, one report has linked the tumor-suppressor function of BAF47 to the presence of BRG1. In this case, the investigators suggested that, in the absence of BAF47, BRG1 may function as an oncogene and drive tumor formation, whereas the loss of both BAF47 and BRG1 together had little effect [82].

BRG1 The potential role for BRG1 as an oncogene is controversial, although this would be in line with the observation that, in normal proliferating stem cells, the SWI/SNF complexes contain BRG1 rather than BRM together with a proliferation-associated BAF subunit composition, whereas differentiated nonproliferating cells rely on BRM as their SWI/SNF core unit [3]. Yet, there are other more compelling data which suggest that BRG1 may act as an oncogene. For example, an increased BRG1 expression was correlated with oral carcinoma development [83], while in prostate carcinoma an increased BRG1 and a reduced BRM expression was found in malignant versus normal tissue, and was associated with high-grade tumors as well as the invasive potential of prostate carcinoma cells [84]. Further, an elevated BRG1 expression was also identified in gastric carcinoma compared to normal mucosa, and increasing BRG1 levels were correlated with tumor progression [85, 86]. Additionally, mechanistic links have been identified to an oncogenic role of BRG1, based on the results of studies of colorectal carcinoma where BRG1 – but not BRM – was overexpressed. In this case, the *PTEN* gene – which encodes a tumor suppressor and phosphatase, which antagonizes the oncogenic phosphatidylinositol 3-kinase (PI3K)–AKT pathway – was shown to be repressed by the BRG1–SWI/SNF complex in colorectal carcinoma cells [87]. In addition, BRG1 was found to interact with the oncogene β-catenin and to promote the expression of β-catenin target genes [88].

In stark contrast to these reports, mounting evidence has emerged in recent years that BRG1 acts as a tumor suppressor. Such a role for BRG1 is highlighted by the fact that BRG1 hemizygous mice are susceptible to tumors (complete BRG1 knockout is developmentally lethal) [89], whilst a total loss of BRG1 has been identified in tumor tissue and also shown to enhance the development of human lung cancer [90]. The loss of BRG1 expression in lung cancer is also associated with a poor prognosis [91, 92], while more recently acquired data have suggested that a significant proportion of primary lung tumors already carry biallelic inactivated BRG1, placing it at an early stage of tumor development [93]. Moreover, BRG1 is silenced or mutated in many human tumor cell lines derived from breast, pancreatic, ovarian, lung, brain, prostate, and colon carcinomas [94–97]. BRG1 has also been shown to be lost in established neuroendocrine carcinomas and adenocarcinomas of the cervix [98]. The tumor-suppressive function of BRG1 is largely attributed to cell-cycle inhibitory roles (these are discussed more fully in Sect. 4.1.1.2).

It is important to emphasize that both oncogenic or tumor-suppressive roles for BRG1 are plausible, based on the assumption that the effect of BRG1 depends largely on the subunit composition of the predominant SWI/SNF complex (e.g., the presence of BAF47, as outlined in the section titled "BAF47"), and that this composition is determined by the biological context [3]. Therefore, even seemingly contradictive observations of increased or decreased levels of BRG1 in the same tumor type may be explained by genetic tumor heterogeneity, since not all tumors – even of the same type – necessarily employ the same pathway alterations to achieve malignant features. For example, as outlined above, the overexpression of BRG1 was correlated with a progression of colorectal and prostate carcinoma, which implied an oncogenic role [84, 87], whereas BRG1 was found to be lost or mutated in cell lines derived from these malignancies, which implied a tumor-suppressive role [97]. Moreover, whereas a common loss of BRG1 was identified in melanoma progression, there was no effect of BRG1 silencing on cell-cycle progression in melanoma cells [99]. Nonetheless, other investigators reported increased BRG1 levels associated with melanoma [100], a discrepancy which may be explained by known genomic melanoma heterogeneity being dependent on ethnic background and exposure to ultraviolet sunlight in an analyzed melanoma cohort [101, 102]. The UK-derived sample set of melanoma is likely to have reduced BAF47 leels, as was found in a melanoma cohort investigated by the same laboratory [78]. However, as this is in line with an oncogenic role for BRG1 [82], it would be interesting to determine whether the Australian melanoma cohort expresses BAF47.

BRM In contrast to BRG1, its homolog BRM appears to be rarely mutated in cancer, although recently a "hotspot" BRM mutation was identified in squamous and basal cell carcinomas [103]. BRM expression is more commonly reduced or lost. In prostate carcinoma, a reduced BRM expression was found in malignant versus normal tissue and also in high-grade lesions [84], while targeted BRM silencing in mice caused prostate hyperplasia. This was in line with the observation that BRM loss was associated with an increased proliferation in human prostate carcinoma cells [104]. BRM is also often lost or reduced in cell lines or tumor tissue such as gastric carcinoma and lung tumors, while a reduced expression of BRM in mice was associated with a carcinogen-induced development of lung carcinoma. Moreover, there is evidence that histone modifications play a role in the loss of BRM expression, since drugs that can inhibit HDAC function (i.e., HDAC inhibitors) may also restore BRM expression [86, 105]. Notably, the expression of BRM may play a role in preventing proliferation in oncogenic RAS-expressing fibroblasts, and this BRM function can also be prevented by exposure to HDAC inhibitors. Interestingly, however, the HDAC inhibitory effect is targeted to the acetylation of BRM rather than acetylation of histone tails. Typically, HDAC inhibitor treatment caused an accumulation of acetylated BRM and prevented its growth-inhibitory functions [106].

BAF57 The SWI/SNF subunit BAF57 mediates interaction with the hormone-dependent estrogen and androgen receptors. BAF57 is essential for estrogen-dependent transcriptional activation [107], and a truncating BAF57 mutation has been identified in a breast ductual carcinoma cell line. Although this mutant BAF57 no longer binds the estrogen receptor, it can still bind to the downstream coactivator SRC1a, rendering SRC1a transcriptional activity independent of estrogen [108]. Additionally, in a subset of prostate carcinoma, an increased BAF57 level was shown to be linked to an androgen-dependent proliferation [109].

BAF155 High levels of BAF155, which normally is a subunit found in proliferating stem cells [3], were associated with cervical carcinoma [110] and also identified in prostate carcinoma versus normal differentiated prostate tissue. Such an elevated BAF155 expression was also correlated with tumor progression and metastasis [111].

BAF180 The novel technology of exome sequencing has identified BAF180 as the second most common gene to be altered by truncating mutations in 41% of renal cell carcinoma [112]. Additionally, BAF180 has been shown, together with BRD7, to cooperate with *p53* in target gene expression during oncogene-induced senescence, and was suggested to act as a tumor suppressor since it was found to be mutated in breast cancer [113, 114].

BAF270A The SWI/SNF subunit BAF270A, which is deficient in approximately 30% of renal carcinomas and 10% of breast carcinomas [115], has been shown to be essential for normal cell cycle arrest, as it has been linked to reduced cyclin D1 and increased *p21* expression [116].

3.2.3 Other ATP-Dependent Chromatin-Remodeling Complexes in Cancer

3.2.3.1 ISWI Amplified Rsf-1, which forms part of the ISWI complex, was identified in prostate carcinoma in comparison to normal tissue [117]. It was also related to ovarian cancer and, more importantly, to the resistance of ovarian carcinoma to the drug paclitaxel; in contrast, the silencing of Rsf-1 in ovarian carcinoma cells led to an increased paclitaxel sensitivity [118]. Additionally, the overexpression of Rsf-1 resulted in DNA damage and chromosomal aberration [119].

3.2.3.2 NuRD/Mi-2/CHD CHD5 has been proposed as a tumor suppressor, since heterozygous mice develop spontaneous tumors, including squamous cell carcinoma [120].

MTA1 (metastasis-associated gene1) or its homologs *MTA2* and *MTA3*, form part of the NuRD/Mi-2/CHD complex. *MTA1*, which is a signal transduction target of the breast cancer-associated oncogene HER2, was found to be overexpressed in breast cancer cell lines and to be tightly linked to invasion in esophageal, colorectal, and gastric carcinomas [121]. The expression of *MTA3* is facilitated by estrogen; moreover, *MTA3* downregulates SNAI1 which, in turn, is a transcriptional repressor of the tumor suppressor and cell–cell adhesion molecule, E-cadherin. Estrogen receptor-negative breast carcinomas thus display a reduced *MTA3* level linked to a loss of E-cadherin and invasiveness [122].

The histone demethylase LSD1 is a component of the NuRD/Mi-2/CHD complex, and is involved in the transcriptional regulation of cell proliferation, survival, and epithelial-to-mesenchymal transition. LSD1 also inhibits the invasion of breast cancer cells *in vitro*, and suppresses breast cancer metastatic potential *in vivo*. In line with this latter finding, LSD1 is downregulated in breast carcinomas [46].

ZIP, a transcriptional repressor, recruits the NuRD/Mi-2/CHD complex to repress expression of the EGFR, inhibits cell proliferation, and suppresses breast carcinogenesis [123].

3.2.3.3 INO80 Recent screenings for proteins, which regulate telomere length, have led to the emergence of INO80 complex proteins, and deletion of the

Ies3 subunit has resulted in an impaired growth and telomeric instability in yeast [124].

The Tip60 HAT, which complexes with INO80, acetylates p53 and is involved in the p53 activation in response to DNA damage; it also mediates the decision between cell cycle arrest and apoptosis [125]. Hence, Tip60 is thought to be a tumor suppressor, and has been proposed to be critical to oncogene-induced DNA damage response, which is impaired by reduced Tip60 levels [126].

The INO80 interacting proteins reptin (Tip48) and pontin (Tip49) are over-expressed in hepatocellular, colorectal, breast, gastric, bladder, and non-small-cell lung carcinoma. They are required for telomerase assembly, and they cooperate with Tip60 in DNA damage recognition and repair processes [127]. However, whether these functions of Tip60, reptin, and pontin are determined by interaction to the INO80 complex is unclear, as these proteins have HAT (Tip60) or ATPase (reptin and pontin) activities, and have been shown to interact with chromatin and other remodeling complexes [127]. Nevertheless, data from yeast are in line with a role for INO80 in DNA double-strand break repair and telomerase assembly [124, 128].

4 Remodeling the Hallmarks of Cancer

As outlined above, the main hallmarks of transformed malignant cells are excessive proliferation, avoidance of cell death, and invasive potential. A number of tumor-suppressor genes and oncogenes have been identified in pathways that affect these hallmarks, because their activity is frequently altered by mutations, genetic amplifications, or loss [1]. Moreover, most – if not all – oncogenes and tumor-suppressor genes are regulated by chromatin-remodeling events linked to malignant transformation. Some of these connections are reviewed in the following subsections.

4.1 Excessive Proliferation

4.1.1 Chromatin-Remodeling Complexes and Cell Cycle-Inhibiting Tumor Suppressor-Proteins

The main function of a broad range of tumor-suppressor proteins is that of cell cycle inhibition and, thus, an appropriate control of cell proliferation. These tumor suppressors are usually inactivated during malignant transformation.

4.1.1.1 Chromatin Remodeling and p53

The tumor suppressor and transcription factor p53 is often referred to as the "*guardian of the genome*" [129], to describe its essential role in responding to genomic stress (e.g., radiation- or carcinogen-induced DNA damage). p53 is mutated or deleted in more than 55% of all human cancers [130]. Remarkably, p53 responds to DNA damage by either cell-cycle arrest and DNA repair. Alternately, if the damage is too severe for an efficient repair, p53 will induce a tightly regulated cell death program (apoptosis) to prevent malignant transformation through the accumulation of genetic changes [129]. p53-induced cell cycle arrest is mainly facilitated by the transcriptional upregulation of the cell-cycle inhibitor $p21^{Waf1}$, and this arrest allows for DNA repair to be completed before replication can continue [131]. Alternately, if the DNA damage is too severe, then p53 may transcriptionally activate the transcription of several pro-apoptotic proteins to induce cell death [129]. Emerging evidence has demonstrated that p53 employs the chromatin-remodeling machinery as part of its normal function as a transcription factor, and to decide between growth arrest and cell death. For instance, various HATs not only transfer acetyl groups to histone residues to allow p53 access to its target promoters; rather, they also transfer acetyl groups to a number of transcription factors or transcriptional repressors, including p53. The latter is directly acetylated by p300, CBP, or PCAF at multiple carboxy-terminal residues, and this correlates to increased transcriptional activity [132]. Moreover, p53 acetylation has been linked to genomic stress, and is generally increased following DNA-damaging insults [133–135]. It has also been shown that, following DNA damage, the interaction between p53 and the HAT subunit hADA3 increased such that, in the presence of hADA3, the p53 transcriptional activity is strongly enhanced [136] and hADA3 is a target of E6 human papilloma virus antigen in cervical carcinoma [137]. Moreover, the acetylation of p53 competes directly with ubiquitylation by the ubiquitin ligase mdm2, thereby contributing to p53 accumulation, as the p53 levels are normally regulated by rapid ubiquitin-dependent proteasomal degradation [138]. HDAC2 was also shown to affect p53 transcriptional activity, though not by modulating the acetylation of p53 itself; rather, a knockdown of HDAC2 increased the p53-DNA binding activity [139].

p53 has been reported to interact with the SWI/SNF chromatin-remodeling complex, and this was critical for the transcriptional regulation of $p21^{Waf1}$. Interaction with SWI/SNF was proposed to occur via BAF60A, and this correlated with an increased $p21^{Waf1}$ transcription [140]. Remarkable other investigations linked BAF53A to p53 interaction with the SWI/SNF complex [141] which, in contrast, correlated with $p21^{Waf1}$ repression, and this was of course in line with BAF53A being a "proliferative" SWI/SNF subunit [3, 142]. Moreover, an N-terminal-truncated BAF53A caused cell death via p53 [143]. SWI/SNF and p53 have also provided another spin to the controversy between oncogenic and tumor-suppressive roles for BRG1. An analysis of the effects of BRG1 silencing on p53 stability showed that the knockdown of BRG1, but not BRM, reduced the ubiquitylation of p53 and thus led to an enhanced p53 accumulation and transcriptional upregulation of $p21^{Waf1}$. The authors of the study concluded that, in tumor cells which have retained wild-type p53, this tumor suppressor was disabled by BRG1 in cooperation with CBP [144]. The SWI/SNF subunit BAF180, on the other hand, has been shown (together with BRD7) to induce a number of *p53* target genes in normal cells during oncogene-induced senescence [113], and this agrees with evidence showing BAF180 to be critical for $p21^{Waf1}$ regulation [114]. Furthermore, the presence of p53 is directly linked to chromatin relaxing during DNA repair, and p53-null cells have changed the patterns of histone acetylation compared to p53-expressing cells; this suggests a more global effect of p53 on chromatin remodeling (for a review, see Ref. [145]). The main functions of p53, and the contribution of chromatin remodeling, are shown schematically in Fig. 3.

4.1.1.2 Chromatin Remodeling and Retinoblastoma Protein (pRb) The retinoblastoma protein (pRb) is often

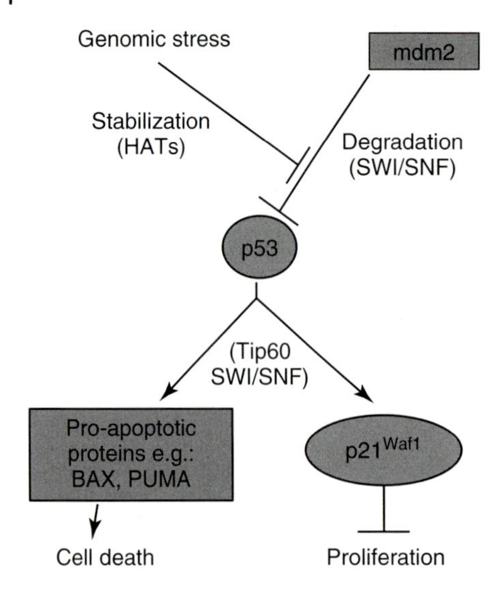

Fig. 3 Chromatin remodeling and p53. p53 has been referred to as the "*guardian of the genome*," and is stabilized in response to genomic stress. Stabilization involves acetylation by HATs and opposes mdm2-dependent p53 degradation, which involves SWI/SNF. p53 transcriptionally upregulates either $p21^{Waf1}$, causing cell cycle arrest, or pro-apoptotic proteins, leading to cell death. The decision involves Tip60, and transcriptional p53 activities depend on SWI/SNF with specific subunit composition.

referred to as the "*master brake*" of cell proliferation, because the main pRb function is to inhibit the E2F transcription factor from expressing a range of proliferation genes [146]. Thus, active pRb causes cell cycle arrest. On the other hand, when cells receive growth factor signaling and are promoted towards cell division, pRb is sequentially phosphorylated by cyclin-dependent kinases (CDKs), which results in pRb inactivation and, consequently, the derepression of proliferation genes. Mutations of pRb have been associated with the childhood cancer, retinoblastoma. Moreover, the cell cycle regulatory pRb pathway is disturbed in over 90% of cancers [1]. During recent years, it has become increasingly clear that pRb relies on chromatin remodeling for transcriptional repression of extended chromatin regions. pRB was shown to recruit HDAC1 to E2F-responsive promoters [147], while HDAC1 assists the pRb-repressive function by the deacetylation of H4 [148]. The phosphorylation of pRb by CDK4, on the other hand, prevents HDAC1–pRb interaction [149, 150]. Furthermore, pRB is thought to engage the SWI/SNF chromatin-remodeling complex for E2F promoter silencing, and this provides an elegant model of the sequential expression of cyclins, which provide activity to CDKs (and thereby pRb inactivation) at specific stages of the cell cycle. The sequential activity of CDK4/cyclin D in G_1 phase of the cell cycle, of CDK2/cyclin E to promote the entry of DNA replication in S-phase, followed by CDK2/cyclin A to complete the DNA replication, requires an initiation by cyclin D, the expression of which is a direct consequence of growth factor signaling. CDK4/cyclin D complexes initiate phosphorylation of the active pRb, which occupies E2F promoters in complex with HDAC1 and SWI/SNF. In turn, this disrupts the pRb–HDAC1 interaction, which then causes derepression of the first E2F target genes, including cyclin E. This leads to active CDK2/cyclin E and further pRb phosphorylation, which dissociates SWI/SNF from the complex and allows for cyclin A expression and traversion through S-phase [150]. There is evidence that both BRG1 and BRM can function in these HDAC1–SWI/SNF–pRb repressive complexes, and it has been proposed

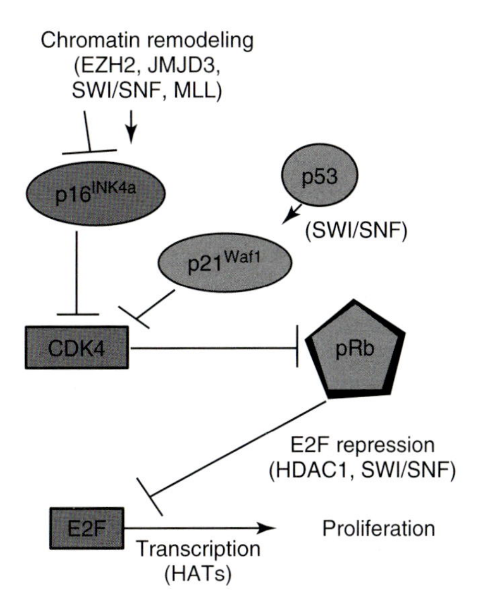

Fig. 4 Chromatin remodeling and pRb. The main function of pRb is to repress E2F from transcriptional activation of proliferation genes, which involves HDAC1 and SWI/SNF. pRb itself can be inhibited by phosphorylation catalyzed by CDKs, while the CDK inhibitors $p21^{Waf1}$ and $p16^{INK4a}$ in turn can keep pRb active by inhibiting the CDKs. Transcriptional regulation of these inhibitors requires SWI/SNF. $p16^{INK4a}$ in particular is strongly regulated by chromatin-remodeling events. The histone methyltransferase EZH2 efficiently represses $p16^{INK4a}$, which is directly opposed by JMJD3, while the histone methyltransferase MLL activates $p16^{INK4a}$ expression.

that pRb requires BRG1 binding for cell-cycle arrest[151, 152], while some cancer-associated BRG1 mutations abrogate pRB-mediated arrest [153]. However, the cell cycle inhibitory effect created by the reintroduction of BRG1 is only seen convincingly in cells that are deficient in both BRG1 and BRM, such as SW-13 adenocarcinoma cells [99]. It should also be noted that this arrest may not involve the direct interaction between pRb and BRG1, as it could be shown that SW-13 cells would arrest upon ectopic BRG1 expression, even when a BRG1 mutant was expressed that was disabled in pRb interaction. Instead, BRG1 led to an upregulation of the CDK inhibitor $p21^{Waf1}$, leading in turn to a hypophosphorylated active pRb and cell cycle arrest [154]. Evidence was also produced that SWI/SNF is involved in the transcriptional upregulation of the CDK inhibitors, $p16^{INK4a}$ and $p15^{INK4b}$ [73, 154], as well as the repression of cyclin D expression [155], and all of these SWI/SNF targets were shown to promote pRb activity. A loss of the SWI/SNF subunit BAF47 impaired $p16^{INK4a}$ and $p21^{Waf1}$ expression [76] (the relevance of BAF47 loss in carcinoma was discussed above; see the section titled "BAF47"). Details of the pRb pathway, and its cooperation with chromatin remodeling, is shown schematically in Fig. 4. pRb also interacts with heterochromatin-forming proteins to induce senescence (as discussed in Sect. 4.1.3).

4.1.1.3 Chromatin Remodeling and $p16^{INK4a}$ The main CDK inhibitors relevant in carcinoma are $p21^{Waf1}$ and $p16^{INK4a}$. $p21^{Waf1}$ is not mutated in carcinoma, but is deregulated via alterations of p53; in contrast, $p16^{INK4a}$ is one of the most frequently mutated, deleted, or silenced genes in cancer [156]. Moreover, individuals with germline mutations in $p16^{INK4a}$ are at risk of developing pancreatic cancer, and carry a highly increased risk of developing melanoma because these mutations generally produce nonfunctional proteins [80, 157, 158]. The main role of $p16^{INK4a}$ is to inhibit the cyclin D-dependent kinases CDK4 (and CDK6), which places $p16^{INK4a}$ at the forefront of cell-cycle inhibition in early G_0/G_1 phase, where it is involved in the central decision of whether to enter proliferation, or not. In normal cells, $p16^{INK4a}$ causes cell cycle arrest associated with differentiation, and it probably plays an even more important role in senescence (see Sect. 4.1.3). Not surprisingly, $p16^{INK4a}$ is therefore efficiently repressed during development and in proliferating progenitor cell populations, whereas the levels are increased when cells differentiate and senesce [17]. $p16^{INK4a}$ regulation on the level of chromatin modifications plays an important role, and alterations in these mechanisms have been associated with cancer. For example, a reversible repression of $p16^{INK4a}$ is achieved by the formation of facultative heterochromatin, which is recognized by the repressor BMI1, a member of the polycomb group of proteins, a well-characterized repressor of $p16^{INK4a}$. More recently, a critical role for the histone methyltransferase EZH2 was identified in $p16^{INK4a}$ repression. In this case, EZH2 cooperates with BMI1 by the trimethylation of H3K27, which attracts BMI1 binding [17]. Moreover, the histone H3 demethylase Ndy1 enhances $p16^{INK4a}$ repression by opposing the EZH2 downregulation, usually in association with senescence, and thereby enhances H3K27 trimethylation and BMI1 binding to the $p16^{INK4a}$ promoter. Consequently, Ndy1 overexpression did cause mouse embryonic fibroblast transformation [159]. On the other hand, EZH2 may be downregulated in concert with an upregulation of its opponent, the H3K27 demethylase JMJD3, during $HRAS^{G12V}$-mediated $p16^{INK4a}$ upregulation and senescence in human fibroblasts. This is in line with the fact that JMJD3 levels are significantly higher in growth-arrested nevi compared to normal skin, while ectopic JMJD3 induces $p16^{INK4a}$ upregulation and senescent features in human fibroblasts [160].

When cells differentiate and reach their finite life span, the SWI/SNF chromatin remodeling complex replaces the BMI1 repressor, which in turn relaxes the chromatin structures around the $p16^{INK4a}$ promoter region. This effect is strictly dependent on the BRG1 and BAF47 subunits, and allows transcription factor access to activate $p16^{INK4a}$ transcription [74].

In contrast to the repressing H3K27me3, H3K4me3 was shown to be required for $p16^{INK4a}$ expression, and was shown to depend on the presence of two components of the H3K4 methyltransferase complex, MLL1 and pRbBP5. Moreover, H3K4me3 status – and thus $p16^{INK4a}$ expression – were reportedly modulated by the pRbBP5 binding partner DDB1, which is a component of the ultraviolet light-damaged DNA-binding protein. DDB1 knockdown reduced H3K4me3 of the $p16^{INK4a}$ promoter and reduced $p16^{INK4a}$ expression. Importantly, in human fibroblasts both MLL1 and DDB1 were crucial in the oncogene-induced upregulation of $p16^{INK4a}$ by $H\text{-}Ras^{G12V}$ [161].

4.1.2 **Chromatin-Remodeling Complexes and Oncogenic Signaling**

The control of oncogenic signaling pathways provides a powerful mechanism to render a cell independent of external growth stimuli, and to ignore anti-proliferative signals as well as to promote resistance to programmed cell death. Hence, such control represents a frequent adaptation to promote malignant transformation.

4.1.2.1 MAPK Pathway The mitogen-activated phosphorylation kinase (MAPK) signaling pathway is frequently aberrantly activated in human cancers, through constitutively activating mutations in RAS or the downstream kinase BRAF, and is thought to confer proliferation and survival mechanisms. There is evidence that MAPK signaling involves the modulation of chromatin. Indeed it has been shown that, upon MAPK signaling, H3S10 is phosphorylated by the H3 kinase MSK1, in association with an increased acetylation of H3 at MAPK target promoters. H1S3 was also found to be phosphorylated in activated RAS expressing mouse fibroblasts, and may destabilize the chromatin structure (for a review, see Ref. [62]). The SWI/SNF core enzyme BRM is proposed to be capable of preventing proliferation in oncogenic RAS-expressing fibroblasts [106]. Consequently, the expression of BRM – but not of BRG1 – is diminished upon the RAS transformation of mouse fibroblasts [162]. A recent search for genes that can cooperate with mutated, oncogenic $BRAF^{V600E}$ in melanoma formation identified SETDB1, which is a histone methyltransferase that facilitates H3K9 trimethylation but opposes $BRAF^{V600E}$-induced senescence, and thereby accelerates melanoma formation in a zebrafish model. SETDB1 is also frequently overexpressed in human melanomas, and also in NSCLC, small-cell lung cancer (SCLC), as well as ovarian, hepatocellular, and breast carcinoma [163].

4.1.2.2 PI3K-AKT Pathway Another important signaling pathway that very frequently is abnormally activated in cancer is the PI3K–AKT pathway, which facilitates increased proliferation and survival, including resistance against chemotherapeutic drugs [164]. Previously, AKT has been shown to interact with SWI/SNF and to phosphorylate the subunit BAF155 (and potentially also BAF47), though the functional outcome of this modification is unclear [165]. Phosphorylation of the acetyltransferase p300 on Ser1834 by AKT was shown to be critical for the p300-mediated histone acetylation and activation of gene expression [166]. Recently, a link between the histone H3K27 methylase EZH2 and the PI3K–AKT pathway was proposed. In this case, the overexpression of EZH2, which is common in a range of carcinomas (see Sect. 3.1.7), was associated with an upregulation of AKT1 and its active, phosphorylated form. Moreover, AKT1 was critical for the cellular localization of the breast cancer-associated BRCA1 tumor suppressor and genomic stability, as regulated by EZH2 [167].

4.1.3 Senescence

Senescence is a form of stable cell cycle arrest that permanently prevents cells from re-entering the cell cycle in response to physiological stimuli. The onset of senescence is a response to various genomic stress features, including telomere shortening, DNA damage, or oxidative stress. Senescence is seen as a major tumor-suppressive mechanism to prevent cells with deleterious DNA damage from dividing and potentially transmitting any transforming genetic changes to their daughter cells. Not surprisingly, senescence is also activated in response to oncogenes such as B-RAF^{V600E} or N-RAS^{Q61K} [168–170]; this may be a last resort to prevent malignant transformation by these oncogenes, as their signaling potently opposes programmed cell death by altering the balance of pro- and anti-apoptotic proteins [171]. pRb is required for an intact senescence program, and the re-introduction of pRb into SAOS osteosarcoma cells that have lost pRb expression may cause senescence [172]. In contrast, the inactivation of pRb with the viral oncoprotein E1A prevents senescence, whereas a mutant E1A, which is impaired in pRb binding, is unable to prevent senescence [173]. Interestingly, the mutant E1A, which is able to bind pRb but unable to interact with the HDAC p300 and p400, leads to a less-efficient development of senescence [173]. These data highlight the role of chromatin remodeling in cellular senescence programs; indeed, during senescence E2F-responsive genomic promoter regions are stably repressed from transcription by the establishment of heterochromatin regions. These regions are visible as microscopic "senescence-associated heterochromatin foci" (SAHF) when cells are stained with certain DNA-intercalating dyes, such as DAPI. SAHF formation coincides with the recruitment of heterochromatin proteins and pRb to E2F-responsive promoters. A number of heterochromatin-associated histone modifications are characteristic for SAHF; these include a lack of H3K9Ac and H3K4me3, but an accumulation of H3K9me3. The latter attracts HP1 proteins, which are required for heterochromatin assembly [173]. Further, SAHF H3K9 trimethylation is thought to be very stable and to prevent HATs from catalyzing histone acetylations. This contributes to a very secure silencing of E2F-responsive proliferation genes, which is not reversible by physiological stimuli [173]. Importantly, although the p53–p21 pathway was shown to be capable of inducing a number of senescent features – and does so in response to DNA damage, telomere dysfunction and during oncogene-induced senescence [174, 175] – E2F promoter silencing via SAHF formation requires an intact p16–pRb pathway. In fact, it has been argued that SAHF formation is critical for the potency of the senescence program in tumor suppression [170, 173]. As expected, SAHF formation is prevented by cancer-associated $p16^{INK4a}$ mutations [176]. Additionally, there is some evidence that the SWI/SNF complex is required for senescence onset, and the introduction of BRG1 into cells induces senescent features, although this is only convincing in cells that also lack the BRM ATPase [154]. Interestingly, it was found that BRG1 can interact with $p16^{INK4a}$, although the absence of BRG1 neither prevents $p16^{INK4a}$-induced growth arrest nor senescence or SAHF formation. It should be noted that the WMM1175 melanoma cells used in these experiments also express BRM, and so still have functional SWI/SNF complexes [99]. This is in line with a suggested role for BRM in melanocyte senescence, since it was shown that BRM was recruited and required – albeit transiently – by the pRb–HDAC1 complex during the initiation of SAHF [177].

H3K9 trimethylation during SAHF formation is thought to be catalyzed by either the SUV39H1 methyltransferase which, together with HP1 interacts with pRb during E2F promoter silencing in senescence [178], or the RIZ1/PRDM2 H3K9 methyltransferase, which was also shown to cooperate in pRb gene repression and is inactivated in colon, breast, and gastric carcinoma [179]. Furthermore, a search for H3K9me3 interacting proteins to identify proteins involved in the senescence program identified JMJD2C, which is a H3K9- and H3K36-specific demethylase. JMJD2C is a direct antagonist of the SUV39H1 H3K9 methyltransferase, and its overexpression in carcinomas is therefore not surprising [48].

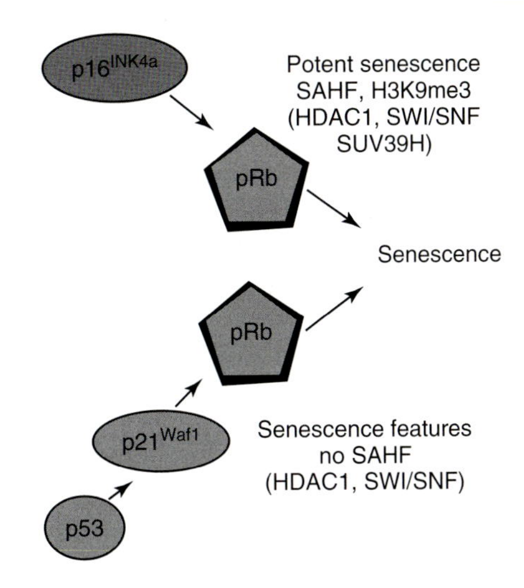

Fig. 5 Chromatin remodeling and senescence. The secure repression of E2F target genes by pRb is central to the cellular senescence program. Regardless of whether this is initiated by $p21^{Waf1}$ or $p16^{INK4a}$, this involves HDAC1 and SWI/SNF action. While p53-$p21^{Waf1}$-induced senescence displays several distinctive features (growth arrest, cell flattening, cell vacuolization, and increased β-galactosidase levels), only $p16^{INK4a}$ action on pRb leads to senescence involving H3K9 trimethylation by SUV39H. This leads to a distinctive formation of SAHF that, arguably, contributes potency to the senescence program.

Consistent with gene silencing and decreased acetylation during senescence is the downregulation of the histone acetylases p300/CBP, as observed in melanocytes reaching their finite lifespan [180]. Chromatin remodeling in senescence is shown schematically in Fig. 5.

4.2
Avoidance of Cell Death

Clearly, oncogenic signaling pathways which employ the chromatin-remodeling machinery (Sect. 4.1.2) can influence apoptotic cell fate. Additionally, as apoptosis involves the fragmentation of DNA it becomes clear that chromatin-remodeling processes must be more intimately involved, to allow the access of endonucleases for DNA cleavage. It is now also recognized that the histone phosphorylations of H2A, H2B, H3, and H4, the dephosphorylation of H1, the acetylation of H3B and H4, and the deacetylations of H4, together with methylations of H3 and H4 and the de-ubiquitylation of H2A, form part of the apoptotic process, and that some of these events are catalyzed by caspases while others involve conventional histone-modifying enzymes (for a review, see Ref. [181]). Furthermore, HDAC inhibitors have shown to have dramatic effects on cultured cells, which includes the onset of apoptosis [40].

There is also some evidence for a contribution of the SWI/SNF complex to the regulation of apoptosis, although this field has not been fully investigated. Although the knockdown of BAF60A was shown to decrease apoptosis, BAF60A was also proposed to interact with p53 and to enhance p53 transcriptional activity [140]; the BAF57 subunit was suggested to induce apoptosis by an upregulation of the pro-apoptotic tumor suppressor, CYLD [182]. On the other hand, N-terminal-truncated BAF53A caused cell death via p53, which suggested a BAF53A function in survival [143], while others proposed a critical role for SWI/SNF complexes to prevent apoptosis in response to DNA damage, as BRG1 knockdown increased apoptosis and this was regulated by a prolonged p53 activity [183]. Furthermore, the activity of the NuRD/Mi-2/CHD subunit MTA1 was shown to repress p53 activity in prostate carcinoma cells, while the inhibition of MTA1 contributed to p53-driven apoptosis [184].

4.3 Invasive Potential

The most important cell adhesion molecule in the normal homeostasis of epithelial cells is arguably E-cadherin. As a cell–cell adhesion molecule, E-cadherin is pivotal for normal cell–cell interaction so as to form a tight epithelial tissue; indeed, the loss of E-cadherin has been linked with an epithelial to mesenchymal transition of tumor cells. Consequently, the loss of E-cadherin would be critical in allowing cells to dislodge from their normal epithelial tissue environment and to invade other tissues [185]. The expression of E-cadherin also parallels that of $p16^{INK4a}$, in a sense that it is not expressed during development (when cells must migrate to find their appropriate tissue location) yet its level of expression is high during differentiation. Not surprisingly, E-cadherin is (similar to $p16^{INK4a}$) repressed by facultative heterochromatin formation involving the H3K27 methyltransferase, EZH2 [186]. Furthermore, it was noted above that MTA3 (which is a subunit of the ATP-dependent chromatin-remodeling complex NuRD/Mi-2/CHD) regulates SNAI1, a transcriptional E-cadherin repressor, and also reduces MTA3, which is linked to a loss of E-cadherin and to breast carcinoma invasiveness [122]. The NuRD/Mi-2/CHD subunit LSD1 was also linked to an inhibition of breast cancer invasiveness, and reduced levels of LSD1 have been found in breast carcinomas [46]. The SWI/SNF ATP-dependent chromatin-remodeling complex is also associated with epithelial to mesenchymal transitions. In fact, the introduction of either BRG1 or BRM into cells deficient of both ATPases led to a reversal of CpG island methylation (a regulative modification at the DNA level) and a derepression of the E-cadherin promoter; subsequently, the SWI/SNF complex was found to engage with the E-cadherin promoter to activate transcription [187]. In contrast, it has been shown recently that BRG1 can interact with the ZEB1 protein and aid the repression of E-cadherin. ZEB1–BRG1 colocalization was also observed in invasive colorectal carcinoma cells [188]. Moreover, BRG1 was shown to be critical for matrix metalloproteinase-2 (MM-2) expression, which is correlated with tumor invasion and angiogenesis. BRG1 was shown to facilitate MM-2 expression by recruiting the transcription factors SP1, SP3, and AP2 to the MM-2 promoter [189] and, as noted above, in prostate carcinoma an increased BRG1 and reduced BRM expression was found in malignant versus normal tissue and associated with the invasive potential of the tumor cells [84]. Thus, it is clear that BRG1 possesses both pro- and anti-invasive properties in carcinoma cells, depending on its binding partners.

5 Conclusion

Currently, an overwhelming amount of evidence exists demonstrating the importance of epigenetic alterations of gene regulation in carcinoma. It is recognized that histone modifications are associated with carcinoma progression and resistance to therapeutic intervention, and that they modulate the expression and function of important tumor suppressors and oncogenes. Furthermore, oncogenic signaling pathways engage the functions of ATP-dependent chromatin-remodeling complexes, which is perhaps not surprising as these complexes have important functions in modulating the expression of tumor suppressors. Thus, chromatin remodeling has been linked to all aspects of tumor evolution, and affects the most important hallmarks of cancer, namely proliferation, survival, and invasion. Not surprisingly, chromatin remodeling has been linked to all frequently studied types of carcinoma.

References

1 Hanahan, D., Weinberg, R.A. (2000) The hallmarks of cancer. *Cell*, **100** (1), 57–70.

2 Cairns, J. (1975) Mutation selection and the natural history of cancer. *Nature*, **255** (5505), 197–200.

3 Hargreaves, D.C., Crabtree, G.R. (2011) ATP-dependent chromatin remodeling: genetics, genomics and mechanisms. *Cell Res.*, **21** (3), 396–420.

4 Frank, S.A., Nowak, M.A. (2004) Problems of somatic mutation and cancer. *BioEssays*, **26** (3), 291–299.

5 Davis, P.K., Brackmann, R.K. (2003) Chromatin remodeling and cancer. *Cancer Biol. Ther.*, **2** (1), 22–29.

6 Luger, K., Mader, A.W., Richmond, R.K., Sargent, D.F., Richmond, T.J. (1997) Crystal structure of the nucleosome core particle at 2.8 Å resolution. *Nature*, **389** (6648), 251–260.

7 King, M.W. © 1996–2011; themedicalbiochemistrypage.org/dna.html.

8 Wang, G.G., Allis, C.D., Chi, P. (2007) Chromatin remodeling and cancer, Part II: ATP-dependent chromatin remodeling. *Trends Mol. Med.*, **13** (9), 373–380.

9 Saha, A., Wittmeyer, J., Cairns, B.R. (2006) Chromatin remodelling: the industrial revolution of DNA around histones. *Nat. Rev. Mol. Cell Biol.*, **7** (6), 437–447.

10 Kouzarides, T. (2007) SnapShot: histone-modifying enzymes. *Cell*, **131** (4), 822.

11 Roth, S.Y., Denu, J.M., Allis, C.D. (2001) Histone acetyltransferases. *Annu. Rev. Biochem.*, **70**, 81–120.

12 Thiagalingam, S., Cheng, K.H., Lee, H.J., Mineva, N., Thiagalingam, A., Ponte, J.F. (2003) Histone deacetylases: unique players in shaping the epigenetic histone code. *Ann. N. Y. Acad. Sci.*, **983**, 84–100.

13 Taddei, A., Roche, D., Bickmore, W.A., Almouzni, G. (2005) The effects of histone deacetylase inhibitors on heterochromatin: implications for anticancer therapy? *EMBO Rep.*, **6** (6), 520–524.

14 Agger, K., Christensen, J., Cloos, P.A., Helin, K. (2008) The emerging functions of histone demethylases. *Curr. Opin. Genet. Dev.*, **18** (2), 159–168.

15 Ng, S.S., Yue, W.W., Oppermann, U., Klose, R.J. (2009) Dynamic protein methylation in chromatin biology. *Cell. Mol. Life Sci.*, **66** (3), 407–422.

16 Taverna, S.D., Li, H., Ruthenburg, A.J., Allis, C.D., Patel, D.J. (2007) How chromatin-binding modules interpret histone modifications: lessons from professional pocket pickers. *Nat. Struct. Mol. Biol.*, **14** (11), 1025–1040.

17 Bracken, A.P., Kleine-Kohlbrecher, D., Dietrich, N., Pasini, D., Gargiulo, G., Beekman, C., Theilgaard-Monch, K., Minucci, S., Porse, B.T., Marine, J.C., Hansen, K.H., Helin, K. (2007) The Polycomb group proteins bind throughout the INK4A-ARF locus and are disassociated in senescent cells. *Genes Dev.*, **21** (5), 525–530.

18 Byvoet, P., Shepherd, G.R., Hardin, J.M., Noland, B.J. (1972) The distribution and turnover of labeled methyl groups in histone fractions of cultured mammalian cells. *Arch. Biochem. Biophys.*, **148** (2), 558–567.

19 Jenuwein, T., Allis, C.D. (2001) Translating the histone code. *Science*, **293** (5532), 1074–1080.

20 Fraga, M.F., Ballestar, E., Villar-Garea, A., Boix-Chornet, M., Espada, J., Schotta, G., Bonaldi, T., Haydon, C., Ropero, S., Petrie, K., Iyer, N.G., Perez-Rosado, A., Calvo, E., Lopez, J.A., Cano, A., Calasanz, M.J., Colomer, D., Piris, M.A., Ahn, N., Imhof, A., Caldas, C., Jenuwein, T., Esteller, M. (2005) Loss of acetylation at Lys16 and trimethylation at Lys20 of histone H4 is

a common hallmark of human cancer. *Nat. Genet.*, **37** (4), 391–400.
21 Seligson, D.B., Horvath, S., Shi, T., Yu, H., Tze, S., Grunstein, M., Kurdistani, S.K. (2005) Global histone modification patterns predict risk of prostate cancer recurrence. *Nature*, **435** (7046), 1262–1266.
22 Elsheikh, S.E., Green, A.R., Rakha, E.A., Powe, D.G., Ahmed, R.A., Collins, H.M., Soria, D., Garibaldi, J.M., Paish, C.E., Ammar, A.A., Grainge, M.J., Ball, G.R., Abdelghany, M.K., Martinez-Pomares, L., Heery, D.M., Ellis, I.O. (2009) Global histone modifications in breast cancer correlate with tumor phenotypes, prognostic factors, and patient outcome. *Cancer Res.*, **69** (9), 3802–3809.
23 Tzao, C., Tung, H.J., Jin, J.S., Sun, G.H., Hsu, H.S., Chen, B.H., Yu, C.P., Lee, S.C. (2009) Prognostic significance of global histone modifications in resected squamous cell carcinoma of the esophagus. *Mod. Pathol.*, **22** (2), 252–260.
24 Barlesi, F., Giaccone, G., Gallegos-Ruiz, M.I., Loundou, A., Span, S.W., Lefesvre, P., Kruyt, F.A., Rodriguez, J.A. (2007) Global histone modifications predict prognosis of resected non small-cell lung cancer. *J. Clin. Oncol.*, **25** (28), 4358–4364.
25 Manuyakorn, A., Paulus, R., Farrell, J., Dawson, N.A., Tze, S., Cheung-Lau, G., Hines, O.J., Reber, H., Seligson, D.B., Horvath, S., Kurdistani, S.K., Guha, C., Dawson, D.W. (2010) Cellular histone modification patterns predict prognosis and treatment response in resectable pancreatic adenocarcinoma: results from RTOG 9704. *J. Clin. Oncol.*, **28** (8), 1358–1365.
26 Cheung, N., Chan, L.C., Thompson, A., Cleary, M.L., So, C.W. (2007) Protein arginine-methyltransferase-dependent oncogenesis. *Nat. Cell Biol.*, **9** (10), 1208–1215.
27 Bai, X., Wu, L., Liang, T., Liu, Z., Li, J., Li, D., Xie, H., Yin, S., Yu, J., Lin, Q., Zheng, S. (2008) Overexpression of myocyte enhancer factor 2 and histone hyperacetylation in hepatocellular carcinoma. *J. Cancer Res. Clin. Oncol.*, **134** (1), 83–91.
28 Shandilya, J., Swaminathan, V., Gadad, S.S., Choudhari, R., Kodaganur, G.S., Kundu, T.K. (2009) Acetylated NPM1 localizes in the nucleoplasm and regulates transcriptional activation of genes implicated in oral cancer manifestation. *Mol. Cell. Biol.*, **29** (18), 5115–5127.
29 Iyer, N.G., Ozdag, H., Caldas, C. (2004) p300/CBP and cancer. *Oncogene*, **23** (24), 4225–4231.
30 Gayther, S.A., Batley, S.J., Linger, L., Bannister, A., Thorpe, K., Chin, S.F., Daigo, Y., Russell, P., Wilson, A., Sowter, H.M., Delhanty, J.D., Ponder, B.A., Kouzarides, T., Caldas, C. (2000) Mutations truncating the EP300 acetylase in human cancers. *Nat. Genet.*, **24** (3), 300–303.
31 Halkidou, K., Gaughan, L., Cook, S., Leung, H.Y., Neal, D.E., Robson, C.N. (2004) Upregulation and nuclear recruitment of HDAC1 in hormone refractory prostate cancer. *Prostate*, **59** (2), 177–189.
32 Choi, J.H., Kwon, H.J., Yoon, B.I., Kim, J.H., Han, S.U., Joo, H.J., Kim, D.Y. (2001) Expression profile of histone deacetylase 1 in gastric cancer tissues. *Jpn. J. Cancer Res.*, **92** (12), 1300–1304.
33 Wilson, A.J., Byun, D.S., Popova, N., Murray, L.B., L'Italien, K., Sowa, Y., Arango, D., Velcich, A., Augenlicht, L.H., Mariadason, J.M. (2006) Histone deacetylase 3 (HDAC3) and other class I HDACs regulate colon cell maturation and p21 expression and are deregulated in human colon cancer. *J. Biol. Chem.*, **281** (19), 13548–13558.
34 Ishihama, K., Yamakawa, M., Semba, S., Takeda, H., Kawata, S., Kimura, S., Kimura, W. (2007) Expression of HDAC1 and CBP/p300 in human colorectal carcinomas. *J. Clin. Pathol.*, **60** (11), 1205–1210.
35 Zhang, Z., Yamashita, H., Toyama, T., Sugiura, H., Ando, Y., Mita, K., Hamaguchi, M., Hara, Y., Kobayashi, S., Iwase, H. (2005) Quantitation of HDAC1 mRNA expression in invasive carcinoma of the breast. *Breast Cancer Res. Treat.*, **94** (1), 11–16.
36 Zhu, P., Huber, E., Kiefer, F., Gottlicher, M. (2004) Specific and redundant functions of histone deacetylases in regulation of cell cycle and apoptosis. *Cell Cycle*, **3** (10), 1240–1242.
37 Huang, B.H., Laban, M., Leung, C.H., Lee, L., Lee, C.K., Salto-Tellez, M., Raju, G.C., Hooi, S.C. (2005) Inhibition of histone deacetylase 2 increases apoptosis and p21Cip1/WAF1 expression, independent

of histone deacetylase 1. *Cell Death Differ.*, **12** (4), 395–404.

38 Song, J., Noh, J.H., Lee, J.H., Eun, J.W., Ahn, Y.M., Kim, S.Y., Lee, S.H., Park, W.S., Yoo, N.J., Lee, J.Y., Nam, S.W. (2005) Increased expression of histone deacetylase 2 is found in human gastric cancer. *APMIS*, **113** (4), 264–268.

39 Zhang, Z., Yamashita, H., Toyama, T., Sugiura, H., Omoto, Y., Ando, Y., Mita, K., Hamaguchi, M., Hayashi, S., Iwase, H. (2004) HDAC6 expression is correlated with better survival in breast cancer. *Clin. Cancer Res.*, **10** (20), 6962–6968.

40 Bolden, J.E., Peart, M.J., Johnstone, R.W. (2006) Anticancer activities of histone deacetylase inhibitors. *Nat. Rev. Drug Discov.*, **5** (9), 769–784.

41 Vire, E., Brenner, C., Deplus, R., Blanchon, L., Fraga, M., Didelot, C., Morey, L., Van Eynde, A., Bernard, D., Vanderwinden, J.M., Bollen, M., Esteller, M., Di Croce, L., de Launoit, Y., Fuks, F. (2006) The Polycomb group protein EZH2 directly controls DNA methylation. *Nature*, **439** (7078), 871–874.

42 Moss, T.J., Wallrath, L.L. (2007) Connections between epigenetic gene silencing and human disease. *Mutat. Res.*, **618** (1–2), 163–174.

43 Kleer, C.G., Cao, Q., Varambally, S., Shen, R., Ota, I., Tomlins, S.A., Ghosh, D., Sewalt, R.G., Otte, A.P., Hayes, D.F., Sabel, M.S., Livant, D., Weiss, S.J., Rubin, M.A., Chinnaiyan, A.M. (2003) EZH2 is a marker of aggressive breast cancer and promotes neoplastic transformation of breast epithelial cells. *Proc. Natl Acad. Sci. USA*, **100** (20), 11606–11611.

44 Berezovska, O.P., Glinskii, A.B., Yang, Z., Li, X.M., Hoffman, R.M., Glinsky, G.V. (2006) Essential role for activation of the Polycomb group (PcG) protein chromatin silencing pathway in metastatic prostate cancer. *Cell Cycle*, **5** (16), 1886–1901.

45 Varambally, S., Dhanasekaran, S.M., Zhou, M., Barrette, T.R., Kumar-Sinha, C., Sanda, M.G., Ghosh, D., Pienta, K.J., Sewalt, R.G., Otte, A.P., Rubin, M.A., Chinnaiyan, A.M. (2002) The polycomb group protein EZH2 is involved in progression of prostate cancer. *Nature*, **419** (6907), 624–629.

46 Wang, Y., Zhang, H., Chen, Y., Sun, Y., Yang, F., Yu, W., Liang, J., Sun, L., Yang, X., Shi, L., Li, R., Li, Y., Zhang, Y., Li, Q., Yi, X., Shang, Y. (2009) LSD1 is a subunit of the NuRD complex and targets the metastasis programs in breast cancer. *Cell*, **138** (4), 660–672.

47 Kahl, P., Gullotti, L., Heukamp, L.C., Wolf, S., Friedrichs, N., Vorreuther, R., Solleder, G., Bastian, P.J., Ellinger, J., Metzger, E., Schule, R., Buettner, R. (2006) Androgen receptor coactivators lysine-specific histone demethylase 1 and four and a half LIM domain protein 2 predict risk of prostate cancer recurrence. *Cancer Res.*, **66** (23), 11341–11347.

48 Cloos, P.A., Christensen, J., Agger, K., Maiolica, A., Rappsilber, J., Antal, T., Hansen, K.H., Helin, K. (2006) The putative oncogene GASC1 demethylates tri- and dimethylated lysine 9 on histone H3. *Nature*, **442** (7100), 307–311.

49 Italiano, A., Attias, R., Aurias, A., Perot, G., Burel-Vandenbos, F., Otto, J., Venissac, N., Pedeutour, F. (2006) Molecular cytogenetic characterization of a metastatic lung sarcomatoid carcinoma: 9p23 neocentromere and 9p23-p24 amplification including JAK2 and JMJD2C. *Cancer Genet. Cytogenet.*, **167** (2), 122–130.

50 Yang, Z.Q., Imoto, I., Fukuda, Y., Pimkhaokham, A., Shimada, Y., Imamura, M., Sugano, S., Nakamura, Y., Inazawa, J. (2000) Identification of a novel gene, GASC1, within an amplicon at 9p23-24 frequently detected in esophageal cancer cell lines. *Cancer Res.*, **60** (17), 4735–4739.

51 Sen, G.L., Webster, D.E., Barragan, D.I., Chang, H.Y., Khavari, P.A. (2008) Control of differentiation in a self-renewing mammalian tissue by the histone demethylase JMJD3. *Genes Dev.*, **22** (14), 1865–1870.

52 Agger, K., Cloos, P.A., Rudkjaer, L., Williams, K., Andersen, G., Christensen, J., Helin, K. (2009) The H3K27me3 demethylase JMJD3 contributes to the activation of the INK4A-ARF locus in response to oncogene- and stress-induced senescence. *Genes Dev.*, **23** (10), 1171–1176.

53 Yamane, K., Tateishi, K., Klose, R.J., Fang, J., Fabrizio, L.A., Erdjument-Bromage, H., Taylor-Papadimitriou, J., Tempst, P., Zhang, Y. (2007) PLU-1 is an H3K4 demethylase involved in transcriptional repression and breast cancer cell proliferation. *Mol. Cell*, **25** (6), 801–812.

54 Scibetta, A.G., Santangelo, S., Coleman, J., Hall, D., Chaplin, T., Copier, J., Catchpole, S., Burchell, J., Taylor-Papadimitriou, J. (2007) Functional analysis of the transcription repressor PLU-1/JARID1B. *Mol. Cell. Biol.*, **27** (20), 7220–7235.

55 Dalgliesh, G.L., Furge, K., Greenman, C., Chen, L., Bignell, G., Butler, A., Davies, H., Edkins, S., Hardy, C., Latimer, C., Teague, J., Andrews, J., Barthorpe, S., Beare, D., Buck, G., Campbell, P.J., Forbes, S., Jia, M., Jones, D., Knott, H., Kok, C.Y., Lau, K.W., Leroy, C., Lin, M.L., McBride, D.J., Maddison, M., Maguire, S., McLay, K., Menzies, A., Mironenko, T., Mulderrig, L., Mudie, L., O'Meara, S., Pleasance, E., Rajasingham, A., Shepherd, R., Smith, R., Stebbings, L., Stephens, P., Tang, G., Tarpey, P.S., Turrell, K., Dykema, K.J., Khoo, S.K., Petillo, D., Wondergem, B., Anema, J., Kahnoski, R.J., Teh, B.T., Stratton, M.R., Futreal, P.A. (2010) Systematic sequencing of renal carcinoma reveals inactivation of histone modifying genes. *Nature*, **463** (7279), 360–363.

56 Roesch, A., Fukunaga-Kalabis, M., Schmidt, E.C., Zabierowski, S.E., Brafford, P.A., Vultur, A., Basu, D., Gimotty, P., Vogt, T., Herlyn, M. (2010) A temporarily distinct subpopulation of slow-cycling melanoma cells is required for continuous tumor growth. *Cell*, **141** (4), 583–594.

57 Sharma, S.V., Lee, D.Y., Li, B., Quinlan, M.P., Takahashi, F., Maheswaran, S., McDermott, U., Azizian, N., Zou, L., Fischbach, M.A., Wong, K.K., Brandstetter, K., Wittner, B., Ramaswamy, S., Classon, M., Settleman, J. (2010) A chromatin-mediated reversible drug-tolerant state in cancer cell subpopulations. *Cell*, **141** (1), 69–80.

58 Spannhoff, A., Sippl, W., Jung, M. (2009) Cancer treatment of the future: inhibitors of histone methyltransferases. *Int. J. Biochem. Cell Biol.*, **41** (1), 4–11.

59 Wang, G.G., Allis, C.D., Chi, P. (2007) Chromatin remodeling and cancer, Part I: Covalent histone modifications. *Trends Mol. Med.*, **13** (9), 363–372.

60 Fernandez-Capetillo, O., Lee, A., Nussenzweig, M., Nussenzweig, A. (2004) H2AX: the histone guardian of the genome. *DNA Repair (Amst.)*, **3** (8–9), 959–967.

61 Terada, Y. (2006) Aurora-B/AIM-1 regulates the dynamic behavior of HP1alpha at the G_2-M transition. *Mol. Biol. Cell*, **17** (7), 3232–3241.

62 Dunn, K.L., Davie, J.R. (2005) Stimulation of the Ras-MAPK pathway leads to independent phosphorylation of histone H3 on serine 10 and 28. *Oncogene*, **24** (21), 3492–3502.

63 Ahn, S.H., Cheung, W.L., Hsu, J.Y., Diaz, R.L., Smith, M.M., Allis, C.D. (2005) Sterile 20 kinase phosphorylates histone H2B at serine 10 during hydrogen peroxide-induced apoptosis in *S. cerevisiae*. *Cell*, **120** (1), 25–36.

64 Cheung, W.L., Ajiro, K., Samejima, K., Kloc, M., Cheung, P., Mizzen, C.A., Beeser, A., Etkin, L.D., Chernoff, J., Earnshaw, W.C., Allis, C.D. (2003) Apoptotic phosphorylation of histone H2B is mediated by mammalian sterile twenty kinase. *Cell*, **113** (4), 507–517.

65 Halliday, G.M., Bock, V.L., Moloney, F.J., Lyons, J.G. (2009) SWI/SNF: a chromatin-remodelling complex with a role in carcinogenesis. *Int. J. Biochem. Cell Biol.*, **41** (4), 725–728.

66 Reisman, D., Glaros, S., Thompson, E.A. (2009) The SWI/SNF complex and cancer. *Oncogene*, **28** (14), 1653–1668.

67 Biegel, J.A., Zhou, J.Y., Rorke, L.B., Stenstrom, C., Wainwright, L.M., Fogelgren, B. (1999) Germ-line and acquired mutations of INI1 in atypical teratoid and rhabdoid tumors. *Cancer Res.*, **59** (1), 74–79.

68 Sevenet, N., Sheridan, E., Amram, D., Schneider, P., Handgretinger, R., Delattre, O. (1999) Constitutional mutations of the hSNF5/INI1 gene predispose to a variety of cancers. *Am. J. Hum. Genet.*, **65** (5), 1342–1348.

69 Rousseau-Merck, M.F., Versteege, I., Legrand, I., Couturier, J., Mairal, A., Delattre, O., Aurias, A. (1999) hSNF5/INI1 inactivation is mainly associated with homozygous deletions and mitotic recombinations in rhabdoid tumors. *Cancer Res.*, **59** (13), 3152–3156.

70 Sevenet, N., Lellouch-Tubiana, A., Schofield, D., Hoang-Xuan, K., Gessler, M., Birnbaum, D., Jeanpierre, C., Jouvet, A., Delattre, O. (1999) Spectrum of hSNF5/INI1 somatic mutations in

human cancer and genotype-phenotype correlations. *Hum. Mol. Genet.*, **8** (13), 2359–2368.

71 Guidi, C.J., Sands, A.T., Zambrowicz, B.P., Turner, T.K., Demers, D.A., Webster, W., Smith, T.W., Imbalzano, A.N., Jones, S.N. (2001) Disruption of Ini1 leads to peri-implantation lethality and tumorigenesis in mice. *Mol. Cell. Biol.*, **21** (10), 3598–3603.

72 Reincke, B.S., Rosson, G.B., Oswald, B.W., Wright, C.F. (2003) INI1 expression induces cell cycle arrest and markers of senescence in malignant rhabdoid tumor cells. *J. Cell. Physiol.*, **194** (3), 303–313.

73 Betz, B.L., Strobeck, M.W., Reisman, D.N., Knudsen, E.S., Weissman, B.E. (2002) Re-expression of hSNF5/INI1/BAF47 in pediatric tumor cells leads to G_1 arrest associated with induction of p16ink4a and activation of RB. *Oncogene*, **21** (34), 5193–5203.

74 Kia, S.K., Gorski, M.M., Giannakopoulos, S., Verrijzer, C.P. (2008) SWI/SNF mediates polycomb eviction and epigenetic reprogramming of the INK4b-ARF-INK4a locus. *Mol. Cell. Biol.*, **28** (10), 3457–3464.

75 Doan, D.N., Veal, T.M., Yan, Z., Wang, W., Jones, S.N., Imbalzano, A.N. (2004) Loss of the INI1 tumor suppressor does not impair the expression of multiple BRG1-dependent genes or the assembly of SWI/SNF enzymes. *Oncogene*, **23** (19), 3462–3473.

76 Chai, J., Charboneau, A.L., Betz, B.L., Weissman, B.E. (2005) Loss of the hSNF5 gene concomitantly inactivates p21CIP/WAF1 and p16INK4a activity associated with replicative senescence in A204 rhabdoid tumor cells. *Cancer Res.*, **65** (22), 10192–10198.

77 Oruetxebarria, I., Venturini, F., Kekarainen, T., Houweling, A., Zuijderduijn, L.M., Mohd-Sarip, A., Vries, R.G., Hoeben, R.C., Verrijzer, C.P. (2004) P16INK4a is required for hSNF5 chromatin remodeler-induced cellular senescence in malignant rhabdoid tumor cells. *J. Biol. Chem.*, **279** (5), 3807–3816.

78 Lin, H., Wong, R.P., Martinka, M., Li, G. (2009) Loss of SNF5 expression correlates with poor patient survival in melanoma. *Clin. Cancer Res.*, **15** (20), 6404–6411.

79 Hayward, N.K. (2003) Genetics of melanoma predisposition. *Oncogene*, **22** (20), 3053–3062.

80 Becker, T.M., Rizos, H., Kefford, R.F., Mann, G.J. (2001) Functional impairment of melanoma-associated p16(INK4a) mutants in melanoma cells despite retention of cyclin-dependent kinase 4 binding. *Clin. Cancer Res.*, **7** (10), 3282–3288.

81 Gallagher, S.J., Thompson, J.F., Indsto, J., Scurr, L.L., Lett, M., Gao, B.F., Dunleavey, R., Mann, G.J., Kefford, R.F., Rizos, H. (2008) p16INK4a expression and absence of activated B-RAF are independent predictors of chemosensitivity in melanoma tumors. *Neoplasia*, **10** (11), 1231–1239.

82 Wang, X., Sansam, C.G., Thom, C.S., Metzger, D., Evans, J.A., Nguyen, P.T., Roberts, C.W. (2009) Oncogenesis caused by loss of the SNF5 tumor suppressor is dependent on activity of BRG1, the ATPase of the SWI/SNF chromatin remodeling complex. *Cancer Res.*, **69** (20), 8094–8101.

83 Gunduz, E., Gunduz, M., Ouchida, M., Nagatsuka, H., Beder, L., Tsujigiwa, H., Fukushima, K., Nishizaki, K., Shimizu, K., Nagai, N. (2005) Genetic and epigenetic alterations of BRG1 promote oral cancer development. *Int. J. Oncol.*, **26** (1), 201–210.

84 Sun, A., Tawfik, O., Gayed, B., Thrasher, J.B., Hoestje, S., Li, C., Li, B. (2007) Aberrant expression of SWI/SNF catalytic subunits BRG1/BRM is associated with tumor development and increased invasiveness in prostate cancers. *Prostate*, **67** (2), 203–213.

85 Sentani, K., Oue, N., Kondo, H., Kuraoka, K., Motoshita, J., Ito, R., Yokozaki, H., Yasui, W. (2001) Increased expression but not genetic alteration of BRG1, a component of the SWI/SNF complex, is associated with the advanced stage of human gastric carcinomas. *Pathobiology*, **69** (6), 315–320.

86 Yamamichi, N., Inada, K., Ichinose, M., Yamamichi-Nishina, M., Mizutani, T., Watanabe, H., Shiogama, K., Fujishiro, M., Okazaki, T., Yahagi, N., Haraguchi, T., Fujita, S., Tsutsumi, Y., Omata, M., Iba, H. (2007) Frequent loss of Brm expression in gastric cancer correlates with histologic features and differentiation state. *Cancer Res.*, **67** (22), 10727–10735.

87 Watanabe, T., Semba, S., Yokozaki, H. (2011) Regulation of PTEN expression by

the SWI/SNF chromatin-remodelling protein BRG1 in human colorectal carcinoma cells. *Br. J. Cancer*, **104** (1), 146–154.

88 Barker, N., Hurlstone, A., Musisi, H., Miles, A., Bienz, M., Clevers, H. (2001) The chromatin remodelling factor Brg-1 interacts with beta-catenin to promote target gene activation. *EMBO J.*, **20** (17), 4935–4943.

89 Bultman, S., Gebuhr, T., Yee, D., La Mantia, C., Nicholson, J., Gilliam, A., Randazzo, F., Metzger, D., Chambon, P., Crabtree, G., Magnuson, T. (2000) A Brg1 null mutation in the mouse reveals functional differences among mammalian SWI/SNF complexes. *Mol. Cell*, **6** (6), 1287–1295.

90 Glaros, S., Cirrincione, G.M., Palanca, A., Metzger, D., Reisman, D. (2008) Targeted knockout of BRG1 potentiates lung cancer development. *Cancer Res.*, **68** (10), 3689–3696.

91 Fukuoka, J., Fujii, T., Shih, J.H., Dracheva, T., Meerzaman, D., Player, A., Hong, K., Settnek, S., Gupta, A., Buetow, K., Hewitt, S., Travis, W.D., Jen, J. (2004) Chromatin remodeling factors and BRM/BRG1 expression as prognostic indicators in non-small cell lung cancer. *Clin. Cancer Res.*, **10** (13), 4314–4324.

92 Reisman, D.N., Sciarrotta, J., Wang, W., Funkhouser, W.K., Weissman, B.E. (2003) Loss of BRG1/BRM in human lung cancer cell lines and primary lung cancers: correlation with poor prognosis. *Cancer Res.*, **63** (3), 560–566.

93 Rodriguez-Nieto, S., Canada, A., Pros, E., Pinto, A.I., Torres-Lanzas, J., Lopez-Rios, F., Sanchez-Verde, L., Pisano, D.G., Sanchez-Cespedes, M. (2011) Massive parallel DNA pyrosequencing analysis of the tumor suppressor BRG1/SMARCA4 in lung primary tumors. *Hum. Mutat.*, **32** (2), E1999–E2017.

94 Muchardt, C., Yaniv, M. (2001) When the SWI/SNF complex remodels...the cell cycle. *Oncogene*, **20** (24), 3067–3075.

95 Decristofaro, M.F., Betz, B.L., Rorie, C.J., Reisman, D.N., Wang, W., Weissman, B.E. (2001) Characterization of SWI/SNF protein expression in human breast cancer cell lines and other malignancies. *J. Cell. Physiol.*, **186** (1), 136–145.

96 Medina, P.P., Romero, O.A., Kohno, T., Montuenga, L.M., Pio, R., Yokota, J., Sanchez-Cespedes, M. (2008) Frequent BRG1/SMARCA4-inactivating mutations in human lung cancer cell lines. *Hum. Mutat.*, **29** (5), 617–622.

97 Wong, A.K., Shanahan, F., Chen, Y., Lian, L., Ha, P., Hendricks, K., Ghaffari, S., Iliev, D., Penn, B., Woodland, A.M., Smith, R., Salada, G., Carillo, A., Laity, K., Gupte, J., Swedlund, B., Tavtigian, S.V., Teng, D.H., Lees, E. (2000) BRG1, a component of the SWI-SNF complex, is mutated in multiple human tumor cell lines. *Cancer Res.*, **60** (21), 6171–6177.

98 Kuo, K.T., Liang, C.W., Hsiao, C.H., Lin, C.H., Chen, C.A., Sheu, B.C., Lin, M.C. (2006) Downregulation of BRG-1 repressed expression of CD44s in cervical neuroendocrine carcinoma and adenocarcinoma. *Mod. Pathol.*, **19** (12), 1570–1577.

99 Becker, T.M., Haferkamp, S., Dijkstra, M.K., Scurr, L.L., Frausto, M., Diefenbach, E., Scolyer, R.A., Reisman, D.N., Mann, G.J., Kefford, R.F., Rizos, H. (2009) The chromatin remodelling factor BRG1 is a novel binding partner of the tumor suppressor p16INK4a. *Mol. Cancer*, **8**, 4.

100 Lin, H., Wong, R.P., Martinka, M., Li, G. (2010) BRG1 expression is increased in human cutaneous melanoma. *Br. J. Dermatol.*, **163** (3), 502–510.

101 Curtin, J.A., Fridlyand, J., Kageshita, T., Patel, H.N., Busam, K.J., Kutzner, H., Cho, K.H., Aiba, S., Brocker, E.B., LeBoit, P.E., Pinkel, D., Bastian, B.C. (2005) Distinct sets of genetic alterations in melanoma. *N. Engl. J. Med.*, **353** (20), 2135–2147.

102 Eide M.J., Weinstock M.A. (2005) Association of UV index, latitude, and melanoma incidence in nonwhite populations – US Surveillance, Epidemiology, and End Results (SEER) Program, 1992 to 2001. *Arch. Dermatol.*, **141** (4), 477–481.

103 Moloney, F.J., Lyons, J.G., Bock, V.L., Huang, X.X., Bugeja, M.J., Halliday, G.M. (2009) Hotspot mutation of Brahma in non-melanoma skin cancer. *J. Invest. Dermatol.*, **129** (4), 1012–1015.

104 Shen, H., Powers, N., Saini, N., Comstock, C.E., Sharma, A., Weaver, K., Revelo, M.P., Gerald, W., Williams, E., Jessen, W.J., Aronow, B.J., Rosson, G., Weissman, B., Muchardt, C., Yaniv, M., Knudsen, K.E. (2008) The SWI/SNF ATPase Brm is a gatekeeper of proliferative control in prostate cancer. *Cancer Res.*, **68** (24), 10154–10162.

105 Glaros, S., Cirrincione, G.M., Muchardt, C., Kleer, C.G., Michael, C.W., Reisman, D. (2007) The reversible epigenetic silencing of BRM: implications for clinical targeted therapy. *Oncogene*, **26** (49), 7058–7066.

106 Bourachot, B., Yaniv, M., Muchardt, C. (2003) Growth inhibition by the mammalian SWI-SNF subunit Brm is regulated by acetylation. *EMBO J.*, **22** (24), 6505–6515.

107 Garcia-Pedrero, J.M., Kiskinis, E., Parker, M.G., Belandia, B. (2006) The SWI/SNF chromatin remodeling subunit BAF57 is a critical regulator of estrogen receptor function in breast cancer cells. *J. Biol. Chem.*, **281** (32), 22656–22664.

108 Kiskinis, E., Garcia-Pedrero, J.M., Villaronga, M.A., Parker, M.G., Belandia, B. (2006) Identification of BAF57 mutations in human breast cancer cell lines. *Breast Cancer Res. Treat.*, **98** (2), 191–198.

109 Link, K.A., Balasubramaniam, S., Sharma, A., Comstock, C.E., Godoy-Tundidor, S., Powers, N., Cao, K.H., Haelens, A., Claessens, F., Revelo, M.P., Knudsen, K.E. (2008) Targeting the BAF57 SWI/SNF subunit in prostate cancer: a novel platform to control androgen receptor activity. *Cancer Res.*, **68** (12), 4551–4558.

110 Shadeo, A., Chari, R., Lonergan, K.M., Pusic, A., Miller, D., Ehlen, T., Van Niekerk, D., Matisic, J., Richards-Kortum, R., Follen, M., Guillaud, M., Lam, W.L., MacAulay, C. (2008) Up regulation in gene expression of chromatin remodelling factors in cervical intraepithelial neoplasia. *BMC Genomics*, **9**, 64.

111 Heeboll, S., Borre, M., Ottosen, P.D., Andersen, C.L., Mansilla, F., Dyrskjot, L., Orntoft, T.F., Torring, N. (2008) SMARCC1 expression is upregulated in prostate cancer and positively correlated with tumour recurrence and dedifferentiation. *Histol. Histopathol.*, **23** (9), 1069–1076.

112 Varela, I., Tarpey, P., Raine, K., Huang, D., Ong, C.K., Stephens, P., Davies, H., Jones, D., Lin, M.L., Teague, J., Bignell, G., Butler, A., Cho, J., Dalgliesh, G.L., Galappaththige, D., Greenman, C., Hardy, C., Jia, M., Latimer, C., Lau, K.W., Marshall, J., McLaren, S., Menzies, A., Mudie, L., Stebbings, L., Largaespada, D.A., Wessels, L.F., Richard, S., Kahnoski, R.J., Anema, J., Tuveson, D.A., Perez-Mancera, P.A., Mustonen, V., Fischer, A., Adams, D.J., Rust, A., Chan-on, W., Subimerb, C., Dykema, K., Furge, K., Campbell, P.J., Teh, B.T., Stratton, M.R., Futreal, P.A. (2011) Exome sequencing identifies frequent mutation of the SWI/SNF complex gene PBRM1 in renal carcinoma. *Nature*, **469** (7331), 539–542.

113 Burrows, A.E., Smogorzewska, A., Elledge, S.J. (2010) Polybromo-associated BRG1-associated factor components BRD7 and BAF180 are critical regulators of p53 required for induction of replicative senescence. *Proc. Natl Acad. Sci. USA*, **107** (32), 14280–14285.

114 Xia, W., Nagase, S., Montia, A.G., Kalachikov, S.M., Keniry, M., Su, T., Memeo, L., Hibshoosh, H., Parsons, R. (2008) BAF180 is a critical regulator of p21 induction and a tumor suppressor mutated in breast cancer. *Cancer Res.*, **68** (6), 1667–1674.

115 Wang, X., Nagl, N.G. Jr, Flowers, S., Zweitzig, D., Dallas, P.B., Moran, E. (2004) Expression of p270 (ARID1A), a component of human SWI/SNF complexes, in human tumors. *Int. J. Cancer*, **112** (4), 636.

116 Nagl, N.G. Jr, Patsialou, A., Haines, D.S., Dallas, P.B., Beck, G.R. Jr, Moran, E. (2005) The p270 (ARID1A/SMARCF1) subunit of mammalian SWI/SNF-related complexes is essential for normal cell cycle arrest. *Cancer Res.*, **65** (20), 9236–9244.

117 Mohamed, M.A., Greif, P.A., Diamond, J., Sharaf, O., Maxwell, P., Montironi, R., Young, R.A., Hamilton, P.W. (2007) Epigenetic events, remodelling enzymes and their relationship to chromatin organization in prostatic intraepithelial neoplasia and prostatic adenocarcinoma. *Br. J. Urol. Int.*, **99** (4), 908–915.

118 Choi, J.H., Sheu, J.J., Guan, B., Jinawath, N., Markowski, P., Wang, T.L., Shih Ie, M. (2009) Functional analysis of 11q13.5 amplicon identifies Rsf-1 (HBXAP) as a gene involved in paclitaxel resistance in ovarian cancer. *Cancer Res.*, **69** (4), 1407–1415.

119 Sheu, J.J., Guan, B., Choi, J.H., Lin, A., Lee, C.H., Hsiao, Y.T., Wang, T.L., Tsai, F.J., Shih Ie, M. (2010) Rsf-1, a chromatin remodeling protein, induces DNA damage and promotes genomic instability. *J. Biol. Chem.*, **285** (49), 38260–38269.

120 Bagchi, A., Papazoglu, C., Wu, Y., Capurso, D., Brodt, M., Francis, D., Bredel, M., Vogel, H., Mills, A.A. (2007) CHD5 is a tumor suppressor at human 1p36. *Cell*, **128** (3), 459–475.

121 Bowen, N.J., Fujita, N., Kajita, M., Wade, P.A. (2004) Mi-2/NuRD: multiple complexes for many purposes. *Biochim. Biophys. Acta*, **1677** (1–3), 52–57.

122 Fujita, N., Jaye, D.L., Kajita, M., Geigerman, C., Moreno, C.S., Wade, P.A. (2003) MTA3, a Mi-2/NuRD complex subunit, regulates an invasive growth pathway in breast cancer. *Cell*, **113** (2), 207–219.

123 Li, R., Zhang, H., Yu, W., Chen, Y., Gui, B., Liang, J., Wang, Y., Sun, L., Yang, X., Zhang, Y., Shi, L., Li, Y., Shang, Y. (2009) ZIP: a novel transcription repressor, represses EGFR oncogene and suppresses breast carcinogenesis. *EMBO J.*, **28** (18), 2763–2776.

124 Yu, E.Y., Steinberg-Neifach, O., Dandjinou, A.T., Kang, F., Morrison, A.J., Shen, X., Lue, N.F. (2007) Regulation of telomere structure and functions by subunits of the INO80 chromatin remodeling complex. *Mol. Cell. Biol.*, **27** (16), 5639–5649.

125 Tang, Y., Luo, J., Zhang, W., Gu, W. (2006) Tip60-dependent acetylation of p53 modulates the decision between cell-cycle arrest and apoptosis. *Mol. Cell*, **24** (6), 827–839.

126 Fazzio, T.G., Huff, J.T., Panning, B. (2008) An RNAi screen of chromatin proteins identifies Tip60-p400 as a regulator of embryonic stem cell identity. *Cell*, **134** (1), 162–174.

127 Grigoletto, A., Lestienne, P., Rosenbaum, J. (2011) The multifaceted proteins Reptin and Pontin as major players in cancer. *Biochim. Biophys. Acta*, **1815** (2), 147–157.

128 Morrison, A.J., Highland, J., Krogan, N.J., Arbel-Eden, A., Greenblatt, J.F., Haber, J.E., Shen, X. (2004) INO80 and gamma-H2AX interaction links ATP-dependent chromatin remodeling to DNA damage repair. *Cell*, **119** (6), 767–775.

129 Lane, D.P. (1992) Cancer. p53, guardian of the genome. *Nature*, **358** (6381), 15–16.

130 Greenblatt, M.S., Bennett, W.P., Hollstein, M., Harris, C.C. (1994) Mutations in the p53 tumor suppressor gene: clues to cancer etiology and molecular pathogenesis. *Cancer Res.*, **54** (18), 4855–4878.

131 el-Deiry, W.S., Tokino, T., Velculescu, V.E., Levy, D.B., Parsons, R., Trent, J.M., Lin, D., Mercer, W.E., Kinzler, K.W., Vogelstein, B. (1993) WAF1, a potential mediator of p53 tumor suppression. *Cell*, **75** (4), 817–825.

132 Prives, C., Manley, J.L. (2001) Why is p53 acetylated? *Cell*, **107** (7), 815–818.

133 Ito, A., Lai, C.H., Zhao, X., Saito, S., Hamilton, M.H., Appella, E., Yao, T.P. (2001) p300/CBP-mediated p53 acetylation is commonly induced by p53-activating agents and inhibited by MDM2. *EMBO J.*, **20** (6), 1331–1340.

134 Liu, L., Scolnick, D.M., Trievel, R.C., Zhang, H.B., Marmorstein, R., Halazonetis, T.D., Berger, S.L. (1999) p53 sites acetylated in vitro by PCAF and p300 are acetylated in vivo in response to DNA damage. *Mol. Cell. Biol.*, **19** (2), 1202–1209.

135 Sakaguchi, K., Herrera, J.E., Saito, S., Miki, T., Bustin, M., Vassilev, A., Anderson, C.W., Appella, E. (1998) DNA damage activates p53 through a phosphorylation-acetylation cascade. *Genes Dev.*, **12** (18), 2831–2841.

136 Wang, T., Kobayashi, T., Takimoto, R., Denes, A.E., Snyder, E.L., el-Deiry, W.S., Brachmann, R.K. (2001) hADA3 is required for p53 activity. *EMBO J.*, **20** (22), 6404–6413.

137 Kumar, A., Zhao, Y., Meng, G., Zeng, M., Srinivasan, S., Delmolino, L.M., Gao, Q., Dimri, G., Weber, G.F., Wazer, D.E., Band, H., Band, V. (2002) Human papillomavirus oncoprotein E6 inactivates the transcriptional coactivator human ADA3. *Mol. Cell. Biol.*, **22** (16), 5801–5812.

138 Brooks, C.L., Gu, W. (2003) Ubiquitination, phosphorylation and acetylation: the molecular basis for p53 regulation. *Curr. Opin. Cell Biol.*, **15** (2), 164–171.

139 Harms, K.L., Chen, X. (2007) Histone deacetylase 2 modulates p53 transcriptional activities through regulation of p53-DNA binding activity. *Cancer Res.*, **67** (7), 3145–3152.

140 Oh, J., Sohn, D.H., Ko, M., Chung, H., Jeon, S.H., Seong, R.H. (2008) BAF60a interacts with p53 to recruit the SWI/SNF complex. *J. Biol. Chem.*, **283** (18), 11924–11934.

141 Lee, D., Kim, J.W., Seo, T., Hwang, S.G., Choi, E.J., Choe, J. (2002) SWI/SNF complex interacts with tumor suppressor p53 and is necessary for the activation of p53-mediated transcription. *J. Biol. Chem.*, **277** (25), 22330–22337.

142 Wang, M., Gu, C., Qi, T., Tang, W., Wang, L., Wang, S., Zeng, X. (2007) BAF53 interacts with p53 and functions in p53-mediated p21-gene transcription. *J. Biochem.*, **142** (5), 613–620.

143 Lee, J.H., Lee, J.Y., Chang, S.H., Kang, M.J., Kwon, H. (2005) Effects of Ser2 and Tyr6 mutants of BAF53 on cell growth and p53-dependent transcription. *Mol. Cell*, **19** (2), 289–293.

144 Naidu, S.R., Love, I.M., Imbalzano, A.N., Grossman, S.R., Androphy, E.J. (2009) The SWI/SNF chromatin remodeling subunit BRG1 is a critical regulator of p53 necessary for proliferation of malignant cells. *Oncogene*, **28** (27), 2492–2501.

145 Allison, S.J., Milner, J. (2004) Remodelling chromatin on a global scale: a novel protective function of p53. *Carcinogenesis*, **25** (9), 1551–1557.

146 Weintraub, S.J., Chow, K.N., Luo, R.X., Zhang, S.H., He, S., Dean, D.C. (1995) Mechanism of active transcriptional repression by the retinoblastoma protein. *Nature*, **375** (6534), 812–815.

147 Brehm, A., Miska, E.A., McCance, D.J., Reid, J.L., Bannister, A.J., Kouzarides, T. (1998) Retinoblastoma protein recruits histone deacetylase to repress transcription. *Nature*, **391** (6667), 597–601.

148 Ferreira, R., Naguibneva, I., Mathieu, M., Ait-Si-Ali, S., Robin, P., Pritchard, L.L., Harel-Bellan, A. (2001) Cell cycle-dependent recruitment of HDAC-1 correlates with deacetylation of histone H4 on an Rb-E2F target promoter. *EMBO Rep.*, **2** (9), 794–799.

149 Takaki, T., Fukasawa, K., Suzuki-Takahashi, I., Hirai, H. (2004) Cdk-mediated phosphorylation of pRB regulates HDAC binding in vitro. *Biochem. Biophys. Res. Commun.*, **316** (1), 252–255.

150 Zhang, H.S., Gavin, M., Dahiya, A., Postigo, A.A., Ma, D., Luo, R.X., Harbour, J.W., Dean, D.C. (2000) Exit from G_1 and S phase of the cell cycle is regulated by repressor complexes containing HDAC-Rb-hSWI/SNF and Rb-hSWI/SNF. *Cell*, **101** (1), 79–89.

151 Dunaief, J.L., Strober, B.E., Guha, S., Khavari, P.A., Alin, K., Luban, J., Begemann, M., Crabtree, G.R., Goff, S.P. (1994) The retinoblastoma protein and BRG1 form a complex and cooperate to induce cell cycle arrest. *Cell*, **79** (1), 119–130.

152 Strobeck, M.W., Knudsen, K.E., Fribourg, A.F., DeCristofaro, M.F., Weissman, B.E., Imbalzano, A.N., Knudsen, E.S. (2000) BRG-1 is required for RB-mediated cell cycle arrest. *Proc. Natl Acad. Sci. USA*, **97** (14), 7748–7753.

153 Bartlett, C., Stammler, T., Rosson, G.S., Weissman, B.E. (2010) BRG1 mutations found in human cancer cell lines inactivate Rb-mediated cell cycle arrest. *J. Cell. Physiol.*, **226** (8), 1989–1997.

154 Kang, H., Cui, K., Zhao, K. (2004) BRG1 controls the activity of the retinoblastoma protein via regulation of p21CIP1/WAF1/SDI. *Mol. Cell. Biol.*, **24** (3), 1188–1199.

155 Rao, M., Casimiro, M.C., Lisanti, M.P., D'Amico, M., Wang, C., Shirley, L.A., Leader, J.E., Liu, M., Stallcup, M., Engel, D.A., Murphy, D.J., Pestell, R.G. (2008) Inhibition of cyclin D1 gene transcription by Brg-1. *Cell Cycle*, **7** (5), 647–655.

156 Sharpless, N.E., DePinho, R.A. (1999) The INK4A/ARF locus and its two gene products. *Curr. Opin. Genet. Dev.*, **9** (1), 22–30.

157 Goldstein, A.M., Chan, M., Harland, M., Gillanders, E.M., Hayward, N.K., Avril, M.F., Azizi, E., Bianchi-Scarra, G., Bishop, D.T., Bressac-de Paillerets, B., Bruno, W., Calista, D., Cannon Albright, L.A., Demenais, F., Elder, D.E., Ghiorzo, P., Gruis, N.A., Hansson, J., Hogg, D., Holland, E.A., Kanetsky, P.A., Kefford, R.F., Landi, M.T., Lang, J., Leachman, S.A., Mackie, R.M., Magnusson, V., Mann, G.J., Niendorf, K., Newton Bishop, J., Palmer, J.M., Puig, S., Puig-Butille, J.A., de Snoo, F.A., Stark, M., Tsao, H., Tucker, M.A., Whitaker, L., Yakobson, E. (2006) High-risk melanoma susceptibility genes and pancreatic cancer, neural system tumors, and uveal melanoma across GenoMEL. *Cancer Res.*, **66** (20), 9818–9828.

158 McKenzie, H.A., Fung, C., Becker, T.M., Irvine, M., Mann, G.J., Kefford, R.F., Rizos,

H. (2010) Predicting functional significance of cancer-associated p16(INK4a) mutations in CDKN2A. *Hum. Mutat.*, **31** (6), 692–701.

159 Tzatsos, A., Pfau, R., Kampranis, S.C., Tsichlis, P.N. (2009) Ndy1/KDM2B immortalizes mouse embryonic fibroblasts by repressing the Ink4a/Arf locus. *Proc. Natl Acad. Sci. USA*, **106** (8), 2641–2646.

160 Barradas, M., Anderton, E., Acosta, J.C., Li, S., Banito, A., Rodriguez-Niedenfuhr, M., Maertens, G., Banck, M., Zhou, M.M., Walsh, M.J., Peters, G., Gil, J. (2009) Histone demethylase JMJD3 contributes to epigenetic control of INK4a/ARF by oncogenic RAS. *Genes Dev.*, **23** (10), 1177–1182.

161 Kotake, Y., Zeng, Y., Xiong, Y. (2009) DDB1-CUL4 and MLL1 mediate oncogene-induced p16INK4a activation. *Cancer Res.*, **69** (5), 1809–1814.

162 Muchardt, C., Bourachot, B., Reyes, J.C., Yaniv, M. (1998) ras transformation is associated with decreased expression of the brm/SNF2alpha ATPase from the mammalian SWI-SNF complex. *EMBO J.*, **17** (1), 223–231.

163 Ceol, C.J., Houvras, Y., Jane-Valbuena, J., Bilodeau, S., Orlando, D.A., Battisti, V., Fritsch, L., Lin, W.M., Hollmann, T.J., Ferre, F., Bourque, C., Burke, C.J., Turner, L., Uong, A., Johnson, L.A., Beroukhim, R., Mermel, C.H., Loda, M., Ait-Si-Ali, S., Garraway, L.A., Young, R.A., Zon, L.I. (2011) The histone methyltransferase SETDB1 is recurrently amplified in melanoma and accelerates its onset. *Nature*, **471** (7339), 513–517.

164 Fresno Vara, J.A., Casado, E., de Castro, J., Cejas, P., Belda-Iniesta, C., Gonzalez-Baron, M. (2004) PI3K/Akt signalling pathway and cancer. *Cancer Treat. Rev.*, **30** (2), 193–204.

165 Foster, K.S., McCrary, W.J., Ross, J.S., Wright, C.F. (2006) Members of the hSWI/SNF chromatin remodeling complex associate with and are phosphorylated by protein kinase B/Akt. *Oncogene*, **25** (33), 4605–4612.

166 Huang, W.C., Chen, C.C. (2005) Akt phosphorylation of p300 at Ser-1834 is essential for its histone acetyltransferase and transcriptional activity. *Mol. Cell. Biol.*, **25** (15), 6592–6602.

167 Gonzalez, M.E., Duprie, M.L., Krueger, H., Merajver, S.D., Ventura, A.C., Toy, K.A., Kleer, C.G. (2011) Histone methyltransferase EZH2 induces Akt-dependent genomic instability and BRCA1 inhibition in breast cancer. *Cancer Res.*, **71** (6), 2360–2370.

168 Campisi, J. (2005) Senescent cells, tumor suppression, and organismal aging: good citizens, bad neighbors. *Cell*, **120** (4), 513–522.

169 Scurr, L.L., Pupo, G.M., Becker, T.M., Lai, K., Schrama, D., Haferkamp, S., Irvine, M., Scolyer, R.A., Mann, G.J., Becker, J.C., Kefford, R.F., Rizos, H. (2010) IGFBP7 is not required for B-RAF-induced melanocyte senescence. *Cell*, **141** (4), 717–727.

170 Haferkamp, S., Scurr, L.L., Becker, T.M., Frausto, M., Kefford, R.F., Rizos, H. (2009) Oncogene-induced senescence does not require the p16(INK4a) or p14ARF melanoma tumor suppressors. *J. Invest. Dermatol.*, **129** (8), 1983–1991.

171 Xue, L., Murray, J.H., Tolkovsky, A.M. (2000) The Ras/phosphatidylinositol 3-kinase and Ras/ERK pathways function as independent survival modules each of which inhibits a distinct apoptotic signaling pathway in sympathetic neurons. *J. Biol. Chem.*, **275** (12), 8817–8824.

172 Xu, H.J., Zhou, Y., Ji, W., Perng, G.S., Kruzelock, R., Kong, C.T., Bast, R.C., Mills, G.B., Li, J., Hu, S.X. (1997) Reexpression of the retinoblastoma protein in tumor cells induces senescence and telomerase inhibition. *Oncogene*, **15** (21), 2589–2596.

173 Narita, M., Nunez, S., Heard, E., Lin, A.W., Hearn, S.A., Spector, D.L., Hannon, G.J., Lowe, S.W. (2003) Rb-mediated heterochromatin formation and silencing of E2F target genes during cellular senescence. *Cell*, **113** (6), 703–716.

174 Beausejour, C.M., Krtolica, A., Galimi, F., Narita, M., Lowe, S.W., Yaswen, P., Campisi, J. (2003) Reversal of human cellular senescence: roles of the p53 and p16 pathways. *EMBO J.*, **22** (16), 4212–4222.

175 Haferkamp, S., Tran, S.L., Becker, T.M., Scurr, L.L., Kefford, R.F., Rizos, H. (2009) The relative contributions of the p53 and pRb pathways in oncogene-induced melanocyte senescence. *Aging (Albany N.Y.)*, **1** (6), 542–556.

176 Haferkamp, S., Becker, T.M., Scurr, L.L., Kefford, R.F., Rizos, H. (2008) p16INK4a-induced senescence is disabled by melanoma-associated mutations. *Aging Cell*, **7** (5), 733–745.

177 Bandyopadhyay, D., Curry, J.L., Lin, Q., Richards, H.W., Chen, D., Hornsby, P.J., Timchenko, N.A., Medrano, E.E. (2007) Dynamic assembly of chromatin complexes during cellular senescence: implications for the growth arrest of human melanocytic nevi. *Aging Cell*, **6** (4), 577–591.

178 Nielsen, S.J., Schneider, R., Bauer, U.M., Bannister, A.J., Morrison, A., O'Carroll, D., Firestein, R., Cleary, M., Jenuwein, T., Herrera, R.E., Kouzarides, T. (2001) Rb targets histone H3 methylation and HP1 to promoters. *Nature*, **412** (6846), 561–565.

179 Kim, K.C., Geng, L., Huang, S. (2003) Inactivation of a histone methyltransferase by mutations in human cancers. *Cancer Res.*, **63** (22), 7619–7623.

180 Bandyopadhyay, D., Okan, N.A., Bales, E., Nascimento, L., Cole, P.A., Medrano, E.E. (2002) Down-regulation of p300/CBP histone acetyltransferase activates a senescence checkpoint in human melanocytes. *Cancer Res.*, **62** (21), 6231–6239.

181 Fullgrabe, J., Hajji, N., Joseph, B. (2010) Cracking the death code: apoptosis-related histone modifications. *Cell Death Differ.*, **17** (8), 1238–1243.

182 Wang, L., Baiocchi, R.A., Pal, S., Mosialos, G., Caligiuri, M., Sif, S. (2005) The BRG1- and hBRM-associated factor BAF57 induces apoptosis by stimulating expression of the cylindromatosis tumor suppressor gene. *Mol. Cell. Biol.*, **25** (18), 7953–7965.

183 Park, J.H., Park, E.J., Hur, S.K., Kim, S., Kwon, J. (2009) Mammalian SWI/SNF chromatin remodeling complexes are required to prevent apoptosis after DNA damage. *DNA Repair (Amst.)*, **8** (1), 29–39.

184 Kai, L., Samuel, S.K., Levenson, A.S. (2010) Resveratrol enhances p53 acetylation and apoptosis in prostate cancer by inhibiting MTA1/NuRD complex. *Int. J. Cancer*, **126** (7), 1538–1548.

185 Kang, Y., Massague, J. (2004) Epithelial-mesenchymal transitions: twist in development and metastasis. *Cell*, **118** (3), 277–279.

186 Fujii, S., Ochiai, A. (2008) Enhancer of zeste homolog 2 downregulates E-cadherin by mediating histone H3 methylation in gastric cancer cells. *Cancer Sci.*, **99** (4), 738–746.

187 Banine, F., Bartlett, C., Gunawardena, R., Muchardt, C., Yaniv, M., Knudsen, E.S., Weissman, B.E., Sherman, L.S. (2005) SWI/SNF chromatin-remodeling factors induce changes in DNA methylation to promote transcriptional activation. *Cancer Res.*, **65** (9), 3542–3547.

188 Sanchez-Tillo, E., Lazaro, A., Torrent, R., Cuatrecasas, M., Vaquero, E.C., Castells, A., Engel, P., Postigo, A. (2010) ZEB1 represses E-cadherin and induces an EMT by recruiting the SWI/SNF chromatin-remodeling protein BRG1. *Oncogene*, **29** (24), 3490–3500.

189 Ma, Z., Chang, M.J., Shah, R., Adamski, J., Zhao, X., Benveniste, E.N. (2004) Brg-1 is required for maximal transcription of the human matrix metalloproteinase-2 gene. *J. Biol. Chem.*, **279** (44), 46326–46334.

29
Pharmaco-Epigenomics to Improve Cancer Therapies

Bart Claes[1,2], Bernard Siebens[1,2], and Diether Lambrechts[1,2]
[1]VIB Vesalius Research Center, Herestraat 49, Box 912, 3000 Leuven, Belgium
[2]KU Leuven Campus Gasthuisberg, Vesalius Research Center, Herestraat 49, Box 912, 3000 Leuven, Belgium

Epigenetic Regulation and Epigenomics: Advances in Molecular Biology and Medicine, First Edition. Edited by Robert A. Meyers.

Keywords

Epigenomics
The study of epigenetic changes on a genome-wide level, where epigenetic refers to heritable changes in gene expression and activity that are not due to changes in the DNA sequence, but are caused by covalent modifications of DNA and histones.

Pharmaco-epigenomics
A recently emerged field of research that uses epigenetic insights to improve pharmacological therapies. The goals are to develop therapies targeted at the epigenome, to identify epigenomic biomarkers that can be used for disease diagnosis, and to predict disease progression as well as the response to a certain therapy.

DNA methylation
The addition of a methyl group to the 5-carbon of the base cytosine in the DNA.

Histone modifications
Post-translational modifications to histones, which are the main protein components of chromatin. Histone modifications include methylation and acetylation.

Cancer therapy
Chemical, physical, or biological treatment of cancer patients in order to inhibit the growth of cancer cells. The most widely used cancer therapies are chemotherapy, radiotherapy, molecularly targeted therapy (which targets a certain protein that is important for cancer growth) and epigenetic therapy.

Angiogenesis
The growth of new blood vessels from pre-existing vessels. Tumors exploit this process to recruit new blood vessels that fuel their growth. Anti-angiogenic therapies represent an effective therapy to slow down tumor growth.

Epigenetic modifications such as aberrant DNA methylation or altered histone modifications play an important role in tumor development and progression. Consequently, therapeutic strategies aimed at reversing epigenetic changes in cancer cells have been developed, and their great promise confirmed in recent studies. Epigenetic modifications can also be used as biomarkers to improve cancer diagnosis or to predict disease prognosis. Importantly, the first examples of how epigenetic alterations can be used as predictive markers for the outcome of conventional chemotherapies or targeted therapies have recently emerged. In this chapter, the recent advances in the field of pharmaco-epigenomics will be highlighted, and a review included of the most recently discovered and promising epigenetic therapies and biomarkers aimed at improving cancer diagnosis and treatment. In particular, attention will be focused on how these epigenetic therapies or biomarkers might change daily clinical practice.

1 Introduction

Despite major advances in its treatment, mortality resulting from cancer remains high, with numbers of cancer-associated deaths worldwide expected to rise to an estimated 12 million in 2030. Much of the recent progress in cancer therapy can be attributed to an improved understanding of cancer as a genetic disease, with several somatic alterations in oncogenes and tumor suppressor genes having been described, functionally validated, and successfully targeted by therapies in the clinic. However, the activity of genes in cancer cells can also be altered by epigenetic mechanisms, which are generally defined as heritable changes in gene expression that are not caused by changes in the DNA sequence itself. Epigenetics encompasses a complex range of reversible changes, which can be categorized into either modifications of the DNA itself or into modifications of the histones. Whilst methylation is the most extensively studied and best understood epigenetic modification of DNA, in the case of histones many different types of post-translational modification have been described, including acetylation, methylation, phosphorylation, ubiquitylation, and sumoylation. In general, histone modifications are less well understood than DNA modifications, and have been reported to occur in different histone proteins, to affect various histone amino acid residues, and to exhibit different degrees of modification (e.g., mono-, di-, and trimethylation).

Besides modifications of the histones and DNA, the microRNAs (miRNAs) represent an additional layer of epigenetic regulation. These small, noncoding RNAs are present endogenously in cells, and function as post-transcriptional silencers of certain sets of target genes [1]. Since the miRNAs induce changes in gene expression that can be heritable, they are also considered to be part of the epigenetic landscape. Moreover, crosstalk between miRNAs and other epigenetic pathways has recently been demonstrated: for example, miRNA expression can be subjected to epigenetic control [2], while in contrast miRNAs can regulate the expression of key components of the epigenetic machinery [3].

An improved understanding of which epigenetic changes occur in tumors, and how they drive tumor growth, have led to numerous attempts at improving existing cancer therapies. In this chapter, attention is focused first on the fundamental role of epigenetic changes in cancer, in terms of the two most well-established mechanisms, namely DNA methylation and histone modifications. The contribution of miRNAs to the deregulation of the epigenetic machinery in cancer has been less clearly established to date, and so will be mentioned only briefly at this point. A description will be provided of how an improved understanding of epigenetic processes has led to the development of epigenetic therapies aimed at reversing the changes that accumulate in cancer cells. On this basis, the first clinical results obtained with DNA methyltransferase

(DNMT) and histone deacetylase (HDAC) inhibitors in the treatment of cancer are described, and suggestions made as to how epigenetic modifications could be used not only to assist in cancer diagnosis but also to predict the prognosis and/or response to a specific therapy. The means by which the methylation status of DNA repair genes might represent a promising biomarker to predict the response to various chemotherapeutic treatments is also discussed. Finally, details are provided of how the tumor stroma can accumulate epigenetic changes and, in contrast, how epigenetic therapies may also target the stroma. In particular, the role of epigenetic changes in blood vessels will be discussed with regards to the anti-angiogenic therapies currently applied on a routine basis in the clinic.

2
The Epigenetic Origin of Cancer

Epigenetic changes occur both early and ubiquitously during the process of carcinogenesis, and continue to accumulate during tumor progression. The two main epigenetic hallmarks of cancer are an altered DNA methylation and histone modifications. Although the exact origin of these alterations is unclear, environmental influences are clearly implicated in their manifestation. For example, whilst monozygotic twins are epigenetically indistinguishable during their early years of life, at an older age they may show remarkable differences in their overall content and distribution of DNA methylation and histone modifications, accounting for the different environmental exposures encountered during their lives [4].

Whereas the role of most epigenetic modifications is poorly understood – especially in the context of cancer – the role of DNA methylation has been investigated more extensively. DNA methylation changes involve both global DNA hypomethylation and local DNA hypermethylation [5]:

- Global DNA *hypomethylation* is frequently observed in tumors, and mostly affects the intergenic regions of the DNA, in particular the repetitive DNA sequences and transposable DNA elements. Hypomethylation is generally believed to result in chromosomal instability and increased mutation events, thereby contributing to tumorigenesis [6], although it has also been associated with the activation of many growth-promoting genes, such as *HRAS*, c-*MYC*, or c-*JUN* [7, 8].
- In contrast, the local DNA *hypermethylation* of CpG islands is found in many promoters of tumor suppressor genes, and is well accepted to repress transcription by inhibiting the binding of specific transcription factors (TFs) and, more indirectly, also by recruiting methyl-CpG-binding proteins and their associated chromatin-remodeling complexes (see Fig. 1) [9].

In cancer, promotor hypermethylation is mostly associated with tumor-suppressor gene silencing, such as the retinoblastoma (*RB*), cyclin-dependent kinase inhibitor 2A (*CDKN2A*), Von Hippel–Lindau (*VHL*), breast cancer 1 (*BRCA1*), or human MutL homolog 1 (*hMLH1*) genes.

CpG island hypermethylation occurs at different stages of cancer development, and affects those genes that are involved in pathways which regulate the cell cycle, apoptosis, DNA repair, cell–cell communication, and angiogenesis, all of which become deregulated during tumor progression [10, 11]. Furthermore,

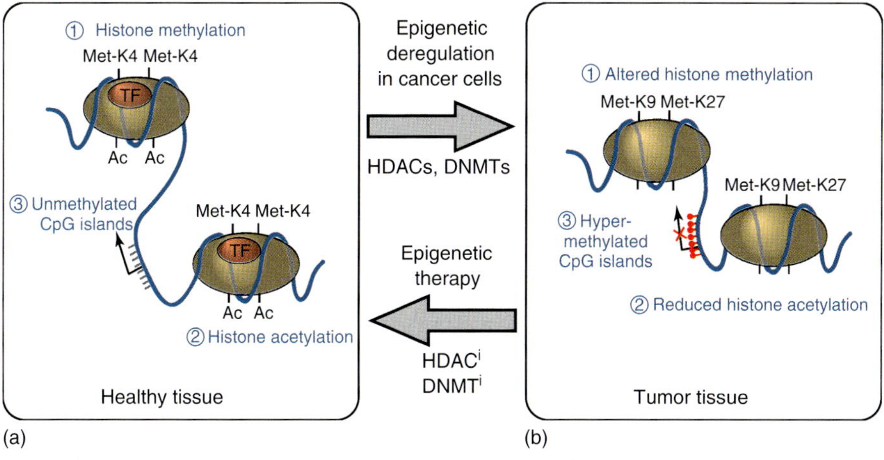

Fig. 1 Restoring epigenetic changes in cancer cells with epigenetic therapy. (a) Chromatin is composed of double-stranded DNA (blue) wrapped around histones (brown). Gene expression is regulated by histone modifications and DNA methylation of CpG islands in the promotor. In normal cells, chromatin surrounding transcriptionally active genes is characterized by specific modifications, such as the methylation of Lys4 of histone H3 ①, histone hyperacetylation ②, and unmethylated CpG islands ③. This leads to an open chromatin state that allows the binding of transcription factors (TFs) and RNA polymerase complexes; (b) In tumor cells, the epigenetic status of chromatin regions becomes altered by the disruption of normal DNA methylation and histone modification patterns. Epigenetic inactivation of tumor suppressor genes is associated with altered histone methylation ①, such as methylation of Lys9 and Lys27 of histone H3, histone hypoacetylation ②, and hypermethylation of CpG islands in promotors ③, leading to a more compact chromatin structure that represses transcription. By interfering with DNA and histone-modifying enzymes, epigenetic drugs can remove inactivation markers, such as DNA methylation, and can induce active markers, such as histone acetylation, leading to the restoration of disrupted epigenetic patterns. AC, acetylation; DNMTs, DNA methyltransferases; HDACs, histone deacetylases; DNMTi, DNMT inhibitor; HDACi, HDAC inhibitor; Met-K4, methylation of Lys4; Met-K9, methylation of Lys9; Met-K27, methylation of Lys27.

CpG island hypermethylation profiles are specific for each cancer type and, similar to the cancer genome, each type can be assigned a cancer methylome [12, 13]. Recently, miRNA genes have also been reported to be inactivated by altered DNA methylation in cancer. For instance, *miR-127* expression is strongly reduced in cancer cells, but can be restored upon treatment with the demethylating agent 5-aza-2′-deoxycytidine. The restoration of a normal expression pattern leads to translational downregulation of the *miR-127* target BCL6, which is a proto-oncogene [14]. In another study, it was shown that the epigenetic inactivation of *miR-124a* in different human tumor types is correlated with the activation of cyclin D kinase 6, a crucial component of complexes that control progression through the cell cycle [15]. These recent findings illustrate how cancer cells can epigenetically silence miRNAs that act as tumor suppressors in order to promote their uncontrolled growth.

Similar to DNA methylation, histone modifications are commonly disrupted in cancer cells. In general, histone modifications determine how tightly or loosely the DNA is wrapped around the histones and, as such, play an important regulatory role in determining gene expression [16]. For instance, acetyl groups neutralize positive charges on the histone tails, leading to weakened electrostatic interactions between the histones and the negatively charged phosphate backbone of DNA [17]. This correlates with an active or open chromatin state, which in turn allows TFs to access the promoters of target genes. By contrast, HDAC-induced deacetylation results in a compaction of the chromatin and an inactivation of genes (see Fig. 1). A global loss of the monoacetylation and trimethylation of histone H4 can be considered a common hallmark of human tumor cells, while altered histone modifications constitute a mechanism for the inactivation of tumor suppressor genes, as illustrated by the hypermethylation of Lys9 in histone H3 of the *CDKN2A* gene [18–20].

Another example of altered HDAC activity is found in acute promyelocytic leukemia (PML). The retinoic acid receptor (RAR) is an important regulator of myeloid cell differentiation, and induces transcriptional repression in the absence of retinoid ligands. Upon the binding of RAR to the retinoic acid response elements (RAREs), the HDACs are recruited to silence the RAR target genes. In PML, translocations cause the production of fusion genes consisting of the *RARα* and *PML* gene (translocation t(15;17)) or *RARα* and *PLZF* (t(11;17)) [21, 22]. The resulting fusion proteins bind to the RAREs and recruit HDACs, but are no longer responsive to retinoids, which results in a permanent and stable silencing of the RAR target genes. These examples demonstrate the powerful oncogenic potential of aberrant HDAC activity. Furthermore, the disruption of histone modifications occurs early during tumorigenesis and accumulates during tumor progression, thus underlining its importance as an essential epigenetic driver of cancer development [5].

Currently, there is emerging evidence that the different epigenetic processes – including DNA methylation and histone modification – are interdependent [23]. Although they involve different chemical reactions and require different sets of enzymes, the biological relationship between these two systems is essential for the modulation of gene expression. It appears that histone methylation can determine specific DNA methylation patterns, while DNA methylation can serve as a template for histone modifications after DNA replication [23]. This mechanistic connection can be accomplished through direct interactions between DNA and the histone-modifying complexes. For instance, methylated DNA has been shown to attract methylated-DNA-binding proteins, such as the transcriptional repressor MeCP2, which recruits chromatin-remodeling complexes to methylated CpG islands. These in turn modify histones and shape the higher-order chromatin structure [16, 24, 25]. Taken together, this interdependency of epigenetic pathways and their deregulation in cancer offers great potential for clinical applications, as will be discussed below.

3 Pharmaco-Epigenomics: Translating Epigenetics into Clinical Practice

As the fundamental role of epigenetic changes in cancer etiology become more clearly understood, the obvious next step would be to translate such knowledge into improved cancer therapies. This novel and emerging research field, which aims at translating epigenetic insights into clinical practice, is referred to as "pharmaco-epigenomics," and is focused on two main areas of interest:

- To develop cancer therapies that reverse the epigenetic changes which accumulate in cancer cells. In contrast to the genetic alterations that accumulate in tumors (such as somatic mutations and chromosomal rearrangements), altered histone modifications and DNA methylation profiles are reversible in nature, and this makes them particularly attractive targets for therapeutic intervention.
- To identify epigenetic biomarkers that could be used to diagnose cancer, as well as to predict disease progression and the response of a specific tumor to a therapy.

Both areas of research, as well as their clinical implications, are discussed in detail in the following sections.

4
Epigenetic Therapies for Cancer

Insights into the fundamental role that epigenetic alterations play during tumorigenesis have led to the development of novel treatment strategies for cancer, which in general are aimed at blocking or reversing the epigenetic alterations that promote malignancy and allow cancer cells to adapt to changes in the microenvironment (see Fig. 1). Whilst the two gene families to be most frequently targeted are the HDACs and the DNMTs, the exact mechanisms underlying the antitumor activity of drugs that target the HDACs or DNMTs have not been elucidated; however, given the vast influences that DNA methylation and histone modifications exert on gene expression, many cellular pathways are likely to be involved. For example, the HDAC inhibitors valproic acid and sodium butyrate have been reported to induce the expression of *CDKN1A*, a cyclin-dependent kinase (CDK) inhibitor that controls cell-cycle arrest and cell differentiation [26]. Valproic acid and sodium butyrate may also activate multiple apoptotic pathways, involving factors such as nuclear factor-κB (NF-κB), c-Jun N-terminal kinase (c-JNK), and B-cell lymphoma 2 (BCL2) [27–29]. Furthermore, HDAC inhibitors have been shown to alter expression of angiogenesis and metastasis-associated genes [30, 31]. Similarly, the DNMT inhibitor decitabine (5-aza-2′-deoxycytidine) reactivates epigenetically silenced tumor suppressor genes, such as *CDKN2A*, or genes implicated in metastasis, such as tissue inhibitor of metalloproteinases 3 (*TIMP3*) [32, 33]. The results of recent studies have also indicated that a DNA-demethylating treatment with decitabine can rescue the growth-inhibitory effects of certain miRNAs [14].

4.1
Epigenetic Therapies Used in Clinical Practice

In the meantime, it has been established that certain tumor types respond well to DNMT and HDAC inhibitor treatments, with the best clinical efficacy seen in hematologic malignancies. The DNMT inhibitor decitabine, for instance, has been approved for the treatment of patients with myelodysplastic syndrome or acute myeloid leukemia [34, 35]. The structure of decitabine mimics that of cytosine, which allows it to replace single cytosine molecules during DNA replication; it is also believed to form covalent adducts with DNMTs, thereby reducing the overall cellular DNMT activity. This interferes with the normal role of several DNMTs, which can no longer reproduce the methylation pattern of the cell in the DNA of daughter cells, and this leads to a reactivation of epigenetically silenced regions [36, 37].

At present, several clinical trials of epigenetic therapy for solid tumors are under way, and although decitabine has not yet been proven effective, it has been reported to stabilize the disease in lung cancer patients [38, 39]. Skepticism regarding the use of decitabine persists, however, since the clinical benefits attributed to it have not been shown to be caused directly by an interference with the epigenetic machinery. The obvious next step would be to examine epigenetic markers and gene expression patterns in patients before and after decitabine treatment, and to evaluate any correlations with benefit from epigenetic therapy. Other DNMT inhibitors, such as the orally active zebularine, are currently also under development [40], while the wide-spectrum HDAC inhibitor trichostatin A (TSA) also appears to be effective against leukemia. A potential synergy between HDAC inhibitors and DNMT inhibitors (e.g., TSA and decitabine) is also currently under investigation [41, 42]. The HDAC inhibitor vorinostat (suberoylanilide hydroxamic acid) has also recently been approved for the treatment of cutaneous T-cell lymphoma in patients with progressive, persistent, or recurrent disease [43]. Another clinically used HDAC inhibitor, phenylbutyrate, has achieved partial responses in hematologic malignancies [44].

An overview of the different DNMT and HDAC inhibitors currently used in clinical practice, or which are still under development, is provided in Table 1.

4.2 Synergism between Epigenetic and Conventional Therapies

Since epigenetic therapy can induce cancer cell reprogramming, it is possible that HDAC and DNMT inhibitors might act synergistically with conventional chemotherapy. This would not only allow the chemotherapy to be applied at lower dosages, but would also result in a reduced toxicity, while the efficacy of the combined therapy would still be increased compared to that of monotherapy. Chemotherapy often induces both genetic and epigenetic alterations that result in the selection of resistant cell clones. Interestingly, however, tumors that have become resistant to the initial treatment with chemotherapy due to epigenetic changes might become sensitive again to the drug when exposed to epigenetic therapies [61]. This principle appears to apply to interferon, which is the standard treatment for melanoma. Although interferon induces tumor cell apoptosis, it has been shown recently that the selection of apoptosis-resistant clones may occur during treatment via epigenetic mechanisms [62]. The epigenetic silencing of genes involved in signaling downstream of interferon, such as interferon regulatory factor 8 (*IRF8*) and XIAP-associated factor 1 (*XAF1*), was found to trigger resistance against interferon therapy [62, 63]. Furthermore, the injection of decitabine into nude mice carrying melanoma xenografts led to a re-sensitization to interferon treatment [62]. A Phase I clinical trial conducted in melanoma patients, in which decitabine was administered together with interleukin-2, further demonstrated an objective response in 31% of the patients [64]. Likewise, preclinical data have suggested that epigenetic therapy can induce radiosensitization and enhance the efficacy

Tab. 1 Clinical trials of epigenetic cancer therapies.

Agent	*Study phase*	*Disease*	*Details*	*Reference(s)*
DNMT inhibitor				
5-Azacytidine	3/FDA-approved	AML	Complete remission in 10–17% Hematological improvement in 23–36%	[45]
	3/FDA-approved	MDS	63% improvement in OS over standard of care	[46]
5-Aza-2′-deoxycytidine	3/FDA-approved	MDS, CML	Safe toxicity profile 34% of patients achieved complete response 73% had objective response	[47]
Arabinosyl-5-azacytidine	1/2	CML, cervical carcinoma		[48, 49]
MG98 (DNMT1 antisense)	2	Metastatic solid tumors		[50]
HDAC inhibitor				
Phenylbutyrate	1	MDS, AML	Well tolerated No partial or complete remissions 15% showed hematological improvement	[51]
Vorinostat (SAHA)	2/FDA-approved	CTCL	30% objective response rate	[52]
	1	AML,CLL, MDS, ALL, and CML	23% showed hematological improvement 6% complete responses	
	1	Advanced solid and hematologic cancers	4% partial responses in CTCL	[53]
Valproic acid (VPA)	1	AML, leukemia		[54, 55]
Belinostat	1	Multiple myeloma		[56]
Romidepsin	1	Lung cancer		[56]
	2	CTCL		

(continued overleaf)

Tab. 1 (*Continued*)

Agent	*Study phase*	*Disease*	*Details*	*Reference(s)*
CI-994	1/2	Lung and pancreatic cancer		[56]
MS-275	1/2	Lung, lymphoma, and leukemia		[56]
MGCD0103	1/2	Solid cancers, lymphoma, and leukemia		[56]
DNMT and HDAC inhibitor combined				
5-Aza-cytidine and VPA	1	Advanced solid cancers	Combination has safe toxicity profile 25% show stable disease (median 6 months)	[57]
5-Aza-cytidine and phenylbutyrate	1	Refractory solid cancers	Combination has safe toxicity profile No clinical benefits observed	[58]
HDAC inhibitor and chemotherapy combined				
Vorinostat, carboplatin, and paclitaxel	1	Advanced NSCLC	2.7-fold improved response, improved PFS (6 vs. 4.1 months) improved OS (13 vs. 9.7 months) compared to chemotherapy alone	[59]
Vorinostat and doxorubicin	1	Solid cancers	8% showed partial response 8% showed stable disease for over eight months	[60]

[a] ALL, acute lymphoblastic leukemia; AML, acute myelogenous leukemia; CLL, chronic lymphocytic leukemia; CTCL, cutaneous T-cell lymphoma; MDS, myelodysplastic syndrome; NSCLC, non-small-cell lung carcinoma; OS, overall survival; PFS, progression-free survival; SAHA, suberoylanilide hydroxamic acid.

of current conventional radiotherapy treatment regimens [65]. Whilst the molecular mechanisms that underlie this radiosensitizing potential are not yet fully understood, they can be partially explained by the silencing of DNA repair genes, such as *Ku70* and *Ku86* [66]. Similarly, treatment with the HDAC inhibitor Vorinostat can prolong the appearance of repair foci identified by phosphorylated histone 2AX (γ-H2AX), which is indicative of a reduced repair efficiency and an increased radiosensitivity [66]. Epigenetic cancer therapy can thus be considered a very promising approach to cancer treatment, based on its synergy with chemotherapy, its resensitization of chemoresistant tumors, and its ability to increase the efficacy of radiotherapy.

4.3 Potential Side Effects of Epigenetic Therapies

Despite its great potential, the nonspecific effects of epigenetic drugs represent an area of concern for clinical application in patients. For instance, given the effect of global DNA hypomethylation on genomic stability, therapy-induced hypomethylation might promote tumor formation in the long run, although this hypothesis requires verification [67, 68]. Epigenetic therapy might also cause an activation of imprinted or silenced genes, and has indeed been shown to be mutagenic [69] and possibly even carcinogenic [70]. These concerns should not be exaggerated, however, as DNMT inhibitors act only on dividing cells while leaving other cells mostly unaffected. Furthermore, current evidence suggests that epigenetic drugs have a tendency to activate genes that have become abnormally silenced [71, 72]. Although no mechanism has yet been demonstrated to explain this, it is possible that the chromatin structure of aberrantly silenced genes is more susceptible to reactivation when compared to genes silenced under normal physiological conditions. However, it should be noted that patients receiving HDAC or DNMT inhibitors in the clinic have not yet suffered from any major toxicities, nor any unexplained long-term adverse effects [73, 74]. Although caution is warranted, the currently available clinical evidence suggests that epigenetic therapy is reasonably safe.

With regards to any potential side effects, there is great promise for specific epigenetic therapies targeted at particular genes via the use of promotor-specific TFs [75]. This strategy has, for instance, been shown specifically to reactivate *Maspin*, a tumor suppressor gene that is silenced by promoter methylation in aggressive epithelial tumors [76]. Consequently, Beltran *et al.* constructed an artificial transcription factor (ATF) that consisted of six zinc-finger domains targeting unique 18 bp sequences in the *Maspin* promoter, linked to an activator domain [76]. This ATF reactivated the epigenetically silenced *Maspin*, induced the apoptosis of cancer cells *in vitro*, and also suppressed tumor growth in a xenograft breast cancer model in nude mice. Hence, despite some concern being expressed regarding long-term safety, epigenetic cancer therapies clearly hold great potential. It is hoped that the next generation of targeted therapies will overcome any possible pitfalls and improve both the clinical efficacy and safety of epigenetic drugs.

5 Epigenetic Biomarkers for Cancer Therapy

Since epigenetic alterations play an important role in determining the behavior of various tumors, it is likely that epigenetic alterations – such as DNA methylation and histone modifications – can also be used for the diagnosis and molecular classification of cancer, and to predict cancer progression or response to therapy. Indeed, although the epigenetic mapping of genes within a clinical research setting is challenging due to the poor preservation of chromatin structure in clinical samples, there exists a tight correlation between methylation patterns, chromatin structure, and gene expression [77]. DNA methylation reflects the chromatin structure of a gene, and can be considered as a stable covalent DNA mark for gene activity [78]. As DNA methylation is better preserved compared to histone modification and chromatin structure (even in poor-quality samples), clinical epigenetic cancer research currently relies on DNA methylation for biomarker identification. However, this could change in the near future with improved sample collection, better storage methods, and novel analytical techniques. Currently, three potential applications may be distinguished for epigenetic markers in cancer management: as complementary diagnostic tools; as prognostic markers of disease progression; and as predictive markers of treatment response.

5.1 Methylated DNA Sequences for Improved Cancer Diagnosis

Epigenetic alterations can be used to complement existing diagnostic tools for cancer detection. Sensitive polymerase chain reaction (PCR)-based methods have been developed to detect hypermethylated CpG islands in DNA from various sources, such as blood, urine, sputum, or tumor biopsies [11]. These approaches have stimulated the discovery of abnormally methylated DNA sequences as tumor markers across multiple cancer types. For instance, the hypermethylation of glutathione *S*-transferase 1 (*GSTP1*) is seen in 80–90% of prostate cancer patients, while benign hyperplastic prostate tissue is not hypermethylated [79, 80]. Consequently, the presence of *GSTP1* methylation in prostate biopsies or urine samples could assist in the diagnosis of malignant prostate cancer [81]. In another study, a panel detecting several hypermethylated genes in breast ductal fluids correctly identified twice as many breast cancers than did classic cytological techniques [82]. More importantly, emerging evidence has indicated that epigenetic alterations occur early in carcinogenesis, before other biomarkers are detectable. For example, a substantial hypermethylation of the tumor suppressor *CDKN2A* can be detected in the bronchial pre-neoplastic epithelium of smokers [83].

Likewise, Melotte *et al.* reported a new biomarker for colorectal cancer (CRC) in stool samples [84]. These authors reported that N-Myc downstream-regulated gene 4 (*NDRG4*), a tumor suppressor candidate, is frequently silenced by promotor hypermethylation in CRC. Thus, by using a methylation-specific PCR assay for *NDRG4*, they could successfully identify 53% of CRC cases, and also correctly predicted which of the individuals were free from cancer [84]. In another recent study, a panel of four genes – namely bone morphogenetic protein 3 (*BMP3*), eyes absent homolog 2 (*EYA2*), aristaless-like homeobox-4 (*ALX4*), and vimentin – was identified that could be methylated with high specificity in colorectal carcinomas and adenomas, but rarely in normal epithelium [85]. These markers would clearly be very attractive for use in diagnostic tests to detect early-stage cancers, where the odds of survival were highest. Another potential plasma biomarker for CRC could lie in the methylation status of the septin 9 (*SEPT9*) gene which, according to the results of a recent study, can identify 72% of CRCs at a specificity of 92% [86].

As diagnostic tests within a routine clinical setting should preferably be noninvasive, the detection of epigenetic biomarkers in blood, stool, or urine would require that methods with improved sensitivity be developed, so as to allow the detection of degraded and diluted epigenetic biomarkers in these body fluids. In this context, Li and colleagues have developed a new technique based on next-generation sequencing, that is capable of detecting one methylated molecule among up to 5000 unmethylated molecules [87]. By using vimentin gene methylation as a plasma biomarker, it was possible to detect 52% of early-stage CRCs, compared to only 14% detection with the classical carcinoembryonic antigen (CEA) test. The myriad of ongoing studies nevertheless clearly demonstrates the enormous potential of epigenetic biomarkers for cancer detection.

5.2
Epigenetic Changes as Prognostic Markers

Epigenetic modifications could also serve as biomarkers to predict the disease prognosis of cancer patients. As patients with histologically similar cancers display a remarkable heterogeneity in their disease prognosis, biomarkers that could identify patients with an increased risk of an aggressive tumor and an associated poor survival would allow the selection of a treatment to achieve the best possible clinical risk : benefit ratio. Although current clinical practice is based mainly on immunohistological analyses, good progress has been made recently to improve risk stratification by using either gene-expression [88] or somatic mutation signatures [89, 90]. Yet, epigenetic biomarkers could possibly complement these existing tools. For instance, hypermethylation of the tumor suppressor genes *APC* and *CDKN2A* were shown to be associated with a poor prognosis in breast cancer and CRC, respectively [91, 92]; similarly, patients with lung cancer also have a poor prognosis if hypermethylation of *CDKN2A* is observed in the tumor [93]. In another study, the prognostic value of APC in breast cancer was confirmed, while the Ras association domain-containing protein 1 (*RASSF1A*) gene was also identified as a potential prognostic marker [94]. Furthermore, the gene for secreted frizzled-related protein 1 (*SFRP1*) has also been associated with a poor overall survival in breast cancer patients. This gene encodes a soluble Wnt antagonist, and defects in the Wnt signaling pathway have been implicated in breast cancer pathogenesis [95]. These examples indicate that future prognostic models will most likely incorporate epigenetic markers for a number of key pathways that are activated in a specific tumor type.

It has also been shown that global histone modification profiles, such as histone lysine methylation and acetylation marks, are correlated with clinical and pathological parameters of prostate cancer, and may serve as a significant predictor of prostate cancer recurrence [96]. Although histone modifications have been less well studied than DNA methylation in a clinical context (due mainly to protein degradation issues in clinical samples), the results of this study showed that histone modification profiles also hold great promise to serve as prognostic markers. Further research and clinical validation is needed, however, before DNA and histone modification profiles can be used to complement current pathological analyses.

5.3
Epigenetic Changes as Predictive Markers for Cancer Therapies

Epigenetic alterations may also function as predictive markers to assess the response to a particular cancer therapy. With the increasing recognition that each tumor has its own genetic profile that requires its own specific therapy, it is to be expected that future cancer therapies will become tailored to individual patients. The paradigm of personalized medicine, as illustrated by the recent approval for a test that determines *KRAS* mutations to predict the response to the epidermal growth factor receptor (EGFR) inhibitor cetuximab in CRC patients, is an example that could also apply to epigenetic markers [97]. It is essential, therefore, to identify any epigenetic differences that might explain any inter-individual variations in response to a certain therapy (see Fig. 2).

A well-established example is the DNA repair gene O^6-methylguanine–DNA methyltransferase (*MGMT*), which reverses guanine alkyl adduct formation induced by alkylating drugs, and thus prohibits the formation of lethal DNA crosslinks [101]; this mechanism explains why *MGMT*-expressing tumors are often resistant to alkylating chemotherapy. There is, however, a substantial variation in *MGMT* expression according to tumor type. For instance, almost half of the gliomas lack *MGMT* expression, and these would be expected to have an increased sensitivity to alkylating drugs. Interestingly, *MGMT* inactivation is rarely the result of a mutation or deletion, but rather results from hypermethylation of the CpG island in its promoter. Two independent studies have shown that the *MGMT* promoter methylation status can predict the response to the alkylating chemotherapies carmustine or temozolomide, and is associated with tumor regression and a prolonged disease-free and overall survival [98, 102]. The predictive value of epigenetic *MGMT* silencing is currently under investigation in several prospective randomized Phase III clinical studies in glioblastoma patients, in which standard chemoradiotherapy is combined with therapies targeted at important angiogenic molecules, such as bevacizumab [an anti-VEGF (vascular endothelial growth factor) antibody] or cilengitide (an integrin inhibitor). It is expected that the *MGMT* status will predict which tumors are sensitive to the standard chemotherapy, and that the addition of anti-angiogenic therapies will enhance the effects of chemotherapy in these tumors. If the results from these trials prove to be positive, then *MGMT* status will become the first predictive epigenetic biomarker to be applied in a clinical setting, hopefully paving the way for many others.

In fact, monitoring the methylation status of the DNA repair genes as biomarkers for chemotherapeutic response might in time become a common theme, with other studies having described the effects of *MGMT* on cyclophosphamide [101], of *hMLH1* on cisplatin [103], of reduced folate carrier (*RFC*) on methotrexate [104, 105], and of Werner syndrome RecQ helicase-like (*WRN*) on irinotecan [105]. Two other DNA repair genes that have been well studied in this respect are the so-called "breast cancer genes," *BRCA1* and *BRCA2*, which are frequently inactivated by genetic and epigenetic mechanisms in several cancer

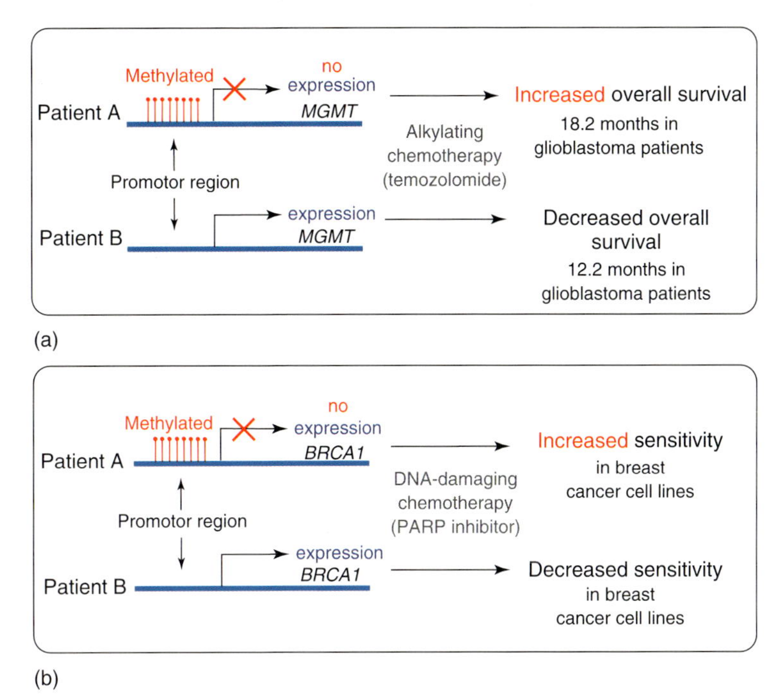

Fig. 2 Prediction of chemotherapeutic response by epigenetic biomarkers. (a) The *MGMT* methylation status can vary remarkably between different cancer patients. About half of glioblastoma patients have CpG island hypermethylation in the *MGMT* promotor (represented by patient A), while the other half does not have *MGMT* promotor methylation and therefore express MGMT (represented by patient B) [98]. The methylation status of *MGMT* in tumors is a predictive biomarker for response to alkylating chemotherapy, since MGMT specifically repairs DNA damage induced by this type of drug. Patients carrying methylated *MGMT* are unable to repair chemotherapeutic damage, which leads to an increased benefit from alkylating drugs and an increased survival compared to unmethylated *MGMT* carriers; (b) Similar heterogeneity exists in *BRCA1* methylation status between breast cancer patients. About 40% of breast cancer patients carry tumors with CpG island hypermethylation in the *BRCA1* promotor (represented by patient A), while the remaining 60% does not and therefore expresses *BRCA1* (represented by patient B) [99]. The methylation status of *BRCA1* in tumors is a potential predictive biomarker for response to drugs that inhibit the DNA repair enzyme poly(ADP)-ribose polymerase (PARP). As both BRCA1 and PARP are involved in DNA repair, *BRCA1*-deficient cells treated with PARP inhibitors no longer have DNA repair activity, accumulate DNA damage and die. Conversely, *BRCA1*-expressing cancer cells treated with PARP inhibitors can still repair DNA damage, and are therefore less sensitive to PARP inhibition. Although this concept has only very recently been proven in breast cancer cell lines [99], it will most likely also be confirmed in cancer patients, as *BRCA1* inactivation via mutation has already been shown to be a predictive biomarker for response to PARP inhibition [100].

types [106–108]. These genes are required for the DNA double-strand break repair processes through homologous recombination. As a consequence, cancer cells carrying inactivated *BRCA1* or *BRCA2* genes are no longer capable of repairing DNA damage induced by, for instance platinum-based compounds. Intriguingly, these cancers have recently been shown to respond well to poly(ADP-ribose) polymerase (PARP) inhibitors [100]. The PARP enzyme is essential for the repair of DNA single-strand breaks; thus, PARP inhibitors may enhance the cytotoxic effects of DNA-damaging agents by selectively targeting cells defective in the *BRCA1/2*-dependent DNA repair pathway and inhibiting their PARP-dependent repair mechanisms [100]. While this concept had already been proven for *BRCA1/2* mutation carriers, the results of a very recent study conducted by Veeck *et al.* showed that the inactivation of *BRCA1* by CpG island hypermethylation is also a powerful predictor of sensitivity towards PARP inhibitors [99]. The data reported by Veeck *et al.* showed clearly the value of epigenetic biomarkers, and further illustrated that future studies should focus on the integration of genomic and epigenomic information, so as to establish a complete picture of *BRCA1/2* loss-of-function, in order that these results might be successfully translated into clinical practice (see Fig. 2).

6
Epigenetic Therapy beyond Cancer Cells

While cancer research initially has focused on the growth autonomy of cancer cells, it is becoming apparent that the stroma that surrounds the cancer cells, such as fibroblasts, endothelial cells, and inflammatory cells, also plays a key role in driving tumor cell proliferation [109] The anti-tumoral properties of novel epigenetic therapies have thus far largely been attributed to the reactivation of silenced tumor suppressor genes in tumor cells. However, given their universal gene regulatory effects, it is likely that epigenetic therapy will also affect cancer stromal cells. There is a variety of stromal cell types that surround and interact with cancer cells, such as endothelial cells, fibroblasts, bone-marrow-derived precursor cells, and infiltrating inflammatory cells, of which several are known to be epigenetically altered under the influence of tumor cells [110–112]. For instance, Hu *et al.* have shown that distinct epigenetic alterations occur in a tumor stage- and cell type-specific manner during breast carcinogenesis, which suggests that epigenetic changes are important for maintenance of the abnormal cellular microenvironment [110]. The results of another study, performed by Chang *et al.*, showed that tumor-associated macrophages (TAMs) display a deacetylation of histones associated with the promotor of *CIITA*, a master regulator of major histocompatability class (MHC)-II expression [113]. This results in a downregulation of *CIITA* expression, a reduced MHC-II-dependent antigen presentation, and thus shows that tumors can induce an immunosuppressive phenotype in TAMs via epigenetic mechanisms.

Within the tumor microenvironment, probably the best-studied cells are the tumor endothelial cells, which not only play a key role during blood vessel formation (angiogenesis) but also respond rapidly to environmental changes, such as tumor hypoxia. Hypoxia is known to induce epigenetic changes, such as a global decrease in H3K9 acetylation, most likely because of HDAC upregulation [114, 115]. Recently, Johnson *et al.* showed that hypoxia-activated genes, such as early growth response 1 (*EGR1*) and *VEGF*, undergo a specific remodeling of the chromatin structure at their core promotors in response to hypoxia, in order to regulate their transcriptional activation [115]. This confirms that epigenetic changes could be important regulators of the response to chronic hypoxia, as often occurs in tumors. Given the critical role of angiogenesis in tumor progression and the widespread clinical application of anti-angiogenic therapies, the epigenetic alterations that occur in tumor vessels, and the means by which epigenetic therapies could target tumor angiogenesis, will now be discussed.

6.1 Epigenetic Changes in the Tumor Vasculature

Angiogenesis is a remarkable dynamic process that is tightly controlled by a balance of stimulatory and inhibitory angiogenic signals. Consequently, an imbalance in these signals will result either in a shortage or an excess of blood vessels, which will contribute to ischemic or malignant disorders, respectively [116]. The first angiogenesis inhibitors have been widely used since 2004 in the first-line treatment of various solid tumors, in combination with chemotherapy [117]. Whilst tumor angiogenesis alone does not initiate malignancy, it does play a critical role in cancer by promoting tumor progression and metastasis [116]. Activation of the so-called "angiogenic switch" is considered as one of the hallmarks of cancer that promotes tumor growth and metastasis [109]. Growing evidence indicates that epigenetic changes of the genes involved in angiogenesis are involved in this switch, and may cause tumors to recruit new blood vessels and sustain their growth [118]. In a recent study reported by Lu *et al.*, it was shown elegantly that cancer cells would induce epigenetic changes in endothelial cells, which in turn promoted tumor angiogenesis. Lu and coworkers identified the methyltransferase EZH2 as a key regulator of tumor angiogenesis, and showed that VEGF produced by cancer cells increased the expression of endothelial EZH2, which methylates and silences vasohibin1, thereby promoting a pro-angiogenic phenotype [119]. It was also shown that the delivery of EZH2 siRNA into tumor and tumor-associated endothelial cells decreased angiogenesis and resulted in a growth inhibition of an orthotopic ovarian cancer model. Taken together, these data confirmed that targeting epigenetic changes in tumor endothelial cells would represent a promising new anti-angiogenic strategy. Several clinical studies in humans have confirmed that epigenetic changes in tumors affect genes involved in the regulation of tumor angiogenesis. Glioblastomas, for instance, are typically characterized by an excessive blood vessel development, and frequently display an epigenetic inactivation of the anti-angiogenic thrombospondin-1 (*THBS-1*) gene [120]. *THBS-1* is also suppressed early in breast carcinogenesis by histone modifications [121], while *THBS-1* silencing via methylation is observed in a significant proportion of primary colorectal adenomas [122]. Interestingly, oxygen–glucose deprivation, which frequently occurs in tumors, was shown to increase *THBS-1* promoter methylation and subsequent silencing; this transcriptional inactivation could be reversed by reoxygenation [123].

6.2 Anti-Angiogenic Effects of Epigenetic Therapies

The fact that epigenetic alterations also occur in tumor endothelial cells, and specifically alter the balance of stimulatory and inhibitory angiogenic signals, suggests that interfering with the epigenetic machinery could be used to reactivate silenced anti-angiogenic factors and inhibit new blood vessel growth. Indeed, several epigenetic therapies have been assigned with anti-angiogenic activities, and some of the mechanisms underlying these have already been elucidated. For instance, the HDAC inhibitor TSA impairs blood vessel formation both *in vitro* and *in vivo* by the downregulation of pro-angiogenic signaling factors, such as VEGF, and the upregulation of anti-angiogenic factors, such as ADAMTS-1 [124–126]. Furthermore, TSA induces the expression of tumor suppressors p53 and VHL, and downregulates HIF-1α, a TF that activates hypoxia-induced angiogenic signaling pathways [127]. However, since p53, VHL, and HIF-1α also have important roles in tumor cells, a major challenge lies in determining whether the effects of HDAC inhibitors act directly on the blood vessels, or indirectly via the tumor cells. Hellebrekers and coworkers recently identified several downregulated genes in tumor-conditioned endothelial cells compared to normal endothelial cells, which included the anti-angiogenic genes *clusterin, fibrillin 1,* and *quiescin Q6* [128]. In this case, it was shown that the expression of these genes could be reactivated by treatment with the HDAC inhibitor, TSA. These findings confirm that the anti-angiogenic effects of HDAC inhibitors can, at least in part, be explained by their direct influence on endothelial gene expression. Another recently identified mechanism of action of HDAC inhibitors appears to be an impairment of endothelial progenitor cell function [129]. Adult progenitor cells possess stem cell-like properties, and are able to differentiate into endothelial cells that assist in the growth of new blood vessels [129, 130]. HDAC inhibitors can block their differentiation into endothelial cells via a repression of the TF HoxA9, a master regulator of expression for endothelial-committed genes, such as *eNOS*, *VEGFR-2,* and *VE-cadherin* [129].

Similar to HDAC inhibitors, the DNMT inhibitors can also reactivate epigenetically silenced genes in tumors, and decrease tumor growth both *in vitro* and *in vivo* [32, 131]. Again, these results cannot be interpreted without considering the effect of DNMT inhibitors on blood vessels. Indeed, the specific inhibitors decitabine and zebularine can decrease vessel formation and inhibit the proliferation of tumor-conditioned endothelial cells by reactivation of growth-inhibiting genes, such as *THBS-1, JUNB,* and *IGFBP3,* which are known to be silenced in tumor-conditioned endothelial cells [33]. Furthermore, these compounds can restore expression of the epigenetically silenced intercellular adhesion molecule-1 (ICAM-1) on tumor-conditioned endothelial cells *in vitro* and *in vivo* by a reversal of histone modifications in the ICAM-1 promoter. This results in a restored leukocyte-endothelial cell adhesion and an enhanced leukocyte infiltration, and also expands the therapeutic effects of DNMTs to the immune system [132]. Together, the HDAC and DNMT inhibitors appear to act on multiple cell types, most notably on tumor and endothelial cells, thereby affecting tumor angiogenesis as well as cancer cell survival.

6.3
Epigenetic Biomarkers for Anti-Angiogenic Therapies

Since epigenetic alterations frequently occur in tumor endothelial cells, it is possible that these epigenetic changes might be a potential predictor of response to the anti-angiogenic treatment. As such, a study by Rini *et al.* investigated the predictive effect of genetic and epigenetic inactivation of the *VHL* gene, a negative regulator of VEGF, in renal cell carcinoma patients treated with the monoclonal anti-VEGF antibody bevacizumab [133]. Patients with *VHL* inactivation had an improved response to the anti-angiogenic therapy, and displayed a strong trend towards a prolonged time to disease progression. Although still speculative, it is possible that *VHL* inactivation in tumors renders these patients more dependent on VEGF for initiating and sustaining angiogenesis, and therefore makes them more susceptible to VEGF inhibition. Although these results require further validation, they clearly show that epigenetic biomarkers could become a valuable predictive tool for anti-VEGF therapy. In a similar study, the predictive value of genetic and epigenetic inactivation of the *VHL* gene was investigated in metastatic renal cell carcinoma patients treated with sunitinib, a tyrosine kinase inhibitor that targets amongst others the VEGF receptors [134]. Patients with *VHL*-inactivating mutations exhibited an increased response rate of 52%, compared to 31% for those with wild-type *VHL*. Likewise, the epigenetic inactivation of *VHL* led to a response rate of 41%, versus 31% for those with wild-type *VHL*. Although the effect of epigenetic inactivation was less pronounced than for *VHL* mutations in this study, the observed effects were very similar to those seen with bevacizumab.

A further anti-angiogenic strategy that is currently being evaluated in clinical trials is that of anti-integrin therapy. The integrins, as cell-surface receptors, are involved in angiogenesis and therefore represent an attractive target for the inhibition of blood vessel growth in tumors. A potential biomarker for anti-integrin therapy might be the *ADAM23* gene, which acts as a negative regulator of the integrin$\alpha_V\beta_{III}$ receptor. *ADAM23* is frequently silenced by promoter hypermethylation and, since its silencing correlates with tumor progression, it might be associated with the acquisition of an angiogenic and metastatic phenotype [135]. Tumors characterized by *ADAM23* silencing might therefore depend more on activated α_V integrin signaling for their blood vessel growth, which would make them more eligible for anti-integrin therapy. These interesting findings warrant further study to evaluate the epigenetic profiles of multiple pro- and anti-angiogenic factors, in order to identify genes that commonly are altered in tumor angiogenesis.

7
Future Perspectives

The field of cancer pharmaco-epigenomics has emerged from an increased understanding of how epigenetic alterations can "drive" cancer growth. Whilst several epigenetic therapies have already been approved for hematological malignancies, it is expected that newer and improved therapies will lead to even better results. Indeed, current epigenetic drugs remain largely unspecific and might be complemented or replaced by more targeted epigenetic therapies with increased specificities for particular key genes. Currently, studies to systematically evaluate epigenetic profiles on a genome-wide level are ongoing, and should lead to an unbiased identification of the most common and relevant epigenetically altered genes for a certain cancer type. These epigenetic drivers of the tumor can then be specifically targeted. A combination of current reference chemotherapy or radiotherapy treatments with epigenetic therapy is also expected to lead to an increased efficacy, while reducing the delivered dose of the chemotherapy and, thereby, reducing also any toxicity. A combination of epigenetic therapies with chemotherapy or radiotherapy might also delay resistance to the chemotherapy, which is frequently observed in high-dose chemotherapeutic treatments. Large-scale efforts to map all of the epigenetic alterations that occur in a given cancer type are ongoing, and it is expected that these will also yield tumor type-specific biomarkers for disease diagnosis, prognosis, or prediction of response to a therapy. For instance, The Cancer Genome Atlas (TCGA) project has identified methylation signatures with a strong prognostic value in glioblastoma, while many other cancer types – such as serous ovarian carcinoma – are currently being investigated. Most importantly, as detection methods to identify the epigenetic changes in tumor DNA are improved, epigenetic biomarkers capable of predicting the efficacy of conventional chemotherapies or targeted therapies will also become applicable in a clinical context. Clearly, pharmaco-epigenomics has the potential to become an essential part of clinical practice in the future. Ultimately, the epigenomic profile of a patient might translate into personalized epigenomic medicine.

References

1 Bartel, D.P. (2004) MicroRNAs: genomics, biogenesis, mechanism, and function. *Cell*, **116**, 281–297.

2 Nomura, T., Kimura, M., Horii, T., Morita, S., Soejima, H., Kudo, S., Hatada, I. (2008) MeCP2-dependent repression of an imprinted miR-184 released by depolarization. *Hum. Mol. Genet.*, **17**, 1192–1199.

3 Szulwach, K.E., Li, X., Smrt, R.D., Li, Y., Luo, Y., Lin, L., Santistevan, N.J., Li, W., Zhao, X., Jin, P. (2010) Crosstalk between microRNA and epigenetic regulation in adult neurogenesis. *J. Cell Biol.*, **189**, 127–141.

4 Fraga, M.F., Ballestar, E., Paz, M.F., Ropero, S., Setien, F., Ballestar, M.L., Heine-Suner, D., Cigudosa, J.C., Urioste, M., Benitez, J., Boix-Chornet, M., Sanchez-Aguilera, A., Ling, C., Carlsson, E., Poulsen, P., Vaag, A., Stephan, Z., Spector, T.D., Wu, Y.Z., Plass, C., Esteller, M. (2005) Epigenetic differences arise during the lifetime of monozygotic twins. *Proc. Natl Acad. Sci. USA*, **102**, 10604–10609.

5 Esteller, M. (2008) Epigenetics in cancer. *N. Engl. J. Med.*, **358**, 1148–1159.

6 Wilson, A.S., Power, B.E., Molloy, P.L. (2007) DNA hypomethylation and human diseases. *Biochim. Biophys. Acta*, **1775**, 138–162.

7 Feinberg, A.P., Tycko, B. (2004) The history of cancer epigenetics. *Nat. Rev. Cancer*, **4**, 143–153.

8 Feinberg, A.P. (2007) Phenotypic plasticity and the epigenetics of human disease. *Nature*, **447**, 433–440.

9 Sasaki, H., Matsui, Y. (2008) Epigenetic events in mammalian germ-cell development: reprogramming and beyond. *Nat. Rev. Genet.*, **9**, 129–140.

10 Esteller, M. (2007) Cancer epigenomics: DNA methylomes and histone-modification maps. *Nat. Rev. Genet.*, **8**, 286–298.

11 Herman, J.G., Baylin, S.B. (2003) Gene silencing in cancer in association with promoter hypermethylation. *N. Engl. J. Med.*, **349**, 2042–2054.

12 Costello, J.F., Fruhwald, M.C., Smiraglia, D.J., Rush, L.J., Robertson, G.P., Gao, X., Wright, F.A., Feramisco, J.D., Peltomaki,

P., Lang, J.C., Schuller, D.E., Yu, L., Bloomfield, C.D., Caligiuri, M.A., Yates, A., Nishikawa, R., Su Huang, H., Petrelli, N.J., Zhang, X., O'Dorisio, M.S., Held, W.A., Cavenee, W.K., Plass, C. (2000) Aberrant CpG-island methylation has non-random and tumour-type-specific patterns. *Nat. Genet.*, **24**, 132–138.

13 Esteller, M., Corn, P.G., Baylin, S.B., Herman, J.G. (2001) A gene hypermethylation profile of human cancer. *Cancer Res.*, **61**, 3225–3229.

14 Saito, Y., Liang, G., Egger, G., Friedman, J.M., Chuang, J.C., Coetzee, G.A., Jones, P.A. (2006) Specific activation of microRNA-127 with downregulation of the proto-oncogene BCL6 by chromatin-modifying drugs in human cancer cells. *Cancer Cell*, **9**, 435–443.

15 Lujambio, A., Ropero, S., Ballestar, E., Fraga, M.F., Cerrato, C., Setien, F., Casado, S., Suarez-Gauthier, A., Sanchez-Cespedes, M., Git, A., Spiteri, I., Das, P.P., Caldas, C., Miska, E., Esteller, M. (2007) Genetic unmasking of an epigenetically silenced microRNA in human cancer cells. *Cancer Res.*, **67**, 1424–1429.

16 Jenuwein, T., Allis, C.D. (2001) Translating the histone code. *Science*, **293**, 1074–1080.

17 Margueron, R., Trojer, P., Reinberg, D. (2005) The key to development: interpreting the histone code? *Curr. Opin. Genet. Dev.*, **15**, 163–176.

18 Fraga, M.F., Ballestar, E., Villar-Garea, A., Boix-Chornet, M., Espada, J., Schotta, G., Bonaldi, T., Haydon, C., Ropero, S., Petrie, K., Iyer, N.G., Perez-Rosado, A., Calvo, E., Lopez, J.A., Cano, A., Calasanz, M.J., Colomer, D., Piris, M.A., Ahn, N., Imhof, A., Caldas, C., Jenuwein, T., Esteller, M. (2005) Loss of acetylation at Lys16 and trimethylation at Lys20 of histone H4 is a common hallmark of human cancer. *Nat. Genet.*, **37**, 391–400.

19 Nguyen, C.T., Weisenberger, D.J., Velicescu, M., Gonzales, F.A., Lin, J.C., Liang, G., Jones, P.A. (2002) Histone H3-lysine 9 methylation is associated with aberrant gene silencing in cancer cells and is rapidly reversed by 5-aza-2′-deoxycytidine. *Cancer Res.*, **62**, 6456–6461.

20 Seligson, D.B., Horvath, S., Shi, T., Yu, H., Tze, S., Grunstein, M., Kurdistani, S.K. (2005) Global histone modification patterns predict risk of prostate cancer recurrence. *Nature*, **435**, 1262–1266.

21 Marks, P., Rifkind, R.A., Richon, V.M., Breslow, R., Miller, T., Kelly, W.K. (2001) Histone deacetylases and cancer: causes and therapies. *Nat. Rev. Cancer*, **1**, 194–202.

22 Johnstone, R.W. (2002) Histone-deacetylase inhibitors: novel drugs for the treatment of cancer. *Nat. Rev. Drug Discov.*, **1**, 287–299.

23 Cedar, H., Bergman, Y. (2009) Linking DNA methylation and histone modification: patterns and paradigms. *Nat. Rev. Genet.*, **10**, 295–304.

24 Turner, B.M. (2007) Defining an epigenetic code. *Nat. Cell Biol.*, **9**, 2–6.

25 Jones, P.L., Veenstra, G.J., Wade, P.A., Vermaak, D., Kass, S.U., Landsberger, N., Strouboulis, J., Wolffe, A.P. (1998) Methylated DNA and MeCP2 recruit histone deacetylase to repress transcription. *Nat. Genet.*, **19**, 187–191.

26 Rocchi, P., Tonelli, R., Camerin, C., Purgato, S., Fronza, R., Bianucci, F., Guerra, F., Pession, A., Ferreri, A.M. (2005) p21Waf1/Cip1 is a common target induced by short-chain fatty acid HDAC inhibitors (valproic acid, tributyrin and sodium butyrate) in neuroblastoma cells. *Oncol. Rep.*, **13**, 1139–1144.

27 Shetty, S., Graham, B.A., Brown, J.G., Hu, X., Vegh-Yarema, N., Harding, G., Paul, J.T., Gibson, S.B. (2005) Transcription factor NF-kappaB differentially regulates death receptor 5 expression involving histone deacetylase 1. *Mol. Cell. Biol.*, **25**, 5404–5416.

28 Dai, Y., Rahmani, M., Dent, P., Grant, S. (2005) Blockade of histone deacetylase inhibitor-induced RelA/p65 acetylation and NF-kappaB activation potentiates apoptosis in leukemia cells through a process mediated by oxidative damage, XIAP down-regulation, and c-Jun N-terminal kinase 1 activation. *Mol. Cell. Biol.*, **25**, 5429–5444.

29 Duan, H., Heckman, C.A., Boxer, L.M. (2005) Histone deacetylase inhibitors down-regulate bcl-2 expression and induce apoptosis in t(14;18) lymphomas. *Mol. Cell. Biol.*, **25**, 1608–1619.

30 Michaelis, M., Michaelis, U.R., Fleming, I., Suhan, T., Cinatl, J., Blaheta, R.A.,

Hoffmann, K., Kotchetkov, R., Busse, R., Nau, H., Cinatl, J., Jr (2004) Valproic acid inhibits angiogenesis in vitro and in vivo. *Mol. Pharmacol.*, **65**, 520–527.

31 Joseph, J., Mudduluru, G., Antony, S., Vashistha, S., Ajitkumar, P., Somasundaram, K. (2004) Expression profiling of sodium butyrate (NaB)-treated cells: identification of regulation of genes related to cytokine signaling and cancer metastasis by NaB. *Oncogene*, **23**, 6304–6315.

32 Suzuki, H., Gabrielson, E., Chen, W., Anbazhagan, R., van Engeland, M., Weijenberg, M.P., Herman, J.G., Baylin, S.B. (2002) A genomic screen for genes upregulated by demethylation and histone deacetylase inhibition in human colorectal cancer. *Nat. Genet.*, **31**, 141–149.

33 Hellebrekers, D.M., Jair, K.W., Vire, E., Eguchi, S., Hoebers, N.T., Fraga, M.F., Esteller, M., Fuks, F., Baylin, S.B., van Engeland, M., Griffioen, A.W. (2006) Angiostatic activity of DNA methyltransferase inhibitors. *Mol. Cancer Ther.*, **5**, 467–475.

34 Hackanson, B., Robbel, C., Wijermans, P., Lubbert, M. (2005) In vivo effects of decitabine in myelodysplasia and acute myeloid leukemia: review of cytogenetic and molecular studies. *Ann. Hematol.*, **84** (Suppl. 1), 32–38.

35 Silverman, L.R., Mufti, G.J. (2005) Methylation inhibitor therapy in the treatment of myelodysplastic syndrome. *Nat. Clin. Pract. Oncol.*, **2** (Suppl. 1), S12–S23.

36 Klose, R.J., Bird, A.P. (2006) Genomic DNA methylation: the mark and its mediators. *Trends Biochem. Sci.*, **31**, 89–97.

37 Goll, M.G., Bestor, T.H. (2005) Eukaryotic cytosine methyltransferases. *Annu. Rev. Biochem.*, **74**, 481–514.

38 Momparler, R.L., Bouffard, D.Y., Momparler, L.F., Dionne, J., Belanger, K., Ayoub, J. (1997) Pilot phase I-II study on 5-aza-2′-deoxycytidine (Decitabine) in patients with metastatic lung cancer. *Anticancer Drugs*, **8**, 358–368.

39 Schrump, D.S., Fischette, M.R., Nguyen, D.M., Zhao, M., Li, X., Kunst, T.F., Hancox, A., Hong, J.A., Chen, G.A., Pishchik, V., Figg, W.D., Murgo, A.J., Steinberg, S.M. (2006) Phase I study of decitabine-mediated gene expression in patients with cancers involving the lungs, esophagus, or pleura. *Clin. Cancer Res.*, **12**, 5777–5785.

40 Cheng, J.C., Matsen, C.B., Gonzales, F.A., Ye, W., Greer, S., Marquez, V.E., Jones, P.A., Selker, E.U. (2003) Inhibition of DNA methylation and reactivation of silenced genes by zebularine. *J. Natl Cancer Inst.*, **95**, 399–409.

41 Shaker, S., Bernstein, M., Momparler, L.F., Momparler, R.L. (2003) Preclinical evaluation of antineoplastic activity of inhibitors of DNA methylation (5-aza-2′-deoxycytidine) and histone deacetylation (trichostatin A, depsipeptide) in combination against myeloid leukemic cells. *Leuk. Res.*, **27**, 437–444.

42 Cameron, E.E., Bachman, K.E., Myohanen, S., Herman, J.G., Baylin, S.B. (1999) Synergy of demethylation and histone deacetylase inhibition in the re-expression of genes silenced in cancer. *Nat. Genet.*, **21**, 103–107.

43 Khan, O., La Thangue, N.B. (2008) Drug Insight: histone deacetylase inhibitor-based therapies for cutaneous T-cell lymphomas. *Nat. Clin. Pract. Oncol.*, **5**, 714–726.

44 Gilbert, J., Baker, S.D., Bowling, M.K., Grochow, L., Figg, W.D., Zabelina, Y., Donehower, R.C., Carducci, M.A. (2001) A phase I dose escalation and bioavailability study of oral sodium phenylbutyrate in patients with refractory solid tumor malignancies. *Clin. Cancer Res.*, **7**, 2292–2300.

45 Silverman, L.R., McKenzie, D.R., Peterson, B.L., Holland, J.F., Backstrom, J.T., Beach, C.L., Larson, R.A. (2006) Further analysis of trials with azacitidine in patients with myelodysplastic syndrome: studies 8421, 8921, and 9221 by the Cancer and Leukemia Group B. *J. Clin. Oncol.*, **24**, 3895–3903.

46 Fenaux, P., Mufti, G.J., Hellstrom-Lindberg, E., Santini, V., Finelli, C., Giagounidis, A., Schoch, R., Gattermann, N., Sanz, G., List, A., Gore, S.D., Seymour, J.F., Bennett, J.M., Byrd, J., Backstrom, J., Zimmerman, L., McKenzie, D., Beach, C., Silverman, L.R. (2009) Efficacy of azacitidine compared with that of conventional care regimens in the treatment of higher-risk myelodysplastic syndromes: a randomised, open-label, phase III study. *Lancet Oncol.*, **10**, 223–232.

47 Kantarjian, H.M., O'Brien, S., Shan, J., Aribi, A., Garcia-Manero, G., Jabbour, E., Ravandi, F., Cortes, J., Davisson, J., Issa, J.P. (2007) Update of the decitabine experience in higher risk myelodysplastic syndrome and analysis of prognostic factors associated with outcome. *Cancer*, **109**, 265–273.

48 Wilhelm, M., O'Brien, S., Rios, M.B., Estey, E., Keating, M.J., Plunkett, W., Sorenson, M., Kantarjian, H.M. (1999) Phase I study of arabinosyl-5-azacytidine (fazarabine) in adult acute leukemia and chronic myelogenous leukemia in blastic phase. *Leuk. Lymphoma*, **34**, 511–518.

49 Manetta, A., Blessing, J.A., Mann, W.J., Smith, D.M. (1995) A phase II study of fazarabine (NSC 281272) in patients with advanced squamous cell carcinoma of the cervix. A Gynecologic Oncology Group study. *Am. J. Clin. Oncol.*, **18**, 439–440.

50 Winquist, E., Knox, J., Ayoub, J.P., Wood, L., Wainman, N., Reid, G.K., Pearce, L., Shah, A., Eisenhauer, E. (2006) Phase II trial of DNA methyltransferase 1 inhibition with the antisense oligonucleotide MG98 in patients with metastatic renal carcinoma: a National Cancer Institute of Canada Clinical Trials Group investigational new drug study. *Invest. New Drugs*, **24**, 159–167.

51 Gore, S.D., Weng, L.J., Zhai, S., Figg, W.D., Donehower, R.C., Dover, G.J., Grever, M., Griffin, C.A., Grochow, L.B., Rowinsky, E.K., Zabalena, Y., Hawkins, A.L., Burks, K., Miller, C.B. (2001) Impact of the putative differentiating agent sodium phenylbutyrate on myelodysplastic syndromes and acute myeloid leukemia. *Clin. Cancer Res.*, **7**, 2330–2339.

52 Garcia-Manero, G., Yang, H., Bueso-Ramos, C., Ferrajoli, A., Cortes, J., Wierda, W.G., Faderl, S., Koller, C., Morris, G., Rosner, G., Loboda, A., Fantin, V.R., Randolph, S.S., Hardwick, J.S., Reilly, J.F., Chen, C., Ricker, J.L., Secrist, J.P., Richon, V.M., Frankel, S.R., Kantarjian, H.M. (2008) Phase 1 study of the histone deacetylase inhibitor vorinostat (suberoylanilide hydroxamic acid [SAHA]) in patients with advanced leukemias and myelodysplastic syndromes. *Blood*, **111**, 1060–1066.

53 Kelly, W.K., O'Connor, O.A., Krug, L.M., Chiao, J.H., Heaney, M., Curley, T., MacGregore-Cortelli, B., Tong, W., Secrist, J.P., Schwartz, L., Richardson, S., Chu, E., Olgac, S., Marks, P.A., Scher, H., Richon, V.M. (2005) Phase I study of an oral histone deacetylase inhibitor, suberoylanilide hydroxamic acid, in patients with advanced cancer. *J. Clin. Oncol.*, **23**, 3923–3931.

54 Chavez-Blanco, A., Segura-Pacheco, B., Perez-Cardenas, E., Taja-Chayeb, L., Cetina, L., Candelaria, M., Cantu, D., Gonzalez-Fierro, A., Garcia-Lopez, P., Zambrano, P., Perez-Plasencia, C., Cabrera, G., Trejo-Becerril, C., Angeles, E., Duenas-Gonzalez, A. (2005) Histone acetylation and histone deacetylase activity of magnesium valproate in tumor and peripheral blood of patients with cervical cancer. A phase I study. *Mol. Cancer*, **4**, 22.

55 Raffoux, E., Chaibi, P., Dombret, H., Degos, L. (2005) Valproic acid and all-trans retinoic acid for the treatment of elderly patients with acute myeloid leukemia. *Haematologica*, **90**, 986–988.

56 Glaser, K.B. (2007) HDAC inhibitors: clinical update and mechanism-based potential. *Biochem. Pharmacol.*, **74**, 659–671.

57 Braiteh, F., Soriano, A.O., Garcia-Manero, G., Hong, D., Johnson, M.M., Silva Lde, P., Yang, H., Alexander, S., Wolff, J., Kurzrock, R. (2008) Phase I study of epigenetic modulation with 5-azacytidine and valproic acid in patients with advanced cancers. *Clin. Cancer Res.*, **14**, 6296–6301.

58 Lin, J., Gilbert, J., Rudek, M.A., Zwiebel, J.A., Gore, S., Jiemjit, A., Zhao, M., Baker, S.D., Ambinder, R.F., Herman, J.G., Donehower, R.C., Carducci, M.A. (2009) A phase I dose-finding study of 5-azacytidine in combination with sodium phenylbutyrate in patients with refractory solid tumors. *Clin. Cancer Res.*, **15**, 6241–6249.

59 Ramalingam, S.S., Maitland, M.L., Frankel, P., Argiris, A.E., Koczywas, M., Gitlitz, B., Thomas, S., Espinoza-Delgado, I., Vokes, E.E., Gandara, D.R., Belani, C.P. (2010) Carboplatin and paclitaxel in combination with either vorinostat or placebo for first-line therapy of advanced non-small-cell lung cancer. *J. Clin. Oncol.*, **28**, 56–62.

60 Munster, P.N., Marchion, D., Thomas, S., Egorin, M., Minton, S., Springett, G., Lee, J.H., Simon, G., Chiappori, A., Sullivan, D., Daud, A. (2009) Phase I trial of vorinostat and doxorubicin in solid tumours: histone

deacetylase 2 expression as a predictive marker. *Br. J. Cancer*, **101**, 1044–1050.

61 Smith, L.T., Otterson, G.A., Plass, C. (2007) Unraveling the epigenetic code of cancer for therapy. *Trends Genet.*, **23**, 449–456.

62 Reu, F.J., Bae, S.I., Cherkassky, L., Leaman, D.W., Lindner, D., Beaulieu, N., MacLeod, A.R., Borden, E.C. (2006) Overcoming resistance to interferon-induced apoptosis of renal carcinoma and melanoma cells by DNA demethylation. *J. Clin. Oncol.*, **24**, 3771–3779.

63 Yang, D., Thangaraju, M., Greeneltch, K., Browning, D.D., Schoenlein, P.V., Tamura, T., Ozato, K., Ganapathy, V., Abrams, S.I., Liu, K. (2007) Repression of IFN regulatory factor 8 by DNA methylation is a molecular determinant of apoptotic resistance and metastatic phenotype in metastatic tumor cells. *Cancer Res.*, **67**, 3301–3309.

64 Gollob, J.A., Sciambi, C.J., Peterson, B.L., Richmond, T., Thoreson, M., Moran, K., Dressman, H.K., Jelinek, J., Issa, J.P. (2006) Phase I trial of sequential low-dose 5-aza-2′-deoxycytidine plus high-dose intravenous bolus interleukin-2 in patients with melanoma or renal cell carcinoma. *Clin. Cancer Res.*, **12**, 4619–4627.

65 Munshi, A., Kurland, J.F., Nishikawa, T., Tanaka, T., Hobbs, M.L., Tucker, S.L., Ismail, S., Stevens, C., Meyn, R.E. (2005) Histone deacetylase inhibitors radiosensitize human melanoma cells by suppressing DNA repair activity. *Clin. Cancer Res.*, **11**, 4912–4922.

66 Munshi, A., Tanaka, T., Hobbs, M.L., Tucker, S.L., Richon, V.M., Meyn, R.E. (2006) Vorinostat, a histone deacetylase inhibitor, enhances the response of human tumor cells to ionizing radiation through prolongation of gamma-H2AX foci. *Mol. Cancer Ther.*, **5**, 1967–1974.

67 Eden, A., Gaudet, F., Waghmare, A., Jaenisch, R. (2003) Chromosomal instability and tumors promoted by DNA hypomethylation. *Science*, **300**, 455.

68 Yang, A.S., Estecio, M.R., Garcia-Manero, G., Kantarjian, H.M., Issa, J.P. (2003) Comment on "Chromosomal instability and tumors promoted by DNA hypomethylation" and "Induction of tumors in nice by genomic hypomethylation". *Science*, **302**, 1153; author reply 1153.

69 Jackson-Grusby, L., Laird, P.W., Magge, S.N., Moeller, B.J., Jaenisch, R. (1997) Mutagenicity of 5-aza-2′-deoxycytidine is mediated by the mammalian DNA methyltransferase. *Proc. Natl Acad. Sci. USA*, **94**, 4681–4685.

70 Carr, B.I., Rahbar, S., Asmeron, Y., Riggs, A., Winberg, C.D. (1988) Carcinogenicity and haemoglobin synthesis induction by cytidine analogues. *Br. J. Cancer*, **57**, 395–402.

71 Karpf, A.R., Peterson, P.W., Rawlins, J.T., Dalley, B.K., Yang, Q., Albertsen, H., Jones, D.A. (1999) Inhibition of DNA methyltransferase stimulates the expression of signal transducer and activator of transcription 1, 2, and 3 genes in colon tumor cells. *Proc. Natl Acad. Sci. USA*, **96**, 14007–14012.

72 Liang, G., Gonzales, F.A., Jones, P.A., Orntoft, T.F., Thykjaer, T. (2002) Analysis of gene induction in human fibroblasts and bladder cancer cells exposed to the methylation inhibitor 5-aza-2′-deoxycytidine. *Cancer Res.*, **62**, 961–966.

73 Yoo, C.B., Jones, P.A. (2006) Epigenetic therapy of cancer: past, present and future. *Nat. Rev. Drug Discov.*, **5**, 37–50.

74 Jones, P.A., Baylin, S.B. (2002) The fundamental role of epigenetic events in cancer. *Nat. Rev. Genet.*, **3**, 415–428.

75 Moore, M., Ullman, C. (2003) Recent developments in the engineering of zinc finger proteins. *Briefings Funct. Genomic. Proteomic.*, **1**, 342–355.

76 Beltran, A., Parikh, S., Liu, Y., Cuevas, B.D., Johnson, G.L., Futscher, B.W., Blancafort, P. (2007) Re-activation of a dormant tumor suppressor gene maspin by designed transcription factors. *Oncogene*, **26**, 2791–2798.

77 Szyf, M. (2004) Toward a discipline of pharmacoepigenomics. *Curr. Pharmacogenomics*, **2**, 357–377.

78 Geiman, T.M., Robertson, K.D. (2002) Chromatin remodeling, histone modifications, and DNA methylation – how does it all fit together? *J. Cell. Biochem.*, **87**, 117–125.

79 Esteller, M., Corn, P.G., Urena, J.M., Gabrielson, E., Baylin, S.B., Herman, J.G. (1998) Inactivation of glutathione *S*-transferase P1 gene by promoter hypermethylation in human neoplasia. *Cancer Res.*, **58**, 4515–4518.

80 Jeronimo, C., Usadel, H., Henrique, R., Oliveira, J., Lopes, C., Nelson, W.G., Sidransky, D. (2001) Quantitation of GSTP1 methylation in non-neoplastic prostatic tissue and organ-confined prostate adenocarcinoma. *J. Natl Cancer Inst.*, **93**, 1747–1752.
81 Cairns, P., Esteller, M., Herman, J.G., Schoenberg, M., Jeronimo, C., Sanchez-Cespedes, M., Chow, N.H., Grasso, M., Wu, L., Westra, W.B., Sidransky, D. (2001) Molecular detection of prostate cancer in urine by GSTP1 hypermethylation. *Clin. Cancer Res.*, **7**, 2727–2730.
82 Fackler, M.J., Malone, K., Zhang, Z., Schilling, E., Garrett-Mayer, E., Swift-Scanlan, T., Lange, J., Nayar, R., Davidson, N.E., Khan, S.A., Sukumar, S. (2006) Quantitative multiplex methylation-specific PCR analysis doubles detection of tumor cells in breast ductal fluid. *Clin. Cancer Res.*, **12**, 3306–3310.
83 Belinsky, S.A., Nikula, K.J., Palmisano, W.A., Michels, R., Saccomanno, G., Gabrielson, E., Baylin, S.B., Herman, J.G. (1998) Aberrant methylation of p16(INK4a) is an early event in lung cancer and a potential biomarker for early diagnosis. *Proc. Natl Acad. Sci. USA*, **95**, 11891–11896.
84 Melotte, V., Lentjes, M.H., van den Bosch, S.M., Hellebrekers, D.M., de Hoon, J.P., Wouters, K.A., Daenen, K.L., Partouns-Hendriks, I.E., Stessels, F., Louwagie, J., Smits, K.M., Weijenberg, M.P., Sanduleanu, S., Khalid-de Bakker, C.A., Oort, F.A., Meijer, G.A., Jonkers, D.M., Herman, J.G., de Bruine, A.P., van Engeland, M. (2009) N-Myc downstream-regulated gene 4 (NDRG4): a candidate tumor suppressor gene and potential biomarker for colorectal cancer. *J. Natl Cancer Inst.*, **101**, 916–927.
85 Zou, H., Harrington, J.J., Shire, A.M., Rego, R.L., Wang, L., Campbell, M.E., Oberg, A.L., Ahlquist, D.A. (2007) Highly methylated genes in colorectal neoplasia: implications for screening. *Cancer Epidemiol. Biomarkers Prev.*, **16**, 2686–2696.
86 de Vos, T., Tetzner, R., Model, F., Weiss, G., Schuster, M., Distler, J., Steiger, K.V., Grutzmann, R., Pilarsky, C., Habermann, J.K., Fleshner, P.R., Oubre, B.M., Day, R., Sledziewski, A.Z., Lofton-Day, C. (2009) Circulating methylated SEPT9 DNA in plasma is a biomarker for colorectal cancer. *Clin. Chem.*, **55**, 1337–1346.
87 Li, M., Chen, W.D., Papadopoulos, N., Goodman, S.N., Bjerregaard, N.C., Laurberg, S., Levin, B., Juhl, H., Arber, N., Moinova, H., Durkee, K., Schmidt, K., He, Y., Diehl, F., Velculescu, V.E., Zhou, S., Diaz, L.A., Jr, Kinzler, K.W., Markowitz, S.D., Vogelstein, B. (2009) Sensitive digital quantification of DNA methylation in clinical samples. *Nat. Biotechnol.*, **27**, 858–863.
88 Sotiriou, C., Pusztai, L. (2009) Gene-expression signatures in breast cancer. *N. Engl. J. Med.*, **360**, 790–800.
89 Pharoah, P.D., Day, N.E., Caldas, C. (1999) Somatic mutations in the p53 gene and prognosis in breast cancer: a meta-analysis. *Br. J. Cancer*, **80**, 1968–1973.
90 Schmidt, M.K., Tollenaar, R.A., de Kemp, S.R., Broeks, A., Cornelisse, C.J., Smit, V.T., Peterse, J.L., van Leeuwen, F.E., Van't, Veer, L.J. (2007) Breast cancer survival and tumor characteristics in premenopausal women carrying the CHEK2*1100delC germline mutation. *J. Clin. Oncol.*, **25**, 64–69.
91 Muller, H.M., Widschwendter, A., Fiegl, H., Ivarsson, L., Goebel, G., Perkmann, E., Marth, C., Widschwendter, M. (2003) DNA methylation in serum of breast cancer patients: an independent prognostic marker. *Cancer Res.*, **63**, 7641–7645.
92 Wettergren, Y., Odin, E., Nilsson, S., Carlsson, G., Gustavsson, B. (2008) p16INK4a gene promoter hypermethylation in mucosa as a prognostic factor for patients with colorectal cancer. *Mol. Med.*, **14**, 412–421.
93 Esteller, M., Gonzalez, S., Risques, R.A., Marcuello, E., Mangues, R., Germa, J.R., Herman, J.G., Capella, G., Peinado, M.A. (2001) K-ras and p16 aberrations confer poor prognosis in human colorectal cancer. *J. Clin. Oncol.*, **19**, 299–304.
94 Muller, H.M., Fiegl, H., Widschwendter, A., Widschwendter, M. (2004) Prognostic DNA methylation marker in serum of cancer patients. *Ann. N. Y. Acad. Sci.*, **1022**, 44–49.
95 Veeck, J., Niederacher, D., An, H., Klopocki, E., Wiesmann, F., Betz, B.,

Galm, O., Camara, O., Durst, M., Kristiansen, G., Huszka, C., Knuchel, R., Dahl, E. (2006) Aberrant methylation of the Wnt antagonist SFRP1 in breast cancer is associated with unfavourable prognosis. *Oncogene*, **25**, 3479–3488.

96 Ellinger, J., Kahl, P., von der, Gathen, J., Rogenhofer, S., Heukamp, L.C., Gutgemann, I., Walter, B., Hofstadter, F., Buttner, R., Muller, S.C., Bastian, P.J., von Ruecker, A. (2009) Global levels of histone modifications predict prostate cancer recurrence. *Prostate*, **70**, 61–69.

97 Normanno, N., Tejpar, S., Morgillo, F., De Luca, A., Van Cutsem, E., Ciardiello, F. (2009) Implications for KRAS status and EGFR-targeted therapies in metastatic CRC. *Nat. Rev. Clin. Oncol.*, **6**, 519–527.

98 Esteller, M., Garcia-Foncillas, J., Andion, E., Goodman, S.N., Hidalgo, O.F., Vanaclocha, V., Baylin, S.B., Herman, J.G. (2000) Inactivation of the DNA-repair gene MGMT and the clinical response of gliomas to alkylating agents. *N. Engl. J. Med.*, **343**, 1350–1354.

99 Veeck, J., Ropero, S., Setien, F., Gonzalez-Suarez, E., Osorio, A., Benitez, J., Herman, J.G., Esteller, M. (2010) BRCA1 CpG island hypermethylation predicts sensitivity to poly(adenosine diphosphate)-ribose polymerase inhibitors. *J. Clin. Oncol.* **28**, e563–e564; author reply e565–e566.

100 Fong, P.C., Boss, D.S., Yap, T.A., Tutt, A., Wu, P., Mergui-Roelvink, M., Mortimer, P., Swaisland, H., Lau, A., O'Connor, M.J., Ashworth, A., Carmichael, J., Kaye, S.B., Schellens, J.H., de Bono, J.S. (2009) Inhibition of poly(ADP-ribose) polymerase in tumors from BRCA mutation carriers. *N. Engl. J. Med.*, **361**, 123–134.

101 Esteller, M., Gaidano, G., Goodman, S.N., Zagonel, V., Capello, D., Botto, B., Rossi, D., Gloghini, A., Vitolo, U., Carbone, A., Baylin, S.B., Herman, J.G. (2002) Hypermethylation of the DNA repair gene O(6)-methylguanine DNA methyltransferase and survival of patients with diffuse large B-cell lymphoma. *J. Natl Cancer Inst.*, **94**, 26–32.

102 Hegi, M.E., Diserens, A.C., Gorlia, T., Hamou, M.F., de Tribolet, N., Weller, M., Kros, J.M., Hainfellner, J.A., Mason, W., Mariani, L., Bromberg, J.E., Hau, P., Mirimanoff, R.O., Cairncross, J.G., Janzer, R.C., Stupp, R. (2005) MGMT gene silencing and benefit from temozolomide in glioblastoma. *N. Engl. J. Med.*, **352**, 997–1003.

103 Strathdee, G., MacKean, M.J., Illand, M., Brown, R. (1999) A role for methylation of the hMLH1 promoter in loss of hMLH1 expression and drug resistance in ovarian cancer. *Oncogene*, **18**, 2335–2341.

104 Ferreri, A.J., Dell'Oro, S., Capello, D., Ponzoni, M., Iuzzolino, P., Rossi, D., Pasini, F., Ambrosetti, A., Orvieto, E., Ferrarese, F., Arrigoni, G., Foppoli, M., Reni, M., Gaidano, G. (2004) Aberrant methylation in the promoter region of the reduced folate carrier gene is a potential mechanism of resistance to methotrexate in primary central nervous system lymphomas. *Br. J. Haematol.*, **126**, 657–664.

105 Agrelo, R., Cheng, W.H., Setien, F., Ropero, S., Espada, J., Fraga, M.F., Herranz, M., Paz, M.F., Sanchez-Cespedes, M., Artiga, M.J., Guerrero, D., Castells, A., von Kobbe, C., Bohr, V.A., Esteller, M. (2006) Epigenetic inactivation of the premature aging Werner syndrome gene in human cancer. *Proc. Natl Acad. Sci. USA*, **103**, 8822–8827.

106 Tapia, T., Smalley, S.V., Kohen, P., Munoz, A., Solis, L.M., Corvalan, A., Faundez, P., Devoto, L., Camus, M., Alvarez, M., Carvallo, P. (2008) Promoter hypermethylation of BRCA1 correlates with absence of expression in hereditary breast cancer tumors. *Epigenetics*, **3**, 157–163.

107 Lee, M.N., Tseng, R.C., Hsu, H.S., Chen, J.Y., Tzao, C., Ho, W.L., Wang, Y.C. (2007) Epigenetic inactivation of the chromosomal stability control genes BRCA1, BRCA2, and XRCC5 in non-small cell lung cancer. *Clin. Cancer Res.*, **13**, 832–838.

108 Birgisdottir, V., Stefansson, O.A., Bodvarsdottir, S.K., Hilmarsdottir, H., Jonasson, J.G., Eyfjord, J.E. (2006) Epigenetic silencing and deletion of the BRCA1 gene in sporadic breast cancer. *Breast Cancer Res.*, **8**, R38.

109 Hanahan, D., Weinberg, R.A. (2000) The hallmarks of cancer. *Cell*, **100**, 57–70.

110 Hu, M., Yao, J., Cai, L., Bachman, K.E., van den Brule, F., Velculescu, V., Polyak, K. (2005) Distinct epigenetic changes in the stromal cells of breast cancers. *Nat. Genet.*, **37**, 899–905.

111 Hanson, J.A., Gillespie, J.W., Grover, A., Tangrea, M.A., Chuaqui, R.F., Emmert-Buck, M.R., Tangrea, J.A., Libutti, S.K., Linehan, W.M., Woodson, K.G. (2006) Gene promoter methylation in prostate tumor-associated stromal cells. *J. Natl Cancer Inst.*, **98**, 255–261.

112 Chung, I., Karpf, A.R., Muindi, J.R., Conroy, J.M., Nowak, N.J., Johnson, C.S., Trump, D.L. (2007) Epigenetic silencing of CYP24 in tumor-derived endothelial cells contributes to selective growth inhibition by calcitriol. *J. Biol. Chem.*, **282**, 8704–8714.

113 Chang, Y.C., Chen, T.C., Lee, C.T., Yang, C.Y., Wang, H.W., Wang, C.C., Hsieh, S.L. (2008) Epigenetic control of MHC class II expression in tumor-associated macrophages by decoy receptor 3. *Blood*, **111**, 5054–5063.

114 Johnson, A.B., Barton, M.C. (2007) Hypoxia-induced and stress-specific changes in chromatin structure and function. *Mutat. Res.*, **618**, 149–162.

115 Johnson, A.B., Denko, N., Barton, M.C. (2008) Hypoxia induces a novel signature of chromatin modifications and global repression of transcription. *Mutat. Res.*, **640**, 174–179.

116 Carmeliet, P. (2005) Angiogenesis in life, disease and medicine. *Nature*, **438**, 932–936.

117 Kerbel, R.S. (2008) Tumor angiogenesis. *N. Engl. J. Med.*, **358**, 2039–2049.

118 Buysschaert, I., Schmidt, T., Roncal, C., Carmeliet, P., Lambrechts, D. (2008) Genetics, epigenetics and pharmaco-(epi)genomics in angiogenesis. *J. Cell. Mol. Med.*, **12**, 2533–2551.

119 Lu, C., Han, H.D., Mangala, L.S., Ali-Fehmi, R., Newton, C.S., Ozbun, L., Armaiz-Pena, G.N., Hu, W., Stone, R.L., Munkarah, A., Ravoori, M.K., Shahzad, M.M., Lee, J.W., Mora, E., Langley, R.R., Carroll, A.R., Matsuo, K., Spannuth, W.A., Schmandt, R., Jennings, N.B., Goodman, B.W., Jaffe, R.B., Nick, A.M., Kim, H.S., Guven, E.O., Chen, Y.H., Li, L.Y., Hsu, M.C., Coleman, R.L., Calin, G.A., Denkbas, E.B., Lim, J.Y., Lee, J.S., Kundra, V., Birrer, M.J., Hung, M.C., Lopez-Berestein, G., Sood, A.K. (2010) Regulation of tumor angiogenesis by EZH2. *Cancer Cell* **18**, 185–197.

120 Li, Q., Ahuja, N., Burger, P.C., Issa, J.P. (1999) Methylation and silencing of the Thrombospondin-1 promoter in human cancer. *Oncogene*, **18**, 3284–3289.

121 Hinshelwood, R.A., Huschtscha, L.I., Melki, J., Stirzaker, C., Abdipranoto, A., Vissel, B., Ravasi, T., Wells, C.A., Hume, D.A., Reddel, R.R., Clark, S.J. (2007) Concordant epigenetic silencing of transforming growth factor-beta signaling pathway genes occurs early in breast carcinogenesis. *Cancer Res.*, **67**, 11517–11527.

122 Rojas, A., Meherem, S., Kim, Y.H., Washington, M.K., Willis, J.E., Markowitz, S.D., Grady, W.M. (2008) The aberrant methylation of TSP1 suppresses TGF-beta1 activation in colorectal cancer. *Int. J. Cancer*, **123**, 14–21.

123 Hu, C.J., Chen, S.D., Yang, D.I., Lin, T.N., Chen, C.M., Huang, T.H., Hsu, C.Y. (2006) Promoter region methylation and reduced expression of thrombospondin-1 after oxygen-glucose deprivation in murine cerebral endothelial cells. *J. Cereb. Blood Flow Metab.*, **26**, 1519–1526.

124 Deroanne, C.F., Bonjean, K., Servotte, S., Devy, L., Colige, A., Clausse, N., Blacher, S., Verdin, E., Foidart, J.M., Nusgens, B.V., Castronovo, V. (2002) Histone deacetylases inhibitors as anti-angiogenic agents altering vascular endothelial growth factor signaling. *Oncogene*, **21**, 427–436.

125 Chou, C.W., Chen, C.C. (2008) HDAC inhibition upregulates the expression of angiostatic ADAMTS1. *FEBS Lett.*, **582**, 4059–4065.

126 Rossig, L., Li, H., Fisslthaler, B., Urbich, C., Fleming, I., Forstermann, U., Zeiher, A.M., Dimmeler, S. (2002) Inhibitors of histone deacetylation downregulate the expression of endothelial nitric oxide synthase and compromise endothelial cell function in vasorelaxation and angiogenesis. *Circ. Res.*, **91**, 837–844.

127 Kim, M.S., Kwon, H.J., Lee, Y.M., Baek, J.H., Jang, J.E., Lee, S.W., Moon, E.J., Kim, H.S., Lee, S.K., Chung, H.Y., Kim, C.W., Kim, K.W. (2001) Histone deacetylases induce angiogenesis by negative regulation of tumor suppressor genes. *Nat. Med.*, **7**, 437–443.

128 Hellebrekers, D.M., Melotte, V., Vire, E., Langenkamp, E., Molema, G., Fuks,

F., Herman, J.G., Van Criekinge, W., Griffioen, A.W., van Engeland, M. (2007) Identification of epigenetically silenced genes in tumor endothelial cells. *Cancer Res.*, **67**, 4138–4148.

129 Rossig, L., Urbich, C., Bruhl, T., Dernbach, E., Heeschen, C., Chavakis, E., Sasaki, K., Aicher, D., Diehl, F., Seeger, F., Potente, M., Aicher, A., Zanetta, L., Dejana, E., Zeiher, A.M., Dimmeler, S. (2005) Histone deacetylase activity is essential for the expression of HoxA9 and for endothelial commitment of progenitor cells. *J. Exp. Med.*, **201**, 1825–1835.

130 Young, P.P., Vaughan, D.E., Hatzopoulos, A.K. (2007) Biologic properties of endothelial progenitor cells and their potential for cell therapy. *Prog. Cardiovasc. Dis.*, **49**, 421–429.

131 Baylin, S.B. (2004) Reversal of gene silencing as a therapeutic target for cancer-roles for DNA methylation and its interdigitation with chromatin. *Novartis Found. Symp.* **259**, 226–233; discussion 234–237, 285–288.

132 Hellebrekers, D.M., Castermans, K., Vire, E., Dings, R.P., Hoebers, N.T., Mayo, K.H., Oude Egbrink, M.G., Molema, G., Fuks, F., van Engeland, M., Griffioen, A.W. (2006) Epigenetic regulation of tumor endothelial cell anergy: silencing of intercellular adhesion molecule-1 by histone modifications. *Cancer Res.*, **66**, 10770–10777.

133 Rini, B.I., Jaeger, E., Weinberg, V., Sein, N., Chew, K., Fong, K., Simko, J., Small, E.J., Waldman, F.M. (2006) Clinical response to therapy targeted at vascular endothelial growth factor in metastatic renal cell carcinoma: impact of patient characteristics and Von Hippel–Lindau gene status. *Br. J. Urol. Int.*, **98**, 756–762.

134 Choueiri, T.K., Vaziri, S.A., Jaeger, E., Elson, P., Wood, L., Bhalla, I.P., Small, E.J., Weinberg, V., Sein, N., Simko, J., Golshayan, A.R., Sercia, L., Zhou, M., Waldman, F.M., Rini, B.I., Bukowski, R.M., Ganapathi, R. (2008) von Hippel-Lindau gene status and response to vascular endothelial growth factor targeted therapy for metastatic clear cell renal cell carcinoma. *J. Urol.* **180**, 860–865; discussion 865–866.

135 Verbisck, N.V., Costa, E.T., Costa, F.F., Cavalher, F.P., Costa, M.D., Muras, A., Paixao, V.A., Moura, R., Granato, M.F., Ierardi, D.F., Machado, T., Melo, F., Ribeiro, K.B., Cunha, I.W., Lima, V.C., Maciel Mdo, S., Carvalho, A.L., Soares, F.F., Zanata, S., Sogayar, M.C., Chammas, R., Camargo, A.A. (2009) ADAM23 negatively modulates alpha(v)beta(3) integrin activation during metastasis. *Cancer Res.*, **69**, 5546–5552.

Part V
Model Organisms

30
Parental Genomic Imprinting in Flowering Plants

Frédéric Berger
Temasek Life Sciences Laboratory (TLL), 1 Research Link, Singapore 117604, Singapore

Epigenetic Regulation and Epigenomics: Advances in Molecular Biology and Medicine, First Edition. Edited by Robert A. Meyers.

Keywords

Parental genomic imprinting
An epigenetic mode of regulation causing preferential gene expression from one of the two parental alleles.

Egg cell
The "true" female gamete that initiates the embryo lineage after double fertilization.

Central cell
The "accessory" female gamete that initiates the endosperm lineage after double fertilization.

Double fertilization
Sexual reproduction in flowering plants requires two parallel fertilizations of the egg cell and the central cell by two sperm cells delivered by the pollen tube.

Endosperm
The embryo-nurturing annex that develops from the fertilized central cell.

Gametophytes
Plant haploid structures that produce gametes. In plants meiosis produces haploid spores that develop as haploid gametophytes. In flowering plants the gametophytes are reduced to a few cells.

Sporophyte
The diploid vegetative generation producing spores. The sporophyte comprises all the vegetative and reproductive plant tissues from the flowering plants.

Pollen
The male gametophyte that contains at maturity two sperm cells and a vegetative cell. The vegetative cell elongates the pollen tube that delivers the two sperm cells to the embryo sac.

Embryo sac
The female gametophyte that contains the egg cell and the central cell.

Ovule integuments
Cell layers from the diploid mother plant, which protect the female gametophyte.

In contrast to most genes, that are expressed equally from both parental alleles, imprinted genes (as identified in flowering plants and mammals) are differentially expressed, depending on their parental origin. In flowering plants, imprinting is regulated by DNA methylation and histone methylation. During vegetative development most imprinted genes are silenced by chromatin modifications. During gametogenesis, however, the male or female allele is activated by the removal of chromatin modifications and remains active after fertilization. The other allele is inherited in a silenced state, leading to an imprinted gene expression. Imprinting mechanisms are conserved across plant species, and to a certain extent there is evidence of a convergent evolution of imprinting mechanisms between plants and mammals. The physiological significance and evolutionary origin of imprinting are still unclear. In flowering plants, imprinting may derive from global epigenetic reprogramming mechanisms that occur in female, but not in male, gametes.

1 General Context of Parental Genomic Imprinting in Plants

1.1 Sexual Reproduction in Flowering Plants: An Overview

In plants, the male and female gametes are produced after meiosis following a series of divisions of a haploid male or female spore (Fig. 1). In flowering plants, male gametogenesis takes place in stamens, leading to the production of pollen containing sperm cells [1] (Fig. 1), while female gametogenesis takes place within the diploid tissues of the ovule [2]. Meiosis results in the production of a haploid megaspore which undergoes a series of three syncytial divisions, followed by cellularization to produce an embryo sac which contains the haploid female gamete or egg cell, and the central cell (Fig. 1).

Plant reproduction is characterized by a double fertilization, when two sperm cells are delivered by the pollen to the egg cell and the central cell. Fertilization of the egg cell leads to embryogenesis (Fig. 2), while the second sperm cell activates division of the central cell, leading to production of the endosperm which develops around the embryo and allows the transfer of maternal nutrients and provides physical protection to the embryo (Fig. 3). In certain species, the endosperm stores reserves in the form of starch, proteins and lipids; in the case of rice and other cereals the endosperm constitutes the edible part of the seed (Fig. 3). In most plant species, the central cell inherits two haploid nuclei from the syncytial gametophyte; thus, the endosperm genome contains two doses of the maternal genome and one dose of the paternal genome. It was this specific parental genomic dosage that attracted much interest during the early

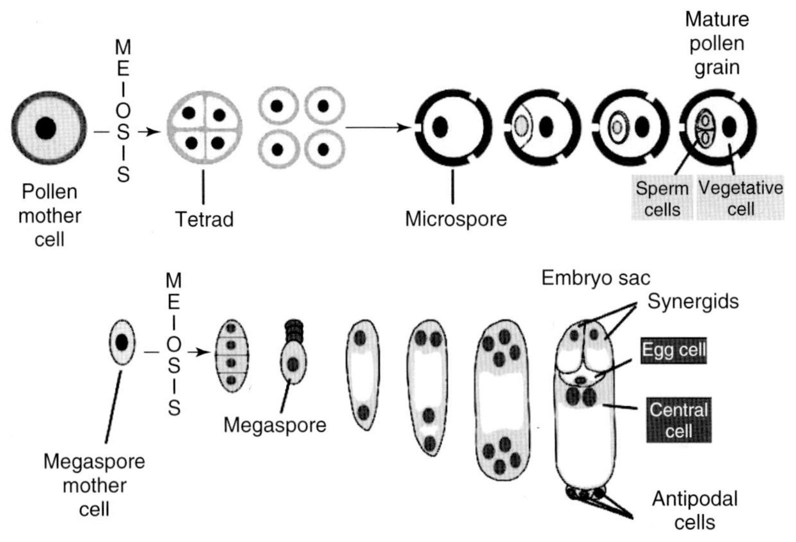

Fig. 1 Flowering plant gametogenesis. Male meiosis produces a tetrad of four haploid microspores, each of which experiences two mitotic divisions to produce a pollen grain comprising a large vegetative cell and two identical haploid sperm cells. Female gametogenesis is also initiated by meiosis in the megaspore mother cell. The megaspore, which is the only surviving meiotic product, experiences three syncytial nuclei divisions, leading to an eight-celled female gametophyte. Cellularization leads to the production of seven cells: three antipodals of unknown function; two synergids, which attract the pollen tube; the egg cell; and the central cell, which contains two nuclei. These seven cells constitute the embryo sac.

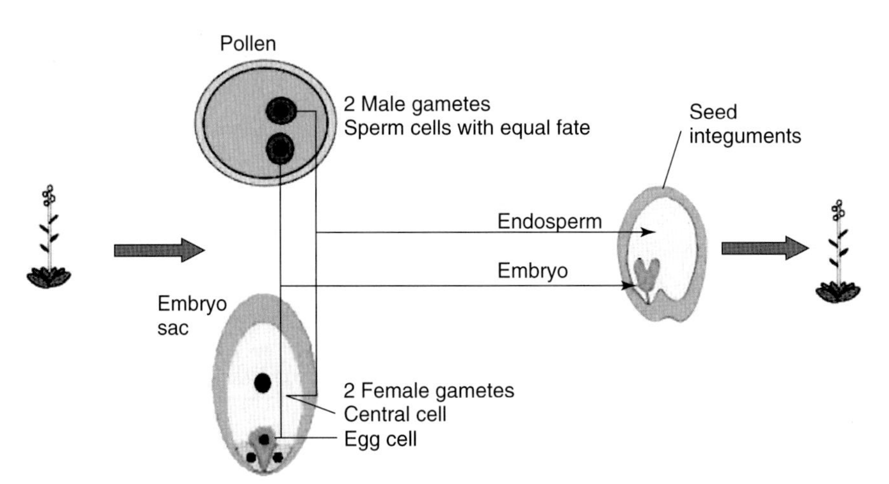

Fig. 2 Double-fertilization in flowering plants. The double-fertilization consists of the parallel fusions of one sperm with the egg cell, and one sperm with the central cell, initiating embryogenesis and endosperm development, respectively. The embryo and the endosperm develop inside the seed integuments.

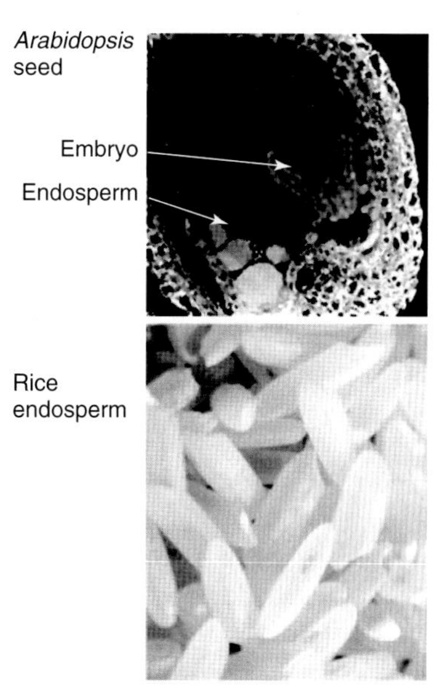

Fig. 3 Developmental features of flowering plant seeds. A confocal section of an *Arabidopsis* seed (at six days after fertilization), showing the embryo surrounded by the endosperm. In *Arabidopsis* the endosperm does not store the seed reserves, in contrast to rice endosperm. At maturity, the endosperm represents the major component of the rice seed, most commonly known as the *edible rice grain*.

studies of plant reproduction, and which led to the discovery of imprinting in plants.

1.2 Historical Discovery of Imprinting

1.2.1 First Reports of Imprinting in Plants

The term *imprinting* was originally adopted to qualify the differential elimination of paternal chromosomes in the mealybug *Sciara* [3]. The first example of the imprinted expression of a gene was identified during studies of the pigmentation of the endosperm outer layers in maize [4], when an irregular anthocyanin pigmentation was linked to certain alleles of the *R* gene and conferred only when the mutation was maternally inherited. The proposal that expression of the *r* allele depended on its parental origin was first made by Kermicle, who subsequently showed that the pigmentation defects associated with the *r* mutation did not depend on the gene dosage [4]. Hence, imprinting was first described in plants several years before the concept was formulated in mice, when it was found that certain chromosomal regions could lead to developmental abnormalities when both copies were exclusively maternally or paternally derived [5]. However, the parent of origin expression in maize was observed only for certain *r* mutant alleles and not others, and the mechanism which causes *r* allele-dependent imprinting currently remains unknown.

1.2.2 The Impact of Interploid Crosses on Imprinting Discovery

Studies of the seeds which develop from crosses between plants with different ploidy have also provided evidence for parental genomic imprinting. As early as the mid-twentieth century, crosses between tetraploid and diploid plants were shown to result in seed abortion due to endosperm failure [6, 7]. It was shown later in maize that a critical maternal : paternal genome dosage in the endosperm was required for seed survival [8], and these experiments were subsequently repeated in *Arabidopsis*, using tetraploid and hexaploid plant lines [9]. An increased paternal contribution caused endosperm enlargement, whereas an increased maternal dosage had the opposite effect. These results could be explained by a differential expression of the paternal alleles and maternal alleles of certain genes which were important for endosperm development. The results of these experiments led to the hypothesis that two sets of maternally and paternally expressed imprinted genes could control endosperm development.

1.2.3 Discovery of the First Imprinted Gene in Arabidopsis

The first imprinted gene in *Arabidopsis* – *MEDEA* (*MEA*) – was identified more than a decade ago, following investigations into the mutations that caused a maternal gametophytic effect [10–12]. In backcrosses with the wild-type plant, the loss-of-function mutant allele *medea* was shown to cause seed abortion only when it was inherited from the mother. The fact

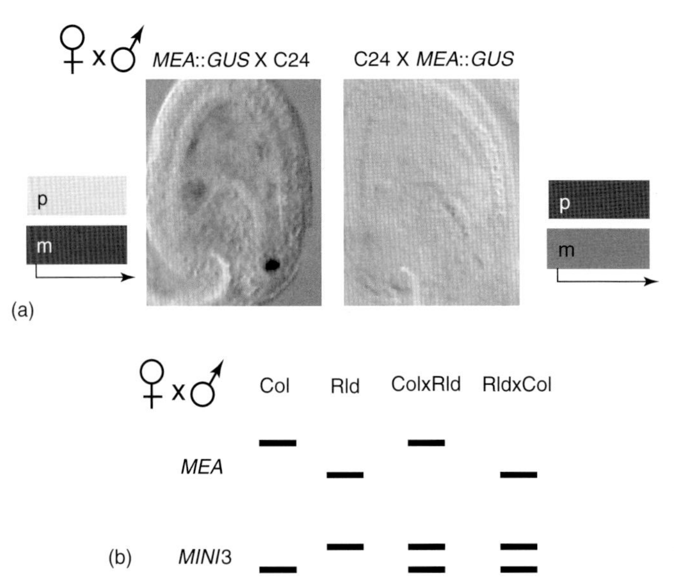

Fig. 4 Analysis of *MEDEA* imprinting. (a) Expression of *MEA::GUS* reporter construct in crosses, which involve the transgenic line carrying the reporter construct as a mother or as a father. *MEA::GUS* is expressed only from the maternal allele (m, box with arrow), while the paternal allele (p) remains silent; (b) Schematic of the analysis of the imprinted expression of *MEA* endogenous locus. A sequence polymorphism is used to distinguish *MEA* mRNA and *MINI3* mRNA between the wild-type strains Col and Rld. Seeds resulting from crosses between the two parents express *MINI3* mRNAs from both parental alleles, and two bands are detected. In contrast, *MEA* mRNA originates only from the maternal allele and a single band is detected.

that 50% of seeds aborted suggested that the maternal effect had originated from a defect in the female gametes [13, 14]. However, it was then shown that *MEA* was actively transcribed after fertilization [14, 15], and that the defects observed in seeds inheriting the *mea* mutant allele could not be explained by defects occurring in the female gametes. The fact that *MEA* is transcribed from the maternal allele in the seed, and not from the paternal allele, was demonstrated by using a polymorphism between two wild-type strains of *Arabidopsis* (Fig. 4) [13]. The maternal expression of *MEA* was further confirmed using transcriptional reporters expressed under the control of the *MEA* promoter (Fig. 4) [13, 16, 17].

2 Imprinted Genes and Their Function

Whereas, in *Arabidopsis*, the endosperm is the only tissue where imprinted gene expression has been identified [18, 19], in both rice [20] and maize [21] the imprinted loci identified are also all expressed in the endosperm. An exception to this is the *Maternally expressed in embryo 1* (*Mee1*) gene in maize, which is imprinted in both the embryo and the endosperm [22] (Table 1).

2.1 Arabidopsis Imprinted Genes

In *Arabidopsis*, amongst the first maternally expressed imprinted genes

Tab. 1 Imprinted genes and their function.

Species	***Name***	***Potential function***
Maize (*Zea mays*)	R (Certain alleles only)	Transcription factor
	Allele MO17 of the *dzr1* locus	Reserve protein
	Fertilization independent endosperm 1 (Fie1)	PcG chromatin remodeling factor
	No apical meristem related protein 1 (Nrp1)	Unknown
	Maize Enhancer of Zeste1 (Mez1)	PcG chromatin remodeling factor
	Maternally expressed gene1 (Meg1)	Cys-rich peptide
Arabidopsis thaliana	*MEDEA (MEA)*	PcG chromatin remodeling factor
	FLOWERING WAGENINGEN (FWA)	Homeobox transcription factor
	PHERES1 (PHE1)	Type1 MADS-box transcription factor
	FERTILIZATION INDEPENDENT SEED 2 (FIS2)	PcG chromatin remodeling factor
	MATERNALLY EXPRESSED PAB C-TERMINAL (MPC) 43	C-terminal domain of poly(A) binding proteins (PABPs); probably controls mRNA stability and translation
	HD-ZIP GENE9 (HDG9)	Transcription factor
	HD-ZIP GENE8 (HDG8)	Transcription factor
	HD-ZIP GENE3 (HDG3)	Transcription factor
	ATMYBR2	Transcription factor
	AT5G62110	Putative transcription factor

identified, *MEA* [13, 14] and *fertilization independent seed 2* (*FIS2*) [16, 23, 24] are core members of the endosperm-specific FIS Polycomb group Repressor Complex 2 (PRC2). The FIS complex also includes *Fertilization-independent endosperm* (*FIE*) [25] and *Multicopy-suppressor of IRA1* (*MSI1*) [26–28], which are not imprinted. PRC2 methylates the Lys27 residue of HISTONE3, and thereby represses transcription [29, 30].

The wild-type endosperm is divided into three major differentiated domains. The chalazal pole (cz) is distinguished from the peripheral and micropylar domains of the endosperm by a multinucleate structure identified as the *cyst* [9, 31, 32] (Fig. 5). The endosperm of *fis* mutants is characterized by multiple defects including an enhanced proliferation, a much enlarged chalazal domain, and an absence of cellularization [12, 24, 27, 33]. This pleiotropic phenotype might be the consequence of the maintenance of a juvenile developmental program [34]. Although some targets of the FIS Polycomb group (PcG) complex have been identified, the pathways downstream of this transcriptional regulation are unknown and the targets whose functions explain the *fis* phenotype have not been fully understood. The function of two targets of the FIS PcG complex, which are themselves imprinted – the *Arabidopsis formin homolog 5* (*AtFH5*) and *PHERES1* (*PHE1*) will be detailed below.

AtFH5 encodes an actin-nucleating agent [35] and is maternally expressed in the endosperm [36]. The posterior endosperm cyst develops from the migration of nuclei from the peripheral endosperm (PE) [27] (Fig. 6). The early endosperm syncytial development ends when cellularization partitions the syncytium into mononucleate cells, but cellularization does not occur in the posterior pole [31, 37]. *AtFH5* expression is confined to the posterior pole, and is

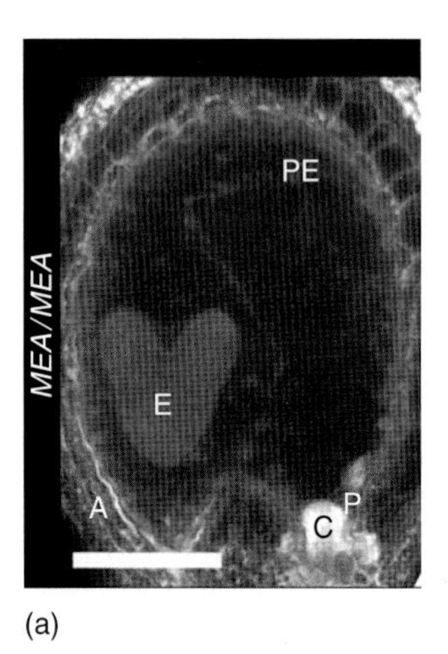

(a)

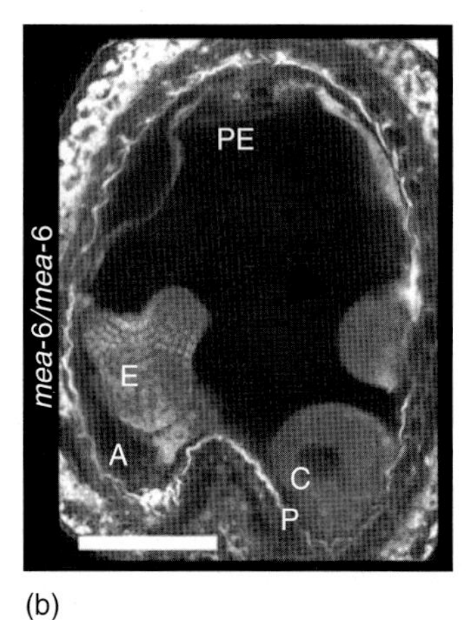

(b)

Fig. 5 Developmental features of endosperm in wild-type and *fis* mutants. (a) In the wild-type at six days after fertilization, the cellular peripheral endosperm (PE) surrounds the embryo (E) at the anterior pole (A). At the posterior pole (P) the chalazal endosperm (C) is a small pocket of cytoplasm containing several nuclei; (b) In the *fis* mutant *medea-6* allele (*mea-6*) the peripheral endosperm does not cellularize and the chalazal endosperm proliferates in ectopic positions. Scale bars = 50 μm.

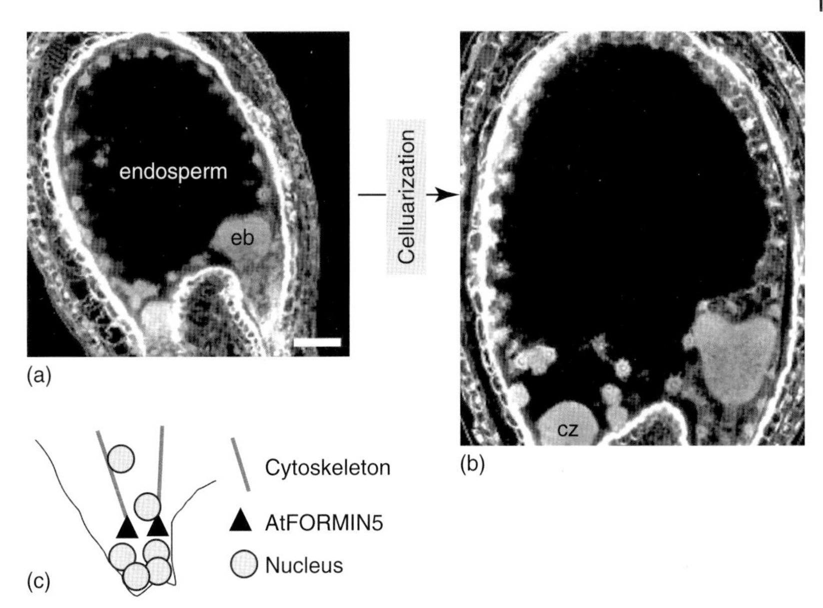

Fig. 6 Role of AtFORMIN5 in endosperm development. (a) The endosperm early development is characterized by nuclei proliferation in absence of cell division, leading to a syncytium surrounding the embryo; (b) After a specific number of nuclei division, each nucleus becomes enclosed in its specific compartment, leading to a cellular endosperm. Cellularization is initiated around the embryo (the micropylar pole) and progresses toward the chalazal pole (cz); (c) The chalazal endosperm does not cellularize, and becomes the site of migration of nuclei along cytoskeletal elements that are organized by the actin polymerization factor AtFORMIN5.

required for nuclear migration to this part of endosperm [35]. The restricted expression of *AtFH5* in the posterior endosperm depends on the FIS PRC2. In the absence of FIS function, *AtFH5* is expressed ectopically, preventing the correct development of the posterior pole [36].

PHE1 is paternally expressed [38, 39] in endosperm, and encodes a type 1 MADS-box transcription factor of the AGAMOUS-LIKE family (AGAMOUS-LIKE37). PHE1 antagonizes the role of FIS PRC2 on endosperm growth, but its mechanism of action remains unclear [33].

Two other imprinted genes identified in *Arabidopsis* have been studied in further details, namely *Flowering wageningen* (*FWA*) and *Maternally expressed PAB C-terminal* (*MPC*). *FWA* encodes an homeodomain leucine zipper (HD-ZIP) protein [40] and is expressed maternally only in endosperm, where its function is not known [41]. When ectopically expressed in vegetative tissues, FWA binds and inhibits the function of *Flowering locus T* (*FT*), causing late flowering [42]. Three other members of the gene family encoding HD-ZIP proteins (HDG3, HDG8, and HDG9) also show imprinted expression in endosperm [43]. HDG8 and HDG9 are maternally expressed, while HDG3 is expressed predominantly from its paternal allele.

MPC encodes the C-terminal region of a poly(A) binding protein (PABP) [44]. At present, the function of MPC is unknown;

however, it is also expressed – but not imprinted – in vegetative tissues and in the embryo.

2.2 Conservation of Polycomb Group Imprinted Genes in Cereals

In maize, only maternally expressed imprinted genes have been identified and, with the exception of homologs of PRC2 members, there is currently only limited evidence available for their function (Table 1) [4, 45–47].

The *Arabidopsis* gene *MEDEA* is homologous to the *Drosophila* PcG protein Enhancer-of-zeste (E(z)). Among the three maize E(z)-like genes – *Mez1* (*Maize enhancer of zeste 1*), *Mez2*, and *Mez3* – only *Mez1* displays a mono-allelic expression pattern in the developing endosperm tissue [48]. The two rice E(z)-like genes, *OsiEZ1* and *OsCLF*, are not imprinted [49].

A stronger conservation of imprinting in cereals was found in the FIE homologs. Notably, the maize FIE2 and sorghum FIE proteins form a monophyletic group, sharing a closer relationship to each other than to the FIE1 protein, which suggests that maize *Fie* genes originated from two different ancestral genomes [50]. The maize *Fie1* gene is maternally expressed exclusively in the endosperm, whereas *Fie2* is maternally expressed in the embryo and also in the endosperm, albeit at lower levels [47, 50].

The rice genome also contains two *FIE* homologs, *OsFIE1* and *OsFIE2*. The former homolog is expressed only in endosperm, with the maternal copy being expressed while the paternal copy remains inactive [49]. At present, the function of *FIE* homologs in the cereal endosperm is unknown.

2.3 Conclusion

Genomic imprinting in plants primarily affects those genes that are expressed in endosperm. The majority of imprinted genes identified in plants are maternally expressed, with only a few paternally expressed imprinted genes having been identified in *Arabidopsis*. The experimental elimination of the paternal genome in fertilized central cells has confirmed that maternal expression is not sufficient for early endosperm development [51]. The requirement for paternal genome expression suggests the existence of as-yet unidentified paternally expressed imprinted genes. Imprinting affects those genes which encode the members of a conserved PcG complex, which plays a key role in the control of several aspects of endosperm development, including polarity, growth, and temporal aspects. The results of studies using interploid crosses have suggested that the overall function of imprinting is related to the control of endosperm growth and to seed size, although a comprehensive picture of the total number of imprinted genes is still lacking. It is likely that the development of deep-sequencing technologies, coupled with the use of polymorphisms, will lead to the rapid discovery of new imprinted genes in a variety of species.

3 Molecular Mechanisms Controlling Imprinting

Parental genomic imprinting originates from epigenetic mechanisms that act during gametogenesis, and which differentiate the transcriptional states of the two prospective parental alleles. Epigenetic regulations include a wide spectrum of mechanisms that regulate and modify phenotypes independently of the genotype [52–54]. Covalent modifications of the chromatin regulate the expression of the genome. In all eukaryotes, the histones are subjected to various types of modification, and in most cases the DNA is methylated at cytosine residues. DNA methylation and certain types of histone methylation can be transmitted through cell division and, as a consequence, may constitute a form of "epigenetic memory."

In plants, most imprinted genes are not expressed in vegetative tissues, and until the completion of gametogenesis, this "silenced state" depending either on DNA methylation or Histone3 lysine 27 (H3K27) methylation. In either the male or female gamete lineage, the chromatin modifications (which silence gene expression) are removed by the end of gametogenesis, whilst after fertilization the difference between the transcriptional status of the two parental alleles persists, leading to a stable imprinted expression in the endosperm.

3.1
Imprinting by DNA Methylation

3.1.1 Maintenance of DNA Methylation on the Silent Alleles

Both alleles of *FIS2*, *MPC*, and *FWA* are silenced throughout the plant life cycle until gametogenesis occurs (Fig. 7). The silencing of *FWA*, *FIS2*, and *MPC* is

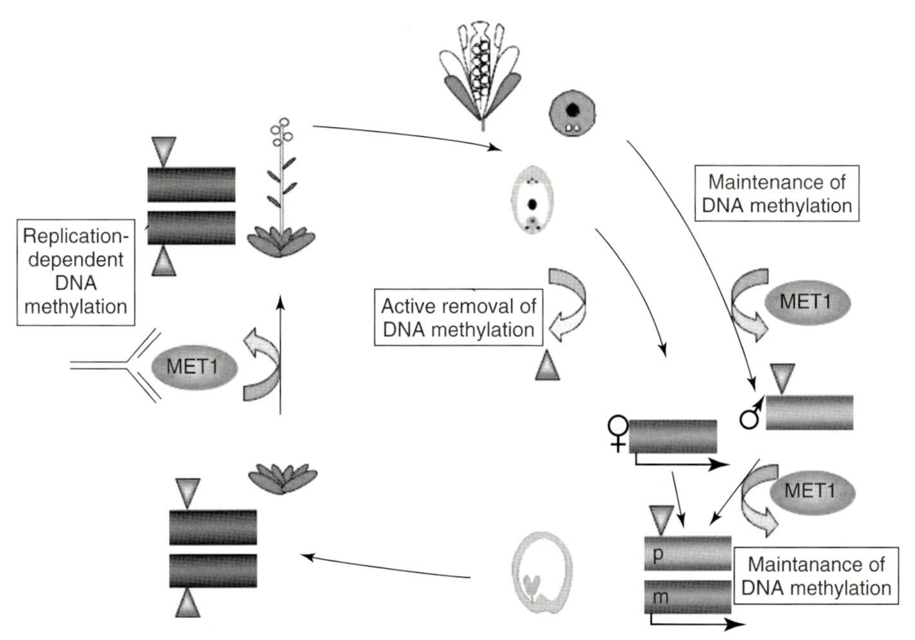

Fig. 7 DNA methylation-dependent mechanisms leading to imprinting of maternally expressed genes in *Arabidopsis*. MET1 maintains CpG methylation silencing marks on the parental alleles of imprinted genes (gray triangle). The two sperm cells fertilize the egg cell and the central cell. During male gametogenesis, CpG methylation is maintained while the gene becomes expressed in the central cell. Hence, the endosperm inherits a silenced paternal allele (p) and an active maternal allele (m), resulting in monoparental imprinted expression in the endosperm.

mediated by the DNA Methyltransferase 1 (MET1), which maintains the DNA methylation of CpG sites [23, 41, 44]. The silenced status of *FWA*, *FIS2*, and *MPC* is maintained during male gametogenesis until the sperm cells are differentiated [23, 44] Fig. 7. During endosperm development, the inherited paternal copy of *FWA*, *FIS2*, and *MPC* remains silenced by MET1, whereas the maternal copy is inherited as transcriptionally active, resulting in monoparental expression [23, 41, 44–56]. This mechanism is conserved in maize, where an analysis of the maternally expressed *Fie1* locus showed that DNA methylation is present in sperm cells, and is specifically removed in the central cell, but not in the egg cell [46, 47]. This direct assessment of DNA methylation in isolated central cells shows that epigenetic marks differ between each gamete, and prefigure the imprinted expression after fertilization in the endosperm.

3.1.2 Two-Step Removal of DNA Methylation in the Central Cell

The removal of DNA methylation marks in the central cell depends on two successive mechanisms that cause passive, and then active, demethylation (Fig. 8). Passive DNA demethylation is caused by the transcriptional repression of MET1 by the *Arabidopsis* Retinoblastoma homolog RBR [57] which, together with its partner, the WD40 domain containing protein MSI1, bind to the promoter of *MET1* and repress *MET1* transcription. This causes a low activity of MET1 during the late phase of female gametogenesis, while the DNA continues to replicate. The Retinoblastoma-dependent repression of MET1 transcription is predicted to cause the production of demethylated DNA.

The activation of expression of *FWA*, *FIS2*, and *MPC* in the central cell also relies on the DNA demethylase DEMETER (DME) [23, 41, 44, 58]. *DME* is expressed in the central cell and encodes a DNA glycosylase that removes methylated cytosine residues [55, 58]. DME creates single-strand DNA breaks that are repaired by the DNA ligase I [59]. The synergistic action of passive demethylation by the repression of MET1 activity, followed by active demethylation by DME, might completely demethylate the *cis* elements in *FIS2* and the *FWA* promoters causing expression of these genes in the central cell. After fertilization, the active maternal allele is inherited with a demethylated *cis* element, while the inactive paternal allele is inherited with a fully methylated *cis* element (Fig. 8). MET1 is active in endosperm and maintains the imbalanced pattern of methylation causing imprinted expression in endosperm [23]. Such a mechanism is likely to apply to all maternally expressed imprinted genes silenced by MET1 in sperm cells.

3.2 Control of Imprinting by Histone Methylation

In vegetative tissues, both alleles of *MEA* are silenced by H3K27 trimethylation, mediated by PcG complexes [55, 56]. Genome-wide arrays of DNA methylation and H3K27 trimethylation have shown that the *MEA* locus is covered with H3K27 trimethylation [60]. Compromising H3K27 trimethylation in mutants for PcG activity causes *MEA* ectopic expression in pollen and vegetative tissues (Fig. 9). The *MEA* imprinted status is lost in mutants for the PcG complex active in endosperm [55, 56], and the silencing of *MEA* by H3K27 methylation implies that

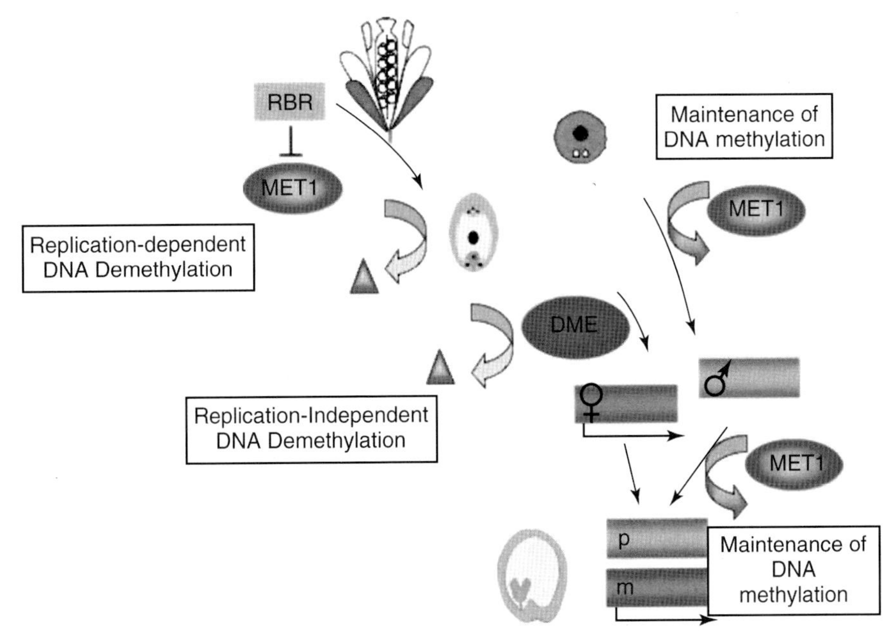

Fig. 8 DNA methylation-dependent mechanisms leading to imprinting of maternally expressed genes in *Arabidopsis*. The DNA methyltransferase MET1 is expressed throughout male gametogenesis and in the sperm cells. MET1 maintains DNA methylation silencing marks on the paternal allele of imprinted genes (gray triangle). During female gametogenesis, the Retinoblastoma (pRB) pathway represses *MET1* expression, causing passive removal of DNA methylation on the maternal allele. This mode of demethylation is not sufficient to cause expression of the target gene. Only in the mature central cell is the glycosylase lyase DEMETER (DME) expressed. This actively removes DNA methylation from the maternal allele provided to the endosperm, resulting in an imprinted expression. DME is not expressed in the egg cell, and both the paternal and maternal alleles remain silenced in the embryo.

the transcriptional activation of *MEA* requires the removal of trimethylated H3K27 during female gametogenesis. Although details of the mechanism causing such a removal remain unknown (Fig. 9), DME is required for *MEA* transcriptional activation [58, 61], and it is possible that MET1 and DME indirectly activate a pathway that removes H3K27 trimethylation marks from *MEA*, leading to its activation. Alternatively, maternal *MEA* expression may require a transcriptional activator that is itself directly controlled by DNA methylation and DME activity.

One of the maize homologs of *MEA*, *Mez1*, is imprinted [48], and the silenced paternal allele carries H3K27 trimethylation [62]. Similar to the self-regulation of *MEA* imprinting, the disruption of PcG function provided by the *Mez1* maternal allele causes expression of the *Mez1* paternal allele [63], suggesting a conservation of the mechanisms that regulate imprinting of *MEA* and its homolog *Mez1*.

PHE1 is a paternally expressed imprinted gene [39]. Whilst silencing of the maternal allele of *PHE1* is mediated by the maternal action of PcG in endosperm [38], the *PHE1* maternal allele is expressed at

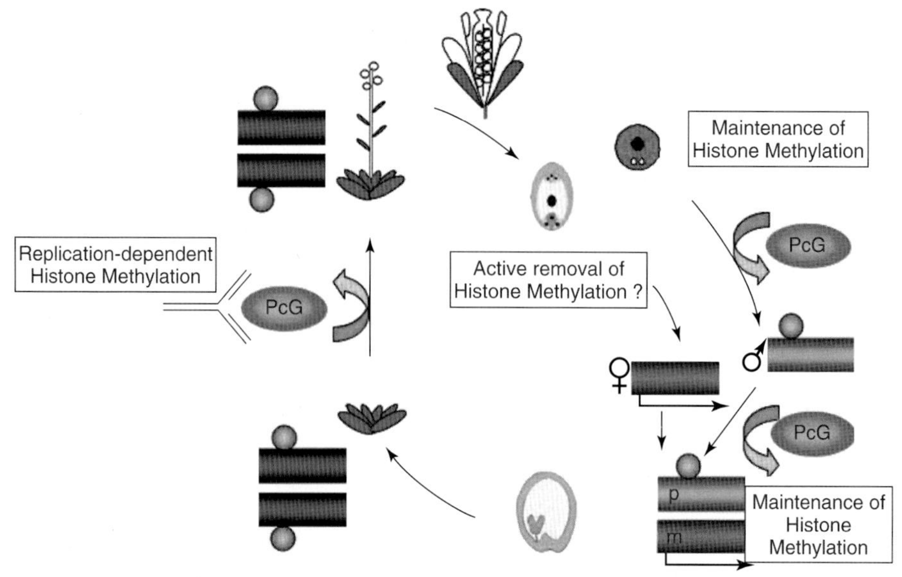

Fig. 9 Polycomb Repressive Complex 2 (PRC2) -dependent mechanisms leading to the imprinting of maternally expressed genes in *Arabidopsis*. PRC2 maintains H3K27 methylation silencing marks on the parental alleles of *MEA* (gray spheres). Each sperm cell fertilizes either the egg cell or the central cell. During male gametogenesis, H3K27 methylation is maintained, while *MEA* becomes expressed in the central cell. Hence, the endosperm inherits a silenced paternal allele (p) and an active maternal allele (m), resulting in an imprinted expression. The origin of *MEA* expression in the central cell remains unknown. PcG, Polycomb group.

variable levels, depending on the natural accessions [64]. The mechanisms causing transcriptional activation of *PHE1* in the male gametes remain unknown, and a PcG independent mechanism regulates *PHE1* (see Sect. 3.3.2).

Transcription of the gene *AtFH5* is also directly controlled by PcG complex activity [36]. *AtFH5* expression is silenced by PcG complexes that are active in vegetative tissues prior to gametogenesis. Unlike *MEA*, the expression of *AtFH5* is not activated in the central cell; rather, only the maternal allele of *AtFH5* is expressed after fertilization. *AtFH5* expression is also confined by PcG activity to the posterior pole of the endosperm, suggesting additional transcriptional controls. The site marked by PcG activity and sufficient for imprinting is contained in a 400 bp domain of the *AtFH5* promoter; however, details of the mechanism which causes the imprinting of *AtFH5* are unknown.

Although histone methylation by PcG is involved in imprinting in mammals [65], it does not appear to act as the essential repressor of the silenced allele of imprinted genes, as has been shown for certain imprinted genes in plants. Consequently, a major challenge is to recognize which mechanisms remove the H3K27 methylation mark from the expressed allele of *MEA*, *AtFH5*, and *PHE1*. An understanding of such mechanisms will further provide means of identifying other imprinted genes controlled by the PcG pathway.

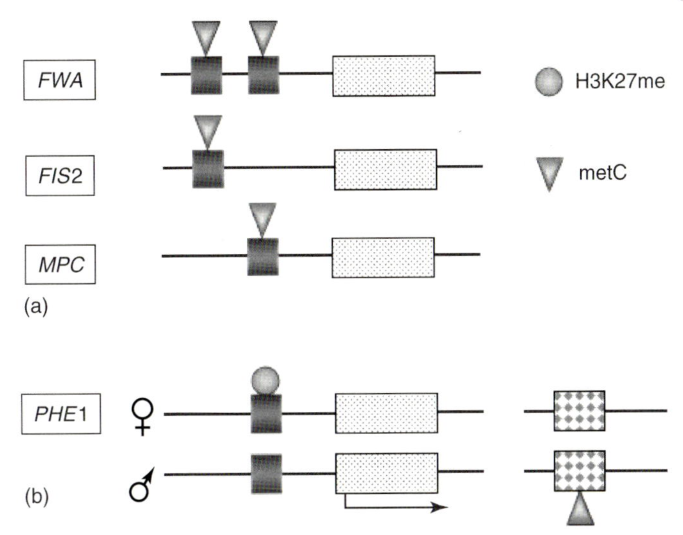

Fig. 10 Cis-elements regulating imprinting. (a) Elements controlling MET1-dependent imprinting; (b) Elements controlling the paternally expressed imprinted gene *PHERES1* (*PHE1*) is silenced by Polycomb Group-dependent histone methylation. In sperm cells, the H3K27me marks (gray spheres) are removed by as-yet unknown mechanisms. A cis-element located at 2.6 kb downstream of the coding sequence of *PHE1* contains repeats, which are methylated on cytosine residues (gray triangles) in vegetative tissues and in sperm cells. The removal of cytosine methylation on the maternal allele in the central cell appears conditional to the proper maintenance of the imprinted status of *PHE1* in endosperm.

3.3
Cis-Elements Controlling Imprinting

H3K27 methylation by PcG activity is widespread over *Arabidopsis* imprinted loci, and very little is known about the polycomb response elements (PREs), which are similar to those described in *Drosophila*. In contrast, DNA methylation at CpG has been localized to well-defined cis-elements in the promoter, and in 3′ of the coding sequence of imprinted genes.

3.3.1 Cis-Elements in the Promoter

FWA imprinting relies in large part on a cis-element in the promoter, consisting of two direct repeats that most likely originated from a duplicated SINE element [40, 66]. These produce small non-coding RNAs that are recognized by the RNA-dependent DNA *de novo* methylation pathway [67] (Fig. 10a). Although 5′ *cis* elements have been also identified for *FIS2* [23] and *MPC* [44], these do not appear to derive from a transposon. Consequently, exactly how DNA methylation is directed towards CG sites on these cis-elements is unknown.

3.3.2 Evidence for Imprinting Regulation by Long-Distance Elements

Long-distance regulatory elements are essential in mammalian imprinting regulation [68]. In plants, a comparable regulatory mechanism affects *PHE1* imprinting [64]. The mechanism responsible for removal of the silencing H3K27 methylation marks from the *PHE1* locus in sperm cells – despite the presence of functional

PcG – remains unknown, although MET1 also appears to regulate *PHE1* imprinting [64]. A methylated repeat region is located 2.6 kb from of the 3′ end of the *PHE1* coding sequence, and the methylation of this is required to maintain the expression of the paternal *PHE1* allele (Fig. 10b).

Similarly, the *MEA* coding region precedes a methylated 3′ cis-element known as *ISR*, which comprises seven 183 bp repeats and is located approximately 500 bp from the end of the *MEA* coding sequence. The ISR is demethylated on the *MEA* maternal allele, and may regulate *MEA* imprinting [5, 24]. It is possible that the methylated *MEA ISR* prevents removal of the silencing H3K27 methylation marks from the *MEA* locus. Thus, the demethylation of *MEA ISR* by DME would be a step required for *MEA* activation in the central cell.

4 Biological Significance and the Evolution of Imprinting

4.1 Parental Conflict

Imprinting arose independently in plants and mammals, both of which are characterized by a mode of reproduction that involves maternal nutrition of the developing embryo through specialized tissues. According to a scenario where a mother produces offspring from different fathers, embryos carrying different paternal genomes compete for resource allocations from the mother. Consequently, the "goal" of the fathers is to derive as much maternal resources as possible for their embryos. When kin selection is considered, it is advantageous for the mother to downregulate such opposing interference from the father to ensure an equitable distribution of nutrients to each offspring. These considerations led to the *parental conflict theory*, which predicts a positive selection of maternally expressed growth inhibitors and paternally expressed growth enhancers [69].

The reproductive scenario associated with the parental conflict theory applies to certain mammalian and outcrossing plant species, but does not apply to self-fertilizing *Arabidopsis*, in which imprinting is nonetheless found to be active. Even if it is assumed that the ancestors of mammals and flowering plants were obligate outbreeders, with no restriction on the numbers of male partners, the parental conflict theory also suggests that imprinted genes should not be under positive selection in species that always self-fertilize. *Arabidopsis thaliana* usually self-fertilizes, in contrast to the closely related species *Arabis lyrata* and *Arabis petraea*. The type of selection exerted on *MEA* in *Arabidopsis* and its homologs in related species has been evaluated in three studies [70–72]. It would appear that *MEA* has not been subjected to a strong positive selection in *Arabidopsis*, and similarly no positive selection was detected for the *MEA* maize homolog, *Mez1* [48]. Although, taken together, the results of these studies failed to provide any strong support for the conflict theory, it is difficult to reach an unambiguous conclusion as no similar data have been gathered for other imprinted genes in plants.

The dramatic result of crosses between plants of different ploidies [9, 73] was considered as a strong support to the parental conflict theory (Fig. 11). An additional dosage of the paternal genome causes an increased endosperm size, while an additional maternal dosage reduces

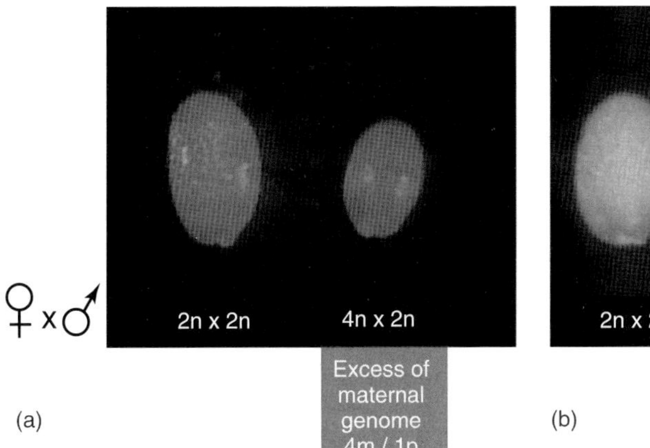

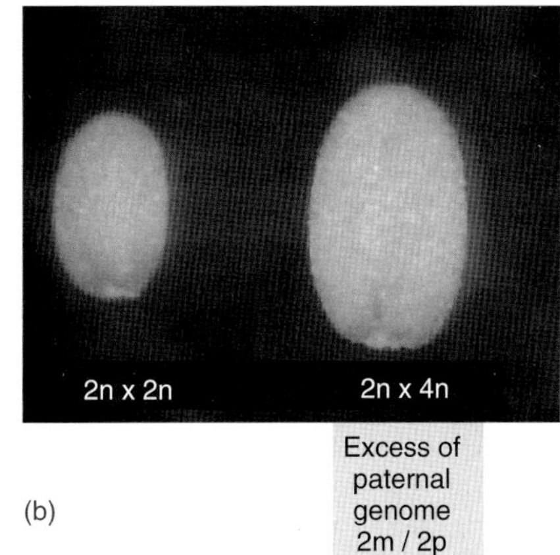

Fig. 11 Consequences of genome dosage imbalance on seed size. An increased parental dosage is achieved by crossing a diploid wild-type *Arabidopsis* strain with a tetraploid strain. At maturity, the seed sizes are compared. An increased maternal dosage causes a reduction of seed size, while an increased paternal dosage has the opposite effect.

the endosperm size. However, the direct measurement of genome dosage on the expression of various imprinted genes showed there to be no direct relationship between the changes in levels of the transcripts of imprinted genes in response to dosage imbalance, and their parent of origin-dependent expression. Whereas, an increased maternal dosage increases the expression level of *FIS2* and *FWA*, it reduces the level of *MEA*. Further analyses have shown that a parental dosage imbalance leads to a complete deregulation of imprinting via PcG and PcG-independent pathways [74–76].

The parental conflict theory also predicts that maternally expressed imprinted genes will suppress endosperm growth, and this is correlated directly with embryo growth and seed size [77]. However, *AtFH5* is an activator of endosperm growth [36] and *FWA* has apparently no direct effect on seed growth [41].

In summary, the parental conflict hypothesis is poorly supported by the persistence of imprinting in self-breeding *Arabidopsis*, the absence of any direct correlation between parental genome dosage imbalance and the expression levels of imprinted genes, and the elusive link between growth control and imprinted genes.

4.2 Maternal Control

The study of the maternal effect of the loss of function of MET1 on seed development highlights the overwhelming influence of the maternal control on endosperm growth. In contrast to the gametophytic paternal effect of *met1* on seed size, the inheritance of *met1* by the female gamete has no effect on endosperm and seed development [61, 78]. However, mother plants deprived of MET1 function produce much larger seeds than their wild-type counterparts [16, 78, 79]. Genetic analyses have shown that this results from the effect of MET1 on the maternally originating integuments, which envelop the endosperm. Seed size is controlled directly not only by the size of the endosperm but also by the growth potential of the maternal integuments cells and their capacity to elongate [77]. MET1 directly prevents the proliferation and elongation of the ovules' integuments, thus controlling seed size [78].

Overall, the coordinated controls exerted by seed integuments on cell proliferation and elongation and endosperm growth depend on maternally expressed genes, and thus determine the level of seed growth. It is possible, therefore, that a dominant matriarchal control was selected to enhance plant reproductive fitness defined in terms of seed growth.

4.3 Imprinting and Speciation

In plant species with an obligate outcrossing reproductive strategy, imprinting might play a role in the prevention of hybridity [80]. Interspecific crosses are possible in related plant species, but often lead to reduced seed fertility. In a study of crosses between *Arabidopsis* relatives, it was shown that *PHE1* plays an important role in the viability of the hybrid seeds [81]. It is possible, therefore, that imprinting has been selected as a mechanism regulating hybridity, and as such would be crucial for speciation. Interestingly, imprinting has also been linked with speciation in mammals [82, 83].

4.4 Imprinting: A Byproduct of Global Epigenetic Changes?

Many theories relating the origin of imprinting have been based on knowledge derived from studies in mammals. To a certain extent, these studies support the concept that the overwhelming imbalance of maternal contribution to embryo development and care after birth was the key event that caused imprinting selection. Although several aspects of imprinting mechanisms are shared between plants and mammals, imprinting in plants is characterized by idiosyncratic features (Fig. 12). The parental conflict theory is poorly supported by imprinting studies in plants, and other alternatives must be envisaged. The maternal alleles of all imprinted genes studied to date undergo demethylation during female gametogenesis on cis-elements in their promoter (*FWA*, *FIS2*, *MPC*, *Fie1*, *Fie2*) or in elements located in 3′ (*MEA* and *PHE1*). Similarly, the maintenance of CG DNA methylation in sperm cells is common to all imprinted genes studied in *Arabidopsis*. This may suggest that demethylation in the central cell represents a common regulatory mechanism for all imprinted genes.

MET1 is expressed strongly in sperm cells, but is repressed in the central cell [57]. This asymmetrical expression of MET1 is reflected by the parent-of-origin genetic effects caused by the loss of function of MET1. The absence of MET1 activity during male gametogenesis inhibits endosperm growth and results in smaller seeds, which is not mirrored by the maternal inheritance of MET1 during female gametogenesis [61, 78].

The reduction of *MET1* expression during female gametogenesis implies a genome-wide DNA demethylation of cytosine residues in CG context in the central cell. Normally, chromatin in the central cell is loosely organized [84], but it becomes compacted in response to the ectopic expression of MET1 seen in the absence of Retinoblastoma [85]. This observation provides indirect evidence for a global demethylation of the central cell chromatin. Demethylation of

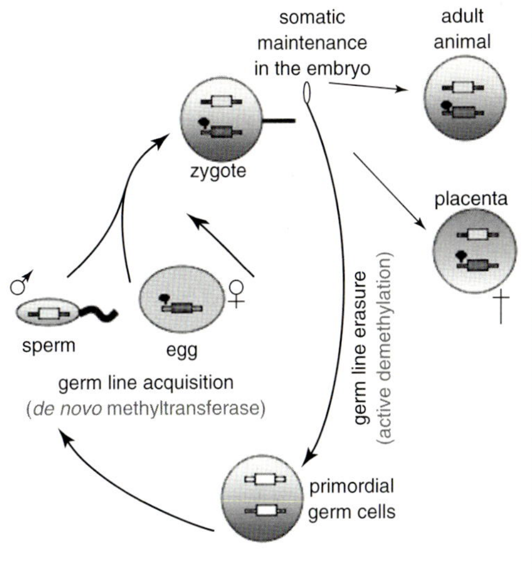

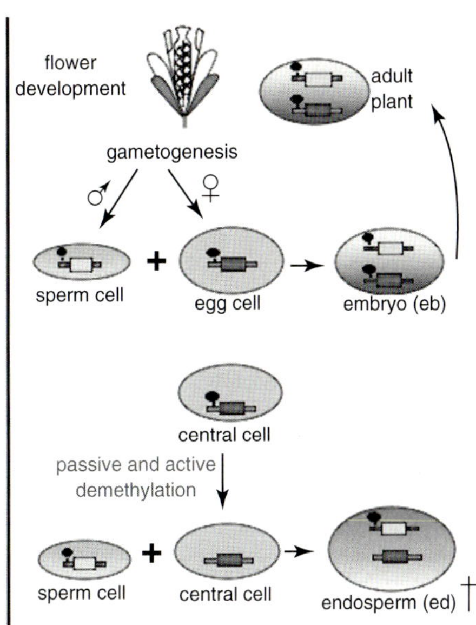

Fig. 12 Comparison between imprinting mechanisms in plants and mammals. Both in plants and mammals, the primary mechanism that controls imprinting relies on DNA methylation (black "lollipops") and to a minor extent on histone methylation. In contrast to plants, imprinted genes are expressed in the embryo and in the adult animal. Hence, in the germline the chromatin methylation needs to be erased. The mechanism involved is still unknown, but likely involves demethylases with an enzymatic action similar to DME. During gametogenesis, DNA methylation marks are deposited differentially in either the male or the female gamete, leading to the inheritance of an active and an inactive allele in the embryo. In contrast to mammals, the marks regulating imprinting in plants are removed in a sex-dependent manner during gametogenesis. This removal results from a global demethylation of DNA during female gametogenesis in the central cell. DNA methylation in maintained during male gametogenesis. The opposite controls of DNA demethylation between male and female gametogenesis leads to the inheritance of an active maternal allele and an inactive paternal allele in the endosperm. As the endosperm dies at the end of seed development, the imprinted status is not passed on to the next generation and there is no imprinting cycle in plants, in contrast to mammals.

the central cell genome is also supported by evidence for genome-wide DNA demethylation in the endosperm [55, 86]. This demethylation most likely originates from the reduced MET1 activity, combined with the active demethylation by DME in the central cell. Parental genomic imprinting could, therefore, be considered to be a byproduct of this global demethylation.

The recent identification of imprinting in three members of the clade of class IV homeodomain transcription factors containing *FWA*, shows that all contain 5′ methylated elements derived from helitrons. This suggests that transposon insertions are tightly linked with the selection of imprinting [43].

One possible scenario for the origin of imprinting in plants is that the differentiation of the endosperm lineage from the embryo lineage involves a global change in chromatin modifications, including a global demethylation of the

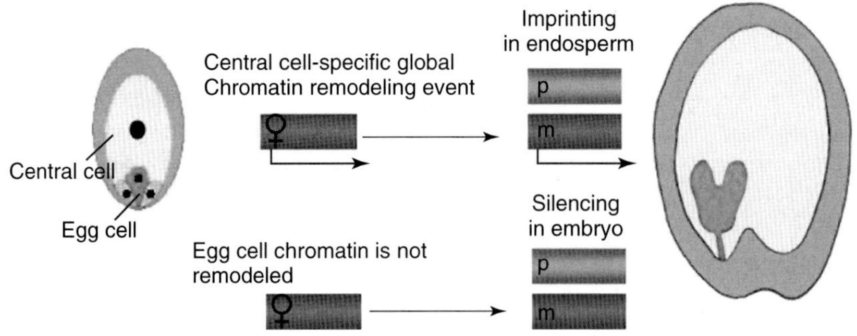

Fig. 13 Model of the origin of imprinting as differential epigenetic reprogramming between the two female gametes. Imprinting appears to be observed mostly in the endosperm. In order to explain this, a global differential reprogramming of the chromatin modifications might take place in the central cell, but not in the egg cell (at least not to the same extent). The two sperm cells are identical in *Arabidopsis* and in most flowering plant species, and thus do not intervene in this model. The central cell genome is remodeled in a global manner, leading to the loss of repressive chromatin modifications. In the egg cell this global remodeling does not take place, or is compensated. After double fertilization the embryo inherits two silent copies, while the endosperm inherits two active copies from the central cell and a silent copy from the sperm cell. If the locus possesses cis-elements that allow expression in the endosperm, the gene is then expressed only from the maternal allele in an imprinted manner.

central cell (Fig. 13). The stochastic insertion of transposons created new elements that are targeted by the RNA-dependent DNA methylation pathway, leading to a new site of DNA methylation. According to this scenario, such transposon insertion sites could create new methylated elements in the genome that would be demethylated in the central cell. As a consequence, the maternal allele would be inherited as demethylated, but the paternal allele would not. If the insert were to be located in a promoter, providing a context favorable for expression in the endosperm, then a new imprinted locus would be created. The selection of such a locus could depend on the potential neutral or beneficial impact on seed fitness.

Acknowledgments

The authors thank Pauline Jullien and Vu Minh Thiet for their comments on the manuscript. The research conducted by F.B. is funded by Temasek Life Sciences Laboratory.

References

1 McCormick, S. (2004) Control of male gametophyte development. *Plant Cell*, (Suppl. 16), S142–S153.

2 Yadegari, R., Drews, G.N. (2004) Female gametophyte development. *Plant Cell*, (Suppl. 16), S133–S141.

3 Crouse, H.V. (1960) The controlling element in the sex chromosome behaviour in *Sciara*. *Genetics*, **45**, 1429–1443.

4 Kermicle, J.L. (1970) Dependence of the R-Mottled aleurone phenotype in maize on mode of sexual transmission. *Genetics*, **66**, 69–85.

5 Cattanach, B.M., Kirk, M. (1985) Differential activity of maternally and paternally derived chromosome regions in mice. *Nature*, **315**, 496–498.

6 Randolph, L. (1935) Cytogenetics of tetraploid maize. *J. Agric. Res.*, **50**, 591–595.

7 Cooper, D. (1951) *Caryopsis* development following matings between diploid and

tetraploid strains of maize. *Am. J. Bot.*, **38**, 702–710.

8 Lin, B.-Y. (1984) Ploidy barrier to endosperm development in maize. *Genetics*, **107**, 103–115.

9 Scott, R.J., Spielman, M., Bailey, J., Dickinson, H.G. (1998) Parent-of-origin effects on seed development in *Arabidopsis thaliana*. *Development*, **125**, 3329–3341.

10 Chaudhury, A.M., Ming, L., Miller, C., Craig, S., Dennis, E.S., Peacock, W.J. (1997) Fertilization-independent seed development in *Arabidopsis thaliana*. *Proc. Natl Acad. Sci. USA*, **94**, 4223–4228.

11 Grossniklaus, U., Vielle-Calzada, J.P., Hoeppner, M.A., Gagliano, W.B. (1998) Maternal control of embryogenesis by *MEDEA*, a polycomb group gene in *Arabidopsis*. *Science*, **280**, 446–450.

12 Kiyosue, T., Ohad, N., Yadegari, R., Hannon, M., Dinneny, J., Wells, D., Katz, A., Margossian, L., Harada, J.J., Goldberg, R.B., Fischer, R.L. (1999) Control of fertilization-independent endosperm development by the MEDEA polycomb gene in *Arabidopsis*. *Proc. Natl Acad. Sci. USA*, **96**, 4186–4191.

13 Kinoshita, T., Yadegari, R., Harada, J.J., Goldberg, R.B., Fischer, R.L. (1999) Imprinting of the MEDEA polycomb gene in the *Arabidopsis* endosperm. *Plant Cell*, **11**, 1945–1952.

14 Vielle-Calzada, J.P., Thomas, J., Spillane, C., Coluccio, A., Hoeppner, M.A., Grossniklaus, U. (1999) Maintenance of genomic imprinting at the *Arabidopsis* medea locus requires zygotic DDM1 activity. *Genes Dev.*, **13**, 2971–2982.

15 Baroux, C., Gagliardini, V., Page, D.R., Grossniklaus, U. (2006) Dynamic regulatory interactions of Polycomb group genes: MEDEA autoregulation is required for imprinted gene expression in *Arabidopsis*. *Genes Dev.*, **20**, 1081–1086.

16 Luo, M., Bilodeau, P., Dennis, E.S., Peacock, W.J., Chaudhury, A. (2000) Expression and parent-of-origin effects for *FIS2*, *MEA*, and *FIE* in the endosperm and embryo of developing *Arabidopsis* seeds. *Proc. Natl Acad. Sci. USA*, **97**, 10637–10642.

17 Wang, D., Tyson, M.D., Jackson, S.S., Yadegari, R. (2006) Partially redundant functions of two SET-domain polycomb-group proteins in controlling initiation of seed development in *Arabidopsis*. *Proc. Natl Acad. Sci. USA*, **103**, 13244–13249.

18 Kinoshita, T., Ikeda, Y., Ishikawa, R. (2008) Genomic imprinting: a balance between antagonistic roles of parental chromosomes. *Semin. Cell Dev. Biol.*, **19**, 574–579.

19 Berger, F., Chaudhury, A. (2009) Parental memories shape seeds. *Trends Plant Sci.*, **14**, 550–556.

20 Rodrigues, J.C., Luo, M., Berger, F., Koltunow, A.M. (2010) Polycomb group gene function in sexual and asexual seed development in angiosperms. *Sex. Plant Reprod.*, **23**, 123–133.

21 Scott, R.J., Spielman, M. (2006) Deeper into the maize: new insights into genomic imprinting in plants. *BioEssays*, **28**, 1167–1171.

22 Jahnke, S., Scholten, S. (2009) Epigenetic resetting of a gene imprinted in plant embryos. *Curr. Biol.*, **19**, 1677–1681.

23 Jullien, P.E., Kinoshita, T., Ohad, N., Berger, F. (2006) Maintenance of DNA methylation during the *Arabidopsis* life cycle is essential for parental imprinting. *Plant Cell*, **18**, 1360–1372.

24 Luo, M., Bilodeau, P., Koltunow, A., Dennis, E.S., Peacock, W.J., Chaudhury, A.M. (1999) Genes controlling fertilization-independent seed development in *Arabidopsis thaliana*. *Proc. Natl Acad. Sci. USA*, **96**, 296–301.

25 Ohad, N., Yadegari, R., Margossian, L., Hannon, M., Michaeli, D., Harada, J.J., Goldberg, R.B., Fischer, R.L. (1999) Mutations in FIE, a WD polycomb group gene, allow endosperm development without fertilization. *Plant Cell*, **11**, 407–416.

26 Guitton, A.E., Berger, F. (2005) Control of reproduction by Polycomb Group complexes in animals and plants. *Int. J. Dev. Biol.*, **49**, 707–716.

27 Guitton, A.E., Page, D.R., Chambrier, P., Lionnet, C., Faure, J.E., Grossniklaus, U., Berger, F. (2004) Identification of new members of Fertilisation Independent Seed Polycomb group pathway involved in the control of seed development in *Arabidopsis thaliana*. *Development*, **131**, 2971–2981.

28 Kohler, C., Hennig, L., Bouveret, R., Gheyselinck, J., Grossniklaus, U., Gruissem, W. (2003) *Arabidopsis* MSI1 is a component of the MEA/FIE Polycomb group complex and required for seed development. *EMBO J.*, **22**, 4804–4814.

29 Hennig, L., Derkacheva, M. (2009) Diversity of Polycomb group complexes in plants: same rules, different players? *Trends Genet.*, **25**, 414–423.

30 Schuettengruber, B., Chourrout, D., Vervoort, M., Leblanc, B., Cavalli, G. (2007) Genome regulation by polycomb and trithorax proteins. *Cell*, **128**, 735–745.

31 Brown, R.C., Lemmon, B.E., Nguyen, H., Olsen, O.-A. (1999) Development of endosperm in *Arabidopsis thaliana*. *Sex. Plant Reprod.*, **12**, 32–42.

32 Boisnard-Lorig, C., Colon-Carmona, A., Bauch, M., Hodge, S., Doerner, P., Bancharel, E., Dumas, C., Haseloff, J., Berger, F. (2001) Dynamic analyses of the expression of the HISTONE::YFP fusion protein in *Arabidopsis* show that syncytial endosperm is divided in mitotic domains. *Plant Cell*, **13**, 495–509.

33 Kohler, C., Hennig, L., Spillane, C., Pien, S., Gruissem, W., Grossniklaus, U. (2003) The Polycomb-group protein MEDEA regulates seed development by controlling expression of the MADS-box gene PHERES1. *Genet. Dev.*, **17**, 1540–1553.

34 Ingouff, M., Haseloff, J., Berger, F. (2005) Polycomb group genes control developmental timing of endosperm. *Plant J.*, **42**, 663–674.

35 Ingouff, M., Fitz Gerald, J.N., Guerin, C., Robert, H., Sorensen, M.B., Van Damme, D., Geelen, D., Blanchoin, L., Berger, F. (2005) Plant formin AtFH5 is an evolutionarily conserved actin nucleator involved in cytokinesis. *Nat. Cell Biol.*, **7**, 374–380.

36 Fitz Gerald, J.N., Hui, P.S., Berger, F. (2009) Polycomb group-dependent imprinting of the actin regulator AtFH5 regulates morphogenesis in *Arabidopsis thaliana*. *Development*, **136**, 3399–3404.

37 Sorensen, M.B., Mayer, U., Lukowitz, W., Robert, H., Chambrier, P., Jurgens, G., Somerville, C., Lepiniec, L., Berger, F. (2002) Cellularisation in the endosperm of *Arabidopsis thaliana* is coupled to mitosis and shares multiple components with cytokinesis. *Development*, **129**, 5567–5576.

38 Makarevich, G., Leroy, O., Akinci, U., Schubert, D., Clarenz, O., Goodrich, J., Grossniklaus, U., Kohler, C. (2006) Different Polycomb group complexes regulate common target genes in *Arabidopsis*. *EMBO Rep.*, **7**, 947–952.

39 Kohler, C., Page, D.R., Gagliardini, V., Grossniklaus, U. (2005) The *Arabidopsis thaliana* MEDEA Polycomb group protein controls expression of PHERES1 by parental imprinting. *Nat. Genet.*, **37**, 28–30.

40 Soppe, W.J., Jacobsen, S.E., Alonso-Blanco, C., Jackson, J.P., Kakutani, T., Koornneef, M., Peeters, A.J. (2000) The late flowering phenotype of fwa mutants is caused by gain-of-function epigenetic alleles of a homeodomain gene. *Mol. Cell*, **6**, 791–802.

41 Kinoshita, T., Miura, A., Choi, Y., Kinoshita, Y., Cao, X., Jacobsen, S.E., Fischer, R.L., Kakutani, T. (2004) One-way control of FWA imprinting in *Arabidopsis* endosperm by DNA methylation. *Science*, **303**, 521–523.

42 Ikeda, Y., Kobayashi, Y., Yamaguchi, A., Abe, M., Araki, T. (2007) Molecular basis of late-flowering phenotype caused by dominant epi-alleles of the FWA locus in *Arabidopsis*. *Plant Cell Physiol.*, **48**, 205–220.

43 Gehring, M., Bubb, K.L., Henikoff, S. (2009) Extensive demethylation of repetitive elements during seed development underlies gene imprinting. *Science*, **324**, 1447–1451.

44 Tiwari, S., Schulz, R., Ikeda, Y., Dytham, L., Bravo, J., Mathers, L., Spielman, M., Guzman, P., Oakey, R.J., Kinoshita, T., Scott, R.J. (2008) MATERNALLY EXPRESSED PAB C-TERMINAL, a novel imprinted gene in *Arabidopsis*, encodes the conserved C-terminal domain of polyadenylate binding proteins. *Plant Cell*, **20**, 2387–2398.

45 Gutierrez-Marcos, J.F., Costa, L.M., Biderre-Petit, C., Khbaya, B., O'Sullivan, D.M., Wormald, M., Perez, P., Dickinson, H.G. (2004) Maternally expressed gene1 is a novel maize endosperm transfer cell-specific gene with a maternal parent-of-origin pattern of expression. *Plant Cell*, **16**, 1288–1301.

46 Gutierrez-Marcos, J.F., Costa, L.M., Dal Pra, M., Scholten, S., Kranz, E., Perez, P., Dickinson, H.G. (2006) Epigenetic asymmetry of imprinted genes in plant gametes. *Nat. Genet.*, **38**, 876–878.

47 Hermon, P., Srilunchang, K.O., Zou, J., Dresselhaus, T., Danilevskaya, O.N. (2007) Activation of the imprinted Polycomb Group Fie1 gene in maize endosperm requires demethylation of the maternal allele. *Plant Mol. Biol.*, **64**, 387–395.

48 Haun, W.J., Laoueille-Duprat, S., O'Connell, M.J., Spillane, C., Grossniklaus, U., Phillips,

A.R., Kaeppler, S.M., Springer, N.M. (2007) Genomic imprinting, methylation and molecular evolution of maize Enhancer of zeste (Mez) homologs. *Plant J.*, **49**, 325–337.

49 Luo, M., Platten, D., Chaudhury, A., Peacock, W.J., Dennis, E.S. (2009) Expression, imprinting, and evolution of rice homologs of the polycomb group genes. *Mol. Plant*, **2**, 711–723.

50 Danilevskaya, O.N., Hermon, P., Hantke, S., Muszynski, M.G., Kollipara, K., Ananiev, E.V. (2003) Duplicated fie genes in maize: expression pattern and imprinting suggest distinct functions. *Plant Cell*, **15**, 425–438.

51 Aw, S.J., Hamamura, Y., Chen, Z., Schnittger, A., Berger, F. (2010) Sperm entry is sufficient to trigger division of the central cell but the paternal genome is required for endosperm development in *Arabidopsis. Development*, **137**, 2683–2690.

52 Roudier, F., Teixeira, F.K., Colot, V. (2009) Chromatin indexing in Arabidopsis: an epigenomic tale of tails and more. *Trends Genet.*, **25**, 511–517.

53 Kouzarides, T. (2007) Chromatin modifications and their function. *Cell*, **128**, 693–705.

54 Law, J.A. Jacobsen, S.E. (2010) Establishing, maintaining and modifying DNA methylation patterns in plants and animals. *Nat. Rev. Genet.* **11**, 204–220.

55 Gehring, M., Huh, J.H., Hsieh, T.F., Penterman, J., Choi, Y., Harada, J.J., Goldberg, R.B., Fischer, R.L. (2006) DEMETER DNA glycosylase establishes MEDEA polycomb gene self-imprinting by allele-specific demethylation. *Cell*, **124**, 495–506.

56 Jullien, P.E., Katz, A., Oliva, M., Ohad, N., Berger, F. (2006) Polycomb group complexes self-regulate imprinting of the Polycomb group gene MEDEA in *Arabidopsis. Curr. Biol.*, **16**, 486–492.

57 Jullien, P.E., Mosquna, A., Ingouff, M., Sakata, T., Ohad, N., Berger, F. (2008) Retinoblastoma and its binding partner MSI1 control imprinting in *Arabidopsis. PLoS Biol.*, **6**, e194.

58 Choi, Y., Gehring, M., Johnson, L., Hannon, M., Harada, J.J., Goldberg, R.B., Jacobsen, S.E., Fischer, R.L. (2002) DEMETER, a DNA glycosylase domain protein, is required for endosperm gene imprinting and seed viability in *Arabidopsis. Cell*, **110**, 33–42.

59 Andreuzza, S., Li, J., Guitton, A.E., Faure, J.E., Casanova, S., Park, J.S., Choi, Y., Chen, Z., Berger, F. (2010) DNA LIGASE I exerts a maternal effect on seed development in *Arabidopsis thaliana. Development* **137**, 73–81.

60 Zhang, X., Germann, S., Blus, B.J., Khorasanizadeh, S., Gaudin, V., Jacobsen, S.E. (2007) The *Arabidopsis* LHP1 protein colocalizes with histone H3 Lys27 trimethylation. *Nat. Struct. Mol. Biol.*, **14**, 869–871.

61 Xiao, W., Brown, R.C., Lemmon, B.E., Harada, J.J., Goldberg, R.B., Fischer, R.L. (2006) Regulation of seed size by hypomethylation of maternal and paternal genomes. *Plant Physiol.*, **142**, 1160–1168.

62 Haun, W.J., Springer, N.M. (2008) Maternal and paternal alleles exhibit differential histone methylation and acetylation at maize imprinted genes. *Plant J.*, **56**, 903–912.

63 Haun, W.J., Danilevskaya, O.N., Meeley, R.B., Springer, N. (2009) Disruption of imprinting by Mu transposon insertions in the 5′ proximal regions of the *Zea mays* Mez1 locus. *Genetics*, **181**, 1229–1237.

64 Makarevich, G., Villar, C.B., Erilova, A., Kohler, C. (2008) Mechanism of PHERES1 imprinting in *Arabidopsis. J. Cell Sci.*, **121**, 906–912.

65 Feil, R., Berger, F. (2007) Convergent evolution of genomic imprinting in plants and mammals. *Trends Genet.*, **23**, 192–199.

66 Kinoshita, Y., Saze, H., Kinoshita, T., Miura, A., Soppe, W.J., Koornneef, M., Kakutani, T. (2007) Control of FWA gene silencing in *Arabidopsis thaliana* by SINE-related direct repeats. *Plant J.*, **49**, 38–45.

67 Chan, S.W., Zhang, X., Bernatavichute, Y.V., Jacobsen, S.E. (2006) Two-step recruitment of RNA-directed DNA methylation to tandem repeats. *PLoS Biol.*, **4**, e363.

68 Bartolomei, M.S. (2009) Genomic imprinting: employing and avoiding epigenetic processes. *Genes Dev.*, **23**, 2124–2133.

69 Haig, D. (2004) Genomic imprinting and kinship: how good is the evidence? *Annu. Rev. Genet.*, **38**, 553–585.

70 Kawabe, A., Fujimoto, R., Charlesworth, D. (2007) High diversity due to balancing selection in the promoter region of the *Medea* gene in *Arabidopsis lyrata. Curr. Biol.*, **17**, 1885–1889.

71 Miyake, T., Takebayashi, N., Wolf, D.E. (2009) Possible diversifying selection in the imprinted gene, MEDEA, in *Arabidopsis*. *Mol. Biol. Evol.*, **26**, 843–857.

72 Spillane, C., Schmid, K.J., Laoueille-Duprat, S., Pien, S., Escobar-Restrepo, J.M., Baroux, C., Gagliardini, V., Page, D.R., Wolfe, K.H., Grossniklaus, U. (2007) Positive Darwinian selection at the imprinted MEDEA locus in plants. *Nature*, **448**, 349–352.

73 Spielman, M., Vinkenoog, R., Dickinson, H.G., Scott, R.J. (2001) The epigenetic basis of gender in flowering plants and mammals. *Trends Genet.*, **17**, 705–711.

74 Jullien, P.E., Berger, F. (2010) Parental genome dosage imbalance deregulates imprinting in *Arabidopsis*. *PLoS Genet.* **6**, e1000885.

75 Walia, H., Josefsson, C., Dilkes, B., Kirkbride, R., Harada, J., Comai, L. (2009) Dosage-dependent deregulation of an AGAMOUS-LIKE gene cluster contributes to interspecific incompatibility. *Curr. Biol.*, **19**, 1128–1132.

76 Erilova, A., Brownfield, L., Exner, V., Rosa, M., Twell, D., Mittelsten Scheid, O., Hennig, L., Kohler, C. (2009) Imprinting of the polycomb group gene MEDEA serves as a ploidy sensor in *Arabidopsis*. *PLoS Genet.*, **5**, e1000663.

77 Garcia, D., Fitz Gerald, J.N., Berger, F. (2005) Maternal control of integument cell elongation and zygotic control of endosperm growth are coordinated to determine seed size in *Arabidopsis*. *Plant Cell*, **17**, 52–60.

78 FitzGerald, J., Luo, M., Chaudhury, A., Berger, F. (2008) DNA methylation causes predominant maternal controls of plant embryo growth. *PLoS One*, **3**, e2298.

79 Adams, S., Vinkenoog, R., Spielman, M., Dickinson, H.G., Scott, R.J. (2000) Parent-of-origin effects on seed development in *Arabidopsis thaliana* require DNA methylation. *Development*, **127**, 2493–2502.

80 de Jong, T.J., Scott, R.J. (2007) Parental conflict does not necessarily lead to the evolution of imprinting. *Trends Plant Sci.*, **12**, 439–443.

81 Josefsson, C., Dilkes, B., Comai, L. (2006) Parent-dependent loss of gene silencing during interspecies hybridization. *Curr. Biol.*, **16**, 1322–1328.

82 Shi, W., Lefebvre, L., Yu, Y., Otto, S., Krella, A., Orth, A., Fundele, R. (2004) Loss-of-imprinting of Peg1 in mouse interspecies hybrids is correlated with altered growth. *Genesis*, **39**, 65–72.

83 Vrana, P.B., Guan, X.J., Ingram, R.S., Tilghman, S.M. (1998) Genomic imprinting is disrupted in interspecific *Peromyscus* hybrids. *Nat. Genet.*, **20**, 362–365.

84 Pillot, M., Baroux, C., Vazquez, M.A., Autran, D., Leblanc, O., Vielle-Calzada, J.P., Grossniklaus, U., Grimanelli, D. (2010) Embryo and endosperm inherit distinct chromatin and transcriptional states from the female gametes in *Arabidopsis*. *Plant Cell*, **22**, 307–320.

85 Johnston, A.J., Matveeva, E., Kirioukhova, O., Grossniklaus, U., Gruissem, W. (2008) A dynamic reciprocal RBR-PRC2 regulatory circuit controls *Arabidopsis* gametophyte development. *Curr. Biol.*, **18**, 1680–1686.

86 Hsieh, T.F., Ibarra, C.A., Silva, P., Zemach, A., Eshed-Williams, L., Fischer, R.L., Zilberman, D. (2009) Genome-wide demethylation of *Arabidopsis* endosperm. *Science*, **324**, 1451–1454.

31
Epigenetics of Filamentous Fungi

Kristina M. Smith, Pallavi A. Phatale, Erin L. Bredeweg, Lanelle R. Connolly, Kyle R. Pomraning, and Michael Freitag
Oregon State University, Department of Biochemistry and Biophysics, and Center for Genome Research and Biocomputing (CGRB), Corvallis, OR 97331-7305, USA

Epigenetic Regulation and Epigenomics: Advances in Molecular Biology and Medicine, First Edition. Edited by Robert A. Meyers.

Keywords

Epigenetics
The study of reversible heritable changes in gene expression in the absence of changes in DNA sequence.

RIP (Repeat-induced point mutation)
A premeiotic genome defense system in some ascomycetes that detects and mutates repeated DNA segments in pairwise fashion by the introduction of C:G to T:A transition mutations.

DNA methylation
In eukaryotes typically the generation of 5-methylcytosine from cytosine by specialized enzymes, DNA methyltransferases (DMTs). Adenine methylation has been detected in some taxa, but has received much less attention.

DCDC
Three proteins termed "Defective In Methylation" (DIMs) that, together with two components of conserved E3 ubiquitin ligase complexes (CUL4 and DDB1), form a complex in which all subunits are required for DNA methylation in *Neurospora crassa*.

Jumonji C (JmjC) domain
A motif found in one of three classes of histone demethylases. JmjC proteins catalyze demethylation through an oxidative reaction, requiring Fe(II) and α-ketoglutarate as cofactors.

Quelling
An RNA-dependent post-transcriptional gene silencing pathway in fungi, first discovered in *N. crassa. Neurospora* quelling-deficient mutants (*qde*) helped to uncover the workings of the conserved RNA interference (RNAi) system.

Meiotic silencing ("MSUD")
A phenomenon that relies on unpaired DNA segments during meiosis to silence gene expression of the unpaired DNA and any transcripts of additional copies, paired or unpaired. Like quelling, this most likely functions through a post-transcriptional, RNA-based mechanism.

LaeA
A putative protein methyltransferase that is important in a variety of developmental and gene regulatory pathways in filamentous fungi. First discovered in *Aspergillus nidulans* because of "loss of *aflR* expression" in the *laeA* mutant.

Epigenetic phenomena are defined by reversible heritable changes in gene expression in the absence of changes in DNA sequence. These include, among others, DNA methylation, position effects, RNA silencing systems, and centromere location. The term *epigenetics* is now also more loosely applied to describe gene regulation via change in chromatin structure, even though such changes are not necessarily heritable (e.g., they may occur in terminally differentiated cells). The filamentous fungi, in particular *Neurospora crassa*, have provided fundamental advances in many of the areas mentioned above. Notably, they share silencing systems that are conserved in higher eukaryotes, for instance RNA interference (RNAi) and DNA methylation. Much can be learned about general mechanisms for these phenomena by comparative biology, for which fungi are especially useful. This is because they are relatively simple organisms with small genomes that often lack redundancy, and they are amenable to rapid genetic manipulations, biochemistry and cytology. At the same time, the manipulation of chromatin structure in fungi promises to unlock previously untapped biochemical potential, for instance in the production of secondary metabolites. In this chapter, a review will be provided of previously conducted studies, notably those reported during the past two years. Areas of research will also be suggested where more depth – or indeed any study – is required to make use of the full potential of filamentous fungi as model organisms.

1
Genome Defense in Filamentous Fungi

Compared to the genomes of the two best-studied fungi, the budding yeast *Saccharomyces cerevisiae* and the fission yeast *Schizosaccharomyces pombe* (both ~12.5 Mb), filamentous fungi have genomes that are larger and typically fall in the range of 35 to 50 Mb [1]. Core genome sizes vary dramatically in different genera and certain gene families may be over-represented, based on adaptation to specific niches. Major differences in the size of non-unique segments of the genome can be caused by three non-exclusive processes: (i) the amplification of repetitive sequences, for example, transposable elements (TEs); (ii) the acquisition of whole dispensable or lineage-specific chromosomes that confer advantageous traits [2, 3]; and (iii) whole-genome duplication. For the first two of these processes, evidence is accumulating as fungal genomes are sequenced at an ever-increasing pace. For example, a comparative study of three species from the genus *Fusarium* suggested the existence of mobile pathogenicity chromosomes in the asexual *Fusarium oxysporum* fs. *lycopersici* [3]. Compared to the two species with sexual reproduction, *Fusarium verticillioides* and

Fusarium graminearum, *F. oxysporum* also has increased numbers and types of TEs.

Perhaps not surprisingly, the study of epigenetic phenomena in filamentous fungi has thus uncovered several pathways that appear primarily evolved to limit the expansion and expression of non-unique genomic segments, in most cases TEs (Fig. 1). These pathways are often referred to as "*genome defense*" mechanisms. In contrast, gene-specific regulation by epigenetic phenomena during development, as has been observed or proposed for plants and animals [4–6], is currently less well understood in filamentous fungi. The recent discoveries of dicer- and argonaute-independent pathways for the generation of small RNA [7], some perhaps with regulatory functions fulfilled by microRNA (miRNA) in plants and animals, and the first report of potential gene- and promoter-specific DNA methylation in *Neurospora* [8] may shed more light on this issue.

2
Epigenetic Phenomena in Filamentous Fungi

Epigenetic phenomena can be divided into DNA-, chromatin-, and RNA-based phenomena (Fig. 1). The first eukaryotic genome defense mechanism was uncovered in *Neurospora crassa* [9]. When crossing two compatible strains, gene-sized duplications of about 1 kb or larger are detected by a genome-wide scanning process, and both copies of pairwise duplications are mutated by C:G to T:A transitions [10–12]. This phenomenon, termed "repeat-induced point mutation" (RIP) is, strictly speaking, a genetic mechanism because it changes the DNA sequence yet generates the AT-rich substrate DNA for subsequent DNA methylation in vegetative *Neurospora* tissue by numerous point mutations restricted to the duplications [13–17]. A related phenomenon, termed "methylation induced premeiotically" (MIP) operates in *Ascobolus immersus*, and shares many hallmarks with RIP [18–21]. In this case, however, mutation does not occur and the duplicated regions are marked by DNA methylation alone.

DNA methylation, which perhaps is the archetype of all epigenetic modifications, is found in many (but not all) filamentous fungi [22]. Both, *Neurospora* and *Ascobolus* have been subjected to extensive studies to uncover both the control and function of DNA methylation in fungi and eukaryotes in general [23–26]. Other well-studied taxa, such as *Aspergillus nidulans* and its close relatives, seem to lack DNA methylation, RIP and MIP, even though proteins with homology to cytosine methyltransferases are found in their genomes [27–29].

Chromatin state-dependent position effects – that is, telomere position effects (TPEs) or position effect variegation (PEV) – have only recently been studied in *Neurospora* and *Aspergillus* [30, 31]. Unlike *S. cerevisiae* and *S. pombe*, filamentous fungi do not possess the ability to switch mating types, such that subtelomeric silencing has received the most attention. Silencing close to the single large rDNA repeat of filamentous fungi has not yet been addressed in any mechanistic detail. As some secondary metabolite pathways tend to be clustered in filamentous fungi, recent studies have been directed at understanding the transcriptional silencing and activation of these fairly substantial chromosomal regions [32, 33]. While these phenomena may not be strictly "epigenetic," they may share some mechanistic aspects with subtelomeric

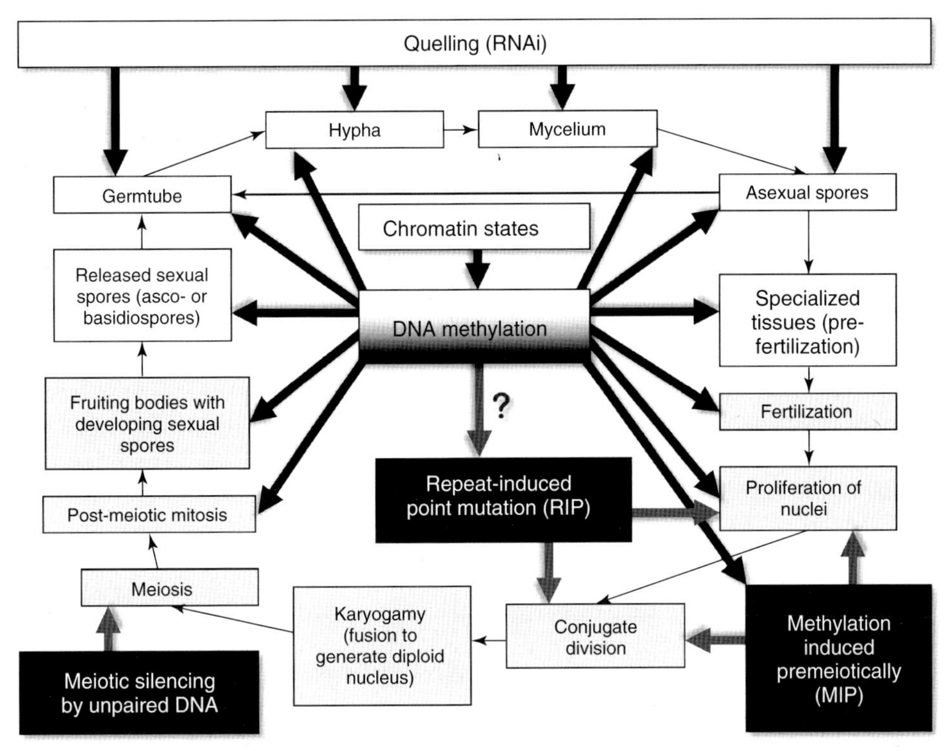

Fig. 1 Epigenetic phenomena acting in an idealized lifecycle for a filamentous fungus, based on *Neurospora*. Indicated are known steps or tissues in the life cycle that are, or may be subject to, "quelling" (RNA interference, "RNAi"), RIP, MIP, and meiotic silencing by unpaired DNA ("MSUD," "meiotic silencing"). In a typical life cycle, the products of meiosis, sexual spores (e.g., ascospores or basidiospores) develop germ tubes that mature into quickly elongating hyphae to form a complicated network of interconnected hyphae, the mycelium. In many fungi, one or many pathways exist to generate asexual spores, which in turn undergo a vegetative reproductive cycle by forming germtubes, hyphae, and mycelia. Under the correct environmental conditions, typically after depleting food sources and induced by nitrogen and/or carbon stress, mating between compatible partners occurs, which requires the production of specialized tissues for fertilization. Nuclei in specialized tissues continue to proliferate and begin a program of conjugated division, which results in karyogamy (nuclear fusion) of nuclei in specialized cells only. This forms the only and rather short diploid life stage of many filamentous fungi (though some fungi are diploid for most of their life cycles). Meiosis ensues, generating four sexual spores. In many species, however, meiosis is followed directly by at least one post-meiotic mitosis that generates eight ascospores or basidiospores. The spores are produced in fruiting bodies that are mostly, if not entirely, composed of specialized maternal tissue. Upon ripening, and under the correct environmental conditions, sexual spores are released to initiate another life cycle. The white and gray boxes indicate tissues or stages during vegetative or sexual development, respectively. Epigenetic phenomena affect discrete stages, with DNA methylation assuming a central role by acting during most stages in the life cycle, and certainly being involved in MIP (and perhaps in RIP). No known interactions occur between RNA-mediated silencing phenomena and DNA- or chromatin-mediated silencing phenomena.

chromatin silencing; hence, they are discussed in a separate section following a description of RNA-based phenomena.

Centromeric chromatin has long been thought of as a constitutively silenced heterochromatic domain. The epigenetic nature of "regional" centromeres, as compared to the single nucleosome "point" centromere of *S. cerevisiae*, has now been established, as specific DNA sequences alone are insufficient – let alone essential – to generate normal centromeres or aberrant "neocentromeres" [34, 35]. While the view of centromeres as constitutively silenced regions has been challenged by cytological and biochemical analyses of rice, fly, and mammalian chromatin [36–38], the original notion still holds in *Neurospora*, where centromeric DNA is associated largely with nucleosomes that are modified by silencing lysine modifications [39].

RNA-based epigenetic phenomena were first discovered in plants (as "cosuppression" or "post-transcriptional gene silencing"; PTGS) and *Neurospora* ("quelling"), before studies in *Caenorhabditis elegans* demonstrated the dependence of "RNA interference" (RNAi) on double-stranded RNA [40–42]. As in plants [43, 44], some fungi employ this system as a defense against viruses, best exemplified by the fungus that causes chestnut blight, *Cryphonectria parasitica* [45–47]. RNA-mediated silencing has also been identified in other fungi, even though details in the pathways vary [48–51]. In addition, *Neurospora* has a meiosis-specific, likely RNA-based mechanism that detects unpaired DNA [52, 53]. If unpaired regions are discovered during this surveillance process, then all regions with homology to the unpaired regions are post-transcriptionally and reversibly silenced, presumably by the degradation of mRNA. While similar "meiotic silencing" phenomena have been discovered in *C. elegans* and mouse, the mechanistic details are different [54].

In the following sections, the aim is to provide an overview of current knowledge of the mechanisms of the above-described epigenetic phenomena, and to discuss in detail those studies conducted during the past two years (though referral will be made to more recent reviews where appropriate). The initial discussion relates to the proteins that participate in epigenetic phenomena in filamentous fungi.

3
Parts of the Machinery: Proteins Involved in Silencing

3.1
DNA Modification

While the capacity to methylate DNA has been lost in many taxa, including many well-studied fungi [55], most filamentous fungi possess DNA methyltransferases (DMTs) that methylate C5 of cytosine to create 5-methylcytosine (5-meC). Adenine methylation has not been observed in any fungus. The fungal DMTs or DMT-like proteins can be grouped into three distinct classes:

- The first class, exemplified by *Neurospora* defective in methylation (DIM-2), is closely related to plant chromomethylases (CMTs) and is responsible for *de novo* and maintenance DNA methylation [56, 57].
- The second class, exemplified by *Ascobolus* Masc2 (Methylase Ascobolus), is more closely related to animal DNA methyltransferase 1 (Dnmt1) and plant MET1 enzymes, and

seems restricted to zygomycetes and basidiomycetes (Fig. 2).

- The third class of DMT-like proteins, exemplified by *Ascobolus* Masc1 [58], *Neurospora* RIP-defective (RID) [59] and *Aspergillus nidulans* DmtA [60], is specific to filamentous fungi in the Pezizomycotina [60–63].

The Masc1/RID/DmtA class of DMT-like proteins, which is most closely related to bacterial methyltransferases, is perhaps the most intriguing group of eukaryotic DMTs. These proteins share similar catalytic domains yet vary drastically in the length and sequence of their N-terminal and C-terminal domains [58–60]. The mutation or deletion of genes encoding these proteins results in distinct phenotypes in various groups of filamentous fungi, and these phenotypes may be linked to the known phylogeny (Fig. 2). In *Ascobolus, masc1* mutants have severe developmental phenotypes in homozygous crosses, while in heterozygous crosses MIP is much reduced; this led to the suggestion that Masc1 might be a *de novo* cytosine methyltransferase [58]; however, *in vitro* methylase activity has not been demonstrated. *Neurospora* RID [59] is required for RIP but is not needed for cytosine methylation in vegetative tissues. RIP may involve either deamination of a methylcytosine intermediate to uracil, or the methylation of cytosines followed by deamination of 5-meC to thymine [12]. RIP occurs during premeiosis, the time at which RID is transcribed [59], although just as for Masc1 no *in vitro* methylation activity by RID alone has been detected.

While *N. crassa* and *A. immersus* have each been subjected to many in-depth studies over the past 30 years, other fungi have only recently been analyzed for active RIP or MIP. The *A. nidulans* homolog of Masc1/RID, DmtA, is required for the production of sexual spores; as RIP/MIP and DNA methylation are absent from this fungus, no effect on these phenomena was observed [60]. The *F. graminearum* genome shows evidence of past RIP activity in the few transposons present in the available genome assembly [64]. Deletion of the *rid* homolog resulted in a loss of RIP (K.R. Pomraning *et al.*, unpublished results). In this species, DNA methylation is either absent or very light when compared to *Neurospora*, even though an apparently functional DIM-2 homolog is present. Both, *Leptosphaeria maculans* [65–67] and *Podospora anserina* [68, 69] can undergo RIP, yet the mutated regions are typically not methylated. Even in *N. crassa* the strength of a methylation signal is roughly proportional to the density of RIP-induced mutations. This allowed the characterization of "maintenance only" signals (where erasure results in a permanent loss of methylation) that are distinct from portable "*de novo* methylation" signals, which become reliably remethylated once the DNA methylation machinery is restored or when reintroduced into the *Neurospora* genome [15, 16, 57, 70]. In *F. graminearum, L. maculans,* and *P. anserina,* mutation frequency and density were relatively low. Thus, it remains possible that the DNA methylation machineries require more heavily mutated alleles as substrates in these fungi.

As with many other *Pezizomycotina, Sordaria macrospora* encodes RID and DIM-2 homologs, although no evidence of RIP or DNA methylation in its repeat-poor genome has been found [71]. More typically, many RIP-like transition mutations are found in regions enriched for TEs or relics of TEs in the genomes of filamentous fungi. While this has often

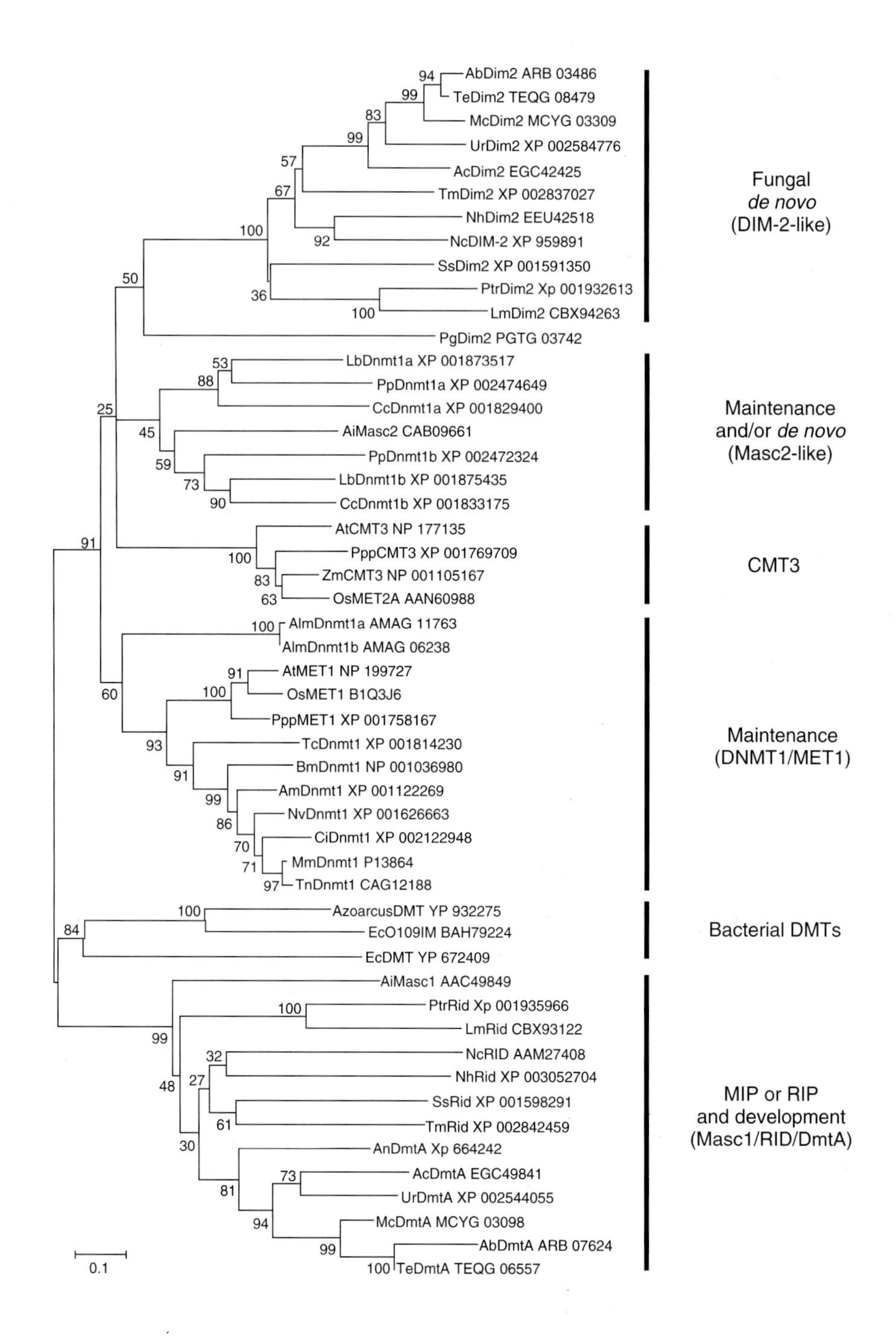
AbDim2 ARB 03486
TeDim2 TEQG 08479
McDim2 MCYG 03309
UrDim2 XP 002584776
AcDim2 EGC42425
TmDim2 XP 002837027
NhDim2 EEU42518
NcDIM-2 XP 959891
SsDim2 XP 001591350
PtrDim2 Xp 001932613
LmDim2 CBX94263
PgDim2 PGTG 03742
LbDnmt1a XP 001873517
PpDnmt1a XP 002474649
CcDnmt1a XP 001829400
AiMasc2 CAB09661
PpDnmt1b XP 002472324
LbDnmt1b XP 001875435
CcDnmt1b XP 001833175
AtCMT3 NP 177135
PppCMT3 XP 001769709
ZmCMT3 NP 001105167
OsMET2A AAN60988
AlmDnmt1a AMAG 11763
AlmDnmt1b AMAG 06238
AtMET1 NP 199727
OsMET1 B1Q3J6
PppMET1 XP 001758167
TcDnmt1 XP 001814230
BmDnmt1 NP 001036980
AmDnmt1 XP 001122269
NvDnmt1 XP 001626663
CiDnmt1 XP 002122948
MmDnmt1 P13864
TnDnmt1 CAG12188
AzoarcusDMT YP 932275
EcO109IM BAH79224
EcDMT YP 672409
AiMasc1 AAC49849
PtrRid Xp 001935966
LmRid CBX93122
NcRID AAM27408
NhRid XP 003052704
SsRid XP 001598291
TmRid XP 002842459
AnDmtA Xp 664242
AcDmtA EGC49841
UrDmtA XP 002544055
McDmtA MCYG 03098
AbDmtA ARB 07624
TeDmtA TEQG 06557
Fungal
de novo
(DIM-2-like)
Maintenance
and/or de novo
(Masc2-like)
CMT3
Maintenance
(DNMT1/MET1)
Bacterial DMTs
MIP or RIP
and development
(Masc1/RID/DmtA)
0.1

been taken as evidence for RIP [72–74], many taxa remain to be directly investigated for active RIP or even DNA methylation.

Based on the phylogenetic trees that can be inferred from alignments of the Masc1/RID/DmtA catalytic domains (Fig. 2), it is tempting to predict phenotypes for deletion alleles of these proteins in the unstudied genera. RID from *Nectria haematococca* and *N. crassa* serve as examples for proteins from the Sordariomycetes, a group which contains the most examples of fungi that have active RIP, including *N. haematococca, F. graminearum* (telomorph: *Gibberella zeae*), *P. anserina,* and *Magnaporthe grisea.* The RID homologs of *L. maculans* and *Pyrenophora tritici-repentis* should be responsible for RIP, which may very well be true for *L. maculans,* one of the few species in which RIP has been experimentally demonstrated. In *P. tritici-repentis,* a homothallic (self-fertile) fungus and its close relative, *Stagonospora nodorum,* relics of RIP exist, but at least in the *P. tritici-repentis* reference genome many nonmutated copies of TEs exist, suggesting the absence of active RIP. Similarly, RID homologs of *Sclerotinia sclerotiorum,* its close relative, *Botryotinia fuckeliana,* and the truffle, *Tuber melanosporum,* may also be responsible for RIP although at present no experimental data are available from this group of fungi.

Although species in the *Aspergillus* group have neither MIP/RIP nor DNA methylation, the DmtA homolog serves an important developmental function, at least in *A. nidulans* [60]. Additional fungi from this group need to be examined experimentally, and the same is true for relatives of *Ajellomyces* (*Histoplasma*) *capsulatus* and *Uncinocarpus reesii,* as well as the dermatophytes, represented here by *Microsporum canis, Arthroderma benhamiae,* and *Trichophyton equitum* (Fig. 2). To uncover the mechanisms for MIP and RIP, and to identify the true function of DmtA homologs in sexual development in this large group of fungi (which includes important human pathogens) remains one of the main challenges of fungal epigenetics investigations.

The two large classes of *bona fide* or putative fungal DMTs are either

←

Fig. 2 Classes of eukaryotic DMTs. A Neighbor-joining tree, based on a Molecular Evolutionary Genetics Analysis (MEGA) alignment of the catalytic domains of the respective DMTs is shown. The same taxa as previously [55] are used, but additional fungal lineages are included. Protein sequences (total of 51) were aligned by sampling 1000 replicate trees. The overall topology of this tree is similar to that constructed by Bayesian approaches [55]. Note that the Masc1/RID/DmtA homologs cluster well with bacterial DMTs. Taxa are abbreviated as follows: Ab, *Arthroderma benhamiae;* Te, *Trichophytum equinum;* Mc, *Microsporum canis;* Ur, *Uncinocarpus reesii;* Ac, *Ajellomyces capsulatus;* Tm, *Tuber melanosporum;* Nh, *Nectria haematococca;* Nc, *Neurospora crassa;* Ss, *Sclerotinia sclerotiorum;* Ptr, *Pyrenophora tritici-repentis;* Lm, *Leptosphaeria maculans;* Pg, *Puccinia graminis;* Lb, *Laccaria bicolor;* Pp, *Postia placenta;* Cc, *Coprinopsis cinerea;* Ai, *Ascobolus immersus;* At, *Arabidopsis thaliana;* Ppp, *Physcomitrella patens;* Zm, *Zea mays;* Os, *Oryza sativa;* Alm, *Allomyces macrogynus;* Tc, *Tribolium castaneum;* Bm, *Bombyx mori;* Am, *Apis mellifera;* Nv, *Nematostella vectensis;* Ci, *Ciona intestinalis;* Mm, *Mus musculus;* Tn *Tetraodon nigroviridis;* Ec, *Escherichia coli.* Accession numbers are indicated. Locus numbers for Te, Ab, Alm, Pg, and Mc are from the Broad Institute web site (http://www.broadinstitute.org/scientific-community/data).

DIM-2-type and most related to plant CMT3, or Masc2-type and most closely resembling mammalian Dnmt1 and plant Met1 DMTs, which are typically thought of as maintenance methyltransferases [62, 63]. That the DIM-2 group of enzymes is more closely related to plant CMT3 enzymes is an attractive hypothesis [63], as both enzymes are part of similar conserved pathways that function through histone modifications [23]. *Neurospora* DIM-2, the only well-studied member of this family, is required for all known DNA methylation, but it is not essential for RIP [56]. In *F. graminearum* and *F. verticillioides*, active DIM-2 enzymes may exist as there is evidence for gene-sized transcripts. Under normal laboratory growth conditions, however, there is very little – if any – DNA methylation observed in *F. graminearum* (K. R. Pomraning *et al.*, unpublished results).

DNA methylation has also been studied in detail in *A. immersus* [75–77]. In addition to *masc1*, a second gene (*masc2*) was discovered by amplification with degenerate primers made to the DMT catalytic domain [78, 79]. Curiously, upon inactivation of the gene, no effect on either MIP or DNA methylation was found [79], but the enzyme had activity *in vitro* [78]. This immediately suggests that *A. immersus* encodes a third DMT gene that is responsible for vegetative DNA methylation or at least is able to complement any *masc2* defects. This would be an unusual situation not found in other filamentous fungi for which there are near-complete genomes available, as these typically have only one Masc1/RID/DmtA and one DIM-2/Dnmt1 enzyme.

Phylogenetic analyses based on the conserved catalytic domains of Dnmt1-type enzymes have suggested that Masc2 and DIM-2 are derived from the same ancestral protein as animal and plant Dnmt1/MET1 and CMT3 [55, 63], respectively, but both reside on their own branches (Fig. 2). This is most obvious for Masc2 and the DMTs from basidiomycete fungi, including the rust *Puccinia graminis*. DMTs from *Laccaria bicolor*, *Coprinopsis cinerea*, and *Postia placenta* are unusual as they cluster together in two closely related Dnmt1-like groups, instead of one Masc1/RID group and one DIM-2/Dnmt1 group. The results of previous studies have suggested that *C. cinerea* has the capacity for MIP, or at least a MIP-like process [80]. The fact that both *C. cinerea* enzymes are considered phylogenetically more distant from the Masc1/RIP group of enzymes than was assumed earlier [60] suggests that MIP may have evolved independently in the two clades (Fig. 2). This proposal deserves further study in additional asco- and basidiomycetes.

Which fungi have lost the genes and, consequently, the capacity to carry out DNA methylation or MIP and RIP? Although, at present, no experimental data on epigenetics in chytrids are available, some chytrids also have two Dnmt1-like DMTs, here exemplified by the two enzymes from *Allomyces macrogynus* (in the Blastocladiomycota). Interestingly, when they are present these DMTs cluster more closely with animal DNMTs than with those of the fungi (Fig. 2). Two species from the Chytridiomycota, however – *Batrachochytrium dendrobatidis* and *Spizellomyces punctatus* – have no genes for DMTs. As with the chytrids, some zygomycetes (e.g., *Phycomyces blakesleeanus*) have a Dnmt1 and DIM-2 but no Masc1/RID enzyme [55], while others (e.g., *Rhizopus oryzae*) have no DMT genes at all (both of these species are in the *Mucorales*). In the basidiomycetes, no genes for DMTs have been found in *Ustilago maydis*, *Phanerochaete chrysosporium*, or *Cryptococcus* spp., and in the ascomycetes the same is true for all *Saccharomyces*, *Candida*, and related taxa, as well as the *Schizosaccharomyces* species.

One group of DMTs that is completely absent from any of the fungi studied thus far is the DRM/Dnmt3 class of *de novo* DMTs found in plants and animals, respectively [63]. This suggests that the DIM-2-like enzymes may carry out both *de novo* and maintenance methylation, as had been found in *N. crassa* [56, 57]. It is possible that the basidiomycete-type Dnmt1 enzymes – most likely generated by gene duplication – have evolved into two groups of enzymes that carry out either function, although no experimental data are currently available to support this notion.

A fourth group of DMT-like enzymes, the DNMT2 homologs, is found in some fungi, namely the genus *Schizosaccharomyces*, the chytrid *Batrachochytrium dendrobatidis*, and many of the basidiomycetes for which genome sequences are available (*Serpula lacrymans*, *Schizophyllum commune*, *L. bicolor*, *C. cinerea*, and *P. placenta*, but curiously not *Phanerochaete chrysosporium*). These proteins share an amazing similarity with DMTs, but have been found to be specialized tRNA methyltransferases instead [63, 81–83].

This introduction to the classes of fungal DMTs underscores the point that epigenetic phenomena in filamentous fungi may be as varied as the lifestyles of members in this large taxonomic unit. While shared themes may be found, details will necessarily vary among this diverse group of organisms.

3.2 Histones of Filamentous Fungi

Epigenetic phenomena can be mediated by post-translational modifications of the core histones H2A, H2B, H3, and H4, or the replacement of canonical histones with histone variants, such as the centromere-specific CENPA (or CenH3) H3 variant, and H2A variants associated with transcription (H2A.Z), repair (H2A.X), or chromosome-wide silencing (macro-H2A) [84, 85]. The linker histone H1 also can affect epigenetic mechanisms, for example, DNA methylation in *Ascobolus* [86], but this may not be a general phenomenon, even in fungi, as this did not hold true in *Neurospora* [87].

Genes for the core histones H3, H2A, and H2B, and the linker histone, H1, are present in single copy in *N. crassa* [88] and *A. nidulans* [89], with the exception of a second, dispensable copy of H4 (*hH4-2*). The relative simplicity of fungal genomes thus provides a unique opportunity to study epigenetic phenomena based on histone structure. Comparisons of all sequenced species of *Fusarium* and *Aspergillus* have revealed that all taxa have the same set of histone genes. Protein sequence alignments derived from the known or predicted histone genes of all available fungal genomes show close to 100% sequence conservation across all filamentous fungi for H3, H4, and H2A. There are minor differences for H2B, for example, the Aspergilli have changes primarily in the N-terminal tail, and thus CLUSTALW [90] pairwise alignment scores between *N. crassa* and *Aspergillus* H2B sequences are low (data not shown).

The replacement of canonical histones with variants can cause epigenetic changes in gene regulation. Among the filamentous fungi, the only variants found

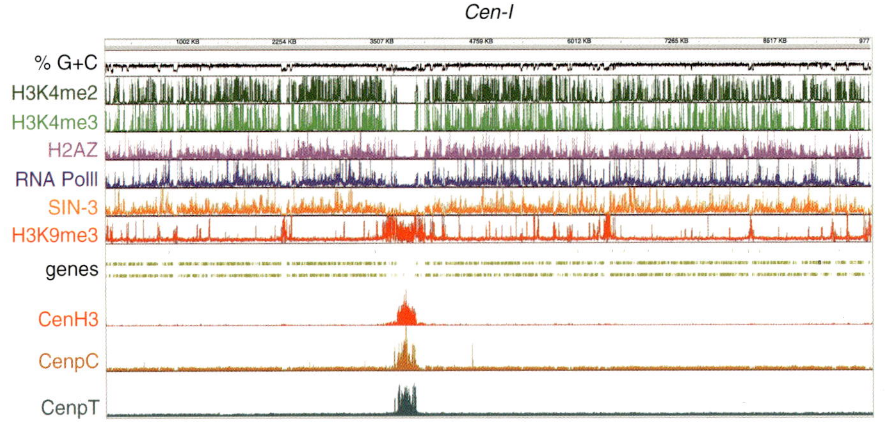

Fig. 3 Example of enrichment for histone modifications, chromatin-associated proteins and centromere proteins in *N. crassa*. ChIP-seq results are shown for *N. crassa* chromosome I with the centromere (*Cen-I*) indicated above and GC content (%G+C, in 1 kb windows) at top in black. Centromeric heterochromatin and dispersed heterochromatin is associated with low GC content and H3K9 methylation (red), but not H3K4 methylation (green) [39]. Centromeric proteins CenH3, CEN-C, and CEN-T are found only at the centromere. Euchromatic regions, where most genes are found, are enriched for marks associated with active transcription, including H3K4 methylation (me2 and me3, green) and RNA PolI II (blue; R. Addison and M.S. Sachs, unpublished results). The H2A.Z histone variant (purple) is associated with promoters of active genes. The SIN-3 corepressor (orange) is a master regulator also found globally in euchromatin, although it is typically associated with the repression of specific genes. For histone modifications, H2A.Z, RNA Pol II, SIN-3, and the centromere proteins, ChIP-seq read counts were mapped with CASHX [94] and binned into 1 kb sliding windows (slide of 200 bp).

are a centromere-specific H3 (CenH3; this is discussed further below) and a variant of H2A, H2A.Z. In yeasts, H2A.Z (referred to as Htz1 in *S. cerevisiae* and $H2A.Z^{Pht1}$ in *S. pombe*) replaces H2A at the promoters of inactive or weakly expressed genes [91, 92]. Fission yeast $H2A.Z^{Pht1}$ also cooperates with the heterochromatin machinery to promote the degradation of antisense read-through transcription from neighboring genes by the exosome [93]. At the present time, the role of H2A.Z in filamentous fungi remains to be explored. We performed ChIP-sequencing with a green fluorescent protein (GFP)-tagged H2A.Z (Fig. 3) in a heterokaryotic transformant that showed slight growth defects. A homokaryotic strain, in which only the tagged copy remains, was not obtained, which suggests that deletion of H2A.Z may be lethal in *Neurospora*. H2A.Z was found to be largely associated with promoters of genes that also showed di- or trimethylation of histone H3 lysine 4 (H3K4me2 or -me3), thus suggesting that *Neurospora* H2A.Z is found in promoters of active genes. H2A.Z was under-represented in pericentric or centromeric regions (Fig. 3 and data not shown).

Extensive investigations have been conducted, primarily in *N. crassa* and more recently in *A. nidulans*, to determine the role of histone modifications in chromatin regulation. The N-terminal tails of H3 and H4 are required to generate transcriptionally repressive heterochromatin and transcriptionally active euchromatin. Typically, lysines in the H3 and H4 tails are hyperacetylated in euchromatin, and H3K4 is trimethylated. In heterochromatin, however, H3K9 is trimethylated and other lysines are hypoacetylated. Mutational studies in *S. cereviseae* and *S. pombe* have revealed a role for specific histone residues in transcription and epigenetic phenomena. For example, changes to H3K9, H3S10, or H3K14 in *S. pombe* caused a loss of centromere silencing and defective chromosome segregation, and a phenotype similar to the deletion of Swi6, a homolog of heterochromatin protein 1 (HP1) [95]. Similar studies are currently under way in *Neurospora.*

3.3 Proteins Involved in Histone Modification

Histone-modifying enzymes are responsible for regulating not only transcription but essentially all DNA processes – that is, recombination, repair, replication, chromatid cohesion, and chromosome segregation. Histones can be modified by lysine acetylation, lysine and arginine methylation, phosphorylation, ubiquitylation, sumoylation, and ADP-ribosylation [96–99]:

- Acetylation is reversible and catalyzed by histone acetyltransferases (HATs) and histone deacetylases (HDACs).
- Lysine methylation is mediated by SET domain-containing histone methyltransferases (HMTs) and a smaller class of proteins homologous to DOT1 [100–103].
- The demethylation of histone lysines is carried out by proteins with Jumonji domains or homologs of LSD-type proteins [100, 104].
- Arginine methylation is catalyzed by protein arginine methyltransferases (PRMTs) [105], and is opposed by peptidylarginine deiminase enzymes [101].
- The phosphorylation of histones is regulated by kinase and phosphatase enzymes, and ubiquitylation by ubiquitin ligase-containing complexes.

In this chapter, the discussion is limited to enzymes that have been well studied in filamentous fungi, and for which roles in epigenetic phenomena have been identified.

Histone modifications themselves can change the conformation of chromatin. More often, specific histone modifications interact with or are "read" by proteins that cause epigenetic changes [106]. For instance, acetylated lysine residues are recognized by bromodomain proteins that are part of complexes involved in transcriptional activation. Methylated lysine residues, however, are bound by chromodomain proteins that can form either transcriptionally silent heterochromatin (e.g., HP1) or may result in transcriptional activation (e.g., chromodomain 1; CHD1). The current knowledge relating to these types of adaptor protein in filamentous fungi, in the context of epigenetic marks and their regulation, are discussed in the following sections.

4 Propagation of Epigenetic Marks in Filamentous Fungi

4.1 Function of DNA Methylation in Fungi

In filamentous fungi, DNA methylation is largely thought of as a genome defense mechanism to block the transcription of "selfish DNA," such as TEs. This notion was proposed following the discovery of DNA methylation and RIP in *Neurospora* [107] and MIP in *A. immersus* [21]. As described above, both mechanisms function to silence duplicated DNA, either by solely epigenetic (MIP) or a combination of epigenetic and mutagenic (RIP) means. As outlined above, in RIP duplicated regions are targets for C→T transition mutations. The remaining cytosines within or directly flanking the duplicated copies are subject to *de novo* DNA methylation during the later stages of the sexual cycle (i.e., in the developing ascospores) and, if the now AT-rich DNA serves as a sufficiently strong signal for *de novo* methylation, as a substrate for the DNA methylation machinery in vegetative tissues [12, 15–17]. In MIP, duplicated DNA segments are methylated, but no mutagenesis occurs [18]. Both, RIP and MIP effectively silence transcription from duplicated regions [19, 108, 109] by their effects on transcript elongation.

Silencing by DNA methylation would, for instance, affect duplications resulting from transposon replication and insertion. In this context, RIP is not always required as a prelude for *de novo* methylation in *N. crassa*, as in-depth studies with portable "methylation signals" have shown [13, 16, 25, 57, 110]. In fact, akin to a DNA-based "immune system," it is sufficient for invading DNA to have a high AT content and for high numbers of ApT dinucleotides to be recognized as "foreign" and be subjected to DNA methylation. The exact consequences of DNA methylation in relation to the activity of the transcriptional apparatus, and whether this mechanism of gene silencing occurs in other fungi, remain to be determined.

Transcriptional silencing by DNA methylation can also spread into neighboring genes [111]. Most recently, this phenotype has been used as the basis for a powerful forward genetic screen to identify additional factors required for DNA methylation [112, 113]. A single *Neurospora* strain contains genes that are reversibly silenced by DNA methylation. These so-called "epialleles," *am::hph*^{me} and *his-3::bar*^{me}, were generated by duplications of the original genes, followed by RIP, and selection and screening for alleles that can be reactivated by the loss of DNA methylation alone. Thus, the absence or altered activity of proteins involved in DNA methylation is expected to result in relief from silencing of the *hph* and *bar* genes. Upon mutagenesis, hygromycin-resistant strains that grow in the presence of Basta should be obtained, as has been illustrated in the first two reports resulting from this selection [112, 113].

In the past, fungi have been fruitful models for DNA methylation research, in part because cytosine methylation is not essential in *N. crassa* and likely *A. immersus*. Nevertheless, the long-term effects of lacking DNA methylation have not yet been examined, and it remains unknown as to how the absence of DNA methylation might affect fungi on an evolutionary timescale. One prediction is that the TEs which are inefficiently mutagenized by RIP during the sexual cycle would still be transcribed. As will be discussed below, such transcripts are likely subject to RNA-mediated silencing. The extent to which DNA methylation contributes to the maintenance or stability of fungal genomes, for example, by potential involvement in DNA recombination or repair, also remains unknown at present.

4.2 Control of DNA Methylation

Intensive studies on the control of DNA methylation in *N. crassa* [112, 114–118] have revealed mechanisms that were also found in plants and mammals [119–124]. Thus, it is now known that several well-studied chromatin factors are essential for the regulation of DNA methylation in *Neurospora* [23, 112], and at least for part of the *de novo* methylation observed in *Arabidopsis* and mammals. The DIM-5 H3K9 HMT is required to place an epigenetic mark on H3, in this case trimethylation of Lys9 (H3K9me3) [114, 115]; this mark is then "read" by the chromodomain of HP1 [116]. HP1 interacts directly with and recruits the DMT, DIM-2 [117], which is the single DMT responsible for all DNA methylation in *Neurospora* [56]. The loss of DIM-2 has little – if any – effect on DIM-5 activity or HP1 localization.

How does DIM-5 know where to go? AT-rich DNA, generated by the action of RIP, is a substrate for DIM-5 [70], which is found in at least one complex, the DCDC (DIM-5, -7, -9, CUL-4, and DDB-1 Complex) that recruits HP1 and DIM-2 [112, 113, 125]. DIM-7 targets DIM-5 to regions of heterochromatin, but the activity of DIM-5 is dependent on the entire DCDC. DIM-7 is also required for the interaction between DIM-5 and DIM-9 (or DDB-1/CUL-4 Associated Factor, DCAF) [112, 126]. Although DDB-1 and CUL-4 form part of a conserved E3 ubiquitin ligase complex [127], the possible substrate in the control of DNA methylation remains unknown.

Whereas the discovery of the DCDC suggested that DIM-5 does not act alone, the characterization of DCDC components still failed to reveal how AT-rich DNA is recognized. One possibility is that additional histone-modifying enzymes play a role in DNA methylation, presumably by acting upstream of DIM-5. This may be accomplished by differentially modifying the H3 tail in a way that DIM-5 or the DCDC complex members would recognize it as a substrate. In fact, some evidence has been obtained to support this proposal, as a partial loss of function allele of *ppp-1*, encoding Protein Phosphatase-1, resulted in increased H3S10 phosphorylation and a selective loss of H3K9 methylation and subsequent DNA methylation from certain regions in the genome [128]. Similarly, mutation of two of the four classical HDAC genes, *hda-1* and *hda-3*, resulted in a loss of DNA methylation from some, but not all, regions of heterochromatin. In this study, a loss of DNA methylation was associated with a loss of H3K9 methylation and increased H3K9 and H3K14 acetylation; this suggested that the HDA proteins might act upstream of DIM-5, or that feedback existed from the HDACs to DIM-5 [129].

The next logical question is how histone kinases and phosphatases or histone acetylases and deacetylases are attracted to AT-rich DNA, which is typically generated as a consequence of the action of RIP or by transformation. An attractive group of proteins involved in this are high-mobility group proteins, the linker histone H1, and other proteins that have "AT-hook" motifs. The treatment of *Neurospora* with distamycin, a compound that competes with AT-hooks for binding to the minor groove of AT-rich DNA, resulted in a decreased DNA methylation [17], while disruption of the *H1* gene had no obvious effect [87]. As there is biochemical evidence for proteins that bind to AT-rich DNA generated by RIP [57], there may be some AT-hook proteins and additional uncharacterized proteins present that recognize DNA mutated by RIP, and that will link the pathway from RIP by RID to DNA methylation by DIM-2.

4.3 Boundaries for DNA Methylation

One of the oldest questions in epigenetics research relates to which types of DNA structures or proteins would serve as a signal to inhibit the spreading of silencing marks, so that silencing would be stopped before it reached euchromatic regions that needed to be transcribed. Recent studies, again with *Neurospora,* have shed light on the activity of a Jumonji C (JmjC) domain and AT-hook containing protein, DMM-1 [118], a relative of the *Arabidopsis* IBM1 protein [123, 124]. DMM-1 is recruited to the edges of heterochromatic regions, and stops the spread of DNA methylation. DMM-1 does not seem to interact with H3 or H4 peptides, but rather functions in a complex with a DNA-binding motif-containing protein, DMM-2. Presumably, the JmjC domain is involved in the demethylation of proteins (perhaps specific histone residues), although even after substantial efforts a substrate has not yet been identified. Growth deficiencies observed by the loss of DMM-1 were relieved by deletion of the gene for the DMT, DIM-2, and also by treatment with a cytosine methylation inhibitor, 5-azacytidine (5AC) [118]. DMM-1 also seems to interact with HP1 independently from the DNA methylation state. This boundary complex is a prime example for the maintenance of chromatin states that are critical for epigenetic inheritance. It suggests that competing chromatin complexes generate a boundary between active and silent regions, rather than specific DNA sequences or protein/DNA structures.

4.4 Chromatin-Based Position Effects

TPEs, which were first demonstrated in *Drosophila melanogaster* [130], occur when a gene is silenced due to its proximity to a telomere. Subtelomeric regions, found proximal to the telomere repeats that cap most eukaryotic genomes, are generally transcriptionally repressive chromatin environments. In the budding yeast, *S. cerevisiae,* a class of proteins with NAD-dependent HDAC activity, Sir2 (Silent information regulator) [131] and the related sirtuins, were shown to be required for telomere silencing [132]. Sir2 functions with the yeast-specific Sir3 and Sir4 proteins [133] to silence transgenes inserted near telomere ends.

As *S. cerevisiae* lacks all components of the pathway to generate heterochromatin via H3K9 methylation, other fungi have become important models to study subtelomeric silencing. In *S. pombe* and *N. crassa,* just as in higher eukaryotes, the sirtuins are conserved but function with H3K9 HMTs to establish repressive environments. The *S. pombe* Sir2p homolog is required to deacetylate H3K9, creating a substrate for the Clr4 H3K9 HMT [134]. Subsequent binding of H3K9me3 by the HP1 homolog, Swi6, is required for silencing at telomeres, mating-type loci, and inner centromeric repeats [134]. A conserved mechanism was demonstrated in *N. crassa* [30], where both NAD-dependent and classical HDACs were shown to cooperate to deacetylate histones in subtelomeric regions [30]. Furthermore, combining multiple sirtuin mutations in *nst* (*Neurospora sir two*) genes caused a more drastic derepression than was seen with single mutations, and demonstrated an overlapping specificity and cooperative functions of the seven NAD-dependent

HDACs and four classical HDACs of *N. crassa* [30, 129, 135]. Derepression of subtelomeric transgenes was even more drastic in *dim-5* or *hpo* mutants than in the combined *nst* mutants, presumably because not all HDAC activity was lost and DIM-5 and HP1 act downstream in the silencing pathway [30].

The role of DNA methylation at *N. crassa* telomeres or subtelomeres is subtle. Whereas light DNA methylation is present at subtelomeric regions, mutation of the DMT gene, *dim-2*, caused only a slight loss of silencing, but not at all genes tested [30]. In this case, repression by H3K9 methylation and HP1 binding is thus independent of their role in establishing DNA methylation.

Subtelomeric gene silencing was also demonstrated in *A. nidulans* [31]. In this case, transgenes inserted near telomeres were silent, and the silencing was relieved by disruption of the genes for HepA (the *A. nidulans* HP1 homolog), ClrD (the Clr4/DIM-5 H3K9 MTase), HdaA (a homolog of *S. cerevisiae* Hda1 and *N. crassa* HDA-1), or NkuA (a homolog of Ku70). Silencing was not affected by disruption of the gene for a Sir2 homolog, *hstA*. A role for NkuA is somewhat surprising, since the Ku70/80 complex is required for telomere silencing in *S. cerevisiae* through an interaction with the Sir2, Sir3, Sir4 complex, which only exists in close relatives [136, 137]. No effect on telomeric silencing resulted from Ku70 mutation in either *S. pombe* [138] or *N. crassa* [30], where *mus-51* encodes the *Neurospora* Ku70 homolog [139].

Other factors involved in telomere silencing in *Neurospora* remain to be discovered. For example, H3K27me3 was enriched at telomeres and reduced in the *nst* mutants [30]; likewise, H4K20me3 was present. The role of these modifications in subtelomeric silencing remains to be explored. Filamentous fungi provide an excellent model for the study of silencing, in that targeted mutations are easy to create and, in general, the genomes lack gene families that create redundancy. There is an untapped potential to further clarify mechanisms of subtelomeric silencing by heterochromatin formation and maintenance, and many additional likely candidate genes exist that have not been investigated. For example, the *Neurospora* genome encodes at least nine SET-domain-containing proteins [135], one of which – the H3K9 HMT DIM-5 – was discussed above. Of the others, only SET-2 has been studied in detail so far (see below). Clearly, a role exists for SET domain proteins that control H3K27 and H4K20 methylation, along with telomere-binding proteins such as the mammalian Telomere Repeat-binding Factors (TRFs; Taz1 in *S. pombe* [140] and Sir3 in *S. cerevisiae* [141]) and the functionally conserved Rap1 protein [142]. While forward genetic screens for mutants are currently still hampered by the leakiness of repression or derepression phenotypes [30], much can be learned from these organisms.

At present, very little is known about the mechanisms of telomere maintenance in filamentous fungi. In both budding and fission yeast and in mammals, the cycle for telomere repeat shrinking and re-elongation and the regulation of telomerase has been the subject of many studies [143–154], but this subject has to date received minimal attention in the filamentous fungi. Forward and reverse genetic screens, as well as biochemical studies, will likely be rewarded with the identification of unknown factors, as *N. crassa* shares DNA methylation with the mammals, a feature lacking in fission and budding yeast. In this context it would be interesting if recent experiments that resulted in chromosomes without telomere repeats in telomerase deficient *S. pombe* could be repeated with filamentous fungi. In this study [146], telomeres were functionally replaced by continually amplifying and rearranging heterochromatic sequences, termed "HAATI" (heterochromatin amplification-mediated and telomerase-independent) that are dependent on the conserved end-protection protein Pot1.

4.5
Centromere Identity and Centromere Silencing

Centromeres form the foundation for kinetochores, the attachment points for spindle microtubules during nuclear division, which is an essential process in all eukaryotes [35]. What constitutes a functional centromere remains unclear, and how centromeres are assembled and maintained remains one of the fundamental questions in cell biology. One constant element across all eukaryotes is the replacement of canonical H3 at centromeric nucleosomes with an H3 variant called *CENPA* in mammals, Cid in *Drosophila*, Cse4 in *S. cerevisiae* and *Candida albicans* [155], and CenH3 in *N. crassa* [39]. CenH3 sequences from fungi show high variability in length and sequence of the N-terminal tail and loop I region within the histone fold domain [156]. Whilst much is known about kinetochore and centromere complexes and their interactions, much less is known about the targeting of centromere proteins to chromatin. Previously, most models have suggested that protein inheritance overrides DNA sequence – that is, the centromere position is determined epigenetically by an inheritance of the parental state, rather than by specific CenH3–centromere DNA interactions [36, 157–169]. Nevertheless, a cause-and-effect relationship between certain histone modifications and centromere identity has not been established. Separating the influence of DNA sequence versus epigenetic modifications on centromere identity is further complicated in most organisms by the difficulty in capturing centromeric DNA sequences in genome assemblies. Among the filamentous fungi, only in *N. crassa* has the centromere DNA been identified both genetically [170] and biochemically [39]. The centromere sequences in *N. crassa* are 150 to 300 kb in length, and consist of RIP-mutated transposon relics. The same classes of TE relics are found at subtelomeric and dispersed heterochromatin regions, but CenH3 only associates with these sequences at the centromeres [39].

Much of what is currently known about centromere assembly and maintenance is based on studies in the yeasts *S. cerevisiae, S. pombe,* and *C. albicans*. Studies with *S. cerevisiae* have uncovered centromeres with well-defined DNA elements recognized by specialized DNA-binding proteins, suggesting a requirement for a specific DNA sequence to determine the centromere position [35, 171]. This "point centromere" model, however, did not hold true for *S. pombe* [172–174], *C. albicans* [175, 176], and other eukaryotes [177–185], which instead form "regional centromeres" that are characterized by varying lengths of usually repeated DNA [38, 157, 166, 186–192].

In the dimorphic *C. albicans*, each regional centromere is composed of a unique 3 to 4.5 kb sequence that is not repetitive [175]. The size and position of these *CEN* sequences is conserved between phylogenetically divergent *C. albicans* strains [193].

Centromere deletion studies show that neocentromeres can form near-repetitive DNA, often near the original centromere [194]. Although Cse4 is preferentially recruited to sites near repetitive DNA, the DNA sequence alone does not establish centromere identity. For example, when naked centromere DNA was introduced on a plasmid Cse4 failed to associate with it, and a centromere was not formed [176].

The much larger centromeric regions of the three *S. pombe* chromosomes each contain the same repeat elements underlying the centromere core (*imr* and *cc*) and surrounding pericentric heterochromatin (*otr* repeats) [195]. To establish CenH3 at the *imr* and *cc*, an RNAi-directed heterochromatin assembly at the *otr* is required. This is only true for the *de novo* assembly of centromeres on plasmid-based minichromosomes, but not for the inheritance of functional, native centromeres [196, 197]. Tethering the Clr4 HMT directly to minichromosomes induced synthetic heterochromatin, which was the only requirement to recruit CenH3 and form a functional artificial centromere [197]; that is, the direct binding of Clr4 abolished the need for the RNAi pathway. The deletion of a centromere caused neocentromere formation near telomeric heterochromatin, or alternatively, intertelomere fusions were seen between the acentric chromosome and another chromosome [198]. The spontaneous formation of neocentromeres in the absence of endogenous centromere repeats is further evidence that regional centromeres can form independently of DNA sequence, although CenH3 has a preference for centromeric DNA. Centromere silencing and chromosome segregation required the presence of the H2A.Z histone variant (H2A.Z^{Pht1}) and the complex that targets it, including the chromatin remodeler Swr1 and JmjC domain protein Msc1 [91, 93]. At the same time, H2A.Z was not found in centromeric nucleosomes, but rather was required for the correct expression of the inner kinetochore protein CenP-C [91]. In the absence of H2A.Z, Swr1, or Msc1, CenP-A localization was normal, but centromere silencing and chromosome segregation were defective [91]. These results provided one of few clues available to distinguish the role of centromeric chromatin from pericentric heterochromatin in centromere silencing.

In *S. pombe, Drosophila* and humans, centromere cores appear to be marked by euchromatic H3K4me2, while pericentric regions contain the heterochromatic H3K9me and HP1 [199]. That centromere cores are associated with H3K4me2 – a modification typically associated with transcriptionally active regions – was unexpected, as centromeres have long been considered constitutively silenced. Indeed, it was found that in *N. crassa*, H3K9me3 coincides with CenH3 [39] (Fig. 3). This result was similar to the situation in mice, where CenH3 has been colocalized with H3K9me2 [200], and also in chicken, where a high-resolution map of kinetochores suggests the coexistence of H3K4me2, H3K9me3, and CenH3 in centromere cores [201]. However, conflicting data make it desirable to further investigate the epigenetic states of centromeric chromatin. For this, *Neurospora* provides an excellent opportunity as it has experimental advantages similar to those of the yeasts (as noted above) and centromeres that are more similar to those of humans. Not only was H3K9me3 found at centromere cores of *N. crassa*, but the H3K9 HMT, DIM-5, and HP1 were required for correct CenH3 distribution at the centromeres [39]. In *dim-5* and *hpo* mutants, CenH3 localization at some (but not all) centromeres was altered. This was in contrast to *S. pombe* centromeres, where the Clr4 H3K9 HMT and the HP1 homolog, Swi6, were required for the *de novo* assembly but not the inheritance of centromeres [196].

One of the themes of future centromere research will be to determine if, what is true for the best-studied filamentous fungus *N. crassa,* also holds for other taxa in this group. Consequently, we have begun to assemble the centromeric DNA sequences of several filamentous fungi in the genera *Aspergillus, Fusarium,* and *Mycosphaerella,* and – as in *Neurospora* – the distribution of centromere proteins in these taxa is currently being mapped with ChIP-seq.

5 RNA-Dependent Silencing Phenomena

Chromatin-dependent gene silencing and DNA methylation may play a major part in the long-term silencing, often of larger chromosome domains. Over the past 20 years, RNA-based silencing mechanisms have been discovered that may permanently or reversibly alter gene expression from invading selfish elements or viruses by small interfering RNA (siRNA), regulate normal gene expression in dedicated pathways by micro RNA (miRNA), or affect long-term silencing by the expression of mRNA-size or larger noncoding transcripts, for example, as found in mammalian X-chromosome inactivation by Xist RNA [202–204]. Examples of the first mechanism were found at an early stage in *N. crassa* and *C. parasitica* [41, 42, 46, 205]. The existence of miRNA-like "milRNA" has suggested that some regulatory function exists for small RNA in filamentous fungi [7], but there is no evidence for long noncoding transcripts of the Xist-type that affect gene silencing. Nonetheless, projects to annotate the transcriptomes of filamentous fungi, similar to recent studies with *C. albicans* [206] and *S. pombe* [207], are currently under way.

5.1 Post-Transcriptional Gene Silencing by Small RNA

One of the first RNA silencing mechanisms to be discovered was "quelling" in *N. crassa* [42]. The insertion of additional ectopic copies of a gene in transformation experiments can cause the production of aberrant RNA, which in turn can result in a reversible silencing of both the ectopic and endogenous copies in vegetative haploid tissue [208]. It has since been shown that this is a post-transcriptional process that relies on the generation of small RNAs that function *in trans* to degrade mRNA from all copies of the target gene and cause gene silencing. Quelling effectively silences TEs [209]. The first mutant in any RNAi pathway, *qde-1* (*quelling defective-1*), was isolated in *N. crassa* and shown to encode an RNA-dependent RNA polymerase (RdRP) [210]. Two additional *qde* mutants were isolated in the initial mutant screen, and encoded an argonaute homolog (QDE-2) [41] and a RecQ DNA helicase homolog (QDE-3) [211, 212]. Small RNAs of about 25 nt, termed siRNAs, were found and their production was shown to be dependent on *qde-1* and *qde-3,* but not *qde-2* [213]. Two dicer-like genes, *dcl-1* and *dcl-2,* have redundant functions in generating the siRNAs from the aberrant RNA generated by QDE-1 and QDE-3, though the *dcl-2* mutant shows greater reduction in siRNA than the *dcl-1* mutant and is therefore most likely the primary dicer involved in quelling [214]. The siRNAs associate with the RNA-induced silencing complex (RISC), in which the argonaute QDE-2 and a QDE-2 interacting protein (QIP) are essential components [41, 213, 215]. The proposed model states that QDE-2 cleaves the passenger strand of the siRNA and the QIP exonuclease removes it, leaving the guide strand that targets mRNA [215].

Mechanisms for the synthesis of aberrant RNA and their recognition remain poorly defined. In a recent study with *N. crassa* [216], DNA damage was shown to induce the expression of QDE-2. The same study also revealed a new class of small RNA, "QDE-2-interacting" RNAs (qiRNAs), which are 20 or 21 nt long, and thus several nucleotides shorter than *Neurospora* siRNAs. The qiRNAs have a strong preference for uridine at the 5′ end, and originate mostly from the ribosomal gene cluster. The production of qiRNAs requires QDE-1, QDE-3, and dicer homologs; their generation also requires DNA-damage-induced aberrant RNA as precursor, a process that is dependent on both QDE-1 and QDE-3, and which suggests that QDE-1 also has DNA-dependent RNA polymerase (DdRP) activity [216]. *Neurospora* quelling (or RNAi) mutants showed an increased sensitivity to DNA damage, which suggested a role for qiRNAs in the DNA damage response, perhaps by inhibiting protein translation [216].

By analyzing small RNAs associated with QDE-2, several additional classes of small RNA were identified in *N. crassa* [7]. Some of these small RNAs share certain hallmarks with miRNAs from animals and plants, but have structural differences and employ different biogenesis pathways when compared to both plant and animal miRNA; these RNAs are referred to as "miRNA-like small RNAs" (milRNAs). In the same study [7], "dicer-independent small interfering RNAs" (disiRNAs) were identified.

milRNAs are produced by at least four different mechanisms, which employ distinct combinations of known and previously unknown factors of the quelling machinery, including dicers, QDE-2, the exonuclease QIP, and a novel RNaseIII domain-containing protein, MRPL3. The disiRNAs originate from loci that generate partially overlapping sense and antisense transcripts, and do not require any of the known RNAi components for their production. The different characteristics of *Neurospora* disiRNAs and animal piRNAs suggest that they are two distinct classes of small RNAs. With the discovery of novel dicer- or argonaute-independent pathways to generate small RNAs [7], it is likely that more components will be identified in the near future [51].

How are small RNAs involved in chromatin-mediated silencing in filamentous fungi? In *S. pombe*, the conserved RNAi pathway directly recruits factors involved in establishing heterochromatin [217–220], and this machinery is involved in the assembly, but not the maintenance, of centromeric regions [196, 197]. A causal role for siRNAs in establishing heterochromatin in *N. crassa* has been difficult to establish. Some studies have shown small RNAs to be produced from heterochromatic regions [7, 221], though mutants in core RNA silencing components retain apparently normal heterochromatin and DNA methylation [70, 222]. Future studies should address these issues in additional species of filamentous fungi.

5.2 Meiotic Silencing by Unpaired DNA ("MSUD" or "Meiotic Silencing")

A second set of core RNA silencing pathway proteins, present in *N. crassa* [135], has been shown to be involved in Meiotic Silencing by Unpaired DNA (MSUD or "meiotic silencing") [52, 53]. Genes in unpaired regions of homologous chromosomes are post-transcriptionally and reversibly silenced throughout meiosis. This has been proposed to be a defense mechanism against TEs [53], as new insertions of TEs that accumulated during vegetative growth and asexual propagation, and which are also often mobilized during meiosis, will likely be unpaired and therefore silenced during meiosis. SAD-1, an RdRP [53] like QDE-1, and SAD-2, a protein that may assist in localizing SAD-1 to the perinuclear region where the actual silencing may occur [223], were the first proteins found to be involved in MSUD. A second argonaute protein, homolog of QDE-2 and encoded by *sms-2* (suppressor of meiotic silencing), was later identified by forward genetics [224]. The dicer protein DCL-1, which was partially redundant for quelling, is expressed predominantly during meiosis and required for MSUD [225].

Although all components of conserved RNAi pathways are required for meiotic silencing, including QIP [226], the biochemistry of meiotic tissues has proven challenging and to date no small RNAs from genes that are subject to meiotic silencing have been isolated. This, and the ordering of a large number of additional components in the system, remains a challenge in this field for the near future.

6 Transcriptional Silencing or Activation by Changes in Chromatin Structure

The realization that some epigenetic phenomena are driven by changes in the post-translational modifications of core histones resulted in the application of the term "*epigenetic*" to many pathways of transcriptional regulation. We consider this incorrect usage, as one requirement for an epigenetic phenomenon is the maintenance of regulatory states through cell division, either mitosis or meiosis [227]. At the same time, mechanisms that have been identified in either transcriptional regulation research or studies in epigenetic phenomena are often useful to advance both fields.

One idea that deserves further study in filamentous fungi is the notion of "transcriptional memory." In *S. cerevisiae*, the GAL (galactose regulon) has been studied in depth to clarify the interdependency of inheritance of expression states during mitosis. Inheritance is dependent on a nonhistone cytoplasmic factor that is diluted during division. Chromatin structure had less of an effect than the canonical protein signaling factors Gal1 and Gal3 [228, 229], and the results of these studies suggested that epigenetic phenomena could be controlled by nonhistone proteins. Additional support for this idea was provided by the variety of additional substrates that are subject to modification by histone-modifying proteins [230, 231] and that aid in transcriptional regulation. How the paths of "transcriptional regulation" and "classical epigenetics" have become intertwined is explained in the following section.

6.1 Chromatin Remodeling and Epigenetic Phenomena

Of the more than two dozen homologs that encode the catalytic subunits of the ATP-dependent chromatin-remodeling factors (CRFs) in *N. crassa* [135], only two – NCU09106 (CRF10-1) and the chromodomain-containing NCU03060 (CRF6-1) – appear to be involved in the circadian clock [8, 232]. NCU03060, which is referred to as *CLOCKSWITCH* (CSW-1), was found to be necessary for the transition from activation to repression at the *frequency (frq)* locus, apparently by altering promoter accessibility [232]. The FRQ protein is the negative element of the *Neurospora* circadian clock, and its expression is induced by the positive elements, the White Collar Complex (WCC), composed of WC-1 and WC-2 [233, 234]. FRQ inhibits *frq* expression by regulating the WCC, which generates a daily oscillation in FRQ levels and other gene products within clock output pathways [235, 236]. The production of asexual spores (conidiation) is controlled by the clock, and creates an easily observable clock phenotype. In Δ*csw-1* mutants, the normal pattern of conidiation is disrupted, and this was accompanied by chromatin changes at the *frq* promoter [232]. CSW-1 localized to the *frq* promoter, and nuclease sensitivity assays, showed a more open chromatin at the *frq* promoter in Δ*csw-1*. In the Δ*csw-1* mutant, the *frq* transcripts and FRQ protein levels were both higher in the evening hours compared to control strains. Thus, it was proposed that closed chromatin generated by CSW-1 is important for the transcriptional silencing of *frq*. Only overall histone acetylation levels were investigated at the *frq* promoter, and little change was observed in the Δ*csw-1* mutant. Consequently, additional studies are required on histone modification changes that underlie this regulation.

The second CRF ATPase subunit with clock phenotypes, CHD-1, is also required for the normal remodeling of chromatin at the *frq* locus, for normal *frq* expression, and for sustained rhythmicity [8]. Astonishingly, DNA sequences within the *frq* promoter appear to be methylated in a DIM-2-dependent way, and the deletion of *chd-1* results in an expansion of this methylated domain, similar to that found after the deletion of *dmm-1* [118]. The DNA methylation of *frq* was also altered in several clock mutants (*frq*, *frh*, *wc-1*, and the *frq* antisense transcript). DNA methylation appears to be ancillary to clock-regulated gene expression, however, as the *dim-2* strains did not exhibit clock phenotypes.

These results are unprecedented since, following decades of research on DNA methylation in *N. crassa*, not a single gene or promoter region had been found that was subject to DNA methylation under "standard" laboratory conditions. Whether the "nonstandard" conditions used in this study (e.g., a lower incubation temperature and a switch from sucrose to glucose as the carbon source) may affect global DNA methylation is unknown, but this is unlikely based on earlier results [237]. In the field, *N. crassa* grows on complex substrates (not sucrose or glucose), thus more detailed studies on "natural" substrates should be conducted to elucidate the true control and function of both DNA methylation and the clock.

Lastly, the *frq* locus is one of a couple of dozen loci in the *Neurospora* genome that has overlapping sense and antisense transcripts [238, 239]. These are produced at roughly the same time, potentially generating the disiRNAs mentioned for RNA-mediated silencing [7]. That DNA methylation is found in this particular gene, suggests that it is possible that some DNA methylation is controlled by the action of small RNA, similar to RNA-dependent DNA methylation in plants. Under typical laboratory conditions, however, the disiRNA loci [7] and DNA methylation patterns obtained by MeDIP-sequencing do not completely overlap (K. M. Smith, K. R. Pomraning, and M. Freitag, unpublished results).

6.2 Silencing and Activation of Secondary Metabolite Gene Clusters

Secondary metabolism can be defined as "the production of ancillary metabolites and 'useful' compounds, initiated after preferred carbon and nitrogen sources have been depleted" [240, 241]. Frequently, such compounds are involved in the virulence of a filamentous fungus, or in its defense against organisms that encroach on its niche. In many cases the precise biological function for secondary metabolites remains unknown.

Genes for the production of secondary metabolites are often clustered and distributed nonrandomly across the genomes of filamentous fungi [3, 31, 242]. Many of the clusters are within several dozen kilobases from the telomere-capped ends of the chromosomes. It has been proposed by several groups that proximity to the telomeres and the capacity for subtelomeric silencing may be important for the silencing and reactivation of these gene clusters. Conceptually, these clusters and many of the similarly subtelomeric pathogenicity genes of filamentous fungi [243, 244] may behave functionally like the switchable mating-type loci in budding or fission yeast.

Several approaches have been taken to alter the balance of DNA or histone modifications in order to stimulate activity from secondary metabolite clusters in the emerging field of "chemical epigenetics" [240, 245]. Wild isolates of *Cladosporium cladosporioides* and *Diatrype disciformis*, when treated with the DNA methylation inhibitor 5AC or an HDAC inhibitor, suberoylanilide hydroxamic acid (SAHA), produced novel secondary metabolite gene products [246]. Similar success was achieved by the treatment of *A. niger* with SAHA and 5AC [247, 248], and the treatment of *Penicillium citreonigrum* with 5AC [245]. The targeted disruption of certain histone-modifying enzymes has yielded additional clues as to the epigenetic regulation of secondary metabolite gene clusters. For example, mutation of the HP1 homolog in *A. nidulans*, HepA, resulted in an induction of silent secondary metabolite gene clusters [249], while HdaA (Hda1) depletion caused a derepression of telomere-proximal secondary metabolite gene clusters [250]. While these approaches yield compounds that can act as scaffolds for organic chemistry, most research groups are more interested in deciphering the underlying pathways, their aim being to create targeted changes in the production of a single or a few compounds.

Transcriptionally active euchromatin is associated with nucleosomes that have trimethylated H3K4, H3K36, and H3K79 residues [251, 252], and that are hyperacetylated. In addition, specific histone variants are used to mark active chromatin regions. At this point, a brief review will be provided of recent revelations concerning the generation of these marks in filamentous fungi, in this case mostly *A. nidulans*.

The Bre2 homolog of *A. nidulans*, CclA, is part of a Set1 protein complex, known as COMPASS (complex proteins associated with Set1). A Set1 homolog, the catalytic H3K4 methyltransferase subunit, is found in all filamentous fungi, and H3K4 di- or trimethylation is necessary for RNA Pol II binding and transcriptional activity in development and differentiation [253]. Chromatin immunoprecipitation (ChIP) experiments have shown that Δ*cclA* mutants have reduced levels of H3K4me2 and -me3, as well as overall reduced H3 acetylation, and surprisingly also reduced levels of both H3K9 di- and trimethylation [32]. The function of COMPASS and heterochromatic marks seems to be conserved in regulating fungal secondary metabolite gene clusters, as the deletion of *hepA*, the HP1 homolog, and *cclA* from the *F. graminearum* genome appear to result in altered chemical profiles [32, 254].

A *N. crassa set-2* null mutant showed absence of H3K36 methylation, poor growth, and conidiation, as well as female sterility [255]. The mutation of H3K36 to an unmodifiable leucine residue phenocopied the *set-2* defect. In concordance with results from budding yeast, H3K36 methylation was enriched in actively transcribed regions of genes [255]. Virtually nothing else is known about H3K36 or H3K79 methylation profiles in filamentous fungi.

The HDACs are required to generate chromatin environments conducive to gene silencing. The *A. nidulans* RpdA homolog of the global repressor, Rpd3, which was first identified in budding yeast, appears necessary for normal growth, conidiation, and gene regulation. In the absence of budding yeast Rpd3, acetylation is increased at H4K5, H4K12, and H3K18 at derepressed genes [256]. As the deletion of *rpdA* is apparently lethal, as is the case in *N. crassa* [129], an inducible *A. nidulans rpdA* silencing strain was constructed which revealed reduced growth and a general increase in H3 and H4 acetylation [257].

Budding yeast Hda1 deacetylates subtelomeric regions as well as the promoters of a set of genes that are largely distinct from those controlled by Rpd3 [256]. The deletion of *hdaA*, the gene for the Hda1 homolog in *A. nidulans*, caused reduced growth under conditions of oxidative stress [258]. This was due at least in part to an inability of the *hdaA* mutant strain to induce expression of the *catB* gene. CatB catalase allows cells to grow in the presence of free radical-producing drugs, and without it the *hdaA* mutant cells are more susceptible. The deletion of *A. nidulans hdaA* also caused derepression of secondary metabolite gene clusters that were relatively close to the telomeres [250]. The *A. fumigates* HdaA homolog, which is required for normal germination and growth, has also been shown to be involved in secondary metabolite gene cluster regulation [259]. Although the mutant demonstrated both up- and downregulation of several non-ribosomal peptide synthase (NRPS) genes, there was no specificity for telomere-linked regions.

Whereas some phenotypic switches in yeasts are clearly epigenetically regulated, studies in filamentous fungi have remained largely phenomenological, for example, in *Aureobasidium pullulans* [260]. Treatment with the HDAC inhibitor trichostatin A (TSA), or deletion of the TSA-sensitive HDAC Hda1, caused increased phenotype switching from white to opaque colonies in *C. albicans* [261]. Subsequently, deletion of the Rpd3 HDAC caused increased switching in both directions (i.e., from white to opaque and opaque to white); changes in the transcription of phase-specific genes were identified as the cause of the increased switching. Both, Rpd3 (class1 HDAC) and Hda1 (class 2) are widely conserved in eukaryotes, while a third class, the Hos3-like HDACs, is specific to fungi [262]. The roles of these HDACs in the epigenetic regulation of secondary metabolite gene clusters is currently being examined in several filamentous fungi, including *A. nidulans*, *F. fujikuroi*, and *F. graminearum*.

Crosstalk between various histone marks is an important regulatory element about which very little is currently known in the filamentous fungi. The role of H3K9, H3S10, and H3K14 modifications in regulating the ability of *Neurospora* DIM-5 to methylate H3K9 was discussed above. A further example of crosstalk is the *S. cerevisiae* system, which relies on the H4R3 arginine methyltransferase, Hmt1, and the HAT Gcn5 [263, 264]. Acetylated H4, especially H4K8ac, creates a preferred substrate for Hmt1p [264]. The Gcn5 HAT is responsible for acetylating H3 and H4, and mutation of *GCN5* prevents activation of its target gene *HIS3*. The mutation of *hmt1* suppresses this phenotype [264], such that Hmt1 transcriptional repression appears to be opposed by Gcn5 transcriptional activation. Hmt1 may recruit the NAD-dependent HDAC Sir2 to transcriptionally silent regions [265]. It is expected that numerous such systems operate to control the specific expression of secondary metabolite gene clusters in filamentous fungi, although to date information on these is lacking.

As noted above, filamentous fungi – like other eukaryotes – express a number of histone variants, including the centromere-specific H3 variant CenH3 and the H2A variant, H2A.Z, which has been associated with both silencing and activating activities. In budding yeast, Htz1 (H2A.Z) acts as an insulator that inhibits the spreading of silencing complexes into euchromatin [266], and is required for genome stability and the recruitment of RNA Pol II [267]. Thus, insulator elements [268] and histone variant exchange, which are known to define heterochromatin–euchromatin borders in other organisms [269, 270], may play a role in defining the boundaries of secondary metabolite gene clusters.

Perhaps the most enigmatic of all proteins affecting secondary metabolite gene clusters is the putative protein methyltransferase LaeA, first identified in *A. nidulans* [271]. Currently, LaeA is known to be a regulator of several secondary metabolite gene clusters in *Aspergillus* species, including those affecting penicillin and aflatoxin production [271, 272]. Whilst the deletion of *laeA* is not lethal, numerous morphological phenotypes are observed, including the reduced production of sclerotia in *A. flavus* [272]. An inhibitor of this general regulator has been shown to reduce the virulence of the human pathogen *A. fumigates* in a murine model, while the deletion of *laeA* in *A. fumigates* results in decreased virulence [273, 274]. One possible explanation for this phenotype stems from studies that showed that *laeA* deletion mutants had lower levels of the hydrophobin RodA, a homolog of the *N. crassa* hydrophobin EAS (or CLOCK CONTROLLED GENE-2, CCG-2; [275]), which apparently increased the uptake of these spores into the host by phagocytosis [276]. These findings also validated the results of additional studies from several laboratories that demonstrated LaeA involvement in coordinating light regulation, carbon metabolism, and secondary metabolite expression in *A. nidulans*, *A. flavus*, and *F. fujikuroi* [277–280].

While lacking several domains found in the SET- or Dot1-like HMTs, LaeA is considered to be a protein methyltransferase, based on the presence of an *S*-adenosylmethionine binding motif, and the fact that point mutations in this motif result in the same phenotype as deletion of the gene [281]. Nevertheless, *in vitro* activity has not yet been established. LaeA also interacts with and controls the activity of major regulators, such as VeA and VelB [282]. Although its regulatory activity may not directly target histones, LaeA has been shown to be instrumental in maintaining the active state of secondary metabolite gene clusters by somehow reversing H3K9 methylation marks [254, 283]. The deletion of LaeA also affects binding of HP1 [249]. Clearly, one of the most important challenges in the near future will be to decipher whether the roles that LaeA plays in multiple pathways emanate from the modification of a single or multiple substrates.

7 Concluding Remarks

When discussing specific subjects within the field of epigenetics in filamentous fungi, we indicated depth of current investigations, or lack of understanding, which will require more detailed analyses. There are, however, several topics that have not yet been introduced.

First, we decided to completely ignore prions as epigenetic principles, even though the infectious shift in protein structure from State A to State B is clearly "epigenetic" in its purest form. This subject is covered more fully in a series of excellent reports and reviews discussing prions from *S. cerevisiae* and the filamentous fungus, *P. anserina* [284–298].

Second, when introducing the idea of "transcriptional memory," mention was made only of the activating modifications, such as histone hyperacetylation and H3K4 methylation. Although, in mammals and flies, ample evidence is available for the inheritance of transcriptional states by the methylation of H3K27, this modification is absent in yeasts, but has been identified in *N. crassa* [30, 255] and *F. graminearum* (L.R. Connolly, K.M. Smith, and M. Freitag, unpublished results). In fact, a lack of H3K27 methylation in *F. graminearum* results in drastic phenotypes, the reasons for which are currently under investigation in our laboratory. Both, *Neurospora* and *Fusarium* offer many advantages for advancing the study of this important chromatin mark. One intriguing aspect is that, while genes for "Enhancer-of-Zeste"-type proteins (these form part of the Polycomb Repressive Complex 2, which generates H3K27 methylation marks) are apparently conserved [269, 299–301], homologs of genes for subunits of protein complexes that read this mark, such as "Polycomb," are not found in any of the filamentous fungi (K.R. Pomraning, K.M. Smith and M. Freitag, unpublished results).

The ever-increasing pace of improvements in mass spectrometry (MS) will soon allow the analysis of exceedingly small samples, so that the spectrum of all histone modifications may be deduced at a particular developmental stage. Whilst some in-roads have been made to that effect in certain fungi [302, 303], a combination of MS and laser dissection microscopy should soon allow the importance of chromatin modifications that occur during development and infection to be addressed. At the very least, such histone modification maps will help to relate chromatin regulation and histone deposition mechanisms from the filamentous fungi to other organisms.

Finally, results of many studies suggest that certain aspects of epigenetic regulation are ultimately defined by the positioning of nucleosomes on specific preferred sequences. This area of research, to which the late Jon Widom and coworkers made lasting and important contributions [304–307], has resulted in the identification of preferred sequences for histone variant insertion [308].

To summarize, filamentous fungi have proved incredibly useful for the discovery of, and mechanistic research on, several epigenetic phenomena, including DNA methylation (*N. crassa, A. immersus*), RNA-mediated silencing (*N. crassa, C. parasitica*), prions (*P. anserina*), and chromatin-mediated gene silencing (*N. crassa, A. nidulans*). Whilst today many such phenomena are studied in depth in more complex organisms, such as plants and animals, the growing number of laboratories engaging in "chemical epigenetics" and starting to decipher the "rules" of reversing silencing to produce bioactive compounds suggests that the future is very bright for epigenetics investigations in fungi. At present, much of what is known of this subject is derived from only a half-dozen organisms. Bearing in mind that there are well over a million species of fungi on Earth, the mechanistic studies of chromatin and epigenetics have clearly only just begun for this large and very diverse group of organisms.

Acknowledgments

The authors apologize to the many authors whose data are neither cited nor discussed for reasons of space limitations, and thank Randolph Addison and Matthew S. Sachs (Texas A&M University) for the unpublished ChIP-seq data. Studies conducted in the Freitag laboratory are supported by grants from the American Cancer Society (RSG-08-030-01-CCG), the National Institutes of Health (P01GM068087 and R01GM097637), and start-up funds from the OSU Computational and Genome Biology Initiative. The authors have no conflicting interests.

References

1 Stajich, J.E., Berbee, M.L., Blackwell, M., Hibbett, D.S., James, T.Y., Spatafora, J.W., Taylor, J.W. (2009) The fungi. *Curr. Biol.*, **19** (18), R840–R845.

2 Miao, V.P., Covert, S.F., VanEtten, H.D. (1991) A fungal gene for antibiotic resistance on a dispensable ("B") chromosome. *Science*, **254** (5039), 1773–1776.

3 Ma, L.J., van der Does, H.C., Borkovich, K.A., Coleman, J.J., Daboussi, M.J., Di Pietro, A., Dufresne, M., Freitag, M., Grabherr, M., Henrissat, B., Houterman, P.M., Kang, S., Shim, W.B., Woloshuk, C., Xie, X., Xu, J.R., Antoniw, J., Baker, S.E., Bluhm, B.H., Breakspear, A., Brown, D.W., Butchko, R.A., Chapman, S., Coulson, R., Coutinho, P.M., Danchin, E.G., Diener, A., Gale, L.R., Gardiner, D.M., Goff, S., Hammond-Kosack, K.E., Hilburn, K., Hua-Van, A., Jonkers, W., Kazan, K., Kodira, C.D., Koehrsen, M., Kumar, L., Lee, Y.H., Li, L., Manners, J.M., Miranda-Saavedra, D., Mukherjee, M., Park, G., Park, J., Park, S.Y., Proctor, R.H., Regev, A., Ruiz-Roldan, M.C., Sain, D., Sakthikumar, S., Sykes, S., Schwartz, D.C., Turgeon, B.G., Wapinski, I., Yoder, O., Young, S., Zeng, Q., Zhou, S., Galagan, J., Cuomo, C.A., Kistler, H.C., Rep, M. (2010) Comparative genomics reveals mobile pathogenicity chromosomes in *Fusarium*. *Nature*, **464** (7287), 367–373.

4 Law, J.A., Jacobsen, S.E. (2010) Establishing, maintaining and modifying DNA methylation patterns in plants and animals. *Nat. Rev. Genet.*, **11** (3), 204–220.

5 Deaton, A.M., Bird, A. (2011) CpG islands and the regulation of transcription. *Genes Dev.*, **25** (10), 1010–1022.
6 Suzuki, M.M., Bird, A. (2008) DNA methylation landscapes: provocative insights from epigenomics. *Nat. Rev. Genet.*, **9** (6), 465–476.
7 Lee, H.C., Li, L., Gu, W., Xue, Z., Crosthwaite, S.K., Pertsemlidis, A., Lewis, Z.A., Freitag, M., Selker, E.U., Mello, C.C., Liu, Y. (2010) Diverse pathways generate MicroRNA-like RNAs and dicer-independent small interfering RNAs in fungi. *Mol. Cell*, **38**, 803–814.
8 Belden, W.J., Lewis, Z.A., Selker, E.U., Loros, J.J., Dunlap, J.C. (2011) CHD1 remodels chromatin and influences transient DNA methylation at the clock gene frequency. *PLoS Genet.*, **7** (7), e1002166.
9 Selker, E.U., Cambareri, E.B., Jensen, B.C., Haack, K.R. (1987) Rearrangement of duplicated DNA in specialized cells of *Neurospora*. *Cell*, **51**, 741–752.
10 Selker, E.U., Garrett, P.W. (1988) DNA sequence duplications trigger gene inactivation in *Neurospora crassa*. *Proc. Natl Acad. Sci. USA*, **85** (18), 6870–6874.
11 Cambareri, E.B., Jensen, B.C., Schabtach, E., Selker, E.U. (1989) Repeat-induced G-C to A-T mutations in Neurospora. *Science*, **244** (4912), 1571–1575.
12 Selker, E.U. (1990) Premeiotic instability of repeated sequences in *Neurospora crassa*. *Annu. Rev. Genet.*, **24**, 579–613.
13 Selker, E.U., Jensen, B.C., Richardson, G.A. (1987) A portable signal causing faithful DNA methylation *de novo* in *Neurospora crassa*. *Science*, **238**, 48–53.
14 Selker, E.U. (1991) Repeat-Induced Point Mutation (RIP) and DNA Methylation, in: Bennet, J.W., Lasure, L. (Eds) *More Gene Manipulations in Fungi*, Academic Press, Inc., New York, pp. 258–265.
15 Singer, M.J., Marcotte, B.A., Selker, E.U. (1995) DNA methylation associated with repeat-induced point mutation in *Neurospora crassa*. *Mol. Cell. Biol.*, **15** (10), 5586–5597.
16 Miao, V.P., Freitag, M., Selker, E.U. (2000) Short TpA-rich segments of the zeta-eta region induce DNA methylation in *Neurospora crassa*. *J. Mol. Biol.*, **300** (2), 249–273.
17 Tamaru, H., Selker, E.U. (2003) Synthesis of signals for de novo DNA methylation in *Neurospora crassa*. *Mol. Cell. Biol.*, **23** (7), 2379–2394.
18 Rossignol, J.-L., Faugeron, G. (1994) MIP: An Epigenetic Gene Silencing Process in Ascobolus immersus, in: Meyer, P. (Ed.) *Gene Silencing in Higher Plants and Related Phenomena in Other Eukaryotes*, Springer-Verlag, Heidelberg, p. 26.
19 Barry, C., Faugeron, G., Rossignol, J.-L. (1993) Methylation induced premeiotically in *Ascobolus*: coextension with DNA repeat lengths and effect on transcript elongation. *Proc. Natl Acad. Sci. USA*, **90**, 4557–4561.
20 Rhounim, L., Rossignol, J.-L., Faugeron, G. (1992) Epimutation of repeated genes in *Ascobolus immersus*. *EMBO J.*, **11** (12), 4451–4457.
21 Faugeron, G., Rhounim, L., Rossignol, J.-L. (1990) How does the cell count the number of ectopic copies of a gene in the premeiotic inactivation process acting in *Ascobolus immersus*? *Genetics*, **124**, 585–591.
22 Freitag, M., Selker, E.U. (2005) Controlling DNA methylation: many roads to one modification. *Curr. Opin. Genet. Dev.*, **15** (2), 191–199.
23 Rountree, M.R., Selker, E.U. (2010) DNA methylation and the formation of heterochromatin in *Neurospora crassa*. *Heredity*, **105** (1), 38–44.
24 Selker, E.U. (2004) Genome defense and DNA methylation in *Neurospora*. *Cold Spring Harbor Symp. Quant. Biol.*, **69**, 119–124.
25 Selker, E.U., Tountas, N.A., Cross, S.H., Margolin, B.S., Murphy, J.G., Bird, A.P., Freitag, M. (2003) The methylated component of the *Neurospora crassa* genome. *Nature*, **422** (6934), 893–897.
26 Selker, E.U. (2002) Repeat-induced gene silencing in fungi. *Adv. Genet.*, **46**, 439–450.
27 Galagan, J.E., Calvo, S.E., Cuomo, C., Ma, L.J., Wortman, J.R., Batzoglou, S., Lee, S.I., Basturkmen, M., Spevak, C.C., Clutterbuck, J., Kapitonov, V., Jurka, J., Scazzocchio, C., Farman, M., Butler, J., Purcell, S., Harris, S., Braus, G.H., Draht, O., Busch, S., D'Enfert, C., Bouchier, C., Goldman, G.H., Bell-Pedersen, D., Griffiths-Jones, S., Doonan, J.H., Yu, J., Vienken, K., Pain, A., Freitag, M., Selker, E.U., Archer, D.B., Penalva, M.A., Oakley,

B.R., Momany, M., Tanaka, T., Kumagai, T., Asai, K., Machida, M., Nierman, W.C., Denning, D.W., Caddick, M., Hynes, M., Paoletti, M., Fischer, R., Miller, B., Dyer, P., Sachs, M.S., Osmani, S.A., Birren, B.W. (2005) Sequencing of *Aspergillus nidulans* and comparative analysis with *A. fumigatus* and *A. oryzae*. *Nature*, **438** (7071), 1105–1115.

28 Galagan, J.E., Selker, E.U. (2004) RIP: the evolutionary cost of genome defense. *Trends Genet.*, **20** (9), 417–423.

29 Galagan, J.E., Henn, M.R., Ma, L.J., Cuomo, C.A., Birren, B. (2005) Genomics of the fungal kingdom: insights into eukaryotic biology. *Genome Res.*, **15** (12), 1620–1631.

30 Smith, K.M., Kothe, G.O., Matsen, C.B., Khlafallah, T.K., Adhvaryu, K.K., Hemphill, M., Freitag, M., Motamedi, M.R., Selker, E.U. (2008) The fungus *Neurospora crassa* displays telomeric silencing mediated by multiple sirtuins and by methylation of histone H3 lysine 9. *Epigen. Chromatin*, **1** (1), 5.

31 Palmer, J.M., Keller, N.P. (2010) Secondary metabolism in fungi: does chromosomal location matter? *Curr. Opin. Microbiol.*, **13** (4), 431–436.

32 Bok, J.W., Chiang, Y.M., Szewczyk, E., Reyes-Dominguez, Y., Davidson, A.D., Sanchez, J.F., Lo, H.C., Watanabe, K., Strauss, J., Oakley, B.R., Wang, C.C., Keller, N.P. (2009) Chromatin-level regulation of biosynthetic gene clusters. *Nat. Chem. Biol.*, **5** (7), 462–464.

33 Shwab, E.K., Keller, N.P. (2008) Regulation of secondary metabolite production in filamentous ascomycetes. *Mycol. Res.*, **112** (Pt 2), 225–230.

34 Black, B.E., Cleveland, D.W. (2011) Epigenetic centromere propagation and the nature of CENP-a nucleosomes. *Cell*, **144** (4), 471–479.

35 Cleveland, D.W., Mao, Y., Sullivan, K.F. (2003) Centromeres and kinetochores: from epigenetics to mitotic checkpoint signaling. *Cell*, **112** (4), 407–421.

36 Sullivan, B.A., Karpen, G.H. (2004) Centromeric chromatin exhibits a histone modification pattern that is distinct from both euchromatin and heterochromatin. *Nat. Struct. Mol. Biol.*, **11** (11), 1076–1083.

37 Ma, J., Wing, R.A., Bennetzen, J.L., Jackson, S.A. (2007) Plant centromere organization: a dynamic structure with conserved functions. *Trends Genet.*, **23** (3), 134–139.

38 Yan, H., Jin, W., Nagaki, K., Tian, S., Ouyang, S., Buell, C.R., Talbert, P.B., Henikoff, S., Jiang, J. (2005) Transcription and histone modifications in the recombination-free region spanning a rice centromere. *Plant Cell*, **17** (12), 3227–3238.

39 Smith, K.M., Phatale, P.A., Sullivan, C.M., Pomraning, K.R., Freitag, M. (2011) Heterochromatin is required for normal distribution of *Neurospora* CenH3. *Mol. Cell. Biol.*, **31**, 2528–2542.

40 Jorgensen, R.A., Que, Q., Stam, M. (1999) Do unintended antisense transcripts contribute to sense cosuppression in plants? *Trends Genet.*, **15** (1), 11–12.

41 Catalanotto, C., Azzalin, G., Macino, G., Cogoni, C. (2000) Gene silencing in worms and fungi. *Nature*, **404** (6775), 245.

42 Romano, N., Macino, G. (1992) Quelling: transient inactivation of gene expression in *Neurospora crassa* by transformation with homologous sequences. *Mol. Microbiol.*, **6** (22), 3343–3353.

43 Baulcombe, D. (2004) RNA silencing in plants. *Nature*, **431** (7006), 356–363.

44 Hamilton, A.J., Baulcombe, D.C. (1999) A species of small antisense RNA in posttranscriptional gene silencing in plants. *Science*, **286** (5441), 950–952.

45 Chen, B., Choi, G.H., Nuss, D.L. (1994) Attenuation of fungal virulence by synthetic infectious hypovirus transcripts. *Science*, **264** (5166), 1762–1764.

46 Choi, G.H., Nuss, D.L. (1992) Hypovirulence of chestnut blight fungus conferred by an infectious viral cDNA. *Science*, **257** (5071), 800–803.

47 Nuss, D.L. (2011) Mycoviruses, RNA silencing, and viral RNA recombination. *Adv. Virus Res.*, **80**, 25–48.

48 Hammond, T.M., Keller, N.P. (2005) RNA silencing in Aspergillus nidulans is independent of RNA-dependent RNA polymerases. *Genetics*, **169** (2), 607–617.

49 Kadotani, N., Nakayashiki, H., Tosa, Y., Mayama, S. (2003) RNA silencing in the phytopathogenic fungus *Magnaporthe oryzae*. *Mol. Plant Microbe Interact.*, **16** (9), 769–776.

50 Nakayashiki, H., Hanada, S., Nguyen, B.Q., Kadotani, N., Tosa, Y., Mayama, S. (2005) RNA silencing as a tool for exploring gene function in ascomycete fungi. *Fungal Genet. Biol.*, **42** (4), 275–283.

51 Dang, Y., Yang, Q., Xue, Z., Liu, Y. (2011) RNA interference in fungi: pathways, functions and applications. *Eukaryotic Cell*, **10**, 1148–1155.

52 Aramayo, R., Metzenberg, R.L. (1996) Meiotic transvection in fungi. *Cell*, **86** (1), 103–113.

53 Shiu, P.K., Raju, N.B., Zickler, D., Metzenberg, R.L. (2001) Meiotic silencing by unpaired DNA. *Cell*, **107** (7), 905–916.

54 Kelly, W.G., Aramayo, R. (2007) Meiotic silencing and the epigenetics of sex. *Chromosome Res.*, **15** (5), 633–651.

55 Zemach, A., McDaniel, I.E., Silva, P., Zilberman, D. (2010) Genome-wide evolutionary analysis of eukaryotic DNA methylation. *Science*, **328** (5980), 916–919.

56 Kouzminova, E.A., Selker, E.U. (2001) Dim-2 encodes a DNA-methyltransferase responsible for all known cytosine methylation in Neurospora. *EMBO J.*, **20** (15), 4309–4323.

57 Selker, E.U., Freitag, M., Kothe, G.O., Margolin, B.S., Rountree, M.R., Allis, C.D., Tamaru, H. (2002) Induction and maintenance of nonsymmetrical DNA methylation in *Neurospora*. *Proc. Natl Acad. Sci. USA*, **99** (Suppl. 4), 16485–16490.

58 Malagnac, F., Wendel, B., Goyon, C., Faugeron, G., Zickler, D., Rossignol, J.L., Noyer-Weidner, M., Vollmayr, P., Trautner, T.A., Walter, J. (1997) A gene essential for de novo methylation and development in *Ascobolus* reveals a novel type of eukaryotic DNA methyltransferase structure. *Cell*, **91** (2), 281–290.

59 Freitag, M., Williams, R.L., Kothe, G.O., Selker, E.U. (2002) A cytosine methyltransferase homologue is essential for repeat-induced point mutation in *Neurospora crassa*. *Proc. Natl Acad. Sci. USA*, **99** (13), 8802–8807.

60 Lee, D.W., Freitag, M., Selker, E.U., Aramayo, R. (2008) A cytosine methyltransferase homologue is essential for sexual development in *Aspergillus nidulans*. *PLoS ONE*, **3** (6), e2531.

61 Colot, V., Rossignol, J.L. (1999) Eukaryotic DNA methylation as an evolutionary device. *BioEssays*, **21** (5), 402–411.

62 Grace Goll, M., Bestor, T.H. (2005) Eukaryotic cytosine methyltransferases. *Annu. Rev. Biochem.*, **74**, 481–514.

63 Zemach, A., Zilberman, D. (2010) Evolution of eukaryotic DNA methylation and the pursuit of safer sex. *Curr. Biol.*, **20** (17), R780–R785.

64 Cuomo, C.A., Guldener, U., Xu, J.R., Trail, F., Turgeon, B.G., Di Pietro, A., Walton, J.D., Ma, L.J., Baker, S.E., Rep, M., Adam, G., Antoniw, J., Baldwin, T., Calvo, S., Chang, Y.L., Decaprio, D., Gale, L.R., Gnerre, S., Goswami, R.S., Hammond-Kosack, K., Harris, L.J., Hilburn, K., Kennell, J.C., Kroken, S., Magnuson, J.K., Mannhaupt, G., Mauceli, E., Mewes, H.W., Mitterbauer, R., Muehlbauer, G., Munsterkotter, M., Nelson, D., O'Donnell, K., Ouellet, T., Qi, W., Quesneville, H., Roncero, M.I., Seong, K.Y., Tetko, I.V., Urban, M., Waalwijk, C., Ward, T.J., Yao, J., Birren, B.W., Kistler, H.C. (2007) The *Fusarium graminearum* genome reveals a link between localized polymorphism and pathogen specialization. *Science*, **317** (5843), 1400–1402.

65 Idnurm, A., Howlett, B.J. (2003) Analysis of loss of pathogenicity mutants reveals that repeat-induced point mutations can occur in the Dothideomycete *Leptosphaeria maculans*. *Fungal Genet. Biol.*, **39** (1), 31–37.

66 Fudal, I., Ross, S., Brun, H., Besnard, A.L., Ermel, M., Kuhn, M.L., Balesdent, M.H., Rouxel, T. (2009) Repeat-induced point mutation (RIP) as an alternative mechanism of evolution toward virulence in *Leptosphaeria maculans*. *Mol. Plant Microbe Interact.*, **22** (8), 932–941.

67 Rouxel, T., Grandaubert, J., Hane, J.K., Hoede, C., van de Wouw, A.P., Couloux, A., Dominguez, V., Anthouard, V., Bally, P., Bourras, S., Cozijnsen, A.J., Ciuffetti, L.M., Degrave, A., Dilmaghani, A., Duret, L., Fudal, I., Goodwin, S.B., Gout, L., Glaser, N., Linglin, J., Kema, G.H., Lapalu, N., Lawrence, C.B., May, K., Meyer, M., Ollivier, B., Poulain, J., Schoch, C.L., Simon, A., Spatafora, J.W., Stachowiak, A., Turgeon, B.G., Tyler, B.M., Vincent, D., Weissenbach, J., Amselem, J., Quesneville,

H., Oliver, R.P., Wincker, P., Balesdent, M.H., Howlett, B.J. (2011) Effector diversification within compartments of the *Leptosphaeria maculans* genome affected by Repeat-Induced Point mutations. *Nat. Commun.*, **2**, 202.

68 Hamann, A., Feller, F., Osiewacz, H.D. (2000) The degenerate DNA transposon Pat and repeat-induced point mutation (RIP) in *Podospora anserina*. *Mol. Gen. Genet.*, **263** (6), 1061–1069.

69 Graia, F., Lespinet, O., Rimbault, B., Dequard-Chablat, M., Coppin, E., Picard, M. (2001) Genome quality control: RIP (repeat-induced point mutation) comes to *Podospora*. *Mol. Microbiol.*, **39**, 1–11.

70 Lewis, Z.A., Honda, S., Khlafallah, T.K., Jeffress, J.K., Freitag, M., Mohn, F., Schubeler, D., Selker, E.U. (2009) Relics of repeat-induced point mutation direct heterochromatin formation in *Neurospora crassa*. *Genome Res.*, **19** (3), 427–437.

71 Nowrousian, M., Stajich, J.E., Chu, M., Engh, I., Espagne, E., Halliday, K., Kamerewerd, J., Kempken, F., Knab, B., Kuo, H.C., Osiewacz, H.D., Poggeler, S., Read, N.D., Seiler, S., Smith, K.M., Zickler, D., Kuck, U., Freitag, M. (2010) De novo assembly of a 40 Mb eukaryotic genome from short sequence reads: *Sordaria macrospora*, a model organism for fungal morphogenesis. *PLoS Genet.*, **6** (4), e1000891.

72 Clutterbuck, A.J. (2011) Genomic evidence of repeat-induced point mutation (RIP) in filamentous ascomycetes. *Fungal Genet. Biol.*, **48** (3), 306–326.

73 Hane, J.K., Oliver, R.P. (2010) In silico reversal of repeat-induced point mutation (RIP) identifies the origins of repeat families and uncovers obscured duplicated genes. *BMC Genomics*, **11**, 655.

74 Hane, J.K., Oliver, R.P. (2008) RIPCAL: a tool for alignment-based analysis of repeat-induced point mutations in fungal genomic sequences. *BMC Bioinformatics*, **9**, 478.

75 Goyon, C., Rossignol, J.L., Faugeron, G. (1996) Native DNA repeats and methylation in *Ascobolus*. *Nucleic Acids Res.*, **24** (17), 3348–3356.

76 Colot, V., Goyon, C., Faugeron, G., Rossignol, J.-L. (1995) Methylation of repeated DNA sequences and genome stability in *Ascobolus immersus*. *Can. J. Bot.*, **73** (Suppl. 1), S221–S225.

77 Colot, V., Maloisel, L., Rossignol, J.L. (1996) Interchromosomal transfer of epigenetic states in *Ascobolus*: transfer of DNA methylation is mechanistically related to homologous recombination. *Cell*, **86** (6), 855–864.

78 Chernov, A.V., Vollmayr, P., Walter, J., Trautner, T.A. (1997) Masc2, a C5-DNA-methyltransferase from *Ascobolus immersus* with similarity to methyltransferases of higher organisms. *Biol. Chem.*, **378** (12), 1467–1473.

79 Malagnac, F., Gregoire, A., Goyon, C., Rossignol, J.L., Faugeron, G. (1999) Masc2, a gene from *Ascobolus* encoding a protein with a DNA-methyltransferase activity in vitro, is dispensable for in vivo methylation. *Mol. Microbiol.*, **31** (1), 331–338.

80 Freedman, T., Pukkila, P.J. (1993) De novo methylation of repeated sequences in *Coprinus cinereus*. *Genetics*, **135**, 357–366.

81 Yoder, J.A., Bestor, T.H. (1998) A candidate mammalian DNA methyltransferase related to pmt1p of fission yeast. *Hum. Mol. Genet.*, **7** (2), 279–284.

82 Goll, M.G., Kirpekar, F., Maggert, K.A., Yoder, J.A., Hsieh, C.L., Zhang, X., Golic, K.G., Jacobsen, S.E., Bestor, T.H. (2006) Methylation of tRNAAsp by the DNA methyltransferase homolog Dnmt2. *Science*, **311** (5759), 395–398.

83 Dong, A., Yoder, J.A., Zhang, X., Zhou, L., Bestor, T.H., Cheng, X. (2001) Structure of human DNMT2, an enigmatic DNA methyltransferase homolog that displays denaturant-resistant binding to DNA. *Nucleic Acids Res.*, **29** (2), 439–448.

84 Ahmad, K., Henikoff, S. (2002) The histone variant H3.3 marks active chromatin by replication-independent nucleosome assembly. *Mol. Cells*, **9** (6), 1191–1200.

85 Talbert, P.B., Henikoff, S. (2010) Histone variants--ancient wrap artists of the epigenome. *Nat. Rev. Mol. Cell Biol.*, **11** (4), 264–275.

86 Barra, J.L., Rhounim, L., Rossignol, J.L., Faugeron, G. (2000) Histone H1 is dispensable for methylation-associated gene silencing in *Ascobolus immersus* and essential for long life span. *Mol. Cell. Biol.*, **20** (1), 61–69.

87 Folco, H.D., Freitag, M., Ramon, A., Temporini, E.D., Alvarez, M.E., Garcia, I.,

Scazzocchio, C., Selker, E.U., Rosa, A.L. (2003) Histone H1 Is required for proper regulation of pyruvate decarboxylase gene expression in *Neurospora crassa. Eukaryot. Cell*, **2** (2), 341–350.

88 Hays, S.M., Swanson, J., Selker, E.U. (2002) Identification and characterization of the genes encoding the core histones and histone variants of *Neurospora crassa. Genetics*, **160** (3), 961–973.

89 Ehinger, A., Denison, S.H., May, G.S. (1990) Sequence, organization and expression of the core histone genes of *Aspergillus nidulans. Mol. Gen. Genet.*, **222** (2-3), 416–424.

90 Thompson, J.D., Higgins, D.G., Gibson, T.J. (1994) CLUSTAL W: improving the sensitivity of progressive multiple sequence alignment through sequence weighting, position specific gap penalties and weight matrix choice. *Nucleic Acids Res.*, **22**, 4673–4680.

91 Hou, H., Wang, Y., Kallgren, S.P., Thompson, J., Yates, J.R. III, Jia, S. (2010) Histone variant H2A.Z regulates centromere silencing and chromosome segregation in fission yeast. *J. Biol. Chem.*, **285** (3), 1909–1918.

92 Wyrick, J.J., Parra, M.A. (2009) The role of histone H2A and H2B post-translational modifications in transcription: a genomic perspective. *Biochim. Biophys. Acta*, **1789** (1), 37–44.

93 Zofall, M., Fischer, T., Zhang, K., Zhou, M., Cui, B., Veenstra, T.D., Grewal, S.I. (2009) Histone H2A.Z cooperates with RNAi and heterochromatin factors to suppress antisense RNAs. *Nature*, **461** (7262), 419–422.

94 Fahlgren, N., Sullivan, C.M., Kasschau, K.D., Chapman, E.J., Cumbie, J.S., Montgomery, T.A., Gilbert, S.D., Dasenko, M., Backman, T.W., Givan, S.A., Carrington, J.C. (2009) Computational and analytical framework for small RNA profiling by high-throughput sequencing. *RNA*, **15** (5), 992–1002.

95 Mellone, B.G., Ball, L., Suka, N., Grunstein, M.R., Partridge, J.F., Allshire, R.C. (2003) Centromere silencing and function in fission yeast is governed by the amino terminus of histone H3. *Curr. Biol.*, **13** (20), 1748–1757.

96 Zhou, V.W., Goren, A., Bernstein, B.E. (2011) Charting histone modifications and the functional organization of mammalian genomes. *Nat. Rev. Genet.*, **12** (1), 7–18.

97 Zaidi, S.K., Young, D.W., Montecino, M., van Wijnen, A.J., Stein, J.L., Lian, J.B., Stein, G.S. (2011) Bookmarking the genome: maintenance of epigenetic information. *J. Biol. Chem.*, **286** (21), 18355–18361.

98 Yun, M., Wu, J., Workman, J.L., Li B. (2011) Readers of histone modifications. *Cell Res.*, **21** (4), 564–578.

99 Voigt, P., Reinberg, D. (2011) Histone tails: ideal motifs for probing epigenetics through chemical biology approaches. *Chembiochem*, **12** (2), 236–252.

100 Li, K.K., Luo, C., Wang, D., Jiang, H., Zheng, Y.G. (2010) Chemical and biochemical approaches in the study of histone methylation and demethylation. *Med. Res. Rev.*, 1–53. doi 10.1002/med.20228.

101 Klose, R.J., Zhang, Y. (2007) Regulation of histone methylation by demethylimination and demethylation. *Nat. Rev. Mol. Cell Biol.*, **8** (4), 307–318.

102 van Leeuwen, F., Gafken, P.R., Gottschling, D.E. (2002) Dot1p modulates silencing in yeast by methylation of the nucleosome core. *Cell*, **109** (6), 745–756.

103 Feng, Q., Wang, H., Ng, H.H., Erdjument-Bromage, H., Tempst, P., Struhl, K., Zhang, Y. (2002) Methylation of H3-lysine 79 is mediated by a new family of HMTases without a SET domain. *Curr. Biol.*, **12** (12), 1052–1058.

104 Whetstine, J.R., Nottke, A., Lan, F., Huarte, M., Smolikov, S., Chen, Z., Spooner, E., Li, E., Zhang, G., Colaiacovo, M., Shi, Y. (2006) Reversal of histone lysine trimethylation by the JMJD2 family of histone demethylases. *Cell*, **125** (3), 467–481.

105 Gary, J.D., Clarke, S. (1998) RNA and protein interactions modulated by protein arginine methylation. *Prog. Nucleic Acid Res. Mol. Biol.*, **61**, 65–131.

106 de la Cruz, X., Lois, S., Sanchez-Molina, S., Martinez-Balbas, M.A. (2005) Do protein motifs read the histone code? *BioEssays*, **27** (2), 164–175.

107 Selker, E.U., Stevens, J.N. (1985) DNA methylation at asymmetric sites is associated with numerous transition mutations. *Proc. Natl Acad. Sci. USA*, **82**, 8114–8118.

108 Rountree, M.R., Selker, E.U. (1997) DNA methylation inhibits elongation but not

initiation of transcription in *Neurospora crassa*. *Genes Dev.*, **11**, 2383–2395.
109 Barra, J.L., Holmes, A.M., Gregoire, A., Rossignol, J.L., Faugeron, G. (2005) Novel relationships among DNA methylation, histone modifications and gene expression in *Ascobolus*. *Mol. Microbiol.*, **57** (1), 180–195.
110 Foss, E.J., Garrett, P.W., Kinsey, J.A., Selker, E.U. (1991) Specificity of repeat induced point mutation (RIP) in *Neurospora*: sensitivity of non-*Neurospora* sequences, a natural diverged tandem duplication, and unique DNA adjacent to a duplicated region. *Genetics*, **127** (4), 711–717.
111 Irelan, J.T., Selker, E.U. (1997) Cytosine methylation associated with repeat-induced point mutation causes epigenetic gene silencing in *Neurospora crassa*. *Genetics*, **146** (2), 509–523.
112 Lewis, Z.A., Adhvaryu, K.K., Honda, S., Shiver, A.L., Knip, M., Sack, R., Selker, E.U. (2010) DNA methylation and normal chromosome behavior in *Neurospora* depend on five components of a histone methyltransferase complex, DCDC. *PLoS Genet.*, **6** (11), e1001196.
113 Lewis, Z.A., Adhvaryu, K.K., Honda, S., Shiver, A.L., Selker, E.U. (2010) Identification of DIM-7, a protein required to target the DIM-5 H3 methyltransferase to chromatin. *Proc. Natl Acad. Sci. USA*, **107** (18), 8310–8315.
114 Tamaru, H., Selker, E.U. (2001) A histone H3 methyltransferase controls DNA methylation in *Neurospora crassa*. *Nature*, **414** (6861), 277–283.
115 Tamaru, H., Zhang, X., McMillen, D., Singh, P.B., Nakayama, J., Grewal, S.I., Allis, C.D., Cheng, X., Selker, E.U. (2003) Trimethylated lysine 9 of histone H3 is a mark for DNA methylation in *Neurospora crassa*. *Nat. Genet.*, **34** (1), 75–79.
116 Freitag, M., Hickey, P.C., Khlafallah, T.K., Read, N.D., Selker, E.U. (2004) HP1 is essential for DNA methylation in *Neurospora*. *Mol. Cells*, **13** (3), 427–434.
117 Honda, S., Selker, E.U. (2008) Direct interaction between DNA methyltransferase DIM-2 and HP1 is required for DNA methylation in *Neurospora crassa*. *Mol. Cell. Biol.*, **28** (19), 6044–6055.
118 Honda, S., Lewis, Z.A., Huarte, M., Cho, L.Y., David, L.L., Shi, Y., Selker, E.U. (2010) The DMM complex prevents spreading of DNA methylation from transposons to nearby genes in *Neurospora crassa*. *Genes Dev.*, **24** (5), 443–454.
119 Lachner, M., O'Carroll, D., Rea, S., Mechtler, K., Jenuwein, T. (2001) Methylation of histone H3 lysine 9 creates a binding site for HP1 proteins. *Nature*, **410** (6824), 116–120.
120 Jackson, J.P., Lindroth, A.M., Cao, X., Jacobsen, S.E. (2002) Control of CpNpG DNA methylation by the KRYPTONITE histone H3 methyltransferase. *Nature*, **416** (6880), 556–560.
121 Lehnertz, B., Ueda, Y., Derijck, A.A., Braunschweig, U., Perez-Burgos, L., Kubicek, S., Chen, T., Li, E., Jenuwein, T., Peters, A.H. (2003) Suv39h-mediated histone h3 lysine 9 methylation directs DNA methylation to major satellite repeats at pericentric heterochromatin. *Curr. Biol.*, **13** (14), 1192–1200.
122 Fuks, F., Hurd, P.J., Deplus, R., Kouzarides, T. (2003) The DNA methyltransferases associate with HP1 and the SUV39H1 histone methyltransferase. *Nucleic Acids Res.*, **31** (9), 2305–2312.
123 Saze, H., Shiraishi, A., Miura, A., Kakutani, T. (2008) Control of genic DNA methylation by a jmjC domain-containing protein in *Arabidopsis thaliana*. *Science*, **319** (5862), 462–465.
124 Miura, A., Nakamura, M., Inagaki, S., Kobayashi, A., Saze, H., Kakutani, T. (2009) An *Arabidopsis* jmjC domain protein protects transcribed genes from DNA methylation at CHG sites. *EMBO J.*, **28** (8), 1078–1086.
125 Zhao, Y., Shen, Y., Yang, S., Wang, J., Hu, Q., Wang, Y., He, Q. (2010) Ubiquitin ligase components Cullin4 and DDB1 are essential for DNA methylation in *Neurospora crassa*. *J. Biol. Chem.*, **285** (7), 4355–4365.
126 Xu, H., Wang, J., Hu, Q., Quan, Y., Chen, H., Cao, Y., Li, C., Wang, Y., He, Q. (2010) DCAF26, an adaptor protein of Cul4-based E3, is essential for DNA methylation in *Neurospora crassa*. *PLoS Genet.*, **6** (9), e1001132.
127 Petroski, M.D., Deshaies, R.J. (2005) Function and regulation of cullin-RING ubiquitin ligases. *Nat. Rev. Mol. Cell Biol.*, **6** (1), 9–20.

128 Adhvaryu, K.K., Selker, E.U. (2008) Protein phosphatase PP1 is required for normal DNA methylation in *Neurospora*. *Genes Dev.*, **22** (24), 3391–3396.

129 Smith, K.M., Dobosy, J.R., Reifsnyder, J.E., Rountree, M.R., Anderson, D.C., Green, G.R., Selker, E.U. (2010) H2B- and H3-specific histone deacetylases are required for DNA methylation in *Neurospora crassa*. *Genetics*, **186** (4), 1207–1216.

130 Weiler, K.S., Wakimoto, B.T. (1995) Heterochromatin and gene expression in *Drosophila*. *Annu. Rev. Genet.*, **29**, 577–605.

131 Imai, S., Armstrong, C.M., Kaeberlein, M., Guarente, L. (2000) Transcriptional silencing and longevity protein Sir2 is an NAD-dependent histone deacetylase. *Nature*, **403** (6771), 795–800.

132 Gottschling, D.E., Aparicio, O.M., Billington, B.L., Zakian, V.A. (1990) Position effect at *S. cerevisiae* telomeres: reversible repression of Pol II transcription. *Cell*, **63** (4), 751–762.

133 Tanny, J.C., Kirkpatrick, D.S., Gerber, S.A., Gygi, S.P., Moazed, D. (2004) Budding yeast silencing complexes and regulation of Sir2 activity by protein-protein interactions. *Mol. Cell. Biol.*, **24** (16), 6931–6946.

134 Shankaranarayana, G.D., Motamedi, M.R., Moazed, D., Grewal, S.I. (2003) Sir2 regulates histone H3 lysine 9 methylation and heterochromatin assembly in fission yeast. *Curr. Biol.*, **13** (14), 1240–1246.

135 Borkovich, K.A., Alex, L.A., Yarden, O., Freitag, M., Turner, G.E., Read, N.D., Seiler, S., Bell-Pedersen, D., Paietta, J., Plesofsky, N., Plamann, M., Goodrich-Tanrikulu, M., Schulte, U., Mannhaupt, G., Nargang, F.E., Radford, A., Selitrennikoff, C., Galagan, J.E., Dunlap, J.C., Loros, J.J., Catcheside, D., Inoue, H., Aramayo, R., Polymenis, M., Selker, E.U., Sachs, M.S., Marzluf, G.A., Paulsen, I., Davis, R., Ebbole, D.J., Zelter, A., Kalkman, E.R., O'Rourke, R., Bowring, F., Yeadon, J., Ishii, C., Suzuki, K., Sakai, W., Pratt, R. (2004) Lessons from the genome sequence of *Neurospora crassa*: tracing the path from genomic blueprint to multicellular organism. *Microbiol. Mol. Biol. Rev.*, **68** (1), 1–108.

136 Aparicio, O.M., Billington, B.L., Gottschling, D.E. (1991) Modifiers of position effect are shared between telomeric and silent mating-type loci in *S. cerevisiae*. *Cell*, **66** (6), 1279–1287.

137 Boulton, S.J., Jackson, S.P. (1998) Components of the Ku-dependent non-homologous end-joining pathway are involved in telomeric length maintenance and telomeric silencing. *EMBO J.*, **17** (6), 1819–1828.

138 Manolis, K.G., Nimmo, E.R., Hartsuiker, E., Carr, A.M., Jeggo, P.A., Allshire, R.C. (2001) Novel functional requirements for non-homologous DNA end joining in *Schizosaccharomyces pombe*. *EMBO J.*, **20** (1–2), 210–221.

139 Ninomiya, Y., Suzuki, K., Ishii, C., Inoue, H. (2004) Highly efficient gene replacements in *Neurospora* strains deficient for nonhomologous end-joining. *Proc. Natl Acad. Sci. USA*, **101** (33), 12248–12253.

140 Cooper, J.P., Watanabe, Y., Nurse, P. (1998) Fission yeast Taz1 protein is required for meiotic telomere clustering and recombination. *Nature*, **392** (6678), 828–831.

141 Hecht, A., Strahl Bolsinger, S., Grunstein, M. (1996) Spreading of transcriptional repressor SIR3 from telomeric heterochromatin. *Nature*, **383** (6595), 92–96.

142 Chen, Y., Rai, R., Zhou, Z.R., Kanoh, J., Ribeyre, C., Yang, Y., Zheng, H., Damay, P., Wang, F., Tsujii, H., Hiraoka, Y., Shore, D., Hu, H.Y., Chang, S., Lei, M. (2011) A conserved motif within RAP1 has diversified roles in telomere protection and regulation in different organisms. *Nat. Struct. Mol. Biol.*, **18** (2), 213–221.

143 Cooper, J.P., Nimmo, E.R., Allshire, R.C., Cech, T.R. (1997) Regulation of telomere length and function by a Myb-domain protein in fission yeast. *Nature*, **385** (6618), 744–747.

144 Miller, K.M., Cooper, J.P. (2003) The telomere protein Taz1 is required to prevent and repair genomic DNA breaks. *Mol. Cells*, **11** (2), 303–313.

145 Ferreira, M.G., Miller, K.M., Cooper, J.P. (2004) Indecent exposure: when telomeres become uncapped. *Mol. Cells*, **13** (1), 7–18.

146 Jain, D., Hebden, A.K., Nakamura, T.M., Miller, K.M., Cooper, J.P. (2010) HAATI survivors replace canonical telomeres with blocks of generic heterochromatin. *Nature*, **467** (7312), 223–227.

147 Biessmann, H., Mason, J.M. (1997) Telomere maintenance without telomerase. *Chromosoma*, **106** (2), 63–69.
148 Huang, Y. (2002) Transcriptional silencing in *Saccharomyces cerevisiae* and *Schizosaccharomyces pombe*. *Nucleic Acids Res.*, **30** (7), 1465–1482.
149 Sadaie, M., Naito, T., Ishikawa, F. (2003) Stable inheritance of telomere chromatin structure and function in the absence of telomeric repeats. *Genes Dev.*, **17** (18), 2271–2282.
150 Chen, X.F., Meng, F.L., Zhou, J.Q. (2009) Telomere recombination accelerates cellular aging in *Saccharomyces cerevisiae*. *PLoS Genet.*, **5** (6), e1000535.
151 Price, C.M., Boltz, K.A., Chaiken, M.F., Stewart, J.A., Beilstein, M.A., Shippen, D.E. (2010) Evolution of CST function in telomere maintenance. *Cell Cycle*, **9** (16), 3157–3165.
152 Pinto, A.R., Li, H., Nicholls, C., Liu, J.P. (2011) Telomere protein complexes and interactions with telomerase in telomere maintenance. *Front. Biosci.*, **16**, 187–207.
153 Takahashi, Y.H., Schulze, J.M., Jackson, J., Hentrich, T., Seidel, C., Jaspersen, S.L., Kobor, M.S., Shilatifard, A. (2011) Dot1 and histone H3K79 methylation in natural telomeric and HM silencing. *Mol. Cells*, **42** (1), 118–126.
154 Tennen, R.I., Chua, K.F. (2011) Chromatin regulation and genome maintenance by mammalian SIRT6. *Trends Biochem. Sci.*, **36** (1), 39–46.
155 Sanyal, K., Carbon, J. (2002) The CENP-A homolog CaCse4p in the pathogenic yeast *Candida albicans* is a centromere protein essential for chromosome transmission. *Proc. Natl Acad. Sci. USA*, **99** (20), 12969–12974.
156 Baker, R.E., Rogers, K. (2006) Phylogenetic analysis of fungal centromere H3 proteins. *Genetics*, **174** (3), 1481–1492.
157 Nagaki, K., Talbert, P.B., Zhong, C.X., Dawe, R.K., Henikoff, S., Jiang, J. (2003) Chromatin immunoprecipitation reveals that the 180-bp satellite repeat is the key functional DNA element of *Arabidopsis thaliana* centromeres. *Genetics*, **163** (3), 1221–1225.
158 Nagaki, K., Cheng, Z., Ouyang, S., Talbert, P.B., Kim, M., Jones, K.M., Henikoff, S., Buell, C.R., Jiang, J. (2004) Sequencing of a rice centromere uncovers active genes. *Nat. Genet.*, **36** (2), 138–145.
159 Jin, W., Melo, J.R., Nagaki, K., Talbert, P.B., Henikoff, S., Dawe, R.K., Jiang, J. (2004) Maize centromeres: organization and functional adaptation in the genetic background of oat. *Plant Cell*, **16** (3), 571–581.
160 Fang, Y., Spector, D.L. (2005) Centromere positioning and dynamics in living *Arabidopsis* plants. *Mol. Biol. Cell*, **16** (12), 5710–5718.
161 Allshire, R.C., Javerzat, J.P., Redhead, N.J., Cranston, G. (1994) Position effect variegation at fission yeast centromeres. *Cell*, **76** (1), 157–169.
162 Nakagawa, H., Lee, J.K., Hurwitz, J., Allshire, R.C., Nakayama, J., Grewal, S.I., Tanaka, K., Murakami, Y. (2002) Fission yeast CENP-B homologs nucleate centromeric heterochromatin by promoting heterochromatin-specific histone tail modifications. *Genes Dev.*, **16** (14), 1766–1778.
163 Partridge, J.F., Scott, K.S., Bannister, A.J., Kouzarides, T., Allshire, R.C. (2002) cis-acting DNA from fission yeast centromeres mediates histone H3 methylation and recruitment of silencing factors and cohesin to an ectopic site. *Curr. Biol.*, **12** (19), 1652–1660.
164 Williams, B.C., Murphy, T.D., Goldberg, M.L., Karpen, G.H. (1998) Neocentromere activity of structurally acentric mini-chromosomes in *Drosophila*. *Nat. Genet.*, **18** (1), 30–37.
165 Maggert, K.A., Karpen, G.H. (2001) The activation of a neocentromere in *Drosophila* requires proximity to an endogenous centromere. *Genetics*, **158** (4), 1615–1628.
166 Blower, M.D., Sullivan, B.A., Karpen, G.H. (2002) Conserved organization of centromeric chromatin in flies and humans. *Dev. Cell*, **2** (3), 319–330.
167 du Sart, D., Cancilla, M.R., Earle, E., Mao, J.I., Saffery, R., Tainton, K.M., Kalitsis, P., Martyn, J., Barry, A.E., Choo, K.H. (1997) A functional neo-centromere formed through activation of a latent human centromere and consisting of non-alpha-satellite DNA. *Nat. Genet.*, **16** (2), 144–153.
168 Lam, A.L., Boivin, C.D., Bonney, C.F., Rudd, M.K., Sullivan, B.A. (2006) Human centromeric chromatin is a dynamic chromosomal domain that can spread over

noncentromeric DNA. *Proc. Natl Acad. Sci. USA*, **103** (11), 4186–4191.
169 Foltz, D.R., Jansen, L.E., Black, B.E., Bailey, A.O., Yates, J.R. III, Cleveland, D.W. (2006) The human CENP-A centromeric nucleosome-associated complex. *Nat. Cell Biol.*, **8** (5), 458–469.
170 Cambareri, E.B., Aisner, R., Carbon, J. (1998) Structure of the chromosome VII centromere region in *Neurospora crassa*: degenerate transposons and simple repeats. *Mol. Cell. Biol.*, **18** (9), 5465–5477.
171 Meraldi, P., McAinsh, A.D., Rheinbay, E., Sorger, P.K. (2006) Phylogenetic and structural analysis of centromeric DNA and kinetochore proteins. *Genome Biol.*, **7** (3), R23.
172 Nakaseko, Y., Adachi, Y., Funahashi, S., Niwa, O., Yanagida, M. (1986) Chromosome walking shows a highly homologous repetitive sequence present in all the centromere regions of fission yeast. *EMBO J.*, **5** (5), 1011–1021.
173 Scott, K.C., Merrett, S.L., Willard, H.F. (2006) A heterochromatin barrier partitions the fission yeast centromere into discrete chromatin domains. *Curr. Biol.*, **16** (2), 119–129.
174 Cam, H.P., Sugiyama, T., Chen, E.S., Chen, X., FitzGerald, P.C., Grewal, S.I. (2005) Comprehensive analysis of heterochromatin- and RNAi-mediated epigenetic control of the fission yeast genome. *Nat. Genet.*, **37** (8), 809–819.
175 Sanyal, K., Baum, M., Carbon, J. (2004) Centromeric DNA sequences in the pathogenic yeast *Candida albicans* are all different and unique. *Proc. Natl Acad. Sci. USA*, **101** (31), 11374–11379.
176 Baum, M., Sanyal, K., Mishra, P.K., Thaler, N., Carbon, J. (2006) Formation of functional centromeric chromatin is specified epigenetically in *Candida albicans*. *Proc. Natl Acad. Sci. USA*, **103** (40), 14877–14882.
177 Richards, E.J., Goodman, H.M., Ausubel, F.M. (1991) The centromere region of *Arabidopsis thaliana* chromosome 1 contains telomere-similar sequences. *Nucleic Acids Res.*, **19** (12), 3351–3357.
178 Copenhaver, G.P., Nickel, K., Kuromori, T., Benito, M.I., Kaul, S., Lin, X., Bevan, M., Murphy, G., Harris, B., Parnell, L.D., McCombie, W.R., Martienssen, R.A., Marra, M., Preuss, D. (1999) Genetic definition and sequence analysis of *Arabidopsis* centromeres. *Science*, **286** (5449), 2468–2474.
179 Zhang, R., Zhang, C.T. (2004) Isochore structures in the genome of the plant *Arabidopsis thaliana*. *J. Mol. Evol.*, **59** (2), 227–238.
180 Dong, F., Miller, J.T., Jackson, S.A., Wang, G.L., Ronald, P.C., Jiang, J. (1998) Rice (*Oryza sativa*) centromeric regions consist of complex DNA. *Proc. Natl Acad. Sci. USA*, **95** (14), 8135–8140.
181 Nagaki, K., Neumann, P., Zhang, D., Ouyang, S., Buell, C.R., Cheng, Z., Jiang, J. (2005) Structure, divergence, and distribution of the CRR centromeric retrotransposon family in rice. *Mol. Biol. Evol.*, **22** (4), 845–855.
182 Yamamoto, M., Miklos, G.L. (1978) Genetic studies on heterochromatin in *Drosophila melanogaster* and their implications for the functions of satellite DNA. *Chromosoma*, **66** (1), 71–98.
183 Sun, X., Le, H.D., Wahlstrom, J.M., Karpen, G.H. (2003) Sequence analysis of a functional *Drosophila* centromere. *Genome Res*, **13** (2), 182–194.
184 Schueler, M.G., Higgins, A.W., Rudd, M.K., Gustashaw, K., Willard, H.F. (2001) Genomic and genetic definition of a functional human centromere. *Science*, **294** (5540), 109–115.
185 Rudd, M.K., Willard, H.F. (2004) Analysis of the centromeric regions of the human genome assembly. *Trends Genet.*, **20** (11), 529–533.
186 Allshire, R.C. (1997) Centromeres, checkpoints and chromatid cohesion. *Curr. Opin. Genet. Dev.*, **7** (2), 264–273.
187 Karpen, G.H., Allshire, R.C. (1997) The case for epigenetic effects on centromere identity and function. *Trends Genet.*, **13** (12), 489–496.
188 Blower, M.D., Karpen, G.H. (2001) The role of *Drosophila* CID in kinetochore formation, cell-cycle progression and heterochromatin interactions. *Nat. Cell Biol.*, **3** (8), 730–739.
189 Sullivan, K.F. (2001) A solid foundation: functional specialization of centromeric chromatin. *Curr. Opin. Genet. Dev.*, **11** (2), 182–188.

190 Sullivan, B.A., Blower, M.D., Karpen, G.H. (2001) Determining centromere identity: cyclical stories and forking paths. *Nat. Rev. Genet.*, **2** (8), 584–596.

191 Shibata, F., Murata, M. (2004) Differential localization of the centromere-specific proteins in the major centromeric satellite of *Arabidopsis thaliana*. *J. Cell Sci.*, **117** (Pt 14), 2963–2970.

192 Black, B.E., Jansen, L.E., Maddox, P.S., Foltz, D.R., Desai, A.B., Shah, J.V., Cleveland, D.W. (2007) Centromere identity maintained by nucleosomes assembled with histone H3 containing the CENP-A targeting domain. *Mol. Cells*, **25** (2), 309–322.

193 Mishra, P.K., Baum, M., Carbon, J. (2007) Centromere size and position in *Candida albicans* are evolutionarily conserved independent of DNA sequence heterogeneity. *Mol. Genet. Genomics*, **278** (4), 455–465.

194 Ketel, C., Wang, H.S., McClellan, M., Bouchonville, K., Selmecki, A., Lahav, T., Gerami-Nejad, M., Berman, J. (2009) Neocentromeres form efficiently at multiple possible loci in *Candida albicans*. *PLoS Genet.*, **5** (3), e1000400.

195 Fishel, B., Amstutz, H., Baum, M., Carbon, J., Clarke, L. (1988) Structural organization and functional analysis of centromeric DNA in the fission yeast *Schizosaccharomyces pombe*. *Mol. Cell. Biol.*, **8** (2), 754–763.

196 Folco, H.D., Pidoux, A.L., Urano, T., Allshire, R.C. (2008) Heterochromatin and RNAi are required to establish CENP-A chromatin at centromeres. *Science*, **319** (5859), 94–97.

197 Kagansky, A., Folco, H.D., Almeida, R., Pidoux, A.L., Boukaba, A., Simmer, F., Urano, T., Hamilton, G.L., Allshire, R.C. (2009) Synthetic heterochromatin bypasses RNAi and centromeric repeats to establish functional centromeres. *Science*, **324** (5935), 1716–1719.

198 Ishii, K., Ogiyama, Y., Chikashige, Y., Soejima, S., Masuda, F., Kakuma, T., Hiraoka, Y., Takahashi, K. (2008) Heterochromatin integrity affects chromosome reorganization after centromere dysfunction. *Science*, **321** (5892), 1088–1091.

199 Allshire, R.C., Karpen, G.H. (2008) Epigenetic regulation of centromeric chromatin: old dogs, new tricks? *Nat. Rev. Genet.*, **9** (12), 923–937.

200 Guenatri, M., Bailly, D., Maison, C., Almouzni, G. (2004) Mouse centric and pericentric satellite repeats form distinct functional heterochromatin. *J. Cell Biol.*, **166** (4), 493–505.

201 Ribeiro, S.A., Vagnarelli, P., Dong, Y., Hori, T., McEwen, B.F., Fukagawa, T., Flors, C., Earnshaw, W.C. (2010) A super-resolution map of the vertebrate kinetochore. *Proc. Natl Acad. Sci. USA*, **107**, 10484–10489.

202 Kim, D.H., Jeon, Y., Anguera, M.C., Lee, J.T. (2011) X-chromosome epigenetic reprogramming in pluripotent stem cells via noncoding genes. *Semin. Cell Dev. Biol.*, **22**, 336–342.

203 Lee, J.T. (2010) The X as model for RNA's niche in epigenomic regulation. *Cold Spring Harbor Perspect. Biol.*, **2** (9), a003749.

204 Leeb, M., Steffen, P.A., Wutz, A. (2009) X chromosome inactivation sparked by non-coding RNAs. *RNA Biol.*, **6** (2), 94–99.

205 Segers, G.C., Zhang, X., Deng, F., Sun, Q., Nuss, D.L. (2007) Evidence that RNA silencing functions as an antiviral defense mechanism in fungi. *Proc. Natl Acad. Sci. USA*, **104** (31), 12902–12906.

206 Sellam, A., Hogues, H., Askew, C., Tebbji, F., van Het Hoog, M., Lavoie, H., Kumamoto, C.A., Whiteway, M., Nantel, A. (2010) Experimental annotation of the human pathogen *Candida albicans* coding and noncoding transcribed regions using high-resolution tiling arrays. *Genome Biol.*, **11** (7), R71.

207 Ni, T., Tu, K., Wang, Z., Song, S., Wu, H., Xie, B., Scott, K.C., Grewal, S.I., Gao, Y., Zhu, J. (2010) The prevalence and regulation of antisense transcripts in *Schizosaccharomyces pombe*. *PLoS ONE*, **5** (12), e15271.

208 Cogoni, C., Irelan, J.T., Schumacher, M., Schmidhauser, T.J., Selker, E.U., Macino, G. (1996) Transgene silencing of the al-1 gene in vegetative cells of *Neurospora* is mediated by a cytoplasmic effector and does not depend on DNA-DNA interactions or DNA methylation. *EMBO J.*, **15** (12), 3153–3163.

209 Nolan, T., Braccini, L., Azzalin, G., De Toni, A., Macino, G., Cogoni, C. (2005) The post-transcriptional gene silencing machinery functions independently of DNA

methylation to repress a LINE1-like retrotransposon in *Neurospora crassa. Nucleic Acids Res.*, **33** (5), 1564–1573.

210 Cogoni, C., Macino, G. (1999) Gene silencing in *Neurospora crassa* requires a protein homologous to RNA-dependent RNA polymerase. *Nature*, **399** (6732), 166–169.

211 Cogoni, C., Macino, G. (1999) Posttranscriptional gene silencing in *Neurospora* by a RecQ DNA helicase. *Science*, **286** (5448), 2342–2344.

212 Kato, A., Akamatsu, Y., Sakuraba, Y., Inoue, H. (2004) The *Neurospora crassa mus-19* gene is identical to the *qde-3* gene, which encodes a RecQ homologue and is involved in recombination repair and postreplication repair. *Curr. Genet.*, **45** (1), 37–44.

213 Catalanotto, C., Azzalin, G., Macino, G., Cogoni, C. (2002) Involvement of small RNAs and role of the *qde* genes in the gene silencing pathway in *Neurospora. Genes Dev.*, **16** (7), 790–795.

214 Catalanotto, C., Pallotta, M., ReFalo, P., Sachs, M.S., Vayssie, L., Macino, G., Cogoni, C. (2004) Redundancy of the two dicer genes in transgene-induced posttranscriptional gene silencing in *Neurospora crassa. Mol. Cell. Biol.*, **24** (6), 2536–2545.

215 Maiti, M., Lee, H.C., Liu, Y. (2007) QIP, a putative exonuclease, interacts with the *Neurospora* Argonaute protein and facilitates conversion of duplex siRNA into single strands. *Genes Dev.*, **21** (5), 590–600.

216 Lee, H.C., Chang, S.S., Choudhary, S., Aalto, A.P., Maiti, M., Bamford, D.H., Liu, Y. (2009) qiRNA is a new type of small interfering RNA induced by DNA damage. *Nature*, **459** (7244), 274–277.

217 Verdel, A., Jia, S., Gerber, S., Sugiyama, T., Gygi, S., Grewal, S.I., Moazed, D. (2004) RNAi-mediated targeting of heterochromatin by the RITS complex. *Science*, **303** (5658), 672–676.

218 Sugiyama, T., Cam, H., Verdel, A., Moazed, D., Grewal, S.I. (2005) RNA-dependent RNA polymerase is an essential component of a self-enforcing loop coupling heterochromatin assembly to siRNA production. *Proc. Natl Acad. Sci. USA*, **102** (1), 152–157.

219 Zhang, K., Fischer, T., Porter, R.L., Dhakshnamoorthy, J., Zofall, M., Zhou, M., Veenstra, T., Grewal, S.I. (2011) Clr4/Suv39 and RNA quality control factors cooperate to trigger RNAi and suppress antisense RNA. *Science*, **331** (6024), 1624–1627.

220 Grewal, S.I. (2010) RNAi-dependent formation of heterochromatin and its diverse functions. *Curr. Opin. Genet. Dev.*, **20** (2), 134–141.

221 Chicas, A., Cogoni, C., Macino, G. (2004) RNAi-dependent and RNAi-independent mechanisms contribute to the silencing of RIPed sequences in *Neurospora crassa. Nucleic Acids Res.*, **32** (14), 4237–4243.

222 Freitag, M., Lee, D.W., Kothe, G.O., Pratt, R.J., Aramayo, R., Selker, E.U. (2004) DNA methylation is independent of RNA interference in *Neurospora. Science*, **304** (5679), 1939.

223 Shiu, P.K., Zickler, D., Raju, N.B., Ruprich-Robert, G., Metzenberg, R.L. (2006) SAD-2 is required for meiotic silencing by unpaired DNA and perinuclear localization of SAD-1 RNA-directed RNA polymerase. *Proc. Natl Acad. Sci. USA*, **103** (7), 2243–2248.

224 Lee, D.W., Pratt, R.J., McLaughlin, M., Aramayo, R. (2003) An argonaute-like protein is required for meiotic silencing. *Genetics*, **164** (2), 821–828.

225 Alexander, W.G., Raju, N.B., Xiao, H., Hammond, T.M., Perdue, T.D., Metzenberg, R.L., Pukkila, P.J., Shiu, P.K. (2008) DCL-1 colocalizes with other components of the MSUD machinery and is required for silencing. *Fungal Genet. Biol.*, **45** (5), 719–727.

226 Lee, D.W., Millimaki, R., Aramayo, R. (2010) QIP, a component of the vegetative RNA silencing pathway, is essential for meiosis and suppresses meiotic silencing in *Neurospora crassa. Genetics*, **186** (1), 127–133.

227 Zhu, B., Reinberg, D. (2011) Epigenetic inheritance: uncontested? *Cell Res.*, **21** (3), 435–441.

228 Kundu, S., Peterson, C.L. (2010) Dominant role for signal transduction in the transcriptional memory of yeast GAL genes. *Mol. Cell. Biol.*, **30** (10), 2330–2340.

229 Kundu, S., Horn, P.J., Peterson, C.L. (2007) SWI/SNF is required for transcriptional memory at the yeast GAL gene cluster. *Genes Dev.*, **21** (8), 997–1004.

230 Brosch, G., Loidl, P., Graessle, S. (2008) Histone modifications and chromatin dynamics: a focus on filamentous fungi. *FEMS Microbiol. Rev.*, **32** (3), 409–439.
231 Sims, R.J. II, Reinberg, D. (2008) Is there a code embedded in proteins that is based on post-translational modifications? *Nat. Rev. Mol. Cell Biol.*, **9** (10), 815–820.
232 Belden, W.J., Loros, J.J., Dunlap, J.C. (2007) Execution of the circadian negative feedback loop in *Neurospora* requires the ATP-dependent chromatin-remodeling enzyme CLOCKSWITCH. *Mol. Cell*, **25** (4), 587–600.
233 Vitalini, M.W., de Paula, R.M., Park, W.D., Bell-Pedersen, D. (2006) The rhythms of life: circadian output pathways in *Neurospora*. *J. Biol. Rhythms*, **21** (6), 432–444.
234 Baker, C.L., Loros, J.J., Dunlap, J.C. (2011) The circadian clock of *Neurospora crassa*. *FEMS Microbiol. Rev.*, e-pub doi 10.1111/j.1574-6976.2011.00288.x.
235 Schafmeier, T., Diernfellner, A., Schafer, A., Dintsis, O., Neiss, A., Brunner, M. (2008) Circadian activity and abundance rhythms of the *Neurospora* clock transcription factor WCC associated with rapid nucleo-cytoplasmic shuttling. *Genes Dev.*, **22** (24), 3397–3402.
236 Cha, J., Yuan, H., Liu, Y. (2011) Regulation of the activity and cellular localization of the circadian clock protein FRQ. *J. Biol. Chem.*, **286** (13), 11469–11478.
237 Roberts, C.J., Selker, E.U. (1995) Mutations affecting the biosynthesis of S-adenosylmethionine cause reduction of DNA methylation in *Neurospora crassa*. *Nucleic Acids Res.*, **23** (23), 4818–4826.
238 Kramer, C., Loros, J.J., Dunlap, J.C., Crosthwaite, S.K. (2003) Role for antisense RNA in regulating circadian clock function in *Neurospora crassa*. *Nature*, **421** (6926), 948–952.
239 Smith, K.M., Sancar, G., Dekhang, R., Sullivan, C.M., Li, S., Tag, A.G., Sancar, C., Bredeweg, E.L., Priest, H.D., McCormick, R.F., Thomas, T.L., Carrington, J.C., Stajich, J.E., Bell-Pedersen, D., Brunner, M., Freitag, M. (2010) Transcription factors in light and circadian clock signaling networks revealed by genomewide mapping of direct targets for *Neurospora* white collar complex. *Eukaryot. Cell*, **9** (10), 1549–1556.
240 Brakhage, A.A., Schroeckh, V. (2011) Fungal secondary metabolites – strategies to activate silent gene clusters. *Fungal Genet. Biol.*, **48** (1), 15–22.
241 Brakhage, A.A., Schuemann, J., Bergmann, S., Scherlach, K., Schroeckh, V., Hertweck, C. (2008) Activation of fungal silent gene clusters: a new avenue to drug discovery. *Prog. Drug Res.*, **66**, 1, 3–12.
242 Fedorova, N.D., Khaldi, N., Joardar, V.S., Maiti, R., Amedeo, P., Anderson, M.J., Crabtree, J., Silva, J.C., Badger, J.H., Albarraq, A., Angiuoli, S., Bussey, H., Bowyer, P., Cotty, P.J., Dyer, P.S., Egan, A., Galens, K., Fraser-Liggett, C.M., Haas, B.J., Inman, J.M., Kent, R., Lemieux, S., Malavazi, I., Orvis, J., Roemer, T., Ronning, C.M., Sundaram, J.P., Sutton, G., Turner, G., Venter, J.C., White, O.R., Whitty, B.R., Youngman, P., Wolfe, K.H., Goldman, G.H., Wortman, J.R., Jiang, B., Denning, D.W., Nierman, W.C. (2008) Genomic islands in the pathogenic filamentous fungus *Aspergillus fumigatus*. *PLoS Genet.*, **4** (4), e1000046.
243 Thon, M.R., Pan, H., Diener, S., Papalas, J., Taro, A., Mitchell, T.K., Dean, R.A. (2006) The role of transposable element clusters in genome evolution and loss of synteny in the rice blast fungus *Magnaporthe oryzae*. *Genome Biol.*, **7** (2), R16.
244 Dean, R.A., Talbot, N.J., Ebbole, D.J., Farman, M.L., Mitchell, T.K., Orbach, M.J., Thon, M., Kulkarni, R., Xu, J.R., Pan, H., Read, N.D., Lee, Y.H., Carbone, I., Brown, D., Oh, Y.Y., Donofrio, N., Jeong, J.S., Soanes, D.M., Djonovic, S., Kolomiets, E., Rehmeyer, C., Li, W., Harding, M., Kim, S., Lebrun, M.H., Bohnert, H., Coughlan, S., Butler, J., Calvo, S., Ma, L.J., Nicol, R., Purcell, S., Nusbaum, C., Galagan, J.E., Birren, B.W. (2005) The genome sequence of the rice blast fungus *Magnaporthe grisea*. *Nature*, **434** (7036), 980–986.
245 Wang, X., Sena Filho, J.G., Hoover, A.R., King, J.B., Ellis, T.K., Powell, D.R., Cichewicz, R.H. (2010) Chemical epigenetics alters the secondary metabolite composition of guttate excreted by an atlantic-forest-soil-derived *Penicillium citreonigrum*. *J. Nat. Prod.*, **73** (5), 942–948.

246 Williams, R.B., Henrikson, J.C., Hoover, A.R., Lee, A.E., Cichewicz, R.H. (2008) Epigenetic remodeling of the fungal secondary metabolome. *Org. Biomol. Chem.*, **6** (11), 1895–1897.

247 Fisch, K.M., Gillaspy, A.F., Gipson, M., Henrikson, J.C., Hoover, A.R., Jackson, L., Najar, F.Z., Wagele, H., Cichewicz, R.H. (2009) Chemical induction of silent biosynthetic pathway transcription in *Aspergillus niger*. *J. Ind. Microbiol. Biotechnol.*, **36** (9), 1199–1213.

248 Henrikson, J.C., Hoover, A.R., Joyner, P.M., Cichewicz, R.H. (2009) A chemical epigenetics approach for engineering the in situ biosynthesis of a cryptic natural product from *Aspergillus niger*. *Org. Biomol. Chem.*, **7** (3), 435–438.

249 Reyes-Dominguez, Y., Bok, J.W., Berger, H., Shwab, E.K., Basheer, A., Gallmetzer, A., Scazzocchio, C., Keller, N., Strauss, J. (2010) Heterochromatic marks are associated with the repression of secondary metabolism clusters in *Aspergillus nidulans*. *Mol. Microbiol.*, **76** (6), 1376–1386.

250 Shwab, E.K., Bok, J.W., Tribus, M., Galehr, J., Graessle, S., Keller, N.P. (2007) Histone deacetylase activity regulates chemical diversity in *Aspergillus*. *Eukaryotic Cell*, **6** (9), 1656–1664.

251 Lachner, M., O'Sullivan, R.J., Jenuwein, T. (2003) An epigenetic road map for histone lysine methylation. *J. Cell Sci.*, **116** (Pt 11), 2117–2124.

252 Sims, R.J. II, Nishioka, K., Reinberg, D. (2003) Histone lysine methylation: a signature for chromatin function. *Trends Genet.*, **19** (11), 629–639.

253 Eissenberg, J.C., Shilatifard, A. (2010) Histone H3 lysine 4 (H3K4) methylation in development and differentiation. *Dev. Biol.*, **339** (2), 240–249.

254 Strauss, J., Reyes-Dominguez, Y. (2011) Regulation of secondary metabolism by chromatin structure and epigenetic codes. *Fungal Genet. Biol.*, **48** (1), 62–69.

255 Adhvaryu, K.K., Morris, S.A., Strahl, B.D., Selker, E.U. (2005) Methylation of histone H3 lysine 36 is required for normal development in *Neurospora crassa*. *Eukaryot. Cell*, **4** (8), 1455–1464.

256 Robyr, D., Suka, Y., Xenarios, I., Kurdistani, S.K., Wang, A., Suka, N., Grunstein, M. (2002) Microarray deacetylation maps determine genome-wide functions for yeast histone deacetylases. *Cell*, **109** (4), 437–446.

257 Tribus, M., Bauer, I., Galehr, J., Rieser, G., Trojer, P., Brosch, G., Loidl, P., Haas, H., Graessle, S. (2010) A novel motif in fungal class 1 histone deacetylases is essential for growth and development of *Aspergillus*. *Mol. Biol. Cell*, **21** (2), 345–353.

258 Tribus, M., Galehr, J., Trojer, P., Brosch, G., Loidl, P., Marx, F., Haas, H., Graessle, S. (2005) HdaA, a major class 2 histone deacetylase of *Aspergillus nidulans*, affects growth under conditions of oxidative stress. *Eukaryot. Cell*, **4** (10), 1736–1745.

259 Lee, I., Oh, J.H., Shwab, E.K., Dagenais, T.R., Andes, D., Keller, N.P. (2009) HdaA, a class 2 histone deacetylase of *Aspergillus fumigatus*, affects germination and secondary metabolite production. *Fungal Genet. Biol.*, **46** (10), 782–790.

260 Slepecky, R.A., Starmer, W.T. (2009) Phenotypic plasticity in fungi: a review with observations on *Aureobasidium pullulans*. *Mycologia*, **101** (6), 823–832.

261 Srikantha, T., Tsai, L., Daniels, K., Klar, A.J., Soll, D.R. (2001) The histone deacetylase genes HDA1 and RPD3 play distinct roles in regulation of high-frequency phenotypic switching in *Candida albicans*. *J. Bacteriol.*, **183** (15), 4614–4625.

262 Trojer, P., Brandtner, E.M., Brosch, G., Loidl, P., Galehr, J., Linzmaier, R., Haas, H., Mair, K., Tribus, M., Graessle, S. (2003) Histone deacetylases in fungi: novel members, new facts. *Nucleic Acids Res.*, **31** (14), 3971–3981.

263 Lacoste, N., Utley, R.T., Hunter, J.M., Poirier, G.G., Cote, J. (2002) Disruptor of telomeric silencing-1 is a chromatin-specific histone H3 methyltransferase. *J. Biol. Chem.*, **277** (34), 30421–30424.

264 Kuo, M.H., Xu, X.J., Bolck, H.A., Guo, D. (2009) Functional connection between histone acetyltransferase Gcn5p and methyltransferase Hmt1p. *Biochim. Biophys. Acta*, **1789** (5), 395–402.

265 Yu, M.C., Lamming, D.W., Eskin, J.A., Sinclair, D.A., Silver, P.A. (2006) The role of protein arginine methylation in the formation of silent chromatin. *Genes Dev.*, **20** (23), 3249–3254.

266 Meneghini, M.D., Wu, M., Madhani, H.D. (2003) Conserved histone variant H2A.Z protects euchromatin from the ectopic spread of silent heterochromatin. *Cell*, **112** (5), 725–736.

267 Adam, M., Robert, F., Larochelle, M., Gaudreau, L. (2001) H2A.Z is required for global chromatin integrity and for recruitment of RNA polymerase II under specific conditions. *Mol. Cell. Biol.*, **21** (18), 6270–6279.

268 Donze, D., Adams, C.R., Rine, J., Kamakaka, R.T. (1999) The boundaries of the silenced HMR domain in *Saccharomyces cerevisiae*. *Genes Dev.*, **13** (6), 698–708.

269 Beck, D.B., Bonasio, R., Kaneko, S., Li, G., Margueron, R., Oda, H., Sarma, K., Sims, R.J. III, Son, J., Trojer, P., Reinberg, D. (2010) Chromatin in the nuclear landscape. *Cold Spring Harbor Symp. Quant. Biol.*, **75**, 11–22.

270 Sarma, K., Reinberg, D. (2005) Histone variants meet their match. *Nat. Rev. Mol. Cell Biol.*, **6** (2), 139–149.

271 Bok, J.W., Keller, N.P. (2004) LaeA, a regulator of secondary metabolism in *Aspergillus* spp. *Eukaryotic Cell*, **3** (2), 527–535.

272 Kale, S.P., Milde, L., Trapp, M.K., Frisvad, J.C., Keller, N.P., Bok, J.W. (2008) Requirement of LaeA for secondary metabolism and sclerotial production in *Aspergillus flavus*. *Fungal Genet. Biol.*, **45** (10), 1422–1429.

273 Bok, J.W., Balajee, S.A., Marr, K.A., Andes, D., Nielsen, K.F., Frisvad, J.C., Keller, N.P. (2005) LaeA, a regulator of morphogenetic fungal virulence factors. *Eukaryotic Cell*, **4** (9), 1574–1582.

274 Perrin, R.M., Fedorova, N.D., Bok, J.W., Cramer, R.A., Wortman, J.R., Kim, H.S., Nierman, W.C., Keller, N.P. (2007) Transcriptional regulation of chemical diversity in *Aspergillus fumigatus* by LaeA. *PLoS Pathog.*, **3** (4), e50.

275 Bell-Pedersen, D., Dunlap, J.C., Loros, J.J. (1992) The *Neurospora* circadian clock-controlled gene, *ccg-2*, is allelic to eas and encodes a fungal hydrophobin required for formation of the conidial rodlet layer. *Genes Dev.*, **6** (12A), 2382–2394.

276 Dagenais, T.R., Giles, S.S., Aimanianda, V., Latge, J.P., Hull, C.M., Keller, N.P. (2010) *Aspergillus fumigatus* LaeA-mediated phagocytosis is associated with a decreased hydrophobin layer. *Infect. Immun.*, **78** (2), 823–829.

277 Bayram, O., Krappmann, S., Ni, M., Bok, J.W., Helmstaedt, K., Valerius, O., Braus-Stromeyer, S., Kwon, N.J., Keller, N.P., Yu, J.H., Braus, G.H. (2008) VelB/VeA/LaeA complex coordinates light signal with fungal development and secondary metabolism. *Science*, **320** (5882), 1504–1506.

278 Amaike, S., Keller, N.P. (2009) Distinct roles for VeA and LaeA in development and pathogenesis of *Aspergillus flavus*. *Eukaryot. Cell*, **8** (7), 1051–1060.

279 Atoui, A., Kastner, C., Larey, C.M., Thokala, R., Etxebeste, O., Espeso, E.A., Fischer, R., Calvo, A.M. (2010) Cross-talk between light and glucose regulation controls toxin production and morphogenesis in *Aspergillus nidulans*. *Fungal Genet. Biol.*, **47** (12), 962–972.

280 Wiemann, P., Brown, D.W., Kleigrewe, K., Bok, J.W., Keller, N.P., Humpf, H.U., Tudzynski, B. (2010) FfVel1 and FfLae1, components of a velvet-like complex in *Fusarium fujikuroi*, affect differentiation, secondary metabolism and virulence. *Mol. Microbiol.*, **77**, 972–974.

281 Bok, J.W., Noordermeer, D., Kale, S.P., Keller, N.P. (2006) Secondary metabolic gene cluster silencing in *Aspergillus nidulans*. *Mol. Microbiol.*, **61** (6), 1636–1645.

282 Sarikaya Bayram, O., Bayram, O., Valerius, O., Park, H.S., Irniger, S., Gerke, J., Ni, M., Han, K.H., Yu, J.H., Braus, G.H. (2010) LaeA control of velvet family regulatory proteins for light-dependent development and fungal cell-type specificity. *PLoS Genet.*, **6** (12), e1001226.

283 Bayram, O., Braus, G.H. (2011) Coordination of secondary metabolism and development in fungi: the velvet family of regulatory proteins. *FEMS Microbiol. Rev.*, e-pub doi 10.1111/j.1574-6976.2011.00285.x.

284 Wickner, R.B., Edskes, H.K., Kryndushkin, D., McGlinchey, R., Bateman, D., Kelly, A. (2011) Prion diseases of yeast: amyloid structure and biology. *Semin. Cell Dev. Biol.*, **22**, 469–475.

285 Saupe, S.J. (2011) The [Het-s] prion of *Podospora anserina* and its role in heterokaryon incompatibility. *Semin. Cell Dev. Biol.*, **22**, 460–468.

286 Halfmann, R., Lindquist, S. (2010) Epigenetics in the extreme: prions and the inheritance of environmentally acquired traits. *Science*, **330** (6004), 629–632.

287 Brown, J.C., Lindquist, S. (2009) A heritable switch in carbon source utilization driven by an unusual yeast prion. *Genes Dev.*, **23** (19), 2320–2332.

288 Benkemoun, L., Sabate, R., Malato, L., Dos Reis, S., Dalstra, H., Saupe, S.J., Maddelein, M.L. (2006) Methods for the in vivo and in vitro analysis of [Het-s] prion infectivity. *Methods*, **39** (1), 61–67.

289 Tyedmers, J., Madariaga, M.L., Lindquist, S. (2008) Prion switching in response to environmental stress. *PLoS Biol.*, **6** (11), e294.

290 Shorter, J., Lindquist, S. (2005) Prions as adaptive conduits of memory and inheritance. *Nat. Rev. Genet.*, **6** (6), 435–450.

291 Baxa, U., Taylor, K.L., Steven, A.C., Wickner, R.B. (2004) Prions of *Saccharomyces* and *Podospora*. *Contrib. Microbiol.*, **11**, 50–71.

292 Sondheimer, N., Lindquist, S. (2000) Rnq1: an epigenetic modifier of protein function in yeast. *Mol. Cells*, **5** (1), 163–172.

293 Wickner, R.B., Edskes, H.K., Maddelein, M.L., Taylor, K.L., Moriyama, H. (1999) Prions of yeast and fungi. Proteins as genetic material. *J. Biol. Chem.*, **274** (2), 555–558.

294 Silar, P., Daboussi, M.J. (1999) Non-conventional infectious elements in filamentous fungi. *Trends Genet.*, **15** (4), 141–145.

295 Marcotte, E.M., Pellegrini, M., Thompson, M.J., Yeates, T.O., Eisenberg, D. (1999) A combined algorithm for genome-wide prediction of protein function. *Nature*, **402** (6757), 83–86.

296 Yool, A., Edmunds, W.J. (1998) Epigenetic inheritance and prions. *J. Evol. Biol.*, **11**, 241–242.

297 Coustou, V., Deleu, C., Saupe, S., Begueret, J. (1997) The protein product of the het-s heterokaryon incompatibility gene of the fungus *Podospora anserina* behaves as a prion analog. *Proc. Natl Acad. Sci. USA*, **94** (18), 9773–9778.

298 Wickner, R.B. (1994) [URE3] as an altered URE2 protein: evidence for a prion analog in Saccharomyces cerevisiae. *Science*, **264** (5158), 566–569.

299 Pasini, D., Malatesta, M., Jung, H.R., Walfridsson, J., Willer, A., Olsson, L., Skotte, J., Wutz, A., Porse, B., Jensen, O.N., Helin, K. (2010) Characterization of an antagonistic switch between histone H3 lysine 27 methylation and acetylation in the transcriptional regulation of Polycomb group target genes. *Nucleic Acids Res.*, **38** (15), 4958–4969.

300 Stewart, S., Tsun, Z.Y., Izpisua Belmonte, J.C. (2009) A histone demethylase is necessary for regeneration in zebrafish. *Proc. Natl Acad. Sci. USA*, **106** (47), 19889–19894.

301 Simon, J.A., Lange, C.A. (2008) Roles of the EZH2 histone methyltransferase in cancer epigenetics. *Mutat. Res.*, **647** (1–2), 21–29.

302 Xiong, L., Adhvaryu, K.K., Selker, E.U., Wang, Y. (2010) Mapping of lysine methylation and acetylation in core histones of *Neurospora crassa*. *Biochemistry*, **49** (25), 5236–5243.

303 Xiong, L., Wang, Y. (2011) Mapping post-translational modifications of histones H2A, H2B and H4 in Schizosaccharomyces pombe. *Int. J. Mass Spectrom.*, **301** (1-3), 159–165.

304 Segal, E., Fondufe-Mittendorf, Y., Chen, L., Thastrom, A., Field, Y., Moore, I.K., Wang, J.P., Widom, J. (2006) A genomic code for nucleosome positioning. *Nature*, **442** (7104), 772–778.

305 Lowary, P.T., Widom, J. (1998) New DNA sequence rules for high affinity binding to histone octamer and sequence-directed nucleosome positioning. *J. Mol. Biol.*, **276** (1), 19–42.

306 Lowary, P.T., Widom, J. (1997) Nucleosome packaging and nucleosome positioning of genomic DNA. *Proc. Natl Acad. Sci. USA*, **94** (4), 1183–1188.

307 Widom, J. (1992) A relationship between the helical twist of DNA and the ordered positioning of nucleosomes in all eukaryotic cells. *Proc. Natl Acad. Sci. USA*, **89** (3), 1095–1099.

308 Le, N.T., Ho, T.B., Ho, B.H. (2010) Sequence-dependent histone variant positioning signatures. *BMC Genomics*, **11** (Suppl. 4), S3.

32
Epigenetic Gene Regulation in Bacteria

Javier López-Garrido, Ignacio Cota, and Josep Casadesús
Universidad de Sevilla, Departamento de Genética, Facultad de Biología, Apartado 1095, 41080 Seville, Spain

Epigenetic Regulation and Epigenomics: Advances in Molecular Biology and Medicine, First Edition. Edited by Robert A. Meyers.

Keywords

DNA adenine methylation
Methylation of the N^6 position of adenine moieties located in specific DNA targets (e.g., 5′GATC3′ for the Dam methylase of γ-proteobacteria, and 5′GANTC3′ for the CcrM methylase of α-proteobacteria).

Dam methylation pattern
Heritable state (methylated, hemimethylated, or nonmethylated) of specific GATC sites or GATC site clusters in the bacterial genome.

Phase variation
Reversible ON–OFF switching of gene expression at high frequency.

Bistability
Bifurcation of a unimodal pattern of gene expression into two patterns, generating two phenotypically distinct bacterial subpopulations.

Noise
Stochastic fluctuation in the level of a cellular product (e.g., mRNA or protein), especially if synthesized in small amounts.

Phenotypic heterogeneity is common in bacteria. Cases of epigenetic formation of bacterial lineages have been known for decades, and more examples have been unveiled by the advent of single-cell analysis. Epigenetic mechanisms establish cell fate in bacterial genera which undergo developmental programs. Lineage formation also occurs during biofilm formation and the colonization of animals by bacterial pathogens, and may be a frequent phenomenon when bacterial populations adapt to harsh environments. Lineage formation can be observed even in the laboratory, which suggests that phenotypic heterogeneity in clonal populations may be intrinsic to the bacterial lifestyle. The underlying mechanisms are diverse, ranging from relatively simple, inheritable feedback loops to complex self-perpetuating DNA methylation patterns.

1
Epigenetic Variation and Lineage Formation in Bacteria

In contrast to the well-known capacity of eukaryotic cells to diversify into lineages, bacteria have been classically viewed as clonal populations of genetically identical cells, the phenotype of which merely reflects their genetic constitution. This view is, however, simplistic as certain bacterial genera undergo developmental programs such as cell dimorphism [1], sporulation [2], the formation of multicellular structures [2, 3], partnership in symbiotic associations [4], and biofilm formation [5]. Furthermore, studies carried out during the past two decades have revealed that phenotypic heterogeneity, rather from being restricted to bacterial differentiation programs, is a common phenomenon among clonal populations of bacteria [6–10].

As in eukaryotes, the formation of bacterial cell lineages is often made possible by the epigenetic control of gene expression, for which some of the underlying mechanisms have been identified. For example, DNA methylation controls lineage formation in certain bacterial species, a phenomenon which has relevant implications in infectious diseases. However, DNA methylation is only one of the mechanisms employed by bacteria to generate epigenetic variation, and many others appear to exist. As discussed below, feedback loops transmissible through cell division generate metastable epigenetic states that are a common cause of phenotypic heterogeneity in clonal bacterial populations. Examples of epigenetic control involving model organisms (e.g., *Escherichia coli, Bacillus subtilis, Caulobacter crescentus*) and bacterial pathogens (e.g., *Salmonella enterica, Haemophilus influenzae, Neisseria meningitidis*) will be described below. The known or hypothetical roles of epigenetic inheritance in bacterial adaptation will be also discussed.

2
DNA Methylation in Bacteria

Three methylated bases are found in bacterial genomes: N^4-methylcytosine; C^5-methylcytosine; and N^6-methyladenine (m6A) [11, 12]. Bacterial base modification by DNA methyltransferases is often associated with the possession of restriction-modification (R-M) systems [13, 14]. In addition, many bacterial genomes contain solitary methyltransferases that do not have a restriction enzyme counterpart. Examples of the latter are the N^6-adenine methylases (Dam and YhdJ) of gamma-proteobacteria, the N^6-adenine methylase (CcrM) of alpha-proteobacteria, and the C^5-cytosine methylase (Dcm) of enteric bacteria [6, 14, 15]. Two of these enzymes – the N^6-adenine methylases Dam and the CcrM – play multiple roles in bacterial physiology, including the regulation of gene expression [6, 14–18]. Both Dam and CcrM methylate adenosine moieties in similar targets (5′GATC3′ and 5′GANTC3′, respectively) using *S*-adenosylmethionine as a methyl donor. Dam methylase is dispensable in many bacterial genera [19, 20], while CcrM is an essential cell function [21].

All known functions of m6A rely on regulating the interaction between DNA-binding proteins and their cognate DNA sequences (Table 1). Steric hindrance of restriction enzyme activity by methylation of a DNA target is a well-known example [13]. Furthermore, methylation of the amino group of

Tab. 1 Examples of DNA-binding proteins sensitive to the Dam methylation state of GATC sites within cognate DNA sequences.

Protein	*Function*	*Methylation state that permits binding and/or function*	*Reference(s)*
MutH	GATC-specific endonuclease	Hemimethylation, nonmethylation	[25]
Lrp	Activation of *traJ* transcription	Strand-specific hemimethylation, nonmethylation	[26, 27]
OxyR	Repression of *agn43* transcription	Hemimethylation (reduced affinity), nonmethylation	[28, 29]
RNA polymerase	Transcription of the IS*10* transposase gene	Strand-specific hemimethylation, nonmethylation	[30]
Fur	Repression of *sci1* transcription	Nonmethylation	[31]
HdfR	Activation of *std* transcription	Nonmethylation	[32] (M Jakomin *et al.*, unpublished results)
DnaA	Initiation of chromosome replication	Methylation	[33]
SeqA	Sequestration of replication origin, nucleoid organization	Methylation (reduced affinity), hemimethylation	[34]

adenine lowers the thermodynamic stability of DNA [22] and alters the DNA curvature [23]. Such structural effects can additionally influence DNA–protein interactions, especially in the case of proteins that recognize their cognate DNA-binding sites by both primary sequence and structure [24].

Bacteria employ m6A as a signal to indicate *when* and *where* a given DNA–protein interaction must occur (or, using the terms coined by Messer and Noyer-Weidner, for "timing and targeting" [35]). For instance, Dam methylation provides signals for the initiation of chromosome replication, chromosome segregation, and DNA strand discrimination during postreplicative repair of nucleotide mismatches [6, 14, 17]. Dam methylation also controls transcription of certain genes, because the methylation state of critical GATC sites can influence binding of RNA polymerase or transcriptional regulators to promoters. Loci whose expression is under the control of DNA adenine methylation fall into two general classes: (i) genes for which expression is coupled with the DNA replication cycle; and (ii) genes for which expression is controlled by the formation of a

DNA methylation pattern [16]. In addition, examples of DNA methylation control, the basis of which remains to be elucidated, have been described.

2.1 Temporal Control of Gene Expression by DNA Adenine Methylation

Methylation of adenine moieties in the bacterial chromosome is postreplicative, and occurs shortly – though not immediately – after passage of the replication fork [6, 14, 16]. Because the bacterial chromosome remains hemimethylated for a brief period of time after DNA replication, the hemimethylated state can provide a signal to couple gene expression to DNA replication. This phenomenon is not epigenetic *sensu stricto*, because the hemimethylated state is not inherited by daughter cells. However, temporal control of gene expression by hemimethylation is epigenetic in the sense that the signal involved is the methylation state of one or more GATC or GANTC sites. Hemimethylation can either activate or repress gene expression, and examples of both types will be described below; however, gene activation by hemimethylation may be more common than repression. This view is consistent with genetic and transcriptomic analyses indicating that DNA adenine methylation is more often a repressor of bacterial gene expression than an activator [36, 37]. Furthermore, *in vitro* studies on the effect of DNA adenine methylation on binding of proteins to DNA have shown that hemimethylation and nonmethylation are often analogous signals, and have similar phenotypic consequences [25, 26, 30]. Hence, the known examples of transcriptional repression by hemimethylation may be exceptional.

2.1.1 IS10 and *traJ*: Two Examples of Activation of Gene Expression by DNA Adenine Hemimethylation

IS*10* is a bacterial insertion element that also forms part of a composite transposon known as *Tn10* [38]. Studies performed in Nancy Kleckner's laboratory during the 1980s indicated that Tn*10* transposition occurs at a higher frequency in *E. coli* mutants lacking Dam methylase [30]. Repression of Tn*10* transposition by Dam methylation is the consequence of two concerted actions: (i) methylation of GATC sites near the ends of the transposon blocks transposase activity; and (ii) methylation of a GATC that overlaps the −10 module of the promoter of the transposase gene prevents transcription, presumably by hindering RNA polymerase binding [30]. When the replication fork passes by the IS*10* element, its GATC sites become hemimethylated, and the hemimethylated state permits transcription of the transposase gene and transposase-mediated cutting of the IS*10* ends; as a consequence, transposition is transiently permitted [30]. The biological significance of these overlapping controls may be tentatively interpreted as part of the self-restraint mechanisms typical of many transposons [39]:

- Hemimethylation restricts transposase synthesis and transposition to a brief lapse of the cell cycle, thus reducing the potential dangers of multiple transposition events and/or transposase-mediated DNA strand breakage.
- Hemimethylation couples transposition to DNA replication, a cell cycle period during which the existence of daughter chromosomes may reduce the chances of a lethal transposition event.

IS*10* transposase synthesis is further reduced by an additional sophistication of the regulatory system. When DNA replication generates two daughter IS*10* elements, transcription of the transposase gene is permitted in one of the hemimethylated IS*10* species only [30]. This subtle control outlines the regulatory capacity of Dam methylation, because the hemimethylated IS*10* species are identical except for the DNA strand that contains a methylated GATC.

Regulation of gene expression by strand-specific DNA hemimethylation has also been described in the *traJ* gene of the *Salmonella enterica* virulence plasmid, also known as pSLT [40]. Plasmid pSLT is a relative of the *E. coli* F sex factor, and promotes conjugation [27, 41]. The structural proteins that form the conjugal pilus and the effectors of the conjugation apparatus are encoded on a single transcriptional unit, the *tra* operon, which contains approximately 35 genes [40]. Transcription of *tra* is controlled by a pSLT-encoded transcription factor, TraJ, and by regulators encoded on the host chromosome [27, 42, 43]. Conjugal transfer of pSLT is inhibited outside the mammalian intestine [44]. Conjugation inhibition involves multiple, overlapping mechanisms, one of which is repression of *traJ* transcription by Dam methylation [20, 27, 45]. Unlike IS*10*, Dam-mediated repression of *traJ* transcription does not occur at the promoter itself but rather at an upstream regulatory sequence (UAS) which contains two binding sites for Lrp, a global bacterial regulator. Lrp activates *traJ* transcription by "coating" the entire UAS [27] (Fig. 1). One of the Lrp binding sites at the *traJ* UAS contains a GATC, the methylation of which has a dual effect on Lrp binding, namely reduction of the overall Lrp binding and alteration of its binding pattern [26]. The combination of both effects prevents *traJ* transcription in the absence of plasmid replication.

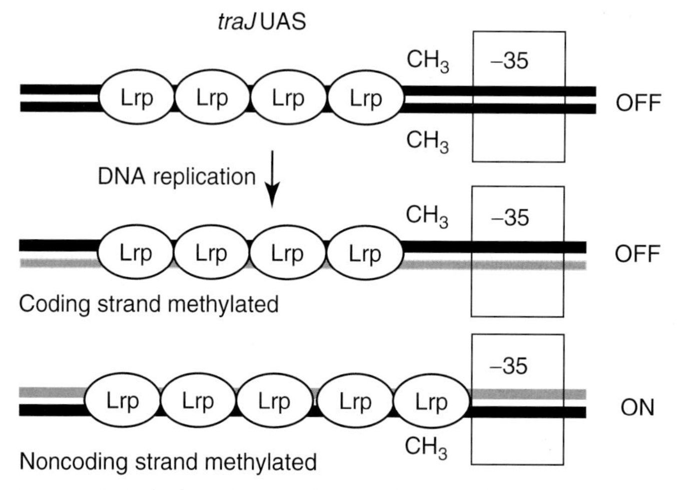

Fig. 1 Control of *traJ* transcription by strand-specific Dam methylation. The transcriptional activator Lrp binds the upstream regulatory sequence (UAS) with different patterns, depending on the DNA strand that contains a methyl group. Only one Lrp binding pattern permits transcription.

Passage of the replication fork renders the site hemimethylated, and transcription is permitted in one of the daughter plasmid molecules [26] (Fig. 1). As in IS*10*, it is remarkable that a single methyl group can create distinct epigenetic states in DNA molecules that otherwise are identical. Another analogy with IS*10* is that activation of *traJ* transcription by strand-specific hemimethylation may be viewed as a strategy to limit the synthesis of potentially dangerous cell products. Excess TraJ might cause a massive synthesis of *tra* operon products, thus representing a burden for the host cell. Excess *tra*-encoded DNA nicking proteins might additionally destabilize the plasmid. The activation of *traJ* transcription by hemimethylation may have an additional adaptive value as a sensor of replication, while Lrp dependence may ensure that transcription occurs when the physiological conditions are appropriate to undertake the energy-consuming process of bacterial mating .

Although the active epigenetic state of *traJ* is not heritable, it may be transmissible to the recipient cell if the daughter plasmid that produces TraJ is chosen for transfer (Fig. 1). Nicking of the transfer origin will occur in the methylated DNA strand, which will be transferred to the recipient cell and immediately replicated. DNA replication in the recipient can be expected to reproduce the active epigenetic state of *traJ* [26]. If sufficient Lrp is available, the recipient might instantly become a donor. This mechanism – which remains hypothetical – might contribute to provide an explanation of an old enigma in the conjugal transfer of F-like plasmids, known as "epidemic spread"; namely, that one limiting factor in plasmid transfer is the number of recipients, which suggests that acquisition of the plasmid is immediately followed by mating [46].

2.1.2 Repression of Gene Expression by DNA Adenine Hemimethylation in the *dnaA* Gene of *E. coli*

Unlike the IS*10* transposase gene and *traJ*, transcription of the *E. coli dnaA* gene, which encodes an essential protein for the initiation of DNA replication, is repressed by Dam hemimethylation [47, 48]. Initiation of chromosome replication requires full methylation of the origin (*oriC*), which contains a high concentration of GATC sites, to permit binding of the active form of DnaA [33, 49]. The *dnaA* gene is transcribed from two promoters; one of them ($dnaA_2$) contains three GATC sites and is only active if these are methylated [47, 48]. Because the *dnaA* gene is located near the origin of replication, the $dnaA_2$ promoter becomes hemimethylated soon after replication initiation. Rapid methylation of the *dnaA2* promoter is prevented by a sequestration mechanism that operates also at the origin of chromosome replication, *oriC*. When replication starts, both *oriC* and the *dnaA* promoter become hemimethylated, while access of the Dam methylase to the *oriC–dnaA* region is prevented by a GATC-binding protein known as *SeqA* [34, 50, 51] and by additional proteins [52]. Sequestration hinders the initiation of a new replication round, and also acts as a timer that delays the start of the next cell cycle until physiological circumstances are adequate [49].

2.1.3 Cell Cycle-Coupled Control of Gene Expression in *Caulobacter crescentus*: Role of CcrM Methylation

Caulobacter crescentus is an alpha-proteobacterium that thrives in oligotrophic environments, such as freshwater lakes and streams. *Caulobacter* divides asymmetrically, giving rise to a stalked cell and a swarmer cell [1]. After a period of motility the swarmer cells differentiate into stalked cells [1, 53]. Chromosome replication occurs only in stalked cells, and requires a full methylation of the GANTC sites at the replication origin, *Cori* [54, 55]. In turn, *Cori* methylation is controlled by a regulatory cascade that uses GANTC hemimethylation as a signal [18, 56, 57]. Synthesis of CcrM methylase itself is activated by GANTC hemimethylation. Because the *ccrM* gene is located closer to the replication terminus than to the *Cori*, CcrM is only synthesized at late stages of chromosome replication [18, 58]. When chromosome replication starts, CcrM is rapidly degraded [18, 58]; as a consequence, DNA replication is not followed by DNA methylation immediately after replication fork passage, and the daughter chromosomes remain hemimethylated until replication reaches the *ccrM* gene. Thus, replication fork progression acts as a timer for coordination of the *Caulobacter* cell cycle. The cascade of transient regulators involved in this temporal control is as follows:

1. Methylation of GANTCs in the *Cori-dnaA* region provides the signal for initiation of the following replication round, and permits synthesis of the replication initiator, DnaA. As in enteric bacteria, the *Caulobacter dnaA* gene is located near the replication origin (*Cori*) and contains two GANTCs. Transcription occurs only when these GANTCs are methylated.
2. Synthesis of DnaA is transient, because its promoter becomes hemimethylated shortly after replication initiation [54]; hence, passage of the replication fork renders the promoter inactive [57]. Because *Caulobacter* stalked cells do not contain CcrM methylase until chromosome replication approaches completion, the *dnaA* promoter remains hemimethylated (and therefore inactive) during most of the cell cycle [18].
3. Before disappearing, DnaA activates transcription of *grcA*, while GrcA activates transcription of the *ctrA* gene, which encodes the next cell-cycle regulator, CtrA [58]. GrcA-mediated activation of *ctrA* transcription occurs upon passage of the replication fork, when a GANTC site near the −35 module of the *ctrA* p_1 promoter becomes hemimethylated [58].
4. CtrA is necessary to activate transcription of *ccrM*, which lies near the replication terminus. However, CtrA-mediated activation of *ccrM* transcription requires hemimethylation of two GANTC sites in the *ccrM* leader region [55]; thus, *ccrM* transcription is delayed until the replication fork passes by. This orderly sequence of events relies on the fact that *ccrM* is transcribed last in the regulatory cascade [58, 59].
5. CcrM synthesis permits the methylation of GANTCs in the *Cori-dnaA* region. As a consequence, *Cori* is activated and the *dnaA* gene is transcribed. Methylation is therefore the signal for initiation of the following replication round.

2.2 Formation of Dam Methylation Patterns in the Bacterial Genome

As a rule, hemimethylation of GATC sites is transient and methylation of the daughter DNA strand restores two-strand GATC methylation shortly after passage of the replication fork [6, 14, 15, 17]. However, the activity of the Dam methylase can be hindered by binding of proteins to specific GATC sites. A phenomenon of this type – sequestration of the origin of chromosome replication by SeqA [51] – was described in Sect. 2.1.2. In addition, studies conducted during the early 1990s showed that the *E. coli* chromosome contains GATC sites that are stably hemimethylated or nonmethylated [60–63]. The methylation state of some such sites was found to vary depending on the growth conditions; this suggests that, in certain cases, undermethylation may be a cellular response to physiological or environmental cues [64–66]. Because active demethylation is not known to occur in bacteria, the formation of stable hemimethylation and nonmethylation was proposed to result from competition between specific DNA-binding proteins and Dam methylase [64–66]. Recent studies have shown that Dam methylase – which usually is a highly processive enzyme – can reduce its processivity at certain GATC sites, thus permitting methylation hindrance by competing DNA-binding proteins [67]. This phenomenon typically occurs at GATC sites which form part of GATC clusters (two or more GATC sites separated by short distances) and are flanked by AT-rich sequences [68]. Nonprocessive GATC sites are found in DNA-binding sequences for Lrp, OxyR, and other regulators of bacterial transcription, and their methylation state is a hallmark of the distinct gene expression patterns found in certain bacterial lineages [6, 16].

2.2.1 The *pap* Operon of Uropathogenic *E. coli*: A Paradigm of Phase Variation Control by Dam Methylation

Phenotypic diversity in clonal populations of bacteria can be generated by the reversible ON–OFF switching of gene expression at high frequencies, a phenomenon known as phase variation [69–71]. In bacterial pathogens, phase variation is often observed at loci encoding surface structures, and may be viewed as a strategy to generate phenotypic polymorphism [69]. Evasion of the host immune system and protection of a bacterial subpopulation against phage infection are well-known advantages of phase variation, though others may exist [71]. The mechanisms that cause phase variation are diverse, but one of the best known involves the formation of heritable Dam methylation patterns. The paradigm for this type of control is the *pap* operon of uropathogenic *E. coli*, a set of genes that encode fimbrial adhesins for adherence to the urinary tract epithelium [72, 73].

Studies performed by David Low and colleagues have shown that a regulatory region upstream the *pap* operon contains six sites for Lrp binding, and two such sites contain GATCs. When the operon is not transcribed (OFF state), Lrp is bound to the downstream three sites, and this may prevent transcriptional initiation by direct hindrance of RNA polymerase binding [74]. In the OFF state, the GATC site located near the *pap* promoter ($GATC_{prox}$) is nonmethylated, while the GATC site located further upstream ($GATC_{dist}$) is methylated [75, 76] (Fig. 2). Nonmethylation of $GATC_{prox}$, which is a GATC site of the non-processive type, is the consequence of Lrp binding, which prevents access of the Dam methylase to the site after DNA replication [77]. Binding of Lrp to the downstream sites reduces its affinity for the upstream sites, thus perpetuating the OFF state [78] (Fig. 2). This DNA methylation pattern is inherited by the daughter cells, and persists endlessly unless a protein referred to as PapI

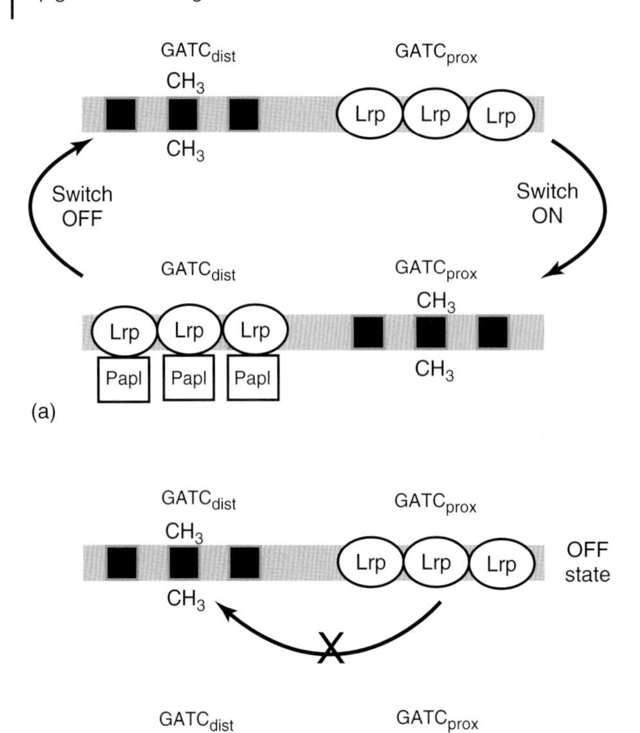

Fig. 2 (a) DNA methylation patterns associated with the OFF and ON states of the *pap* operon. DNA hemimethylation states during switching are not shown; (b) Feedback loops that maintain the OFF and ON state of the *pap* operon. Binding to the downstream sites reduces Lrp affinity for the upstream sites, and propagates the OFF state. Expression of *pap* yields PapB protein, which activates *papI* transcription. A high PapI level maintains Lrp + PapI bound at the distal sites, propagating the ON state.

is present [78, 79]. The transcription of *papI* occurs at very low levels, and may undergo stochastic fluctuation or "noise" [73, 78]. Whenever PapI reaches a level above a critical threshold – an event which, under laboratory conditions, occurs at an average frequency of 10^{-4} per cell and generation – PapI directs the translocation of Lrp to the upstream three binding sites in the *pap* UAS [74, 78]. As a consequence, Lrp and PapI prevent the methylation of $\mathrm{GATC_{dist}}$, which becomes nonmethylated after two replication rounds [72, 73, 76]. In turn, $\mathrm{GATC_{prox}}$ is no longer bound by Lrp, and is quickly methylated by the Dam methylase. Under this configuration ($\mathrm{GATC_{dist}}$ methylated, $\mathrm{GATC_{prox}}$ nonmethylated), the *pap* operon is transcriptionally active, and Pap fimbriae are synthesized [73]. A positive feedback loop sustains the ON state: transcriptional activation of the *pap* operon yields PapB, a *pap* operon product that increases transcription of *papI* [73] (Fig. 2). The ON state is inherited by the bacterial progeny and, under laboratory conditions, is perpetuated during 10–12 generations on average, probably with significant fluctuation. Return to the OFF state requires a decrease in the concentration of PapI, and the mechanism that reduces the PapI level remains to be identified. When PapI is not present, Lrp is unable to bind the upstream regulatory sites, and translocates to the downstream sites [72, 76]. Lrp abandon of the upstream sites permits methylation of $GATC_{dist}$, while Lrp translocation to the downstream sites hinders methylation of $GATC_{prox}$ [72, 76, 80]. Consequently, the OFF configuration of the *pap* regulatory region ($GATC_{dist}$ methylated, $GATC_{prox}$ nonmethylated) is restored after two replication rounds [73] (Fig. 2).

Transcription of the *pap* operon is additionally controlled by host-encoded factors such as CRP, H-NS, CpxAR, and RimJ [81–84]. These regulators may skew the switching frequencies in response to environmental or physiological cues, thus introducing deterministic elements in the stochastic mechanism that controls *pap* phase variation.

2.2.2 Dam Methylation-Dependent Control of Phase Variation in the *E. coli agn43* Gene

Antigen 43 (Agn43) is an *E. coli* outer membrane protein involved in cell aggregation and biofilm formation [85, 86]. Expression of *agn43* is subjected to phase variation, and the underlying mechanism involves transcriptional control by Dam methylation and OxyR, an *E. coli* sensor of oxidative stress that also acts as a transcriptional regulator [28, 85]. Regulation of *agn43* by OxyR is, however, independent of the redox state of the OxyR protein [87]. In the OFF state, OxyR represses *agn43* transcription by binding a regulatory region that contains three GATC sites and, as a consequence of OxyR binding, the GATCs located within the region become nonmethylated after two replication rounds [29, 88]. Nonmethylation facilitates OxyR binding, thus generating a positive feedback loop that propagates the OFF state. Switching to the ON state occurs when OxyR is excluded from the *agn43* promoter, and two-strand DNA methylation occurs [6]. The ON state is perpetuated because methylation of the GATCs prevents OxyR binding [29, 88]. As in the *pap* operon, the OFF and ON states of *agn43* are metastable, and switching occurs at relatively constant frequencies, which vary slightly depending on the culture conditions [89]. OxyR cannot bind the *agn43* regulatory region if its GATCs are methylated, but does show a residual affinity for hemimethylated *agn43* DNA [89]. Such residual affinity may be crucial to permit competition between OxyR and Dam methylase during DNA replication. Hence, switching to the OFF state may require OxyR binding to hemimethylated *agn43*, followed by Dam methylation hindrance during two consecutive DNA replication rounds.

2.2.3 Other Phase Variation Systems under Dam Methylation Control

A variety of phase variation systems that are controlled by Dam methylation and share features with *pap* and *agn43* have been described in pathogenic *E. coli* strains, *Salmonella*, as well as in *Salmonella* bacteriophage P22. In all cases, phase variation causes changes in the bacterial surface, and may play a role in the interaction of the pathogen with its host. One such system, encoded on the glycosyltransferase operon (*gtr*) of phage P22, controls modification of the lipopolysaccharide O antigen [90]. Because the O antigen is the receptor for P22 and other *Salmonella* bacteriophages, *gtr* phase variation in P22 lysogens may permit infection by related phages. This might in part explain the variable prophage assortment typical of *Salmonella* strains [91], and might have consequences for virulence since *Salmonella* phages often carry virulence determinants [92]. Phase variation of *gtr* is controlled by OxyR in a Dam-dependent manner. The *gtr* UAS contains two pairs of closely linked GATC sites. Binding of OxyR to the upstream GATC pair permits *gtr* transcription, and renders both GATCs nonmethylated. In contrast, OxyR binding to the promoter-proximal GATC pair represses *gtr* transcription, and renders both GATCs nonmethylated [90]. The OFF and ON states are thus characterized by specific, heritable DNA methylation patterns, and switching must involve an OxyR translocation mechanism that remains to be identified. Another phase variation system that is controlled by OxyR binding to two pairs of GATC sites has been found in *STM2209-STM2208*, a *Salmonella*-specific locus. *STM2209* and *STM2208* form a bicistronic transcriptional unit, and encode cytoplasmic membrane proteins that are putatively involved in lipopolysaccharide modification, phage P22 sensitivity, and *Salmonella*–macrophage interaction (I. Cota *et al.*, unpublished results).

While *gtr* and *STM2209-STM2208* may be classified as complex variants of the *agn43* phase variation system, the *clp* operon of enterotoxicogenic *E. coli* and the *pef* operon of the *Salmonella* virulence plasmid show similarity to *pap* [93–95]. In both *clp* and *pef*, phase variation in the synthesis of fimbriae is controlled by binding of Lrp to a regulatory region which contains two GATCs arranged in a configuration reminiscent of *pap*. As in *pap*, Lrp binding creates heritable Dam methylation patterns. Switching to the ON state involves translocation of Lrp from $GATC_{prox}$ region to the $GATC_{dist}$ region, whereas switching to the OFF state occurs when Lrp translocates to the $GATC_{prox}$ region [94, 95]. One difference, however, is that the PapI homologs involved in the regulation of *clp* and *pef* do not participate in switching to ON, but rather in switching to OFF [94, 95]. In other words, PapI is a *pap* activator while the Pap-like proteins that regulate *clp* and *pef* are repressors. The PapI-like protein PefI is encoded near the *pef* gene cluster in the *Salmonella* virulence plasmid, while *clp* may be regulated by a chromosomal PapI homolog [6].

The Dam methylation-dependent phase variation systems so far characterized in *E. coli* and *Salmonella* may be examples of a widespread phenomenon. A tentative list of Dam-dependent phase variation genes in the genomes of enteric bacteria may include *gtr* clusters (similar to P22 *gtr*) found in the *Salmonella* chromosome [90], as well as some of the >10 fimbrial operons found in many *Salmonella* strains [96]. Furthermore, bioinformatic searches in *E. coli* and *Salmonella* have indicated the presence of additional GATC sites, the arrangement and flanking sequence of which might confer reduced processivity to Dam methylase (J. Casadesús, unpublished results). If these GATC sites were to form part of phase variation systems, the contribution of Dam methylation to phenotypic heterogeneity in the bacterial population may be even more relevant than currently considered. A simple calculus may suffice to illustrate this point: the existence of 10 phase variation loci with independent switching might be able to generate 2^{10} bacterial cell variants. Such phenotypic heterogeneity may have an adaptive value during host–pathogen interaction, as well as in the environment. The results of recent studies have shown, for instance, that subpopulations are formed during the colonization of animals by *Salmonella* (see below). Dam methylation is essential for *Salmonella* infection [97, 98] and regulates the synthesis of virulence determinants [37, 99]. In addition, the formation of Dam-dependent bacterial lineages may facilitate colonization of the animal host by increasing phenotypic heterogeneity, especially in the bacterial envelope.

2.2.4 **Phasevarions: The Formation of Bacterial Cell Lineages by Phase Variation of DNA Methylase Genes**

Certain restriction–modification (R-M) systems of types I and III show phase variation [100, 101], and a common mechanism for switching between OFF and ON states is the alteration of nucleotide repeats [102]. Expansion or contraction of repeats is usually caused by DNA polymerase slippage during DNA replication. Phase variation of R-M systems may generate subpopulations of bacterial cells that differ in their susceptibility to phage infection, and also in their ability to acquire foreign DNA by transformation. In addition, DNA adenine methylation by certain phase-variable type III R-M systems regulates the expression of specific genes [101, 103–106]. As a result, these systems conserve their R-M activity but have additionally acquired epigenetic regulatory capacity. One notable observation is that, in some such systems, the gene encoding the restriction enzyme is inactivated by mutation, while the modification gene (*mod*) remains active [106]. In these mutant type III R-M systems, the Mod enzyme is therefore a functional analog of solitary methyltransferases such as Dam and CcrM. A difference, however, is that synthesis of the type III Mod DNA methylase is phase-variable, and generates two subpopulations of bacterial cells, one of which contains N^6-methyladenine in the genome while the other lineage does not (Fig. 3). As a consequence, each lineage shows a distinct pattern of gene expression that affects all DNA methylation-sensitive loci. DNA adenine methylation by type III Mod enzymes has been shown to regulate gene expression in the human pathogens *Haemophilus influenzae, Neisseria meningitidis,* and *Neisseria gonorrhoeae,* while the loci under Mod control include genes with roles in envelope structure, virulence, and stress responses [106].

Phasevarions may be viewed as a remarkable evolutionary achievement. Individual phase variation systems such as *pap* and *agn43* generate heterogeneity of a single phenotypic trait, while the cell lineages under phasevarion control differ in multiple phenotypic traits (Fig. 3). An additional *tour de force* in the capacity of phasevarions to generate bacterial lineages is found in bacterial species that contain multiple *mod* alleles, each with slightly different DNA-binding domains [105]. Independent switching in the synthesis of several Mod proteins can be expected to generate multiple gene expression patterns, thus increasing the phenotypic heterogeneity of the population [106].

2.2.5 **Formation of Dam Methylation Patterns upon Deterministic Switching**

In the phase variation systems described in previous sections, ON–OFF switching is reversible, and relies on stochastic mechanisms. In other systems, however,

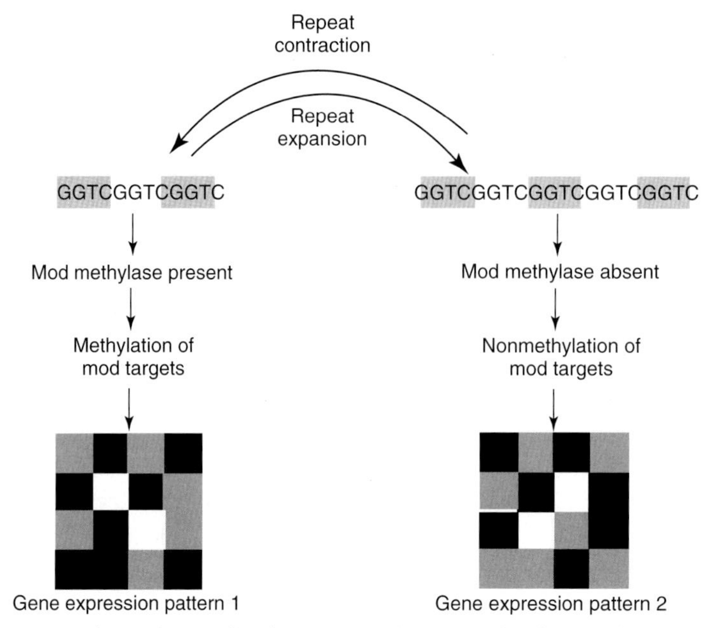

Fig. 3 The workings of a phasevarion. Phase variation of the DNA methylase gene, *mod*, generates two cell lineages. In one lineage the Mod DNA targets are methylated; in the other lineage the Mod DNA targets are nonmethylated. As a consequence, each lineage shows a distinct expression pattern of Mod methylation-sensitive loci.

DNA adenine methylation patterns are generated in a deterministic manner. Aside from this difference, stochastic and deterministic epigenetic switches are both propagated by feedback loops which make DNA methylation patterns transmissible to daughter cells. For instance, transcription of the *sci1* gene cluster of enteroaggregative *E. coli*, which encodes a type VI secretion system, is under the control of the ferric uptake regulator, Fur [31]. The *sci1* promoter region contains two Fur binding sites, one of which overlaps the −10 module and contains a GATC site. In the presence of Fe^{2+}, Fur binds its cognate sites at *sci1*, and prevents transcription. Fur binding protects the downstream site from the Dam methylase, and renders the GATC nonmethylated [31]. However, because nonmethylation increases the affinity of Fur for the downstream binding site, a feedback loop propagates repression. If the intracellular concentration of ferrous ion decreases, Fur leaves the *sci1* promoter region, and the GATC site within the downstream Fur-binding site can be immediately methylated. GATC methylation contributes to sustain the ON state because Fur shows low affinity for the downstream GATC site when it is methylated [31]. Thus, reestablishment of repression may require relatively high levels of Fe^{2+} as well as GATC hemimethylation during DNA replication.

Another Dam-dependent deterministic switch is found in the *S. enterica std* fimbrial operon. Synthesis of Std fimbriae is tightly repressed under laboratory conditions [37], and has been detected only in the intestine of animals [107]. One crucial factor to prevent *std* operon expression outside the animal milieu is Dam methylation [32, 108], which prevents binding of HdfR, a poorly known transcription factor of the LysR family, to the *std* control region [32]. Methylation and hemimethylation of three GATC sites in the *std* UAS prevent HdfR binding (M Jakomin Cota *et al.*, unpublished results). Transcriptional activation of the operon may require, besides HdfR, unidentified factors which may be synthesized by *Salmonella* in the animal intestine only. When *std* activation occurs, binding of HdfR prevents methylation of two GATCs in the UAS (M Jakomin Cota *et al.*, unpublished results). Nonmethylation facilitates HdfR binding, thus creating a positive feedback loop that maintains *std* expression active.

The examples of *sci1* and *std* involve DNA methylation patterns with opposite physiological significance: in *sci1*, nonmethylation is associated with transcriptional repression, and in *std* to transcriptional activation. In both cases, however, the methylation state of GATCs in the regulatory region creates a feedback loop that propagates either the active or the inactive state. Switching is deterministic, and requires ancillary molecules such as ferric ion in *sci1* and an unknown factor in *std*. Dam-dependent, deterministic switches may also operate in *pap*-like fimbrial operons that do not undergo phase variation. One such example may be the *fae* fimbrial operon of enterotoxicogenic *E. coli*, which is repressed by a PapI-like factor known as *FaeA* and by the global regulator Lrp [109, 110].

2.3 Other Bacterial Loci under Dam Methylation Control

Genetic screens and transcriptomic and proteomic analyses have identified multiple cell functions, the synthesis of which is altered in Dam^- mutants of *E. coli* and *Salmonella* [36, 37]. Changes in gene expression in Dam^- mutants do not necessarily indicate Dam-sensitive transcription, because *dam* mutations are pleiotropic and can alter gene expression in an indirect manner [19, 20]. An example is found in the DNA-damage responsive SOS regulon of *E. coli* and *Salmonella*, which shows elevated expression in Dam^- mutants [20, 111]. However, transcription of SOS genes is not under Dam methylation control, and their elevated expression in Dam^- mutants is a consequence of DNA double-strand breakage by the MutHLS mismatch repair system in the absence of methyl-directed DNA strand discrimination [25, 112]. Bioinformatic searches to identify genes of which the transcription is controlled by Dam methylation are also difficult [113–115], and their predictions turn out often to be naïve or simplistic – if not wrong – when validated by experimental analysis. It is not always obvious where the search for relevant GATC sites should be performed, because Dam methylation can regulate a promoter from distant regulatory sites: for instance, the $GATC_{dist}$ of the *pap* operon is located more than 100 bp away from the transcription start site [72]. In the *E. coli* chromosome, the average distance between GATC neighbor sites is 214 bp [14], with the obvious consequence that GATC sites at distances potentially relevant for transcriptional control are found in many promoters. Furthermore, the presence of a GATC site at a critical position in the genome is largely uninformative, as exemplified by the P1 *cre* of bacteriophage P1, which possesses two promoters that contain GATC sites, though only one of the two promoters is regulated by

Dam methylation [116]. An additional complication arises from the fact that GATC-less genes can be controlled by Dam methylation if their transcription is controlled by a cell factor under direct Dam methylation control. An example of this type is found in *Salmonella* pathogenicity island 1 (SPI-1), a cluster of virulence genes the expression of which is reduced in Dam^- mutants [37]. However, only one SPI-1 gene, *hilD*, is necessary to transmit Dam methylation dependence to the entire SPI-1 [99].

The *hilD* gene of *Salmonella* provides also an example of an additional twist in the intricacies of Dam-dependent regulation of gene expression: Dam methylation does not regulate *hilD* transcription but *hilD* mRNA stability [99]. The effect must be indirect, because the Dam methylase is not known to methylate RNA. Other examples of post-transcriptional regulation of gene expression by Dam methylation have been described in *E. coli* [117, 118] and *Yersinia enterocolitica* [119]. Although these enigmatic cases remain to be explained at the molecular level, their existence has interesting implications. Many bacterial genes under Dam methylation control form part of genomic islands, transposable elements and plasmids [6, 14, 15], which suggests that Dam methylation might play a role in the control of horizontally acquired genetic elements, in a fashion reminiscent of the eukaryotic control of transposon activity by 5-methylcytosine. However, the existence of Dam-dependent cellular functions involved in mRNA stability and/or mRNA translation suggests that bacterial evolution has also placed certain elements of the gene expression machinery under Dam methylation control. This is not surprising, given that Dam methylation also controls housekeeping processes such as chromosome replication and segregation [14].

A different type of enigma is found in the *finP* gene of the *Salmonella* virulence plasmid. Transcription of *finP* requires Dam methylation to prevent repression by the nucleoid protein H-NS [42]. However, the effect does not rely on a GATC site found at the *finP* promoter, nor on other GATC sites in the region [42]. A tentative explanation may be that methylation of GATC sites in the genome (around 20 000 in *E. coli* K-12 and *S. enterica*) might contribute to shape the structure of the bacterial nucleoid by a cumulative effect. However, methylation of individual GATCs would make a small, perhaps irrelevant contribution.

3
Bacterial Lineage Formation by Hereditary Transmission of Feedback Loops

It is possible that hereditary DNA methylation patterns permit the formation of especially robust feedback loops. The constancy of switching rates in Dam-dependent phase variation systems [6] argues in favor of this possibility. However, bacterial lineages can also be generated without DNA modification, and many interesting examples have been deciphered during the past few decades. The absence of epigenetic marks (e.g., methylation patterns) in the genome makes these epigenetic systems largely elusive to classical molecular biology. Fortunately enough, flow cytometry, fluorescence microscopy, and microfluidics have provided methods to detect subpopulation formation in clonal bacterial populations, and have revealed that phenotypic heterogeneity is extremely common [7, 9, 10, 120, 121]. The Experimental analysis of epigenetic heterogeneity is especially feasible in systems that exhibit a reduced number of epigenetic states (e.g., two states; a phenomenon known as *bistability*) and a low frequency of switching between states, so that stable cell subpopulations are formed.

Bistable gene expression occurs when a unimodal pattern of gene expression becomes bimodal, bifurcating into two distinct patterns. Bistability can be generated either by a positive feedback loop or by a double-negative feedback loop [122]. A classical example of bistability generated by a positive feedback loop was described six decades ago in the *E. coli lac* operon by Novick and Weiner [123]. IPTG (isopropyl-D-thio-β-galactopyranoside is a gratuitous inducer of the *lac* operon, and is noncatabolizable by *E. coli*. When added at high concentrations, IPTG fully derepresses the *lac* operon, but at low concentrations it is unable to induce a naïve (uninduced) culture. However, if a fully induced culture is transferred to a medium containing low concentrations of IPTG, a subpopulation of cells is able to maintain the *lac* operon induced [123]. The mechanism of maintenance is simple: fully induced cells have a high level of β-galactoside permease in their membrane, while the permease (which can transport IPTG) provides a high internal concentration of inducer, thus maintaining full induction (Fig. 4). The positive feedback loop in this system is that a high permease level is required to concentrate the inducer in the cell, and high internal inducer levels are required for high levels of permease synthesis [123]. In other cells, however, a decrease in the internal concentration of inducer (which may easily occur, for example during cell elongation and division) will reduce permease synthesis, which in turn will cause a further decrease in the internal concentration of inducer, driving the cell towards *lac* repression (Fig. 4). The overall consequence is that a fully induced population bifurcates into two bistable states: fully induced and non-induced [123–125].

Bistability generated by a double-negative feedback loop can be illustrated with another classical example. The infection of *E. coli* by bacteriophage lambda can have two different outcomes, namely lysis or lysogeny. Although the decision is influenced by environmental conditions, and also by the physiological state of the cell, the fate of individual phage infections is unpredictable and may be considered stochastic. Phage lambda has two repressors, known as *cI* and *Cro*, each of which represses the expression of the other [126]. At the onset of infection, both repressors are produced and the lysis–lysogeny decision may be viewed as a repressor race. The repressor that first occupies specific regulatory DNA sites in lambda DNA will repress the synthesis of its antagonist. If the winner is cI, synthesis of Cro will be repressed and lambda will lysogenize the host cell. However, if the winner is Cro then the synthesis of cI will be repressed and lambda will lyse the cell. It should be noted that the outcomes of a positive feedback loop and a double-negative feedback loop are analogous [122]. In the case of lambda, for instance, shutting off the synthesis of Cro is equivalent to positive autoregulation of cI, and *vice versa* [124, 126].

3.1
DNA Uptake Competence in *Bacillus subtilis*

The literature on bacterial bistable switches has been enriched with interesting cases during the past two decades. Because comprehensive, insightful reviews on bistability are available

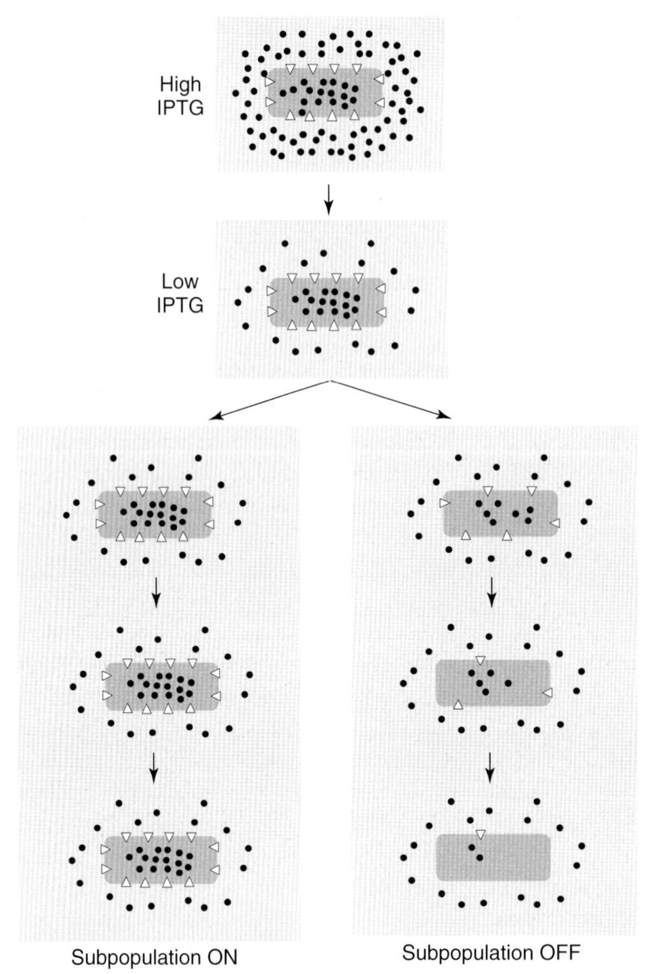

Fig. 4 Bifurcation in *lac* operon expression when an induced culture is transferred to lower, intermediate levels of the inducer, IPTG. Black dots represent IPTG molecules; white triangles represent Lac permease.

[7, 9, 10, 120, 121], only a few examples will be mentioned here. One remarkable case is the acquisition of competence for DNA uptake by *Bacillus subtilis*.

When *B. subtilis* cells enter the stationary phase, about 10% of them become competent while the remainder of the cell population remains noncompetent [127]. The key factor for competence development is ComK, a transcriptional activator of genes necessary for DNA uptake [128]. In addition, ComK shows positive autoregulation [129]. During exponential growth, ComK is synthesized but rapidly degraded. When the culture approaches stationary phase a quorum sensing-related factor stabilizes ComK [130, 131]. At that moment, a competition starts between several repressors and ComK itself to bind the *comK* promoter

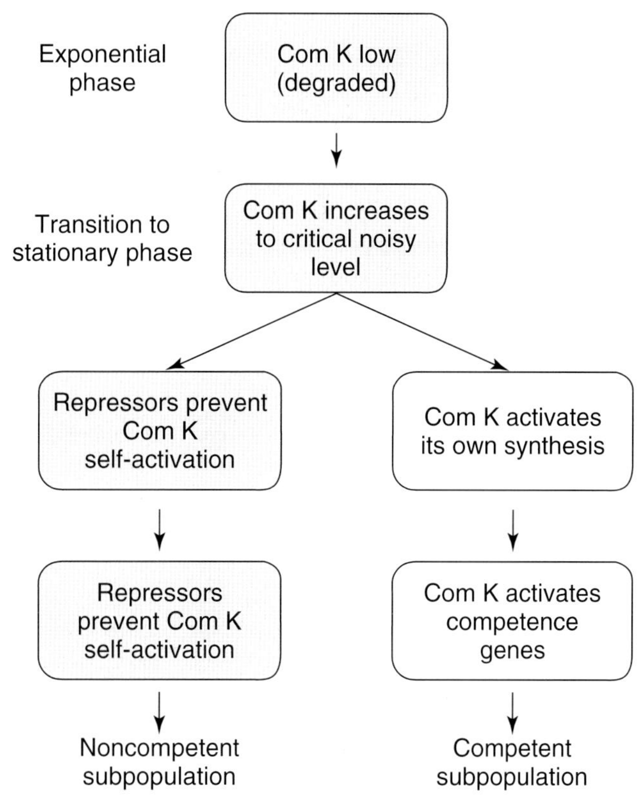

Fig. 5 Subpopulation formation during the development of competence in *Bacillus subtilis*. The fate of individual cells is decided at a critical moment in which Com K levels are intermediate and noisy. Above a threshold level, Com K drives the cell towards competence. Below the threshold, repressors prevent Com K synthesis and the cell does not become competent.

regulatory region [132, 133]. Binding of ComK will create a positive feedback loop, leading to increased synthesis of ComK and subsequent transcription of competence genes (Fig. 5). Binding of the repressors will prevent competence development. The level of ComK in individual cells is noisy, in the sense that it shows a relatively broad, unimodal distribution. This property, which is typical of many cellular products synthesized at low levels, is crucial for bifurcation of the population into two subpopulations. When the ComK level reaches a threshold in a *B. subtilis* cell, a quantitative difference will become qualitative, whereby the ComK positive feedback loop will be activated and competence will develop [134]. In turn, those cells in which the ComK level remains below the threshold will permit the repression of *comK* expression, and will not become competent. Thus, the development of competence occurs in cells that undergo a small, but critical, increase in ComK concentration; the amplification of this signal by the ComK feedback loop propagates the competent state to daughter cells [10].

3.2
Epigenetic Control of Spore Formation in *Bacillus subtilis*

In *B. subtilis* and other Gram-positive bacteria, nutrient scarcity triggers a complex process that generates spores; these are dormant cells with an adamant resistance to physico-chemical injury. Under nutrient limitation, however, only a subpopulation of *B. subtilis* cells enter the developmental program of spore formation [7]; this heterogeneity reflects cell-to-cell differences in the activation of the master regulator of sporulation, Spo0A [135]. Activation of Spo0A requires phosphorylation, and is controlled by a complex phosphorelay [136]. The switch that triggers Spo0A activation in certain cells may be a component of the regulatory cascade that undergoes noisy expression, and converts a quantitative difference into a qualitative state above a given threshold. While this component remains to be identified, a complex positive feedback loop that sustains Spo0A accumulation has been identified [7]. Spo0A stimulates the synthesis of a sigma factor, σ^H, which in turn activates transcription of the *spo0A* gene, and also of genes encoding components of the phosphorelay that activates Spo0A [137]. Hence, Spo0A boosts both its synthesis and its activation [7].

The occurrence of bistability in spore formation is at first sight paradoxical, as the potential benefits of reduced spore formation are difficult to understand. A attractive hypothesis is that maintenance of a subpopulation of nonsporulating cells may be advantageous if nutrient levels increase and the decision to sporulate turns out to be a false alarm. Furthermore, activation of Spo0A triggers lysis of nonsporulating cells, a phenomenon known as bacterial cannibalism [138, 139]. Lysis of nonsporulating siblings releases nutrients that slow down or even arrest spore formation in Spo0A-ON cells. Delayed commitment to the sporulation program may have selective value because sporulation is an energy-consuming process [7], and because a sudden increase in nutrient availability might permit vegetative division, which obviously is more proficient as a reproductive strategy.

3.3
Epigenetic Control of Virulence in *Pseudomonas aeruginosa*

Several examples of epigenetic control of virulence have been proposed in the opportunistic pathogen *Pseudomonas aeruginosa* [140]. A positive feedback loop involving the transcriptional regulator BexR activates expression of the so-called "BexR regulon," which includes the virulence-related *aprA* gene and other loci of unknown function [141]. In addition, BexR shows positive autoregulation [141]. As in the *B. subtilis* ComK system, bistability in BexR expression is the consequence of noisy, low-level BexR synthesis, followed by amplification of the BexR level in cells that produce BexR above a critical threshold [141]. One difference here, however, is that competence is acquired by 10% of *B. subtilis* cells [7], while the BexR feedback loop is activated in only 0.004% of cells [141]. This difference indicates that bistable or multistable epigenetic switches can split clonal populations into subpopulations of diverse sizes. This concept will be used below to discuss the epigenetic mechanisms that may permit the formation of rare cell variants in bacterial populations.

Epigenetic switches may also control mucoidy and cytotoxicity during lung colonization by *P. aeruginosa* in cystic fibrosis patients. An elegant model involving interlocked positive and negative feedback circuits has been proposed to control the synthesis of alginate, an extracellular polymer that renders *P. aeruginosa* mucoid and may play a role in biofilm formation [142]. In turn, a positive feedback loop has been proposed to induce cytotoxic capacity to a *P. aeruginosa* subpopulation during lung infection [143]. Although neither model has been dissected at the molecular level, the potential involvement of epigenetic mechanisms in the adaptation of *P. aeruginosa* to the human lung dissents from the fashionable view that adaptation of pathogens to animal environments often involves increased mutation rates. Indeed, hypermutation has been demonstrated in *P. aeruginosa* isolates from cystic fibrosis patients [144], and may likewise occur during the colonization of harsh environments by other pathogens [145]. However, theoretical calculations have indicated that increased mutation rates often involve a mutational burden incompatible with adaptation [146, 147]. Given this limitation, epigenetic mechanisms may be a less risky bet than mutation for the generation of phenotypic diversity in bacterial populations. If this view were to be correct, many more epigenetic mechanisms for bacterial lineage formation can be expected to be discovered in the future.

3.4 Error-Based Epigenetic Switches

Noise, which is an intrinsic property of biological systems, reflects fluctuations in the efficiency of the cellular machines [148]. In addition to noise, all biological systems exhibit an intrinsic degree of inaccuracy [149, 150]. A well-known example of found in DNA replication, the fidelity of which is checked during base selection and also by proofreading and by postreplicative mismatch correction systems, resutting in error rates as low as 10^{-10} per nucleotide incorporated [151]. However early DNA replication studies, recognized that the accuracy of existing DNA polymerases could be readily improved, as illustrated by the existence of antimutator alleles of T4 DNA polymerase [152]. This gave rise to the notion that a certain degree of DNA replication inaccuracy may have been selected during evolution to create genetic polymorphism [153]. An optimal mutation rate may exist to provide variation, which is the raw material for natural selection and subsequent adaptation. Both transcription and translation are likewise error-prone, putatively reflecting a trade-off between efficiency and accuracy [154–157].

A study conducted by Christophe Herman and colleagues has provided evidence that errors made during transcription can provide signals for epigenetic switching in the *E. coli lac* operon [158]. The experiments were carried out in the presence of intermediate concentrations of a nonmetabolizable inducer, which in this case was not the classical IPTG but thiomethyl-galactoside (TMG). An increased error rate during transcription, caused either by RNA polymerase mutations which reduce transcriptional fidelity or by a lack of transcription fidelity factors, increased switching of the *lac* operon from the OFF to the ON state [158]. The interpretation of this observation was that errors in *lacI* mRNA synthesis caused a transient decrease in the Lac repressor level, which permitted a switching to the ON state [158, 159]. It should be noted that an uninduced *E. coli* cell contains approximately 10 molecules of Lac repressor, an amount which is small enough to make the system noisy and therefore metastable. Perturbation of this delicate equilibrium by transcriptional inaccuracy switches the system to the ON state. Even though the decrease in Lac repressor concentration is transient, synthesis of permease will create a positive feedback loop that will maintain the ON state in certain cells, as in the Novick and Weiner experiment [123]. Transient errors in transcription – and perhaps also in other cellular transactions – may therefore be a source of phenotypic heterogeneity whenever two conditions are met. One condition is that the cellular function involved exists in small amounts, which makes the system intrinsically noisy. The second condition is that a feedback loop can be formed, using a transient signal to generate a self-perpetuating epigenetic state.

4 Phenotypic Heterogeneity of Bacteria in Natural Environments

Experimental use of batch cultures to study bacterial physiology has concealed during decades cell-to-cell variation in bacterial populations. Although bacterial phenotypic variation can be easily observed under laboratory conditions [160, 161], its widespread occurrence and biological significance may be best appreciated in natural environments. An example is the phenomenon known as bacterial persistence, which may be one of the causes why tuberculosis and other bacterial infections are refractory to antibiotics [7–9]. Antibiotic treatment – either in the laboratory or in infected animals – will kill the majority of the bacterial population. However, the kinetics of bacterial killing is biphasic: after an initial period of rapid killing, the rate of killing slows down such that a bacterial subpopulation which is refractory to the antibiotic survives the treatment [8]. The antibiotic resistance shown by persisters is generally transient, which rules out mutation as a common cause and suggests the involvement of epigenetic mechanisms [162].

The results of a recent study have proposed that toxin–antitoxin modules may be involved in persistent formation in *E. coli*, causing population diversification into dormant and growing cells [163]. Fluctuations in the toxin–antitoxin levels may render the system noisy, and may trigger dormancy when the toxin–antitoxin imbalance exceeds a threshold. Furthermore, the degree of toxin–antitoxin imbalance may determine the duration of dormancy. Persisters may be cells which are kept dormant for long periods, and their dormancy state may confer antibiotic resistance [163]. Because persisters are rare, however, switching must occur at very low frequencies, which is in contrast to other epigenetic mechanisms that generate larger subpopulations. Toxin-induced dormancy is unlikely to be the only cause of persistence, because evidence exists that persisters of a given species often belong to several phenotypic classes [8].

In *S. enterica*, phenotypic heterogeneity is observed at several stages of animal colonization. For example, during intestinal infection flagella are necessary for swimming and also facilitate the invasion of intestinal cells; yet, intestinal populations of *Salmonella* include a mixture of flagellated and nonflagellated bacteria [164]. The formation of a nonflagellated subpopulation may be viewed as a stealthy strategy, because flagellin is highly immunogenic. Another phenotypic bifurcation during *Salmonella* infection is observed in systemic infection: upon entry into macrophages, the *Salmonella* population will split into two subpopulations, one of which replicates while the other enters a dormant-like state [165]. This quiescent state is reversible, which suggests that phenotypic bifurcation may involve an epigenetic switch. The colonization of the gallbladder by *Salmonella* provides another example of lineage formation. Although the bile-laden gallbladder is a harsh environment for bacteria, because bile salts are bactericidal [166], *Salmonella* can survive and multiply in this environment. A spectacular manifestation of this capacity is the so-called "carrier state" that is found in 10–20% of humans who survive typhoid fever [167]; such individuals are asymptomatic carriers of *Salmonella typhi*, which resides mainly in the gallbladder, and act as typhoid transmitters upon fecal shedding [167]. Because treatments other than surgical extirpation of the gallbladder are not currently available for the treatment of *Salmonella* carriage, the development of pharmacological treatments to expel *Salmonella* from the gallbladder might represent a major breakthrough in public health. This seems difficult, however, because colonization of the gallbladder by *Salmonella* is accompanied by lineage formation, whereby one subpopulation invades the gallbladder epithelium [168] while another remains in the gallbladder lumen. Further diversification occurs in the presence of gallstones, on which *Salmonella* is able to form biofilms [169, 170].

Biofilms are bacterial communities formed on a surface, and constitute the prevailing lifestyle in bacteria in most natural environments [5]. Biofilms can be formed by a single bacterial species, or by multiple species [5]. Biofilm formation is a rudimentary form of cell differentiation, and involves sequential changes in gene expression [171–173]. In a biofilm, the bacterial cells are embedded in a matrix of exopolysaccharides, which protects the bacterial community from antibacterial agents, including antibiotics [5]. In a wide range of bacterial species, biofilm formation is accompanied by phenotypic diversification [174], with certain such changes perhaps being caused by mutation [174]. Other changes may be the consequence of physiological adaptation to environmental conditions, which can vary greatly from one biofilm area to another [174]. In addition, cases of phenotypic heterogeneity potentially caused by epigenetic switching have been described in biofilms. In *B. subtilis*, production of the extracellular matrix is carried out by a subset of cells which do not occupy any special location but show a distinct gene expression pattern [175]. Marine bacteria of the genus *Pseudoalteromonas* form biofilms on chitin, and chitinase synthesis occurs only in certain cells. Typically, $ChiA^+$ and ChiA cells are intermingled in the biofilm, which suggests that phenotypic differences might be caused by a bistable switch [176, 177]. Furthermore, many biofilms contain non dividing cells with the properties described above for bacterial persisters [178, 179].

5
Perspectives

Single cell analysis facilitates the study of phenotypic heterogeneity in bacteria, and has unveiled the involvement of a variety of epigenetic mechanisms. More examples of epigenetic inheritance in bacteria can be expected to be discovered in the further. In certain developmental programs, the regulatory networs involved are known in detail but the epigenetic switches that establish cell fate are only beginning to be understood. Examples are fruiting body development in *Myxococcus* [2] and differentiation of *Rhizobium* into nitrogen-fixing bacteroids [4]. In *Caulobacter*, DNA methylation is known to be involved in the epigenetic control of the cell cycle [1, 18], but the epigenetic switch(es) that permit the formation of distinct cell types (swarmers and stalked cells) remain to be identified. Because of their complexity and medical relevance, biofilms are also fascinating systems for the study of phenotypic heterogeneity in bacteria [5]. Persister formation and subpopulation differentiation during animal colonization are also medically relevant phenomena whose epigenetic basis is beginning to be understood [162, 163] or remains to be deciphered [164, 165]. Lastly, it may be worth to mentioned that the formation of bacterial lineages does not occur in natural environments only, and can be observed under laboratory conditions [160, 161]. Epigenetic regulation of gene expression may thus be intrinsical to the bacterial lifestyle.

References

1 Laub, M.T., Shapiro, L., McAdams, H.H. (2007) Systems biology of *Caulobacter*. *Annu. Rev. Genet.*, **41**, 429–441.

2 Kroos, L. (2007) The *Bacillus* and *Myxococcus* developmental networks and their transcriptional regulators. *Annu. Rev. Genet.*, **41**, 13–39.

3 Kaiser, D. (2008) *Myxococcus* – from single-cell polarity to complex multicellular patterns. *Annu. Rev. Genet.*, **42**, 109–130.

4 Gibson, K.E., Kobayashi, H., Walker, G.C. (2008) Molecular determinants of a symbiotic chronic infection. *Annu. Rev. Genet.*, **42**, 413–441.

5 Watnick, P., Kolter, R. (2000) Biofilm, city of microbes. *J. Bacteriol.*, **182**, 2675–2679.

6 Casadesus, J., Low, D. (2006) Epigenetic gene regulation in the bacterial world. *Microbiol. Mol. Biol. Rev.*, **70**, 830–856.

7 Dubnau, D., Losick, R. (2006) Bistability in bacteria. *Mol. Microbiol.*, **61**, 564–572.

8 Dhar, N., McKinney, J.D. (2007) Microbial phenotypic heterogeneity and antibiotic tolerance. *Curr. Opin. Microbiol.*, **10**, 30–38.

9 Davidson, C.J., Surette, M.G. (2008) Individuality in bacteria. *Annu. Rev. Genet.*, **42**, 253–268.

10 Smits, W.K., Kuipers, O.P., Veening, J.W. (2006) Phenotypic variation in bacteria: the role of feedback regulation. *Nat. Rev. Microbiol.*, **4**, 259–271.

11 Cheng, X. (1995) Structure and function of DNA methyltransferases. *Annu. Rev. Biophys. Biomol. Struct.*, **24**, 293–318.

12 Cheng, X. (1995) DNA modification by methyltransferases. *Curr. Opin. Struct. Biol.*, **5**, 4–10.

13 Bickle, T.A., Kruger, D.H. (1993) Biology of DNA restriction. *Microbiol. Rev.*, **57**, 434–450.

14 Wion, D., Casadesus, J. (2006) N^6-methyladenine: an epigenetic signal for DNA–protein interactions. *Nat. Rev. Microbiol.*, **4**, 183–192.

15 Marinus, M.G., Casadesus, J. (2009) Roles of DNA adenine methylation in host–pathogen interactions: mismatch repair, transcriptional regulation, and more. *FEMS Microbiol. Rev.*, **33**, 488–503.

16 Low, D.A., Casadesus, J. (2008) Clocks and switches: bacterial gene regulation by DNA adenine methylation. *Curr. Opin. Microbiol.*, **11**, 106–112.

17 Løbner-Olesen, A., Skovgaard, O., Marinus, M.G. (2005) Dam methylation: coordinating cellular processes. *Curr. Opin. Microbiol.*, **8**, 154–160.

18 Collier, J. (2009) Epigenetic regulation of the bacterial cell cycle. *Curr. Opin. Microbiol.*, **12**, 722–729.

19 Marinus, M.G., Morris, N.R. (1973) Isolation of deoxyribonucleic acid methylase mutants of *Escherichia coli* K-12. *J. Bacteriol.*, **114**, 1143–1150.

20 Torreblanca, J., Casadesus, J. (1996) DNA adenine methylase mutants of *Salmonella typhimurium* and a novel dam-regulated locus. *Genetics*, **144**, 15–26.

21 Stephens, C., Reisenauer, A., Wright, R., Shapiro, L. (1996) A cell cycle-regulated bacterial DNA methyltransferase is essential for viability. *Proc. Natl Acad. Sci. USA*, **93**, 1210–1214.

22 Engel, J.D., von Hippel, P.H. (1978) Effects of methylation on the stability of nucleic acid conformations. Studies at the polymer level. *J. Biol. Chem.*, **253**, 927–934.

23 Diekmann, S. (1987) DNA methylation can enhance or induce DNA curvature. *EMBO J.*, **6**, 4213–4217.

24 Polaczek, P., Kwan, K., Campbell, J.L. (1998) GATC motifs may alter the conformation of DNA depending on sequence context and N^6-adenine methylation status: possible implications for DNA-protein recognition. *Mol. Gen. Genet.*, **258**, 488–493.

25 Modrich, P. (1989) Methyl-directed DNA mismatch correction. *J. Biol. Chem.*, **264**, 6597–6600.

26 Camacho, E.M., Casadesús, J. (2005) Regulation of *traJ* transcription in the *Salmonella* virulence plasmid by strand-specific DNA adenine hemimethylation. *Mol. Microbiol.*, **57**, 1700–1718.

27 Camacho, E.M., Casadesus, J. (2002) Conjugal transfer of the virulence plasmid of *Salmonella enterica* is regulated by the leucine-responsive regulatory protein and DNA adenine methylation. *Mol. Microbiol.*, **44**, 1589–1598.

28 Haagmans, W., van der Woude, M. (2000) Phase variation of Ag43 in *Escherichia coli*: dam-dependent methylation abrogates OxyR binding and OxyR-mediated repression of transcription. *Mol. Microbiol.*, **35**, 877–887.

29 Waldron, D.E., Owen, P., Dorman, C.J. (2002) Competitive interaction of the OxyR DNA-binding protein and the Dam methylase at the antigen 43 gene regulatory region in *Escherichia coli*. *Mol. Microbiol.*, **44**, 509–520.

30 Roberts, D., Hoopes, B.C., McClure, W.R., Kleckner, N. (1985) IS10 transposition is regulated by DNA adenine methylation. *Cell*, **43**, 117–130.

31 Brunet, Y., Bernard, C.S., Gavioli, M., Lloubès, R., Cascales, E. (2011) A phase variation regulatory mechanism involving overlapping Fur and DNA methylation controls the expression of the enteroaggregative *Escherichia coli* sci1 type 6 secretion gene cluster. *PLoS Genet.*, **7**, e1002205.

32 Jakomin, M., Chessa, D., Baumler, A.J., Casadesus, J. (2008) Regulation of the *Salmonella enterica* std fimbrial operon by DNA adenine methylation, SeqA and HdfR. *J. Bacteriol.*, **190**, 7406–7413.

33 Messer, W., Bellekes, U., Lother, H. (1985) Effect of dam methylation on the activity of the *E. coli* replication origin, oriC. *EMBO J.*, **4**, 1327–1332.

34 Lu, M., Campbell, J.L., Boye, E., Kleckner, N. (1994) SeqA: a negative modulator of replication initiation in *E. coli*. *Cell*, **77**, 413–426.

35 Messer, W., Noyer-Weidner, M. (1988) Timing and targeting: the biological functions of Dam methylation in *E. coli*. *Cell*, **54**, 735–737.

36 Oshima, T., Wada, C., Kawagoe, Y., Ara, T., Maeda, M., Masuda, Y., Hiraga, S., Mori, H. (2002) Genome-wide analysis of deoxyadenosine methyltransferase-mediated control of gene expression in *Escherichia coli*. *Mol. Microbiol.*, **45**, 673–695.

37 Balbontin, R., Rowley, G., Pucciarelli, M.G., Lopez-Garrido, J., Wormstone, Y., Lucchini, S., Garcia-Del Portillo, F., Hinton, J.C., Casadesus, J. (2006) DNA adenine methylation regulates virulence gene expression in *Salmonella enterica* serovar *typhimurium*. *J. Bacteriol.*, **188**, 8160–8168.

38 Mahillon, J., Chandler, M. (1998) Insertion sequences. *Microbiol. Mol. Biol. Rev.*, **62**, 725–774.

39 Doolittle, W.F., Kirkwood, T.B., Dempster, M.A. (1984) Selfish DNAs with self-restraint. *Nature*, **307**, 501–502.

40 Rotger, R., Casadesus, J. (1999) The virulence plasmids of *Salmonella*. *Int. Microbiol.*, **2**, 177–184.

41 Ahmer, B.M.M., Tran, M., Heffron, F. (1999) The virulence plasmid of *Salmonella typhimurium* is self-transmissible. *J. Bacteriol.*, **181**, 1364–1368.

42 Camacho, E.M., Serna, A., Madrid, C., Marques, S., Fernandez, R., de la Cruz, F., Juarez, A., Casadesus, J. (2005) Regulation of finP transcription by DNA adenine methylation in the virulence plasmid of *Salmonella enterica*. *J. Bacteriol.*, **187**, 5691–5699.

43 Serna, A., Espinosa, E., Camacho, E.M., Casadesus, J. (2010) Regulation of bacterial conjugation in microaerobiosis by host-encoded functions ArcAB and sdhABCD. *Genetics*, **184**, 947–958.

44 Garcia-Quintanilla, M., Ramos-Morales, F., Casadesus, J. (2008) Conjugal transfer of the *Salmonella enterica* virulence plasmid in the mouse intestine. *J. Bacteriol.*, **190**, 1922–1927.

45 Torreblanca, J., Marques, S., Casadesus, J. (1999) Synthesis of FinP RNA by plasmids F and pSLT is regulated by DNA adenine methylation. *Genetics*, **152**, 31–45.

46 Dempsey, W.B. (1993) Key Regulatory Aspects of Transfer of F-Related Factors, in: Clewell, D.B. (Ed.) *Bacterial Conjugation*, Plenum Press, New York, pp. 53–73.

47 Braun, R.E., Wright, A. (1986) DNA methylation differentially enhances the expression of one of the two *E. coli* dnaA promoters *in vivo* and *in vitro*. *Mol. Gen. Genet.*, **202**, 246–250.

48 Kucherer, C., Lother, H., Kolling, R., Schauzu, M.A., Messer, W. (1986) Regulation of transcription of the chromosomal dnaA gene of *Escherichia coli*. *Mol. Gen. Genet.*, **205**, 115–121.

49 Zakrzewska-Czerwinska, J., Jakimowicz, D., Zawilak-Pawlik, A., Messer, W. (2007) Regulation of the initiation of chromosomal replication in bacteria. *FEMS Microbiol. Rev.*, **31**, 378–387.

50 Boye, E., Stokke, T., Kleckner, N., Skarstad, K. (1996) Coordinating DNA replication initiation with cell growth: differential roles for DnaA and SeqA proteins. *Proc. Natl Acad. Sci. USA*, **93**, 12206–12211.

51 Waldminghaus, T., Skarstad, K. (2009) The *Escherichia coli* SeqA protein. *Plasmid*, **61**, 141–150.

52 Riber, L., Fujimitsu, K., Katayama, T., Lobner-Olesen, A. (2009) Loss of Hda activity stimulates replication initiation from I-box, but not R4 mutant origins in *Escherichia coli*. *Mol. Microbiol.*, **71**, 107–122.

53 Lawler, M.L., Brun, Y.V. (2007) Advantages and mechanisms of polarity and cell shape determination in *Caulobacter crescentus*. *Curr. Opin. Microbiol.*, **10**, 630–637.

54 Marczynski, G.T., Shapiro, L. (2002) Control of chromosome replication in *Caulobacter crescentus*. *Annu. Rev. Microbiol.*, **56**, 625–656.

55 Reisenauer, A., Kahng, L.S., McCollum, S., Shapiro, L. (1999) Bacterial DNA methylation: a cell cycle regulator? *J. Bacteriol.*, **181**, 5135–5139.

56 Kahng, L.S., Shapiro, L. (2001) The CcrM DNA methyltransferase of *Agrobacterium tumefaciens* is essential, and its activity is cell-cycle regulated. *J. Bacteriol.*, **183**, 3065–3075.

57 Collier, J., McAdams, H.H., Shapiro, L. (2007) A DNA methylation ratchet governs progression through a bacterial cell cycle. *Proc. Natl Acad. Sci. USA*, **104**, 17111–17116.

58 Collier, J., Murray, S.R., Shapiro, L. (2006) DnaA couples DNA replication and the expression of two cell cycle master regulators. *EMBO J.*, **25**, 346–356.

59 Reisenauer, A., Shapiro, L. (2002) DNA methylation affects the cell cycle transcription of the CtrA global regulator in *Caulobacter*. *EMBO J.*, **21**, 4969–4977.

60 Blyn, L.B., Braaten, B.A., Low, D.A. (1990) Regulation of pap pilin phase variation by a mechanism involving differential dam methylation states. *EMBO J.*, **9**, 4045–4054.

61 Wang, M.X., Church, G.M. (1992) A whole genome approach to in vivo DNA-protein interactions in *E. coli*. *Nature*, **360**, 606–610.

62 Ringquist, S., Smith, C.L. (1992) The *Escherichia coli* chromosome contains specific, unmethylated dam and dcm sites. *Proc. Natl Acad. Sci. USA*, **89**, 4539–4543.

63 Hale, W.B., van der Woude, M.W., Low, D.A. (1994) Analysis of nonmethylated GATC sites in the *Escherichia coli* chromosome and identification of sites that are differentially methylated in response to environmental stimuli. *J. Bacteriol.*, **176**, 3438–3441.

64 Charlier, D., Gigot, D., Huysveld, N., Roovers, M., Pierard, A., Glansdorff, N. (1995) Pyrimidine regulation of the *Escherichia coli* and *Salmonella typhimurium* carAB operons: CarP and integration host factor (IHF) modulate the methylation status of a GATC site present in the control region. *J. Mol. Biol.*, **250**, 383–391.

65 van der Woude, M., Hale, W.B., Low, D.A. (1998) Formation of DNA methylation patterns: nonmethylated GATC sequences in gut and pap operons. *J. Bacteriol.*, **180**, 5913–5920.

66 Tavazoie, S., Church, G.M. (1998) Quantitative whole-genome analysis of DNA–protein interactions by in vivo

methylase protection in *E. coli*. *Nat. Biotechnol.*, **16**, 566–571.

67 Peterson, S.N., Reich, N.O. (2006) GATC flanking sequences regulate Dam activity: evidence for how Dam specificity may influence pap expression. *J. Mol. Biol.*, **355**, 459–472.

68 Peterson, S.N., Reich, N.O. (2008) Competitive Lrp and Dam assembly at the pap regulatory region: implications for mechanisms of epigenetic regulation. *J. Mol. Biol.*, **383**, 92–105.

69 van der Woude, M.W., Baumler, A.J. (2004) Phase and antigenic variation in bacteria. *Clin. Microbiol. Rev.*, **17**, 581–611 (table of contents).

70 van der Woude, M.W. (2006) Re-examining the role and random nature of phase variation. *FEMS Microbiol. Lett.*, **254**, 190–197.

71 van der Woude, M.W. (2011) Phase variation: how to create and coordinate population diversity. *Curr. Opin. Microbiol.*, **14**, 205–211.

72 van der Woude, M., Braaten, B., Low, D. (1996) Epigenetic phase variation of the pap operon in *Escherichia coli*. *Trends Microbiol.*, **4**, 5–9.

73 Hernday, A., Krabbe, M., Braaten, B., Low, D. (2002) Self-perpetuating epigenetic pili switches in bacteria. *Proc. Natl Acad. Sci. USA*, **29**, 29.

74 Weyand, N.J., Low, D.A. (2000) Regulation of Pap phase variation. Lrp is sufficient for the establishment of the phase off pap DNA methylation pattern and repression of pap transcription in vitro. *J. Biol. Chem.*, **275**, 3192–3200.

75 Braaten, B.A., Blyn, L.B., Skinner, B.S., Low, D.A. (1991) Evidence for a methylation-blocking factor (mbf) locus involved in pap pilus expression and phase variation in *Escherichia coli*. *J. Bacteriol.*, **173**, 1789–1800.

76 Braaten, B.A., Nou, X., Kaltenbach, L.S., Low, D.A. (1994) Methylation patterns in pap regulatory DNA control pyelonephritis-associated pili phase variation in *E. coli*. *Cell*, **76**, 577–588.

77 van der Woude, M.W., Kaltenbach, L.S., Low, D.A. (1995) Leucine-responsive regulatory protein plays dual roles as both an activator and a repressor of the *Escherichia coli* pap fimbrial operon. *Mol. Microbiol.*, **17**, 303–312.

78 Hernday, A.D., Braaten, B.A., Low, D.A. (2003) The mechanism by which DNA adenine methylase and PapI activate the pap epigenetic switch. *Mol. Cell*, **12**, 947–957.

79 Kaltenbach, L.S., Braaten, B.A., Low, D.A. (1995) Specific binding of PapI to Lrp-pap DNA complexes. *J. Bacteriol.*, **177**, 6449–6455.

80 Nou, X., Braaten, B., Kaltenbach, L., Low, D.A. (1995) Differential binding of Lrp to two sets of pap DNA binding sites mediated by Pap I regulates Pap phase variation in *Escherichia coli*. *EMBO J.*, **14**, 5785–5797.

81 White-Ziegler, C.A., Black, A.M., Eliades, S.H., Young, S., Porter, K. (2002) The N-acetyltransferase RimJ responds to environmental stimuli to repress pap fimbrial transcription in *Escherichia coli*. *J. Bacteriol.*, **184**, 4334–4342.

82 White-Ziegler, C.A., Villapakkam, A., Ronaszeki, K., Young, S. (2000) H-NS controls pap and daa fimbrial transcription in *Escherichia coli* in response to multiple environmental cues. *J. Bacteriol.*, **182**, 6391–6400.

83 Hernday, A.D., Braaten, B.A., Broitman-Maduro, G., Engelberts, P., Low, D.A. (2004) Regulation of the pap epigenetic switch by CpxAR: phosphorylated CpxR inhibits transition to the phase ON state by competition with Lrp. *Mol. Cell*, **16**, 537–547.

84 Weyand, N.J., Braaten, B.A., van der Woude, M., Tucker, J., Low, D.A. (2001) The essential role of the promoter-proximal subunit of CAP in pap phase variation: Lrp- and helical phase-dependent activation of papBA transcription by CAP from -215. *Mol. Microbiol.*, **39**, 1504–1522.

85 Henderson, I.R., Owen, P. (1999) The major phase-variable outer membrane protein of *Escherichia coli* structurally resembles the immunoglobulin A1 protease class of exported protein and is regulated by a novel mechanism involving Dam and oxyR. *J. Bacteriol.*, **181**, 2132–2141.

86 Danese, P.N., Pratt, L.A., Dove, S.L., Kolter, R. (2000) The outer membrane protein, antigen 43, mediates cell-to-cell interactions within *Escherichia coli* biofilms. *Mol. Microbiol.*, **37**, 424–432.

87 Wallecha, A., Correnti, J., Munster, V., van der Woude, M. (2003) Phase variation of

Ag43 is independent of the oxidation state of OxyR. *J. Bacteriol.*, **185**, 2203–2209.
88 Wallecha, A., Munster, V., Correnti, J., Chan, T., van der Woude, M. (2002) Dam- and OxyR-dependent phase variation of agn43: essential elements and evidence for a new role of DNA methylation. *J. Bacteriol.*, **184**, 3338–3347.
89 Correnti, J., Munster, V., Chan, T., Woude, M. (2002) Dam-dependent phase variation of Ag43 in *Escherichia coli* is altered in a seqA mutant. *Mol. Microbiol.*, **44**, 521–532.
90 Broadbent, S.E., Davies, M.R., van der Woude, M.W. (2010) Phase variation controls expression of *Salmonella* lipopolysaccharide modification genes by a DNA methylation-dependent mechanism. *Mol. Microbiol.*, **77**, 337–353.
91 Figueroa-Bossi, N., Coissac, E., Netter, P., Bossi, L. (1997) Unsuspected prophage-like elements in *Salmonella typhimurium*. *Mol. Microbiol.*, **25**, 161–173.
92 Bossi, L., Fuentes, J.A., Mora, G., Figueroa-Bossi, N. (2003) Prophage contribution to bacterial population dynamics. *J. Bacteriol.*, **185**, 6467–6471.
93 Martin, C. (1996) The clp (CS31A) operon is negatively controlled by Lrp, ClpB, and L- alanine at the transcriptional level. *Mol. Microbiol.*, **21**, 281–292.
94 Crost, C., Garrivier, A., Harel, J., Martin, C. (2003) Leucine-responsive regulatory protein-mediated repression of clp (encoding CS31A) expression by L-leucine and L-alanine in *Escherichia coli*. *J. Bacteriol.*, **185**, 1886–1894.
95 Nicholson, B., Low, D. (2000) DNA methylation-dependent regulation of pef expression in *Salmonella typhimurium*. *Mol. Microbiol.*, **35**, 728–742.
96 Humphries, A.D., Raffatellu, M., Winter, S., Weening, E.H., Kingsley, R.A., Glansdorff, N. (2003) The use of flow cytometry to detect expression of subunits encoded by 11 *Salmonella enterica* serotype *typhimurium* fimbrial operons. *Mol. Microbiol.*, **48**, 1357–1376.
97 Heithoff, D.M., Sinsheimer, R.L., Low, D.A., Mahan, M.J. (1999) An essential role for DNA adenine methylation in bacterial virulence [see comments]. *Science*, **284**, 967–970.
98 Garcia-Del Portillo, F., Pucciarelli, M.G., Casadesus, J. (1999) DNA adenine methylase mutants of *Salmonella typhimurium* show defects in protein secretion, cell invasion, and M cell cytotoxicity. *Proc. Natl Acad. Sci. USA*, **96**, 11578–11583.
99 Lopez-Garrido, J., Casadesus, J. (2010) Regulation of *Salmonella enterica* pathogenicity island 1 by DNA adenine methylation. *Genetics*, **184**, 637–649.
100 Zaleski, P., Wojciechowski, M., Piekarowicz, A. (2005) The role of Dam methylation in phase variation of *Haemophilus influenzae* genes involved in defence against phage infection. *Microbiology*, **151**, 3361–3369.
101 Fox, K.L., Srikhanta, Y.N., Jennings, M.P. (2007) Phase variable type III restriction-modification systems of host-adapted bacterial pathogens. *Mol. Microbiol.*, **65**, 1375–1379.
102 Moxon, R., Bayliss, C., Hood, D. (2006) Bacterial contingency loci: the role of simple sequence DNA repeats in bacterial adaptation. *Annu. Rev. Genet.*, **40**, 307–333.
103 Srikhanta, Y.N., Maguire, T.L., Stacey, K.J., Grimmond, S.M., Jennings, M.P. (2005) The phasevarion: a genetic system controlling coordinated, random switching of expression of multiple genes. *Proc. Natl Acad. Sci. USA*, **102**, 5547–5551.
104 Fox, K.L., Dowideit, S.J., Erwin, A.L., Srikhanta, Y.N., Smith, A.L., Jennings, M.P. (2007) *Haemophilus influenzae* phasevarions have evolved from type III DNA restriction systems into epigenetic regulators of gene expression. *Nucleic Acids Res.*, **35**, 5242–5252.
105 Srikhanta, Y.N., Dowideit, S.J., Edwards, J.L., Falsetta, M.L., Wu, H.J., Harrison, O.B., Fox, K.L., Seib, K.L., Maguire, T.L., Wang, A.H., Maiden, M.C., Grimmond, S.M., Apicella, M.A., Jennings, M.P. (2009) Phasevarions mediate random switching of gene expression in pathogenic *Neisseria*. *PLoS Pathog.*, **5**, e1000400.
106 Srikhanta, Y.N., Fox, K.L., Jennings, M.P. (2010) The phasevarion: phase variation of type III DNA methyltransferases controls coordinated switching in multiple genes. *Nat. Rev. Microbiol.*, **8**, 196–206.
107 Chessa, D., Winter, M.G., Jakomin, M., Baumler, A.J. (2009) *Salmonella enterica* serotype *typhimurium* Std fimbriae bind

terminal alpha(1,2)fucose residues in the cecal mucosa. *Mol. Microbiol.*, **71**, 864–875.
108 Chessa, D., Winter, M.G., Nuccio, S.P., Tukel, C., Baumler, A.J. (2008) RosE represses Std fimbrial expression in *Salmonella enterica* serotype *typhimurium*. *Mol. Microbiol.*, **68**, 573–587.
109 Huisman, T.T., Bakker, D., Klaasen, P., de Graaf, F.K. (1994) Leucine-responsive regulatory protein, IS1 insertions, and the negative regulator FaeA control the expression of the fae (K88) operon in *Escherichia coli*. *Mol. Microbiol.*, **11**, 525–536.
110 Huisman, T.T., de Graaf, F.K. (1995) Negative control of fae (K88) expression by the 'global' regulator Lrp is modulated by the 'local' regulator FaeA and affected by DNA methylation. *Mol. Microbiol.*, **16**, 943–953.
111 Peterson, K.R., Wertman, K.F., Mount, D.W., Marinus, M.G. (1985) Viability of *Escherichia coli* K-12 DNA adenine methylase (dam) mutants requires increased expression of specific genes in the SOS regulon. *Mol. Gen. Genet.*, **201**, 14–19.
112 Glickman, B., van den Elsen, P., Radman, M. (1978) Induced mutagenesis in dam-mutants of *Escherichia coli*: a role for 6-methyladenine residues in mutation avoidance. *Mol. Gen. Genet.*, **163**, 307–312.
113 Henaut, A., Rouxel, T., Gleizes, A., Moszer, I., Danchin, A. (1996) Uneven distribution of GATC motifs in the *Escherichia coli* chromosome, its plasmids and its phages. *J. Mol. Biol.*, **257**, 574–585.
114 Riva, A., Delorme, M.O., Chevalier, T., Guilhot, N., Henaut, C., Henaut, A. (2004) The difficult interpretation of transcriptome data: the case of the GATC regulatory network. *Comput. Biol. Chem.*, **28**, 109–118.
115 Seshasayee, A.S. (2007) An assessment of the role of DNA adenine methyltransferase on gene expression regulation in *E. coli*. *PLoS ONE*, **2**, e273.
116 Sternberg, N., Sauer, B., Hoess, R., Abremski, K. (1986) Bacteriophage P1 cre gene and its regulatory region. Evidence for multiple promoters and for regulation by DNA methylation. *J. Mol. Biol.*, **187**, 197–212.
117 Bell, D.C., Cupples, C.G. (2001) Very-short-patch repair in *Escherichia coli* requires the dam adenine methylase. *J. Bacteriol.*, **183**, 3631–3635.
118 Campellone, K.G., Roe, A.J., Lobner-Olesen, A., Murphy, K.C., Magoun, L., Brady, M.J., Donohue-Rolfe, A., Tzipori, S., Gally, D.L., Leong, J.M., Marinus, M.G. (2007) Increased adherence and actin pedestal formation by dam-deficient enterohaemorrhagic *Escherichia coli* O157:H7. *Mol. Microbiol.*, **63**, 1468–1481.
119 Falker, S., Schilling, J., Schmidt, M.A., Heusipp, G. (2007) Overproduction of DNA adenine methyltransferase alters motility, invasion, and the lipopolysaccharide O-antigen composition of *Yersinia enterocolitica*. *Infect. Immun.*, **75**, 4990–4997.
120 Veening, J.W., Smits, W.K., Kuipers, O.P. (2008) Bistability, epigenetics, and bet-hedging in bacteria. *Annu. Rev. Microbiol.*, **62**, 193–210.
121 Veening, J.W., Stewart, E.J., Berngruber, T.W., Taddei, F., Kuipers, O.P., Hamoen, L.W. (2008) Bet-hedging and epigenetic inheritance in bacterial cell development. *Proc. Natl Acad. Sci. USA*, **105**, 4393–4398.
122 Ferrell, J.E. Jr (2002) Self-perpetuating states in signal transduction: positive feedback, double-negative feedback and bistability. *Curr. Opin. Cell Biol.*, **14**, 140–148.
123 Novick, A., Weiner, M. (1957) Enzyme induction as an all-or-none phenomenon. *Proc. Natl Acad. Sci. USA*, **43**, 553–566.
124 Casadesus, J., D'Ari, R. (2002) Memory in bacteria and phage. *BioEssays*, **24**, 512–518.
125 Laurent, M., Charvin, G., Guespin-Michel, J. (2005) Bistability and hysteresis in epigenetic regulation of the lactose operon. Since Delbruck, a long series of ignored models. *Cell. Mol. Biol. (Noisy-le-grand)*, **51**, 583–594.
126 Johnson, A.D., Poteete, A.R., Lauer, G., Sauer, R.T., Ackers, G.K., Ptashne, M. (1981) Lambda repressor and cro-components of an efficient molecular switch. *Nature*, **294**, 217–223.
127 Chen, I., Christie, P.J., Dubnau, D. (2005) The ins and outs of DNA transfer in bacteria. *Science*, **310**, 1456–1460.
128 van Sinderen, D., Luttinger, A., Kong, L., Dubnau, D., Venema, G., Hamoen, L. (1995) comK encodes the competence transcription factor, the key regulatory protein for competence development in *Bacillus subtilis*. *Mol. Microbiol.*, **15**, 455–462.

129 van Sinderen, D., Venema, G. (1994) comK acts as an autoregulatory control switch in the signal transduction route to competence in *Bacillus subtilis*. *J. Bacteriol.*, **176**, 5762–5770.

130 Magnuson, R., Solomon, J., Grossman, A.D. (1994) Biochemical and genetic characterization of a competence pheromone from *B. subtilis*. *Cell*, **77**, 207–216.

131 Turgay, K., Hahn, J., Burghoorn, J., Dubnau, D. (1998) Competence in *Bacillus subtilis* is controlled by regulated proteolysis of a transcription factor. *EMBO J.*, **17**, 6730–6738.

132 Hoa, T.T., Tortosa, P., Albano, M., Dubnau, D. (2002) Rok (YkuW) regulates genetic competence in *Bacillus subtilis* by directly repressing comK. *Mol. Microbiol.*, **43**, 15–26.

133 Hamoen, L.W., Kausche, D., Marahiel, M.A., van Sinderen, D., Venema, G., Serror, P. (2003) The *Bacillus subtilis* transition state regulator AbrB binds to the −35 promoter region of comK. *FEMS Microbiol. Lett.*, **218**, 299–304.

134 Smits, W.K., Eschevins, C.C., Susanna, K.A., Bron, S., Kuipers, O.P., Hamoen, L.W. (2005) Stripping *Bacillus*: ComK auto-stimulation is responsible for the bistable response in competence development. *Mol. Microbiol.*, **56**, 604–614.

135 Chung, J.D., Stephanopoulos, G., Ireton, K., Grossman, A.D. (1994) Gene expression in single cells of *Bacillus subtilis*: evidence that a threshold mechanism controls the initiation of sporulation. *J. Bacteriol.*, **176**, 1977–1984.

136 Burbulys, D., Trach, K.A., Hoch, J.A. (1991) Initiation of sporulation in *B. subtilis* is controlled by a multicomponent phosphorelay. *Cell*, **64**, 545–552.

137 Veening, J.W., Hamoen, L.W., Kuipers, O.P. (2005) Phosphatases modulate the bistable sporulation gene expression pattern in *Bacillus subtilis*. *Mol. Microbiol.*, **56**, 1481–1494.

138 Gonzalez-Pastor, J.E., Hobbs, E.C., Losick, R. (2003) Cannibalism by sporulating bacteria. *Science*, **301**, 510–513.

139 Ellermeier, C.D., Hobbs, E.C., Gonzalez-Pastor, J.E., Losick, R. (2006) A three-protein signaling pathway governing immunity to a bacterial cannibalism toxin. *Cell*, **124**, 549–559.

140 Gonzalez-Pastor, J.E. (2011) Cannibalism: a social behavior in sporulating *Bacillus subtilis*. *FEMS Microbiol. Rev.*, **35**, 415–424.

141 Turner, K.H., Vallet-Gely, I., Dove, S.L. (2009) Epigenetic control of virulence gene expression in *Pseudomonas aeruginosa* by a LysR-type transcription regulator. *PLoS Genet.*, **5**, e1000779.

142 Guespin-Michel, J.F., Bernot, G., Comet, J.P., Merieau, A., Richard, A., Hulen, C., Polack, B. (2004) Epigenesis and dynamic similarity in two regulatory networks in *Pseudomonas aeruginosa*. *Acta Biotheor.*, **52**, 379–390.

143 Filopon, D., Merieau, A., Bernot, G., Comet, J.P., Leberre, R., Guery, B., Polack, B., Guespin-Michel, J. (2006) Epigenetic acquisition of inducibility of type III cytotoxicity in *P. aeruginosa*. *BMC Bioinformatics*, **7**, 272.

144 Oliver, A., Canton, R., Campo, P., Baquero, F., Blazquez, J. (2000) High frequency of hypermutable *Pseudomonas aeruginosa* in cystic fibrosis lung infection. *Science*, **288**, 1251–1254.

145 Saint-Ruf, C., Matic, I. (2006) Environmental tuning of mutation rates. *Environ. Microbiol.*, **8**, 193–199.

146 Pettersson, M.E., Andersson, D.I., Roth, J.R., Berg, O.G. (2005) The amplification model for adaptive mutation: simulations and analysis. *Genetics*, **169**, 1105–1115.

147 Roth, J.R., Kugelberg, E., Reams, A.B., Kofoid, E., Andersson, D.I. (2006) Origin of mutations under selection: the adaptive mutation controversy. *Annu. Rev. Microbiol.*, **60**, 477–501.

148 Raser, J.M., O'Shea, E.K. (2005) Noise in gene expression: origins, consequences, and control. *Science*, **309**, 2010–2013.

149 Ninio, J. (1997) The evolutionary design of error-rates, and the fast fixation enigma. *Orig. Life Evol. Biosph.*, **27**, 609–621.

150 D'Ari, R., Casadesus, J. (1998) Underground metabolism. *BioEssays*, **20**, 181–186.

151 Drake, J.W. (1991) A constant rate of spontaneous mutation in DNA-based microbes. *Proc. Natl Acad. Sci. USA*, **88**, 7160–7164.

152 Drake, J.W., Allen, E.F. (1968) Antimutagenic DNA polymerases of bacteriophage T4. *Cold Spring Harbor Symp. Quant. Biol.*, **33**, 339–344.

153 Bernardi, F., Ninio, J. (1978) The accuracy of DNA replication. *Biochimie*, **60**, 1083–1095.
154 Libby, R.T., Gallant, J.A. (1991) The role of RNA polymerase in transcriptional fidelity. *Mol. Microbiol.*, **5**, 999–1004.
155 Ruusala, T., Andersson, D., Ehrenberg, M., Kurland, C.G. (1984) Hyper-accurate ribosomes inhibit growth. *EMBO J.*, **3**, 2575–2580.
156 Savageau, M.A., Freter, R.R. (1979) On the evolution of accuracy and cost of proofreading tRNA aminoacylation. *Proc. Natl Acad. Sci. USA*, **76**, 4507–4510.
157 Freter, R.R., Savageau, M.A. (1980) Proofreading systems of multiple stages for improved accuracy of biological discrimination. *J. Theor. Biol.*, **85**, 99–123.
158 Gordon, A.J., Halliday, J.A., Blankschien, M.D., Burns, P.A., Yatagai, F., Herman, C. (2009) Transcriptional infidelity promotes heritable phenotypic change in a bistable gene network. *PLoS Biol.*, **7**, e44.
159 Satory, D., Gordon, A.J., Halliday, J.A., Herman, C. (2011) Epigenetic switches: can infidelity govern fate in microbes? *Curr. Opin. Microbiol.*, **14**, 212–217.
160 Shapiro, J.A., Higgins, N.P. (1989) Differential activity of a transposable element in *Escherichia coli* colonies. *J. Bacteriol.*, **171**, 5975–5986.
161 Aguilar, C., Vlamakis, H., Losick, R., Kolter, R. (2007) Thinking about *Bacillus subtilis* as a multicellular organism. *Curr. Opin. Microbiol.*, **10**, 638–643.
162 Balaban, N.Q., Merrin, J., Chait, R., Kowalik, L., Leibler, S. (2004) Bacterial persistence as a phenotypic switch. *Science*, **305**, 1622–1625.
163 Rotem, E., Loinger, A., Ronin, I., Levin-Reisman, I., Gabay, C., Shoresh, N., Biham, O., Balaban, N.Q. (2010) Regulation of phenotypic variability by a threshold-based mechanism underlies bacterial persistence. *Proc. Natl Acad. Sci. USA*, **107**, 12541–12546.
164 Cummings, L.A., Wilkerson, W.D., Bergsbaken, T., Cookson, B.T. (2006) In vivo, fliC expression by *Salmonella enterica* serovar *typhimurium* is heterogeneous, regulated by ClpX, and anatomically restricted. *Mol. Microbiol.*, **61**, 795–809.
165 Helaine, S., Thompson, J.A., Watson, K.G., Liu, M., Boyle, C., Holden, D.W. (2010) Dynamics of intracellular bacterial replication at the single cell level. *Proc. Natl Acad. Sci. USA*, **107**, 3746–3751.
166 Merritt, M.E., Donaldson, J.R. (2009) Effect of bile salts on the DNA and membrane integrity of enteric bacteria. *J. Med. Microbiol.*, **58**, 1533–1541.
167 Gonzalez-Escobedo, G., Marshall, J.M., Gunn, J.S. (2011) Chronic and acute infection of the gall bladder by *Salmonella typhi*: understanding the carrier state. *Nat. Rev. Microbiol.*, **9**, 9–14.
168 Menendez, A., Arena, E.T., Guttman, J.A., Thorson, L., Vallance, B.A., Vogl, W., Finlay, B.B. (2009) *Salmonella* infection of gallbladder epithelial cells drives local inflammation and injury in a model of acute typhoid fever. *J. Infect. Dis.*, **200**, 1703–1713.
169 Prouty, A.M., Schwesinger, W.H., Gunn, J.S. (2002) Biofilm formation and interaction with the surfaces of gallstones by *Salmonella* spp. *Infect. Immun.*, **70**, 2640–2649.
170 Crawford, R.W., Rosales-Reyes, R., Ramirez-Aguilar Mde, L., Chapa-Azuela, O., Alpuche-Aranda, C., Gunn, J.S. (2010) Gallstones play a significant role in *Salmonella* spp. gallbladder colonization and carriage. *Proc. Natl Acad. Sci. USA*, **107**, 4353–4358.
171 Prigent-Combaret, C., Vidal, O., Dorel, C., Lejeune, P. (1999) Abiotic surface sensing and biofilm-dependent regulation of gene expression in *Escherichia coli*. *J. Bacteriol.*, **181**, 5993–6002.
172 Dorel, C., Vidal, O., Prigent-Combaret, C., Vallet, I., Lejeune, P. (1999) Involvement of the Cpx signal transduction pathway of *E. coli* in biofilm formation. *FEMS Microbiol. Lett.*, **178**, 169–175.
173 Watnick, P.I., Kolter, R. (1999) Steps in the development of a *Vibrio cholerae* El Tor biofilm. *Mol. Microbiol.*, **34**, 586–595.
174 Stewart, P.S., Franklin, M.J. (2008) Physiological heterogeneity in biofilms. *Nat. Rev. Microbiol.*, **6**, 199–210.
175 Chai, Y., Chu, F., Kolter, R., Losick, R. (2008) Bistability and biofilm formation in *Bacillus subtilis*. *Mol. Microbiol.*, **67**, 254–263.

176 Baty, A.M. III, Eastburn, C.C., Diwu, Z., Techkarnjanaruk, S., Goodman, A.E., Geesey, G.G. (2000) Differentiation of chitinase-active and non-chitinase-active subpopulations of a marine bacterium during chitin degradation. *Appl. Environ. Microbiol.* **66**, 3566–3573.

177 Baty, A.M. III, Eastburn, C.C., Techkarnjanaruk, S., Goodman, A.E., Geesey, G.G. (2000) Spatial and temporal variations in chitinolytic gene expression and bacterial biomass production during chitin degradation. *Appl. Environ. Microbiol.*, **66**, 3574–3585.

178 Lewis, K. (2008) Multidrug tolerance of biofilms and persister cells. *Curr. Top. Microbiol. Immunol.*, **322**, 107–131.

179 Singh, R., Ray, P., Das, A., Sharma, M. (2009) Role of persisters and small-colony variants in antibiotic resistance of planktonic and biofilm-associated *Staphylococcus aureus*: an in vitro study. *J. Med. Microbiol.*, **58**, 1067–1073.

33
Epigenetics of Ciliates

Jason A. Motl, Annie W. Shieh, and Douglas L. Chalker
Washington University in St Louis, Biology Department, 1 Brookings Drive, St Louis, MO 63130, USA

Epigenetic Regulation and Epigenomics: Advances in Molecular Biology and Medicine, First Edition. Edited by Robert A. Meyers.

Keywords

Nuclear dimorphism

Containing two different types of nuclei, like ciliates.

Micronucleus (Mic)

The smaller, germline nucleus of nuclear dimorphic ciliates.

Macronucleus (Mac)

The larger, somatic nucleus of nuclear dimorphic ciliates.

Conjugation

Sexual reproduction process of ciliates that involves cross-fertilization and genetic exchange between mating partners to produce progeny.

Autogamy

Sexual reproduction process of ciliates during which one individual self-fertilizes to produce progeny with a completely homozygous genome.

Internal eliminated sequences (IESs)
Sequences ranging from 26 bp to 22 kb, which necessitate removal from introns, exons, and noncoding DNA sequences during sexual reproduction to produce a functional zygotic macronucleus.

DNA elimination
Process of removing repetitive sequences and IESs from the somatic, zygotic macronucleus in ciliates, which occurs during sexual reproduction.

Transposons
DNA elements that can "jump" or transpose around the genome when active.

RNA interference
Process through which ncRNAs and the products of their cleavage, sRNAs, affect transcriptional and post-transcriptional regulation in cells.

Heterochromatin
Chromatin state defined molecularly by histone hypoacetylation and methylation of H3K9 and/or H3K27, which causes condensation of chromatin and gene silencing.

Genetic studies of ciliated protozoa delivered some of the earliest evidence that epigenetic mechanisms play profound roles in determining phenotype. The nuclear dimorphism of these unconventional unicellular organisms has provided a rich context within which to uncover epigenetic mechanisms that regulate genome activities. Comparisons of the chromatin of the transcriptionally active somatic genome and the silent germline have revealed that histone modifications and specialized variants are important regulatory mechanisms, allowing homologous sequences to exist in different states. However, these genomes do not just differ in epigenetic characteristics; they have major structural differences, the result of developmentally programmed DNA rearrangements that occur during nuclear differentiation. These rearrangements eliminate between 15% and 95% of a ciliate's germline-derived DNA to create a streamlined genome that is devoid of most repetitive elements. More recent investigations have revealed that homologous noncoding RNAs (ncRNAs) and RNA interference mechanisms play essential roles in guiding these DNA rearrangements by mediating a comparison of the genome content of the current somatic genome to that in the germline. Continuing research into the process of DNA elimination in ciliates shows promise to provide new insights into the potential of ncRNAs to remodel genomes during development.

1 Ciliate Biology

1.1 Historical Perspective

The concept of "epigenetics" was largely formulated by Conrad Waddington to provide a framework to describe the development of multicellular organisms, and to explain how cells with the same genetic composition can differentiate into functionally distinct types. During these early days of genetics research, the chromosome theory of inheritance was viewed to bridge the observations of Mendelian inheritance and microscopic description of chromosome behavior in cells. However, this genetic theory was somewhat inadequate to account for development of different tissues within an individual, where all cells had the same chromosomes. It was difficult to envision how the apparently static chromosomes (genes) could by themselves manifest phenotypic differences – the fundamental basis of cellular differentiation.

Epigenetic theory thus arose from the need to bridge the gap between genotype and observed phenotypes that could not be accounted for by the behavior of chromosomes. The gap was quite apparent in single-celled organisms, most notably the ciliate *Paramecium* in studies by Tracy Sonneborn [1]. Sonneborn and his colleagues described several examples of phenotypic traits – for example, serotype and mating type – which did not follow conventional Mendelian inheritance, but instead appeared to be passed on through cytoplasmic inheritance. Thus, while these traits were encoded by genes, clonal lines with identical genotypes arose with persistently different phenotypes. Through these studies, Sonneborn and others revealed that the cytoplasm was an important supplement to chromosomes in transmitting heritable information.

While studies of ciliate genetics largely started with those of Sonneborn, research using these organisms has continued to provide important understanding of epigenetic phenomena and their underlying mechanisms that help explain unexpected patterns of inheritance. In this chapter, some early examples of non-Mendelian inheritance observed in ciliates are described to provide a historical context, even though the exact mechanisms that account for these phenomena still await discovery. Nonetheless, research efforts aimed at describing the intricate biology of this fascinating group of microbes have provided new ways to consider epigenetics that stretch well beyond ciliates. Fundamental discoveries of the role of chromatin modification in gene regulation, and the role of noncoding RNAs (ncRNAs) in gene silencing, have secured the place of ciliates as pioneering model systems for epigenetic studies. Much of the utility of these organisms for this research stems from their unique biology, with both germline and somatic copies of the genome maintained in a single cell. Below, the germline and somatic dichotomy of ciliates are described, followed by details of the process of their differentiation, in order to provide the necessary background for describing these epigenetic discoveries.

1.2 Life Cycle and Genetics

The ciliated protozoa belong to the superphylum of Alveolates, which is a lineage that diverged from the ancestors of plants and animals more than a billion years ago [2]. They have evolved into a diverse array of species that have adapted to different environments and strategies for life. Members of the phylum Ciliophora (i.e., ciliates) are commonly found in fresh water, but can also exist in many water-rich environments as free-living organisms, symbiotes, or even parasites. Ciliates have elaborate cellular architectures, most noticeably the organized arrays of cilia that cover their exteriors. An anterior oral apparatus or "mouth," constant swimming enabled by their cilia, and relatively large size give the ciliates animal-like qualities, despite their being unicellular. Ciliates are capable of both asexual and sexual reproduction:

- *Asexual reproduction* (or vegetative population growth) occurs by binary fission, and is the means through which ciliates amplify their populations clonally (Fig. 1a).
- *Sexual reproduction* occurs upon the conjugation of two cells, and involves the exchange of genetic information between each partner and new somatic genome differentiation, without an increase in cell number (Fig. 1b).

The most important feature of ciliates to consider in regards to inheritance is their nuclear dimorphism. In each single cell, ciliates organize two copies of their genome in nuclei that are structurally and functionally distinct. These two different genomes serve the analogous roles to that of germline and somatic cells in metazoans. The germline copy of the genome is contained in the smaller nuclear compartment, the *micronucleus*. The micronuclei are diploid, but interestingly are transcriptionally silent during vegetative growth, serving only to maintain and transmit the genome to progeny cells upon sexual reproduction. The much larger *macronuclei*, on the other hand, carry the somatic genome and, as such, are responsible for all gene expression necessary for vegetative growth. Macronuclei are polyploid, with different ciliate species having widely different copy numbers in their somatic genomes. For example, *Tetrahymena* retain approximately 50 copies of each macronuclear chromosome, whereas *Paramecium* macronuclei contain several hundred copies. During sexual reproduction, the macronucleus – like the soma of metazoa – is lost when a new one is formed from a zygotic nucleus, which is derived from the germline genomes of the parental cells after meiosis.

While all ciliates exhibit nuclear dimorphism, the actual number of germline micronuclei and somatic macronuclei in each cell differs between species. In many of the figures in the chapter, it has been elected to illustrate a single micronucleus and macronucleus per cell, to simplify the discussion. The key nuclear events that occur throughout the ciliate life cycle are presented in a generalized representation in Fig. 1. Vegetative growth involves clonal amplification of the cell's population, during which the micronucleus is duplicated by closed mitosis (i.e., without dissociation of the nuclear envelope), thus ensuring an accurate maintenance of the germline genome (Fig. 1a). The polyploid macronucleus divides amitotically, splitting its nuclear content into roughly equal halves so as to partition its centromere-less chromosomes into each progeny cell. Exactly how the macronuclei maintain the correct copy number of somatic chromosomes is not well understood, but the results of studies conducted in *Oxytricha* and *Stylonychia* have indicated that the copy number can be regulated epigenetically [3, 4]. Nevertheless, high ploidy and the endoreplication of somatic chromosomes appear to maintain the correct DNA content and prevent lethal gene loss.

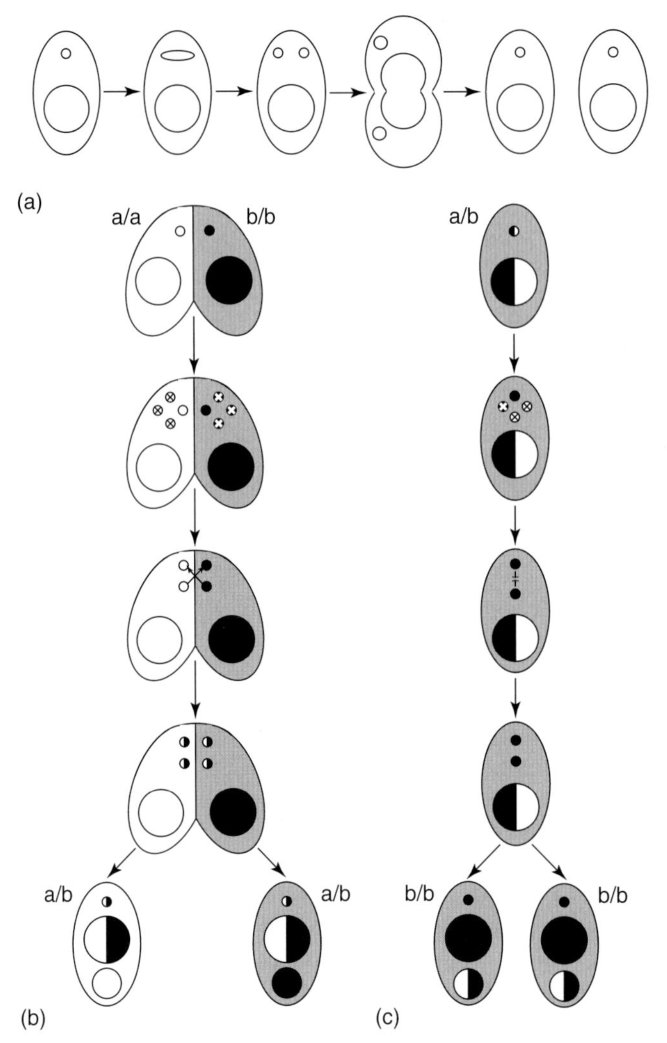

Fig. 1 Ciliates reproduce both vegetatively by (a) binary fission and sexually via (b) conjugation or (c) autogamy.

The micronucleus and macronucleus of a ciliate are replaced after each round of sexual reproduction. *Conjugation,* which can be induced in the laboratory by nutrient starvation, begins with the pairing of two mating compatible cells (see Fig. 1b). The micronuclei of the mating partners undergo meiosis, where a single haploid meiotic product in each partner is selected to be passed on to their progeny; the nonselected meiotic products are then degraded. The chosen haploid nuclei replicate their chromosomes and then undergo an additional nuclear division to produce two haploid nuclei with identical genomes, one of which is exchanged with the mating partner. The exchanged haploid nucleus then fuses with the partner's stationary haploid nucleus to form the zygotic nuclei of the mating pair.

This nuclear cross-fertilization produces identical heterozygous, diploid genomes in each partner. In the case where mating-compatible partners are unavailable, some species will undergo *autogamy*; this is a form of self-fertilization, where two genetically identical haploid nuclei fuse with each other, producing a homozygous diploid genome (Fig. 1c).

When the haploid "gametic" nuclei have fused (*karyogamy*) to give rise to the zygotic genome, additional rounds of DNA replication and nuclear division produce the precursors of the new micronucleus and macronucleus. As development proceeds, these progenitors (which often are called *anlagen*) differentiate into the new germline and somatic nuclei. Whereas, the cross-fertilization that occurs during conjugation generates genetically identical progenitor nuclei, the individual progeny cells of a mating pair can differentiate with distinct phenotypes (e.g., different mating types) in non-Mendelian inheritance patterns. In some cases, specific phenotypes can be traced through a particular cytoplasmic lineage. It is important to note that new somatic nuclei differentiate within the cytoplasms of the two parental cells, such that the DNA is replaced while many existing cellular structures are preserved. This feature of ciliate biology is a major contributor to the non-Mendelian inheritance phenomena described in the following sections.

1.3
Differentiation of Somatic and Germline Genomes

Macronuclear differentiation is an extreme example of genome reprogramming, as the cells start with a genome that is transcriptionally silent and remodel it into one that supports regulated gene expression during vegetative growth. In addition to switching the genome from a silent to an active state, this reprogramming involves a transition from mitotic to amitotic division, accompanied by chromosome breakage and extensive DNA rearrangements (Fig. 2). Research efforts to understand the differences between the transcriptional activity of micro- and macronuclei have uncovered regulatory systems that have solidified the ciliates" place as major models for elucidating epigenetic mechanisms. Before discussing these discoveries further, it is important to briefly touch upon the structural rearrangements

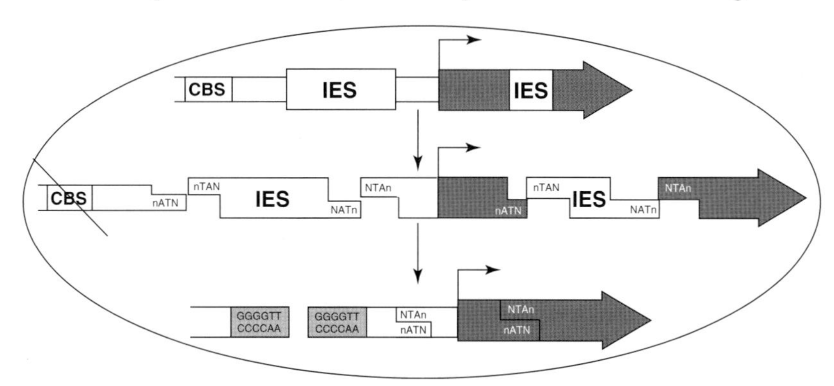

Fig. 2 The somatic genome undergoes extensive DNA rearrangements, including chromosome fragmentation and internal DNA elimination.

that streamline and selectively amplify the genome in differentiating macronuclei. These are important, not only for providing an understanding of some of the historical examples of non-Mendelian inheritance, but also providing – through their study – new avenues by which epigenetic regulation can be further explored.

Ciliates streamline their somatic, macronuclear genome through massive genome rearrangements that fragment the germline-derived chromosomes and eliminate large portions of their genomic complexity (Fig. 2). The fraction of the germline genome removed from the macronucleus ranges from 15% to as much as 95% (for a review, see Ref. [5]). Fragmentation of the developing macronuclear chromosomes is coupled with *de novo* telomere addition, which stabilizes the newly formed termini. The degree of chromosome fragmentation varies widely among the ciliate lineages. For Stichotrichs such as *Oxytricha* and *Euplotes*, this fragmentation is so extensive that the average macronuclear chromosome is only a few kilobase pairs that typically contains a single gene. At the other end of the spectrum, the Oligohymenophora, which include *Tetrahymena* and *Paramecium*, break their developing macronuclear chromosomes at just tens to hundreds of sites to produce chromosomes that are typically several hundred kilobase pairs in size. Following chromosome fragmentation, these small chromosomes are amplified to their final high copy number in the polyploid macronucleus.

In addition to chromosome fragmentation, ciliates eliminate many DNA segments from internal sites. These germline-limited, internal eliminated sequences (IESs) are numerous in all ciliate genomes that have been studied. They are removed from thousands of loci and, in some species, from up to tens of thousands of loci. In some ciliates, such as *Tetrahymena*, essentially all of the IESs are found within intergenic regions, whereas in most other ciliates studied, they are also common within genes. When IESs are present in coding regions, they are precisely excised during macronuclear differentiation. A common class of IESs found in diverse ciliates species is characterized by flanking 5′-TA-3′ dinucleotides, one copy of which is retained upon excision (for a review, see Ref. [6]). The sequences eliminated from somatic macronuclei represent most of the repetitive sequences residing in the germline genome, including transposable elements [7]. The majority of IESs may actually be the remnants of transposons, or be otherwise derived from the activity of transposable elements ([8–10]; see also a review in Ref. [11]). Intriguingly, recent evidence has suggested that the excision of IESs utilizes domesticated transposases [12–14]. As will be discussed below, the mechanisms that ciliates use to identify IESs are related to RNA interference (RNAi), which is used by many eukaryotes as a surveillance system to limit the activity of transposons in the genome [15]. These mechanisms will be described in detail, as they reveal important insights into the use of homologous, ncRNAs in epigenetic regulation.

1.4 Micro- and Macronuclei: Models for Silent and Active Chromatin

The recognition that the micro- and macronuclei of ciliates have opposite activity states promoted the development of these organisms as models with which to examine cellular mechanisms that differentially regulate identical sequences – the

Tab. 1 Histone modifications found in the nuclei of ciliates.

Nucleus	***Histone composition***	***Histone modifications***[a]
Micronucleus	H2A, H2B, H3, H4, micH1	H3K27me, H3S10ph, micH1ph
Macronucleus	H2A, H2B, H3, H4, hv1, hv2, macH1	H3K4me, H3K9me[b], H3K27me, H2Aac, H2Bac, H3ac, H4ac, H2Aph, macH1ph

The histone composition and modifications of the micro- and macronucleus are listed above. Most of these histones and modifications are found throughout the life cycle of *T. thermophila*, but one (H3K9me) is restricted to developing macronuclei during conjugation.
[a]There is no distinction between mono-, di-, and tri-modifications of each histone in this table.
[b]This modification is only found during sexual reproduction in the developing zygotic macronucleus.

very definition of epigenetics. The most significant contributions in this area have been made by groups investigating the chromatin structure of the different nuclei of *Tetrahymena*. Such efforts began about four decades ago, and helped to establish a number of paradigms of epigenetic control, including the importance of histone variants and the role of histone acetylation in transcriptional regulation. A summary of the histone variants and modifications found in the micro- and macronucleus is listed in Table 1.

1.4.1 Differential Histone Composition of Micro- and Macronuclei

The core histones form the largest fraction of chromatin in both the micro- and macronuclei; however, a comparison of the chromatin proteins found in each type of nucleus led to the characterization of some of the first known histone variants. The histone variants, Hv1 and Hv2, were identified as forms of Histone H2A and H3, respectively, that are localized specifically within the transcriptionally active macronucleus [16, 17]; these proteins represent the equivalent of the widely conserved variants H2A.Z and H3.3. While Hv1 (H2A.Z) is essential in *Tetrahymena* [18]. In addition to its presence in the macronucleus, this variant has been observed in micronuclei during early conjugation, when these nuclei first exhibit transcriptional activity [19–21]. Hv2 (H3.3) has properties consistent with its role as a replacement histone. This variant was shown to be constitutively expressed during the cell cycle, in contrast to core histone H3.1, which is expressed only during early S-phase [22]. This led to the hypothesis that H3.1 is only deposited into chromatin during DNA replication, whereas Hv2 is deposited outside of S-phase. The exclusive presence of these two histone variants in the macronucleus (or meiotic micronuclei) provided some of the first evidence that specific variants are preferentially associated with transcriptionally active chromatin.

In addition to core histone variants, the micro- and macronuclei also have distinct linker histones. Although neither linker histone is essential [23], when the genes for the micronuclear and macronuclear linker histones were disrupted, the nucleus in which they normally reside was increased in volume. These results were interpreted to mean that, in the absence of the linker histones, the chromosomes exhibited lower degrees of chromatin compaction. In addition, cells lacking the macronuclear linker histone showed altered gene expression profiles, a finding that providing some of the first evidence that linker histones have roles outside of maintaining general chromosome structure [24].

1.4.2 Differential Histone Modifications of Micro- and Macronuclei

The finding that histones in the macronucleus were hyperacetylated relative to those in the micronucleus provided evidence which corroborated Allfrey's observations, namely that acetylated histones were important for transcriptional activity in animals [25]. The ability to make targeted mutations in *Tetrahymena thermophila* allowed Martin Gorovsky and coworkers to test whether acetylation of the H2A.Z tail was critical for transcription, and to further assess whether specific sites needed to be acetylated [26]. In fact, Gorovsky's group found that the mutation of all normally acetylated lysines in the H2A.Z tail to arginines, which were not able to be acetylated, was lethal. However, the mutant phenotype could be rescued by H2A.Z proteins containing a single acetylated lysine. In addition, the Hv1 tail could be substituted for by the core H2A tail, thus demonstrating that the overall histone tail charge density was more important than the modification of particular tail lysine residues [27].

Arguably, one of the landmark discoveries in epigenetics research was the cloning of the first nuclear histone acetyltransferase (HAT). C. David Allis and coworkers had set out to identify the protein responsible for the hyperacetylation of macronuclear chromatin, by employing an in-gel histone acetylation assay [28]. For this, the histones were first polymerized directly into the denaturing protein gels used to fractionate the *Tetrahymena* extracts. After renaturing the proteins in the polyacrylamide matrix, the gels were incubated with radiolabeled acetyl-CoA. Subsequently, the group identified, and then purified, a 55 kDa protein that shared significant similarity with the yeast GCN5 transcriptional regulator. It was this discovery which established the paradigm that transcriptional regulators act by modifying chromatin [28, 29].

Other histone modifications enriched in either micro- or macronuclei hinted at their biological function. Histone H3 methylated on Lys4 was found exclusively in the macronucleus, thus providing the early evidence that this modification was associated with active chromatin [30]. This modification is absent from micronuclei, but is rapidly established on the bulk of the genome soon after developing macronuclei are formed. In contrast, the methylation of histone H3 on Lys9 is found exclusively during conjugation on the chromatin of IESs in developing macronuclei [31]. This modification is lost from macronuclei as the IESs are removed from the genome. While the methylation of histone H3 on Lys9 was already known to be associated with silent heterochromatin in *Schizosaccharomyces pombe* and other eukaryotes, its linkage to IES excision – which was found concurrently to be controlled by a RNAi-related mechanism – provided one of the first examples (along with studies conducted in *S. pombe*) that RNAi-directed transcriptional gene silencing targeted the chromatin modifications to specific genomic regions [31–33].

While most chromatin modifications are enriched in macronuclei, the phosphorylation of histone H3 on Ser10 was found to be highly enriched in micronuclei undergoing mitosis or meiosis, indicating that this modification may be involved in chromosome condensation [34]. The mutation of Ser10 to alanine resulted in chromosome segregation defects, which further supported the importance of phosphorylation of this position on histone H3 in chromatin compaction during nuclear division [35]. These structural and functional differences between the micro- and macronuclei provided a rich biological context by which to start unraveling the role of chromatin proteins and their post-translational modifications for controlling epigenetic phenomena in ciliates.

2
Epigenetic Phenomena in Ciliates

Ciliates had been firmly established as genetic models for uncovering epigenetic phenomena long before many research groups began to use the differentiation of micro- and macronuclei as a means of resolving the molecular basis of epigenetic control. The many classical examples of non-Mendelian inheritance and other epigenetic phenomena that are described in the following sections have been included on the basis that it is useful to revisit these early observations in light of more recent molecular studies. Such examples of structural and cytoplasmic inheritance have a common feature, notably that the pre-existing phenotypic state of the parent cells is able somehow to "template" the phenotype that emerges in the next generation. These phenomena challenge many of the preconceived ideas of simple genetic inheritance, and beg for further investigation to decipher their underlying mysteries.

2.1
Structural Inheritance

In addition to nuclear dualism, ciliates are characterized by the extraordinary complexity and asymmetry of their cellular structures. The ciliate cortex is comprised of a matrix of cytoskeletal and membranous components, while organized within the cortex are organelles with specialized functions, such as the anteriorly positioned oral apparatus (a mouth-like phagocytic structure) and a posterior cytoproct. The elaborate ciliate body plan is faithfully reconstructed after each round of binary fission. The anterior daughter cell must reform the posterior structures, and the posterior daughter must generate a new mouth and other anterior components. Both, genetic and physical manipulations of the cortex have revealed that the cellular structure of ciliates is largely organized by the pre-existing structures, thereby demonstrating that a cell's phenotype is not determined solely by genotype.

These cells' numerous cilia, which are used primarily for locomotion and feeding, project from arrays of cortical units, aligned into rows that are organized along the anteroposterior axis. Each cortical unit assumes a distinct anterior–posterior and left–right orientation that is crucial for the correct function of the cilia. During each cell cycle, the units are duplicated to ensure that each daughter cell inherits a complete set of structures that assumes the correct orientations. An early scientific question was whether this cortical organization was determined by the action of genes; subsequently, it was revealed that the structural organization of daughter cells is not established purely by the cells genotype, but rather is templated by the geometry of the pre-existing units (i.e., it is inherited through a non-genic mechanism). One of the earliest studies on cortical inheritance was performed using doublet cells. The "doublet" phenotype arises from a failure of pair separation at the end of conjugation, which leads to a fusion of the progeny. This phenotype is fairly stable, and can be propagated such that the vegetative progeny inherits a duplicated set of cortical structures. Genetic crosses demonstrated that the heredity of the doublet phenotype was not determined by genes or the cytoplasm, but rather was communicated through the architecture of the cortex itself [36].

Cortical inversion, a condition in which the cells have one or more ciliary rows rotated 180° in the plane of the cell surface, further illustrates the phenomenon of structural inheritance. In this case, an inverted patch of cilia is produced that results in the cells exhibiting an abnormal "twisting" swimming phenotype. As with the "doublet" phenotype, the progeny of cells with inverted patches inherits the inverted orientation of cilia, as the new cortical organization is templated by the parental cortical organization [37]. What this and other experiments show, in the case of ciliary orientation, is that whilst the genes supply the building blocks, the assembly into a functional organelle is determined by the structure of the pre-existing cortex. The ciliate cortex thus provides an example of structural memory, and reveals that genes are not the only cellular component that can pass on heritable information to the next generation.

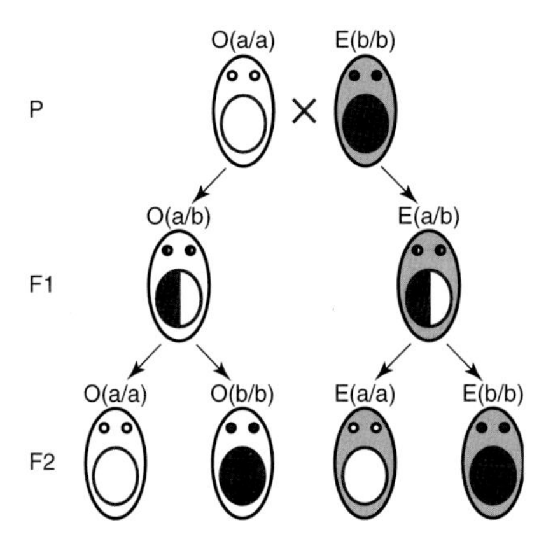

Fig. 3 Cytoplasmic inheritance. The mating type of *Paramecium* is determined by the parental cytoplasm, not the genotype of the progeny. P, Parent; F1/F2, generations.

2.2 Cytoplasmic Inheritance

The inheritance of pre-existing cellular structures is a specialized example of epigenetic influence on the phenotype. A more general non-nuclear medium for transmission of heritable information is the cytoplasm, the role of which as a director for epigenetic information is well documented in ciliates, notably in sexually reproducing *Paramecium aurelia* and related species. One reason for this is that, unlike some ciliates (e.g., *T. thermophila*), the conjugation of *P. aurelia* involves almost no cytoplasmic exchange between the mating pairs. Therefore, while cross-fertilization produces identical zygotic nuclei, these identical genomes develop in the different cytoplasmic environments of their respective parental cells. The interesting observation here is that these progeny – which are genetic twins – commonly express different phenotypes as determined by the cytoplasm in which their macronuclei develop.

Cytoplasmic inheritance in ciliates is most easily illustrated by determination of the mating type trait (Fig. 3) [38, 39]. *Paramecium* exist as two mating types: Even (E) and Odd (O). When two cells of opposite type mate, the progeny that arise from the E parent almost always assume the E mating type, whereas those from the O parent almost always assume the O mating type, despite each having received identical genotypes. This observation suggests that something other than genes is directing the determination of the mating phenotype. A comparison of progeny mating types from crosses that do, and do not, exchange cytoplasm during conjugation further implicated cytoplasm as a key component in mating type determination [40–42]. If cytoplasmic exchange occurred between the mating pairs during conjugation, then the progeny of the O cell would often be switched to the E mating type. Furthermore, an injection of cytoplasm from the E mating partner into the O partner was found to transform the progeny's mating type from O to E. No effect was observed upon the transfer of O cytoplasm into E cells, which suggested that the cytoplasmic factor(s) must exist in the E cell to determine the E mating type, and that the E mating type is dominant over O [43].

Similar to mating type, the serotype of the *Paramecium* progeny can be strongly influenced by the cytoplasm in which a new somatic genome differentiates. Serotype is determined by the specific surface antigen protein that is expressed and displayed on the cell surface. Although several genes encode the different antigen proteins, only one gene is expressed in any given cell. Upon conjugation, the sexual progeny typically express the parental serotype. For instance, when cells of serotype A are crossed with serotype B, the progeny of both types will emerge expressing the serotype of the parent in which their nuclei developed [44]. The inheritance of mating type and serotype is, therefore, specified by the cytoplasmic environment rather than purely as genetic traits.

2.3
Epigenetic Control of Traits Converge with the Regulation of DNA Rearrangements

As noted above, the differentiation of a developing somatic macronucleus from its zygotic precursor involves an extensive streamlining of its germline-derived genome by removing extraneous "junk" DNA (see Fig. 2). Thus, the process of genome rearrangement directs major changes to the overall DNA sequence in the somatic macronucleus relative to the input from the germline. As the DNA removed is primarily noncoding, the suggestion that this DNA reorganization may or may not affect gene expression has not been extensively studied. For many ciliates, which have IESs imbedded within their coding regions, DNA elimination must occur to generate an expressible protein-coding region. It has been postulated – and supported by several experimental observations – that the epigenetic control of these DNA rearrangements may underlie at least some of the examples of non-Mendelian inheritance that have been discovered. The proposal that ciliates may differentially eliminate DNA sequences as a mechanism to alter the phenotype expressed by their progeny, is discussed in the following sections.

A genetic screen that initially was aimed at elucidating the molecular basis for mating-type expression eventually uncovered an intriguing link between this trait and the control of DNA rearrangement. A genetic mutation, mtF^E, was isolated in a cell line that produces only mating type E [45]. As noted above, *Paramecium* sexual progeny almost always assume the mating type of the parent (i.e., O parent, O progeny; E parent, E progeny). Hence, when an E individual that carries the mtF^E mutation (mtF^E/mtF^E) is

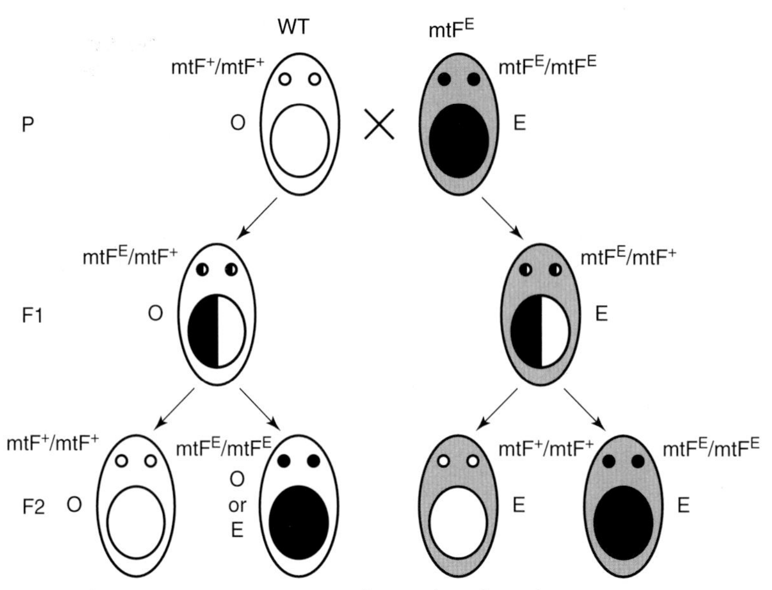

Fig. 4 The mtFE mutation: a genetic lesion that alters the cytoplasmic inheritance of mating type in *Paramecium*. P, Parent; F1/F2, generations; WT, wild-type.

crossed with a wild-type O individual (mtF^+/mtF^+), the mating type of the F1 progeny still follows the cytoplasmic inheritance typical of a wild-type mating (Fig. 4). However, homozygous mtF^E F2 progeny produced from autogamy of F1 O individuals (mtF^+/mtF^E) frequently switch to mating type E (compare Figs 3 and 4). Although the gene mutated in mtF^E strains has not been identified, a detailed study of the mtF^E/mtF^E strains showed that this mutation also led to a failure to eliminate an IES located in the G surface antigen gene. In turn, this observation led to the hypothesis that the gene mutated in mtF^E strains was involved in DNA rearrangement and by extension, that DNA rearrangement may be involved with mating type determination [46].

However, the mtF^E mutation studies provided more than just a link between DNA rearrangement and mating type; rather, they uncovered a means by which the epigenetic regulation of DNA rearrangements could alter the expression of specific traits. Further studies – not of mating type, but of G gene expression – revealed that the IES+ state of the G gene (apparently caused by the mtF^E mutation) became the heritable state of the G gene that was propagated through subsequent generations, even after reintroduction of the wild-type mtF^+ allele. Given the observed cytoplasmic inheritance patterns of both mating type and serotype traits in *Paramecium*, this finding offered an intriguing connection between alternative rearrangements and altered phenotypes.

The propagation of the IES+ state in the mtF^+ progeny showed that it was not a genetic lesion or other alteration to the germline genome that limited expression of the G gene. It was, in fact, the IES+ state itself that was present in the parental macronucleus and which elicited the transmittable influence of the "cytoplasm" during development. This was demonstrated more conclusively by directly injecting the IES+ version of the G coding sequence into the maternal macronucleus, and showing that this alone was able to block the elimination of the homologous IES from the newly developed macronucleus after autogamy [47]. It is important to note that the injected DNA is destroyed along with the maternal macronucleus, so the IES+ state must be communicated to the developing macronucleus through the cytoplasm. The injection of plasmid DNA containing just the one IES, without any flanking G gene coding sequence, was found to be sufficient to block the elimination of this IES, while the remaining IESs within the G gene were excised efficiently. Thus, particular IES sequences present in the maternal macronucleus are able to communicate their presence to the zygotic macronucleus, and alter the normally efficient removal of the homologous sequence.

However, not all IESs were found to be subject to this form of homology-dependent regulation. When ten different IESs were microinjected into parental macronuclei to test their ability to block the excision of the homologous sequence, only four were able to inhibit DNA rearrangement. Whilst it was difficult to see why only some IESs in the zygotic macronucleus could sense the presence of homologous copies in the parental macronuclei, the clear implication here was that many characteristics could be reproducibly inherited in a non-Mendelian fashion, every time a new macronucleus is formed.

The serotype genes of *Paramecium* have proven to be fertile ground for uncovering epigenetic phenomena relating to genome rearrangements. One early and particularly interesting example was revealed by studies of a mutant strain called *d48*, that lacked the ability to express the surface antigen A gene [48]. Subsequent carefully conducted genetic studies showed that the d48 micronucleus contained a wild-type copy of the A gene; but that the macronucleus was missing the A gene-coding region [49]. The remarkable discovery was that the progeny of d48 strains reproducibly eliminated the A gene from their developing macronuclei during conjugation, making these progeny unable to express the A serotype.

The results of a series of microinjection and nuclear transplantation experiments confirmed that the presence of the A gene in the parental macronucleus was necessary for it to be retained in the progeny. Subsequently, microinjection of the A gene into the macronucleus of strains lacking the A gene in both the micro- and the macronuclei was sufficient to restore A gene expression during vegetative growth; however, this expression was lost during sexual reproduction when the microinjected parental macronuclei were fragmented and destroyed [50]. On the other hand, in the d48 strain – which lacks the A gene only in the macronucleus – microinjection of the A gene was sufficient to rescue A gene expression during vegetative growth, both in the parental strain and also in progeny cells following sexual reproduction [51–53]. Strains missing the surface antigen B gene have also been observed and rescued in a similar fashion [54, 55].

The rescue of A gene expression in the *Paramecium* d48 strain was found to be sequence-specific. Microinjection of the A gene or an allele of the A gene that has 97% identity resulted in A gene retention in the newly formed macronuclei of progeny. In contrast, introduction of the G surface antigen gene – which shares approximately 80% similarity with the A gene – failed to rescue the A gene deficiency in the progeny [56, 57]. Thus, the DNA sequence of the parental macronucleus was again shown to have the ability to dramatically influence the types of sequence retained during the development of new macronuclei.

The observations made with d48 strains share intriguing parallels with both the inheritance of the IES+ state in the mtF progeny, and with the examples of cytoplasmic inheritance described above. In each case, the trait (or sequence) propagated is that which was expressed from the parental macronucleus. Thus, for ciliates the regulation of DNA rearrangements allows for somatic states of gene expression to be transmitted to the next generation. Recent studies of the mechanisms that guide DNA rearrangements have shown that homologous RNAs and chromatin-based regulatory schemes are key components. Studies of ciliate DNA elimination during macronuclear development have revealed that ncRNAs may also be the molecules responsible for many of the cytoplasmic and homology-dependent inheritance phenomena observed previously. These mechanisms will be described in more detail in the following subsections, as they offer many unique insights into how ncRNAs can pattern the genome and influence chromatin structure.

3
RNA-Mediated Epigenetic Mechanisms

3.1
Homology-Dependent Gene Silencing

Homology-dependent epigenetic phenomena have been observed widely, with the introduction of transgenes into plant cells often leading to a silencing of the endogenous copy. One of the most-often cited such examples resulted from an effort to create petunias that had darker flower petals, by adding exogenous copies of the chalcone synthase gene that generates the purple pigment [58]. However, instead of producing the expected increase in petal pigmentation, the transgenic petunias showed a decrease in coloration, in conjunction with an overall reduction in the mRNA level of chalcone synthase; this phenomenon was termed *co-suppression.* Similarly, the introduction of transgenes into the fungi *Neurospora crassa* induced a phenomenon known as *quelling*, which involved a silencing of the homologous endogenous gene [59]. Co-suppression has also been observed in the ciliate, *Paramecium tetraurelia*, upon high-copy microinjection of transgenes that lack 5′ and 3′ regulatory regions (i.e., lacking either promoters or transcription terminators), which resulted in a silencing of the endogenous homologous genes [60, 61].

The mysterious mechanism underlying these phenomena was discovered to be RNAi. A mechanistic insight into homology-dependent phenomena in ciliates has likewise been provided via connections to RNAi. In general, RNAi refers to a diverse collection of cellular mechanisms that employ RNA molecules to regulate the expression of genes (for reviews, see Refs [62–64]). In this case, the triggering molecule is typically double-stranded RNA (dsRNA) that is recognized by a ribonuclease known as *Dicer*, which cleaves dsRNA into fragments of approximately 20–30 nt. These so-called small RNA (sRNA) species serve as the specificity factors that guide an associated protein complex to a target mRNA or gene, where these effector RNA–protein complexes can promote silencing, either transcriptionally or post-transcriptionally.

RNAi appears to be an integral part of a variety of processes in ciliates. An examination of the bulk sRNA species in either *Paramecium* or *Tetrahymena* revealed distinct size classes, thus suggesting the existence of at least two different RNAi pathways [33, 65, 66]. The larger species (ca. 25 nt in *Paramecium* and 27–30 nt in *Tetrahymena*) were shown to be produced exclusively during conjugation, and to guide the extensive DNA rearrangements that occur in the differentiating somatic macronucleus (this RNA-guided genome reorganization is discussed in detail in Sect. 4).

A second class of ca. 23 nt RNAs is produced in growing cells, as well as during conjugation in *Paramecium* and *Tetrahymena*. This size class mediates post-transcriptional gene silencing (PTGS), and also the transgene co-suppression introduced above. In addition, the introduction of dsRNA aimed to experimentally induce gene silencing, either through feeding or direct injection into *Paramecium* cells, or by hairpin RNA expression in *Tetrahymena*, resulted in the production of these ca. 23 nt RNAs [61, 66–70]. Thus, these sRNAs are similar in function to the small interfering RNAs (siRNAs) discovered initially in plants by Baulcombe and colleagues, in that they carry out PTGS [71]. In *Tetrahymena*, these sRNAs are produced by Dcr2p from presumed pseudogenes or defective endogenous genes, which triggered the production of dsRNA precursors necessary for siRNA production [65, 72]. They are anti-sense to these predicted open-reading frames (ORFs), and depend on the activity of RNA-dependent RNA polymerase (RdRP), Rdr1p, which is found in a common complex with Dcr2p [72]. In *Paramecium*, a subclass of these smaller sRNAs is only anti-sense to mRNA transcripts, and is produced by a secondary amplification that involves the RdRPs, Rdr1p and Rdr2p [66, 73]. Although RNAi is clearly an important mechanism during the vegetative life of ciliates, its critical role has yet to be carefully examined. On the other hand, the function of RNAi pathways during development of the zygotic macronucleus has promoted new considerations regarding epigenetic programming of the genome.

3.2
RNA-Guided Genome Reorganization

Both, ncRNAs and RNAi-related mechanisms provide much more than a gene-silencing role in ciliates, as these organisms employ RNAs as guides to extensively remodel their genomes during sexual differentiation. Investigations aimed at elucidating the molecular mechanisms associated with the reorganization of the somatic genome of several ciliates have uncovered the involvement of ncRNAs [69, 74–77]. Indeed, the mechanisms identified have been shown to vary substantially among the different ciliate species studied, such that the data relating to *Paramecium*, *Tetrahymena*, and *Oxytricha* will be described separately in the following sections. Nevertheless, a common theme has emerged, in that these RNAs can serve as potent mediators capable of transmitting sequence-specific information between generations. The examples of homology-dependent regulation of phenotypes (particularly those described earlier in *Paramecium*; see Sect. 2.3) hinted that the mechanism(s) guiding genome rearrangements utilized some form of nucleic acid to transmit sequence-specific information between the somatic macronucleus of one generation and the developing macronucleus of the next. These phenomena require that the state of the DNA in the parental macronucleus serves as a "template" for the traits expressed from the genome of the progeny.

Studies of *Paramecium* and *Tetrahymena* DNA rearrangements have identified two types of sequence-specific mediator RNAs – one which is produced from the germline genome, and a second produced from the parental somatic genome [69, 74, 75]. The germline-specific RNAs are in the form of sRNAs (known as scan RNAs; scnRNAs), that are produced during meiosis and act to identify the IESs as germline-limited sequences to be eliminated from the developing somatic genome [33, 66]. The second type of mediator RNA consists of longer transcripts produced from the parental macronucleus, and which appear to antagonize the action of the scnRNAs [69, 75]. It is these macronuclear transcripts that are the key epigenetic regulators that may explain the non-Mendelian inheritance of specific traits. In *Oxytricha*, analogous transcripts created from the parental somatic genome are postulated to serve as templates to directly guide the rearrangements, while a role for sRNAs is, as yet, unknown [76].

Genome scanning is a term used to describe the mechanism by which RNAs from the germline and somatic genomes can communicate the existing genomic content of the parental nucleus to the next generation [33]. Scanning occurs by a comparison of the germline-derived scnRNAs, with long ncRNA transcripts produced by the parental macronucleus [69, 75]. Such scanning assures that those scnRNAs made to regions of the genome which are not IESs, are removed from the pool of scnRNAs that target specific sequences for elimination. Scanning not only allows a "proofreading" of the sRNA pool to prevent any inadvertent elimination of sequences that should be retained, but also permits the retention of IESs that were maintained in the macronucleus of the previous generation and which offered some advantage or specified an alternative phenotype. The mechanisms of RNA-guided genome reorganization and genome scanning are described in the following sections, as these studies reveal the power of homologous RNAs to direct the programming of the somatic genome.

4 Small RNA-Mediated DNA Rearrangements

4.1 RNAi-Dependent DNA Elimination in *Paramecium*

The germline genome of *Paramecium tetraurelia* contains approximately 60 000 IESs that range in size from 26 to 886 bp [8, 78]. Many of these are found within coding sequences, and must be identified and excised with precision from the developing macronuclear chromosomes. Furthermore, during this genome maturation in *P. tetraurelia*, the more than 50 micronuclear chromosomes are fragmented into an unknown number of mini-chromosomes, amplified to 800*n* [79, 80]. The elimination of IESs occurs during both self-mating and sexual reproduction, at which time the parental macronucleus is destroyed and a new zygotic macronucleus is generated.

The results of studies performed over the past decade have revealed that the IESs are identified through the actions of homologous RNAs via an RNAi-related mechanism in *Paramecium*, and support the model shown in Fig. 5 [69]. A class of sRNAs each of ca. 25 nt, produced only

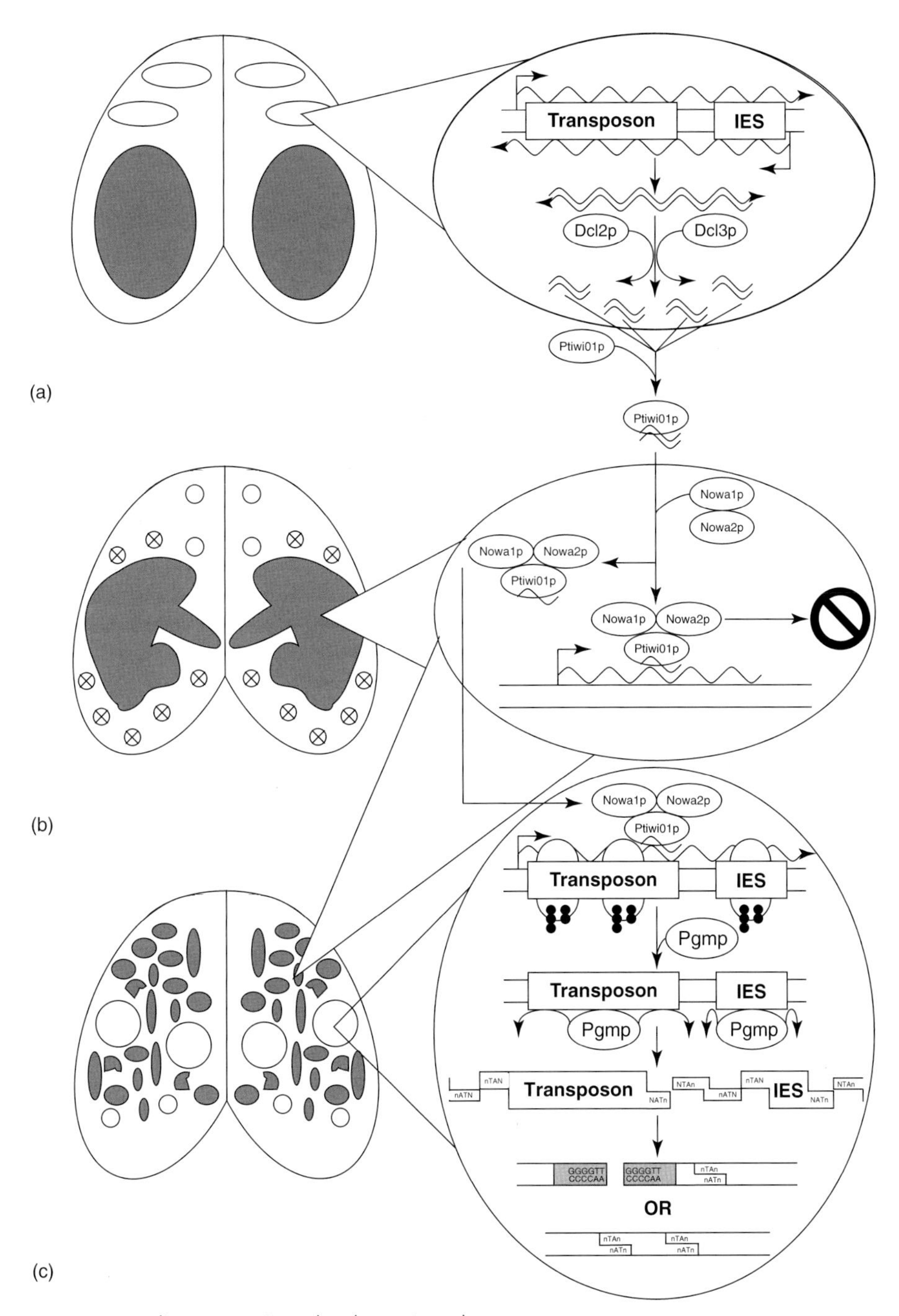

Fig. 5 *P. tetraurelia* uses RNAi and a domesticated transposase, Pgmp, for programmed DNA elimination.

during meiosis, has been shown to be necessary and sufficient to trigger the DNA elimination of IES sequences [66]. These were found to be homologous to a variety of DNA sequences throughout the genome, and likely function in similar manner to the *T. thermophila* scnRNAs described below [33, 66, 69]. These *Paramecium* scnRNAs have 2 bp 3′ overhangs, consistent with cleavage by an RNase III homolog [66, 81–83] (also see review in Ref. [84]). Seven RNase III homologs are present in *P. tetraurelia*, together with three Dicer (DCR) and four Dicer-like (DCL) homologs [66]. Although a single knockdown of the DCL genes has no effect on scnRNA production, double knockdowns of DCL2 and DCL3 will cause it to be abolished. The localization of Dcl2p in the crescent micronucleus early in meiosis indicates that the production of scnRNAs only takes place there at this early time point of conjugation. Double knockdowns of DCL2 and DCL3 also caused a failure of DNA elimination and produced non-viable progeny, further supporting the conclusion that the scnRNAs which they produce target the IESs for excision.

The scnRNAs produced by Dcl2p and Dcl3p cleavage in the crescent micronucleus are transported by the Piwi homologs, Ptiwi01p and Ptiwi09p, into the parental macronucleus to carry out genome scanning [85]. The scnRNAs that match the parental macronuclear genome are removed from the population that will be transported to the developing macronucleus later in development, to participate in genome restructuring. This scanning occurs by comparison of these germline-derived scnRNAs with a second type of regulatory RNA (long ncRNA transcripts produced in the maternal macronucleus), and ensures that scnRNAs made to regions of the genome that are not IESs are not inadvertently excised [69, 75].

Only a few proteins are known to play a role in the genome-scanning process in *P. tetraurelia*. Two glycine-tryptophan (GW) repeat proteins, Nowa1p and Nowa2p [86], have been identified as playing a role in this process; these were found initially to localize within the parental macronucleus during pre-zygotic development, and then to move to the developing macronucleus after its formation. A deletion analysis of Nowa1p showed that the N-terminal portion of the protein has nucleic acid-binding capabilities, particularly for RNA/DNA duplexes. The dimerization of Nowa1p, either with itself or perhaps with Nowa2p, appears to be essential for the nucleic acid-binding function. The double knockdown of NOWA1 and NOWA2 caused a failure of the DNA elimination of a specific class of IESs in *P. tetraurelia*; this was referred to as a maternally controlled internal eliminated sequences (mcIESs) [78, 86]. The failure of DNA elimination was complete in some cases, but incomplete in others [86]. A double knockdown of NOWA1 and NOWA2 also produced non-viable progeny, which indicated an essential function for the completion of autogamy or conjugation.

The question then was, "How might the NOWA proteins contribute to the epigenetic control of IES excision?", and "What RNAs might they interact with?" Previously, long ncRNA has been shown to have a role in several epigenetic phenomena in higher eukaryotes, including dosage compensation and genomic imprinting [87–93]. Data derived from *P. tetraurelia* have provided strong support for an interaction between the maternal long ncRNA and meiotic scnRNAs, and revealed exactly why this interaction is likely to be fundamental to genome programming [69].

Reverse-transcription polymerase chain reaction (RT-PCR) studies of RNA isolated early in autogamy demonstrated the production of ncRNA without IESs, which were thought to be transcribed from the parental macronucleus. When a strain of *P. tetraurelia* containing a mcIES in the parental macronucleus was fed bacteria producing dsRNA prior to autogamy, or were directly injected with 23 nt siRNAs or 25 nt scnRNAs early during autogamy against this mcIES, the latter was removed from the developing macronucleus later in autogamy. These results indicated that genome scanning could be affected by degrading the long ncRNA in the parental macronucleus through bacterial feeding to produce 23 nt siRNAs, or by the direct injection of 23 nt siRNAs, as well as directly injecting the biologically active 25 nt scnRNAs to allow removal of an mcIES that normally would be retained on the completion of autogamy.

Long ncRNA also plays a role in the developing macronucleus by directing the remaining scnRNAs to sequences of DNA that are to be eliminated. In *P. tetraurelia*, the transport of these remaining scnRNA complexes to the developing macronucleus is mediated by the Piwi homologs, Ptiwi01 and Ptiwi09, where the production of long, ncRNA containing IESs has been detected using RT-PCR [69, 85]. Injection of the 25 nt scnRNAs in the same *P. tetraurelia* strain containing a mcIES in the parental macronucleus later during autogamy also causes removal of the mcIES, but the simultaneous injection of 23 nt siRNAs failed to cause DNA elimination [69]. In this case, it seemed likely that the 23 nt siRNAs actually promoted a failure of DNA elimination by targeting the long ncRNA needed for DNA elimination for degradation, while the 25 nt scnRNAs were able to recruit the necessary proteins for the DNA elimination of this mcIES.

4.2
The Role of a Domesticated PiggyBac Transposase in DNA Elimination and Chromosome Breakage in the Developing Somatic Nucleus of *Paramecium*

Each of the different varieties of RNA that are seen only during autogamy or conjugation in *P. tetraurelia* are all directed to one goal, namely the elimination of IESs and repetitive sequences. The removal of any of these types of RNA during the reproductive process causes nonviability [66, 69]. In order to eliminate IESs and repetitive sequences from the genome, these scnRNAs must recruit an excisase, a role for which recent data have implicated the domesticated piggyBac transposase, Pgmp [12]. In order to understand the role of Pgmp in DNA elimination, a brief description of IESs is called for. In *P. tetraurelia*, each IES is flanked by terminal inverted repeats, the consensus sequence of which is 5′-tggTAYAGYNR-3′ [8, 94]. Subsequently, cleavage occurs between the two guanosines in the consensus sequence, to produce a 5′ 4 bp overhang centered around the TA dinucleotide [95]. Mutations in either the T, A, or G in the third, fourth, and eighth position, respectively, of the above consensus sequence are sufficient to block cleavage [96–99]. Cleavage of the consensus sequence, 5′-TTAA-3′, by piggyBac transposases to produce a 5′ 4 bp overhang is somewhat similar to the *P. tetraurelia* consensus IES sequence and cleavage product [100, 101]. An analysis of the *P. tetraurelia* genome identified a piggyBac homolog, called piggyMac (PGM) [12]. Localization of the green fluorescent protein (GFP)–Pgmp was found only in the developing macronucleus late in autogamy. The knockdown of PGM late in conjugation resulted in a failure to produce any viable progeny, a failure of IES excision and chromosome breakage, and an overexpression of IES-containing ncRNA from the developing macronucleus. These knockdown phenotypes implicated Pgmp as having an essential role in the completion of DNA elimination and chromosome breakage in *P. tetraurelia*, most likely through Pgmp-mediated dsDNA breakage to remove IESs and other repetitive sequences. The repair of these dsDNA breaks is mediated by the DNA ligase IV homologs, LIG4a and LIG4b [102].

The removal of these IESs and other repetitive sequences in *P. tetraurelia* and other ciliates is the ultimate epigenetic action. Unlike most other eukaryotes, which heterochromatize their repetitive and noncoding sequences, the ciliates excise and degrade these sequences from their somatic macronucleus, and then amplify the remaining sequences so as to create a streamlined genome that allows a greater cell size than most other eukaryotes and a growth rate comparable to that of yeast. As discussed earlier in brief, a removal of IESs and other repetitive elements occurs in completion of sexual reproduction (the actual removal of these two types of sequence may differ slightly, and even impact the final state of the genome after sexual reproduction). Two different classes of IESs have been identified – namely mcIESs and non-mcIESs – which are small, are found throughout the genome, and eliminated in a precise fashion [47, 78, 95]. The mcIESs are capable of having their excision blocked by the insertion of a copy of the mcIES into the parental macronucleus [47, 78]. The mcIESs tend to be larger in general, and it has been hypothesized that their elimination is dependent on chromatin modifications directed by genome scanning [103]. In contrast, non-mcIESs are smaller, with most being shorter than the amount of DNA wrapped around a nucleosome, which would necessitate a different targeting method for DNA elimination. It seems possible that their elimination could take place through a directed binding of Pgmp, or through guidance of Pgmp via a nucleotide modification to their cleavage sequences. Repetitive sequences are removed with much less precise methods, and this results in either variable cleavage or fragmentation of the chromosome [104]. Both types of DNA elimination depend on the action of Pgmp [12].

Despite all that has been learned regarding the epigenetic phenomenon of RNAi-directed DNA elimination in *P. tetraurelia*, many questions remain to be answered:

- "How are these ncRNAs produced in any of the nuclei?"
- "What is the difference between mcIESs and non-mcIESs, and how does that affect their DNA elimination?"
- "How does DNA elimination, RNAi, and heterochromatin function in related ciliates, and in general how is this biological process related to other epigenetic processes in other eukaryotes?"

Investigations into the RNAi-directed DNA elimination process in a related ciliate, *T. thermophila*, have provided additional insights into many of these questions.

4.3
RNAi-Dependent DNA Elimination in *Tetrahymena*

Like *P. tetraurelia*, the ciliate *T. thermophila* also undergoes massive DNA elimination and chromosome breakage during sexual reproduction or conjugation. During conjugation in *T. thermophila*, the developing zygotic macronucleus is fragmented into approximately 200 minichromosomes from five chromosomes, while 30% of the overall DNA content is removed and the remaining DNA content is amplified to 50*n* [7, 105–110] (for a review, see Ref. [5]). Similar to *P. tetraurelia*, the mechanism of this process was poorly understood until the discovery of conjugation-specific, long ncRNAs and a class of sRNA (termed *scnRNAs*) that are derived from the ncRNAs, and which has led to the model shown in Fig. 6 [33, 74, 75].

The scnRNA model of RNAi-dependent DNA elimination in *T. thermophila* can effectively be broken into two parts: (i) production and selection of the scnRNAs by conventional RNAi-associated proteins; and (ii) transduction of the scnRNA signal into heterochromatin formation, which subsequently triggers DNA elimination of the heterochromatic DNA in the developing zygotic macronucleus. For each of these parts, the experimental data supporting the model, how that data can be used to further elucidate the mechanism of RNAi-dependent DNA elimination, and how the results obtained relate to epigenetics in ciliates and other eukaryotes, are discussed in the following subsections.

4.4 RNAi Apparatus and Genome Scanning in DNA Elimination

The role of RNA during the development of many eukaryotes has been well documented [33, 74, 87–92, 111–116]. For example, *T. thermophila*, like *P. tetraurelia*, has been shown to possess two classes of sRNAs that range from 23 to 24 nt and from 28 to 30 nt in size [33, 65, 117, 118], where the larger class – the scnRNAs – is restricted to conjugation [33]. These appear to be functionally similar to piRNAs that have been described in a variety of organisms, and which are known to act to protect the germline genome in the micronucleus against possible deleterious effects that active transposons can inflict, such as gene inactivation, chromosome translocation, and chromosome breakage [118–125]. Unlike piRNAs, which are Dicer-independent, scnRNA production in both *P. tetraurelia* and *T. thermophila* is totally dependent on a group of DCL proteins [66, 117, 118, 120, 123]. If the DCL genes are either knocked out or knocked down, the scnRNAs are not produced during conjugation, and this triggers a developmental arrest [66, 117, 118]. Whilst it is intriguing that these scnRNAs in *P. tetraurelia* and *T. thermophila* exhibit properties of both piRNAs and siRNAs, further studies of the scnRNA pathway may contribute to a fundamental understanding of how both the piRNA and siRNA pathways arose in higher eukaryotes.

4.5 Bidirectional Transcription of Long dsRNAs

The production of scnRNAs depends on the synthesis of long dsRNA precursors [117, 118]. At an early stage during conjugation, the micronucleus detaches from a groove in the parental macronucleus and elongates to form a crescent that is approximately the length of two cells [127, 128]. During vegetative growth in *T. thermophila*, the micronucleus is transcriptionally silent, although some decades ago it had been observed that early during conjugation (starting after micronuclear detachment from the parental macronucleus and peaking just prior to full crescent elongation) there was copious

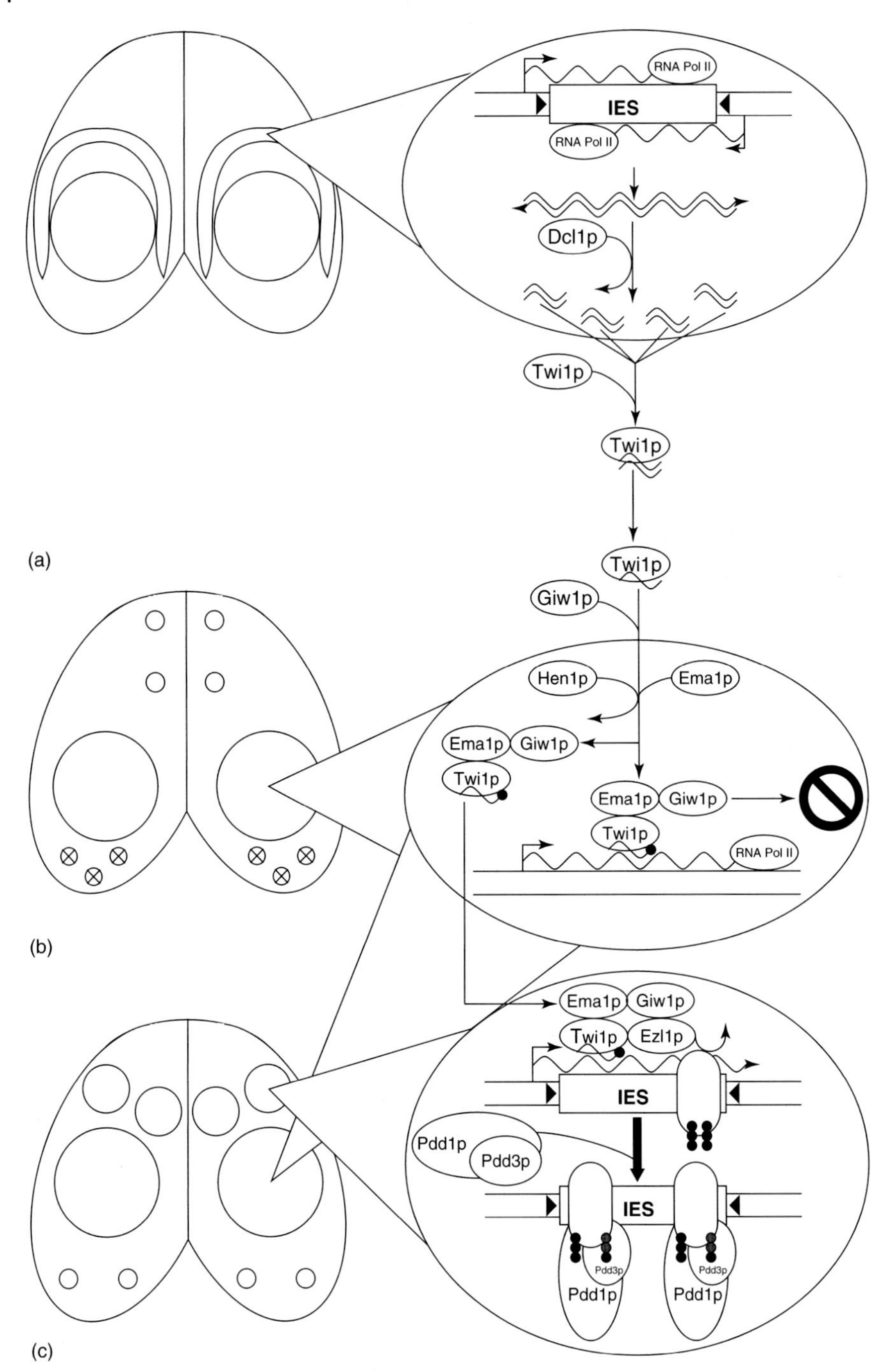

Fig. 6 Meiotic germline transcription and a RNAi pathway direct DNA elimination in *T. thermophila*.

transcription from the micronucleus [19, 20]. The results of later studies conducted in *T. thermophila* showed that, just prior to this period of micronuclear transcription, the *T. thermophila* H2A.Z homolog was deposited in the micronucleus, despite normally being found only in the macronucleus [21, 129]. Other studies also showed that RNA polymerase components, including a putative TATA-binding protein, TBP1, and a RNA polymerase II subunit, RPB3, were localized to the micronucleus during this burst of transcription [130, 131]. This implied that the RNA polymerase responsible for this early micronuclear transcription was RNA polymerase II. Investigations performed on an IES, the M element, showed that the transcription of both strands was markedly increased early in conjugation during the same time period that general micronuclear transcription was increased [74]. These transcripts produced early in conjugation were also heterogeneous at the 5′ and 3′ ends and, unlike RNA polymerase II mRNA transcripts, lacked 3′ polyadenylation. Further studies of the transcription of other known IESs indicated that this is a general characteristic of RNAs produced during this time point in conjugation in *T. thermophila*, which meant that the burst of transcription seen in the micronucleus produced the long, IES-specific dsRNA precursors required for scnRNA production.

4.5.1 Processing of Long dsRNAs into scnRNAs, and Their Subsequent Nuclear Localization

The long, IES-specific dsRNA transcripts are scnRNA precursors, which are processed by Dicer proteins [74, 117, 118]. An analysis of the sequence of the *T. thermophila* macronuclear genome indicated the presence of three putative Dicer proteins [117, 118], two of which were expressed throughout the *T. thermophila* life cycle, while the third Dicer protein, Dicer-like protein 1 (Dcl1p), was expressed exclusively during conjugation. Although the DCL proteins, such as DCL1 in *T. thermophila* and DCL2 and DCL3 in *P. tetraurelia*, lack the conserved RNA helicase domain, they have been shown to play an important role in epigenetic phenomena in other organisms besides ciliates, including *Arabidopsis thaliana* [132]. Knockouts of DCL1 caused a massive increase in these long, IES-specific dsRNA transcripts, yet at the same time they caused the abrogation of scnRNAs [33, 117, 118]; this verified that the long, IES-specific dsRNA transcripts produced early in conjugation are precursors for scnRNAs [117, 118]. Knockouts of DCL1 also failed to complete conjugation and, more importantly, failed to undergo DNA elimination similar to the DCL2/DCL3 double knockdown in *P. tetraurelia* [66, 117, 118]. The localization of Dcl1p, like Dcl2p in *P. tetraurelia*, showed that it was exclusively a micronuclear protein, which meant that the long dsRNAs produced in the micronucleus were processed into scnRNAs in the micronucleus itself, and not exported for cleavage.

Studies of the scnRNA structure itself showed that they were phosphorylated at the 5′ end, and also contained a 3′ hydroxyl group, which was consistent with cleavage by the ribonuclease III family member Dcl1p [33, 117, 118, 133–136]. Hybridization of these scnRNAs to micronuclear and macronuclear genomic DNA preparations from early to late in conjugation (2–10 h) showed a gradual increase in the ratio of scnRNAs hybridizing to micronuclear DNA when compared to macronuclear DNA, thus indicating the existence of a scnRNA sorting mechanism [33, 137]. At 2 h, the ratio of micronuclear DNA to macronuclear DNA binding was approximately threefold [137], but as conjugation proceeded this ratio gradually increased to a maximum of approximately 30-fold at 10 h [33]. Further analysis of some of these scnRNAs showed that they were homologous to the M and long terminal repeat (LTR) IES sequences, consistent with their production from long, IES-specific dsRNAs [75, 117].

Argonaute proteins have been shown to be essential effector proteins in sRNA pathways [138]. The same is true for *T. thermophila* as an Argonaute homolog, TWI1, was shown to bind scnRNAs [33]. A phylogenetic analysis of Twi1p indicated that it was homologous to the *Drosophila melanogaster* Piwi protein, and belonged to the Piwi subfamily of Argonaute proteins. TWI1 was predicted to contain functional PAZ and PIWI domains, which facilitate nucleic acid binding and "Slicer" or ribonuclease activity, respectively. The immunoprecipitation of Twi1p shortly after the production of scnRNAs at 5 h into conjugation demonstrated Twi1p/scnRNAs interaction [137]. The localization of Twi1p showed that the protein was predominantly macronuclear with some cytoplasmic localization, but was excluded completely from the crescent micronucleus; this indicated that the scnRNAs would have to undergo active or passive transport into the cytoplasm to interact with Twi1p [33]. Mutation of the DDH motif in the PIWI domain of TWI1 abolishes ribonuclease activity in Twi1p, and prevents removal of the passenger strand in Twi1p/scnRNA complexes found in the cytoplasm [139]. Mutation of the DDH motif also blocks the import of the Twi1p/scnRNA complexes into the parental macronucleus, which leads to scnRNA instability and degradation over a similar time course when compared to TWI1 knockouts [137, 139].

Like Argonaute proteins in other organisms, Twi1p does not act alone during RNAi-dependent DNA elimination in *T. thermophila*. In order for import into the parental macronucleus of the Twi1p/scnRNA complexes to occur, Twi1p must also interact with an accessory protein called *Giw1p* [139]. Although GIW1 shows no homology to any known domains of any gene, Giwi1p coimmunoprecipitates with full-length Twi1p, interacting with the PAZ and PIWI domains of Twi1p along several discrete protein sequences. Mutation of the DDH motif in Twi1p, which blocks cleavage of the double-stranded scnRNA and also prevents binding of Giw1p to Twi1p, ensures Twi1p/scnRNA complex activation prior to parental macronuclear import. Localization of Giw1p is seen generally in the parental macronucleus and the cytoplasm early in conjugation, where it is capable of participation with Twi1p/scnRNA complexes before importing them into the parental macronucleus. Giw1p also localizes to the developing zygotic macronucleus later in conjugation, although its function there at that time is not known. Knockouts of GIW1 cause failure of Twi1p/scnRNA complex import into the parental macronucleus, but do not affect scnRNA cleavage or unwinding of the scnRNA passenger strand which, along with the Twi1p/scnRNA complex binding data, indicates the activation of Twi1p/scnRNA complexes before Giw1p-dependent import. Like the DCL1 knockout, knockouts of TWI1 and GIW1, as well as the TWI1 PIWI domain mutation, fail to complete conjugation and block the DNA elimination of IESs [33, 137, 139].

4.5.2 Genome Scanning via Comparison of scnRNA Complexes to the Parental Genome

Localization of the Twi1p/scnRNA complexes into the parental macronucleus sets the stage for one of the unique aspects of DNA elimination in *T. thermophila*. As noted above, there is an increase in hybridization levels of scnRNAs to micronuclear genomic DNA when compared to macronuclear genomic DNA as conjugation proceeds, indicating the presence of a sorting mechanism [33, 137]. The sorting process through which micronuclear-specific scnRNA enrichment occurs is referred to as *genome scanning* [33]; this is similar to the situation in *P. tetraurelia*, and involves comparing each Twi1p/scnRNA complex to ncRNA transcribed from the parental macronucleus. Those Twi1p/scnRNA complexes which bind to the parental macronuclear ncRNA are removed from the biologically active Twi1p/scnRNA complex pool through unknown means, although a handful of proteins have been identified that play a role in this genome-scanning process.

Emphasizing the connection of scnRNAs with piRNAs, a homolog of HEN1 (the protein which is known to stabilize piRNAs through methylation) has also been found to have the same role in *T. thermophila* with scnRNAs [140]. The homolog in *T. thermophila*, which is also called *HEN1*, is a RNA methyltransferase that adds a methyl group to the terminal 2′ hydroxyl group of scnRNAs and has homologs in *A. thaliana*, *D. melanogaster*, and *Mus musculus* [140–144]. Hen1p colocalizes with Twi1p in the parental macronucleus early in conjugation during meiosis of the micronucleus; indeed, *in vitro* experiments with recombinant Hen1p and Twi1p have shown that Hen1p also coimmunoprecipitates with Twi1p during this period of development [140]. Knocking out HEN1 causes a loss of 2′-O-methylation in scnRNAs, and decreases scnRNA stability in a similar fashion to the TWI1 knockout and TWI1 PIWI domain mutant [137, 139, 140]. However unlike the TWI1, GIW1, and DCL1 knockouts, knockouts of HEN1 do not show a complete failure of conjugation and blockage of DNA elimination [117, 118, 137, 139, 140]. HEN1 knockouts are able to produce only 3% of possible progeny, but are able to undergo a complete rearrangement of the IESs tested on 67.8% (38/56) of occasions [140]. It is possible that, since scnRNA destabilization is not as extreme as in a TWI1 knockout or PIWI domain mutant, the sheer number of scnRNAs remaining is able to facilitate DNA elimination of IESs and the completion of conjugation.

Several Argonaute proteins that associate with piRNAs in other organisms have also been found to associate with RNA helicases [126, 145–147] (for a review, see Ref. [148]). An RNA helicase in *T. thermophila*, Ema1p, interacts with Twi1p/scnRNA complexes and plays a pivotal role in genome scanning by facilitating the Twi1p/scnRNA/ncRNA interaction [75]. Ema1p colocalizes with Twi1p in the parental macronucleus early in conjugation and later in the developing zygotic macronucleus, where the proteins have also been found to interact through coimmunoprecipitation [33, 75]. Ema1p localization is unaffected in TWI1 or GIW1 knockouts, which indicates that it is imported into the parental macronucleus either by itself, or by the same group of proteins that imports Giw1p/Twi1p/scnRNA complexes [75, 139]. Knockouts of EMA1 logically do not inhibit scnRNA cleavage or import of Twi1p/scnRNA complexes into the parental macronucleus, since it is never seen to accumulate in the cytoplasm where these processes occur. However, chromatin-spreading experiments and RNA immunoprecipitation followed by RT-PCR, have shown that in EMA1 knockouts the Twi1p/scnRNA complexes are no longer able to interact with chromatin and ncRNA when compared to wild-type. This was especially significant since chromatin was thought to be the site of ncRNA production, and that Twi1p/scnRNA/ncRNA interaction was required for genome scanning. As conjugation proceeds, the EMA1 knockouts also displayed an increase in macronuclear-specific scnRNAs compared to wild-type matings. These data implied that Ema1p would facilitate genome scanning by coupling Twi1p/scnRNA complexes with the ncRNA produced in the parental macronucleus, and also through an unknown mechanism which negatively selected against those Twi1p/scnRNA complexes capable of binding successfully to the ncRNA. Finally, EMA1 knockouts failed to complete conjugation yet, curiously, only showed a failure of DNA rearrangement in a select set of IESs. This may point towards the existence of different classes of IESs in *T. thermophila* (as occurs in *P. tetraurelia*) that do not undergo this selection process [47, 75, 78].

Although relatively few proteins are known to play a role in the RNAi-dependent DNA elimination process, there exist a few situations in this process where homologs in one ciliate are found to play the same or similar role in another ciliate [12, 14, 33, 66, 85, 86, 117, 118, 149]. One of these sets of homologs is the GW repeat proteins Nowa1p and Nowa2p in *P. tetraurelia*, and Wag1p and CnjBp in *T. thermophila* [86, 149]. The GW repeat proteins have been found to interact with Argonaute family proteins in *A. thaliana*, *D. melanogaster*, and *Homo sapiens*, and to play a role in sRNA effector function [150–152]. Although, Nowa1p and Nowa2p appear to have RNA-binding capabilities, the function of their homologs, Wag1p and CnjBp, in *T. thermophila*, is unclear [86, 149]. Subsequent colocalization and coimmunoprecipitation experiments with Wag1p and CnjBp demonstrated a protein–protein interaction with Twi1p [75, 149]. CnjBp was also shown to localize to the crescent micronucleus during meiosis (unlike Twi1p and Wag1p), although its role there is currently unknown [149]. Double knockouts of WAG1 and CNJB caused the retention of macronuclear-specific scnRNAs compared to wild-type matings, as conjugation proceeded in a similar fashion to the EMA1 knockout [75, 149]. Unlike the EMA1 knockout, the double WAG1/CNJB knockout also showed a slight increase in the retention of micronuclear-specific scnRNAs. This may entail a more general function of these two GW repeat proteins in the genome-scanning process, for the Twi1p/scnRNA complexes that need to be sequestered in the parental macronucleus, and for those complexes that need eventually to be transported to the developing zygotic macronucleus [148]. Although double knockouts of WAG1/CNJB show an increased retention of scnRNAs, the Twi1p/scnRNA complexes are able to interact with ncRNA through Ema1p normally, indicating that their biological function lies downstream of the initial binding of Twi1p/scnRNA complexes with ncRNA. Like many of the proteins involved in RNAi-directed DNA elimination, the double knockouts of WAG1/CNJB failed to complete conjugation, but failed DNA elimination in a specific set of IESs only (much like EMA1 knockouts) [75, 149]. Curiously, this set of IESs was slightly different from those in EMA1 knockouts [149], and although GW repeat proteins have been shown to affect Argonaute function, the actual mechanism remains a mystery [149–152]. In fact, even among ciliates there is no clear mode of action for these GW repeat proteins [86, 149]. Nonetheless, as more information becomes available regarding the RNAi-dependent DNA elimination pathway in both *P. tetraurelia* and *T. thermophila*, it will be interesting to see whether Nowa1p and Nowa2p in *P. tetraurelia* function similarly to Wag1p and CnjBp in *T. thermophila*, through sorting Argonaute/scnRNA complexes. Likewise, the proof of RNA binding by Wag1p and CnjBp (which has already been demonstrated in Nowa1p and Nowa2p) could help to define a common mode of action for GW repeat proteins in ciliates, and possibly in other eukaryotes in general.

Long ncRNA has been shown to play a vital role in a variety of epigenetic phenomena, as noted above [74, 75, 87–93]. In both *P. tetraurelia* and *T. thermophila* there appear to be three sources of long ncRNA during sexual reproduction: the crescent micronucleus; the parental macronucleus; and the developing zygotic macronucleus [74, 75]. The ncRNA produced in the parental macronucleus is vital to the genome scanning process, and was initially detected in *T. thermophila* alongside the bidirectional transcribed long, IES-specific dsRNA scnRNA precursors, and the ncRNA produced in the developing macronucleus [74]. PCR-based assays devised to further examine ncRNA transcription during conjugation showed that the long, IES-specific dsRNA scnRNA precursor transcription peaked at 3 h, ncRNA transcription from the parental macronucleus necessary for genome scanning peaked at 6 h, and ncRNA transcription from the developing zygotic macronucleus for IES targeting peaked at 10 h [75]. Blocking the transcription of parental macronuclear ncRNA by treatment with actinomycin D during the peak hours of genome scanning (4–6 h into conjugation) caused a significant increase in the failure of IES excision and DNA elimination [74]. Besides using actinomycin D, it is also possible to block the excision of individual IESs by inserting the IES sequence into the parental macronucleus prior to conjugation, similar to the blockade of mcIES excision in *P. tetraurelia* [47, 78, 153, 154]. For example, in *T. thermophila*, an insertion of the M element IES into the parental macronucleus causes a massive increase in M element long dsRNAs, but with no change in the level of scnRNAs [153]. This indicates that the excess long dsRNAs were not being processed into scnRNA, but were most likely acting as ncRNAs in the parental macronucleus, thereby removing M element scnRNA/Twi1p complexes from the biological active pool of Twi1p/scnRNA complexes.

4.6 DNA Elimination of DNA Sequences from the Developing Somatic Nucleus

When initially discovered, the phenomenon of DNA elimination in ciliates appeared to be an aberration in the world of biology, that was focusing increasingly on genetic processes. However, the rise of epigenetics has facilitated a clearer view of how DNA elimination relates to other biological processes. Whilst the link between scnRNAs and piRNAs was discussed in Sect. 4.5, this is not the only biologically relevant link that DNA elimination in ciliates has to other organisms. Just as RNAi was shown to direct heterochromatin formation in *A. thaliana* and *S. pombe*, it was also shown that a correct heterochromatin formation in the developing zygotic macronucleus through H3K9 and H3K27 methylation would depend on the normal function of RNAi components in *T. thermophila* [112, 117, 118, 155–157]. Thus, DNA elimination depends on an establishment of heterochromatin to control the glut of repetitive elements in its genome [31, 155, 158]. As with other eukaryotes, the initial methylation of histones associated with repetitive elements precipitates heterochromatin formation and the compaction of these sequences. Typically, *T. thermophila* and other ciliates take the additional step of removing these heterochromatic sequences out of their somatic genome, in order to create a streamlined genome (not unlike many simple eukaryotes) to optimize their fitness. This streamlining process begins when the Twi1p/scnRNA complexes have been transported to the developing zygotic macronucleus to target the H3K9 and H3K27 methylation of IESs [31, 137, 158]. These methylated histones then act to recruit chromodomain and other accessory proteins, which ultimately promote IES excision and DNA elimination by the domesticated piggyBac transposase, Tpb2p [14, 159–163]. The link between RNAi and heterochromatin, IES-specific chromatin modifications, heterochromatin readers, and the nature of IESs and DNA elimination in *T. thermophila*, will be described in the following section, together with details of relevant experiments to determine each of these steps.

4.6.1 Targeting of scnRNA Complexes and Modification of Chromatin of DNA Sequences to be Eliminated

Like RNAi-directed heterochromatin formation in *A. thaliana* and *S. pombe*, RNAi-dependent DNA elimination in *T. thermophila* requires the production of ncRNA [74, 75, 112, 156, 164]. This ncRNA (which is created in the developing zygotic macronucleus) is necessary for targeting IESs, and interacts with the remaining Ema1p/Twi1p/scnRNA complexes, which are transported there once the developing macronucleus has moved to the anterior of the cell and has begun to enlarge [33, 75, 137, 139]. The Twi1p accessory proteins involved in genome scanning, Ema1p, Wag1p, and CnjBp, are also transported to the developing macronucleus, although it is unclear whether this occurs in a greater complex with Twi1p, or independently [75, 149]. The Ema1p/Twi1p/scnRNA/ncRNA complex interaction facilitates the binding of this complex with another group of proteins referred to as the *Ezl1p complex* (S.D. Taverna *et al.*, unpublished data) [75].

In the RNAi-directed heterochromatin formation pathways in *A. thaliana* and *S. pombe*, heterochromatin formation is directed by H3K9me2, which is catalyzed by the Su(var) 3-9 homologs, Kryptonite (KYP), and Clr4, respectively [165–167]. RNAi-dependent DNA elimination in *T. thermophila* is dependent instead on Ezl1p, an E(z) homolog, and other associated proteins (S.D. Taverna *et al.*, unpublished data) [158]. The Ezl1p complex, which consists of Ezl1p, Esc1p, Rnf1p, Rnf2p, and Nud1p, contains homologs from two protein complexes, PRC1 and PRC2, as found in higher eukaryotes. These complexes are known to play a fundamental role in the developmental regulation of heterochromatin through histone methylation and gene silencing in many organisms, which the Ezl1p complex has subsumed in *T. thermophila* (S.D. Taverna *et al.*, unpublished data) [158, 168–170]. Immunoprecipitations of Ezl1p, Nud1p and Rnf1p are able to pull-down Ema1p, thus demonstrating a protein–protein interaction between the Ema1p/Twi1p/scnRNA complex and the Ezl1p complex (S.D. Taverna *et al.*, unpublished data). Nud1p, Rnf1p, Rnf2, and Esc1p of the Ezl1p complex appear to have no catalytic function themselves, unlike other homologs found in PRC1 and PRC2 complexes, but instead act to enhance targeting of Ezl1p to IESs and Ezl1p methylase activity at the IESs (S.D. Taverna, unpublished data) [171–175] (see review in Ref. [176]). Ezl1p, which is the effector component of the Ezl1p complex, is an E(z) homolog and contains the SET domain, which is capable of trimethylation of H3K9 and H3K27 (S.D. Taverna *et al.*, unpublished data) [158, 177–180]. The coimmunoprecipitation of Ezl1p is able to pull-down the other members of the Ezl1p complex, Nud1p, Rnf1p, Rnf2, and Esc1p (S.D. Taverna *et al.*, unpublished data); reciprocal pulldowns using tagged-Nud1p and -Rnf1p are also able to immunoprecipitate Ezl1p. The colocalization of H3K9me3 and H3K27me3 with Rnf1p of the Ezl1p complex shows that it is capable of histone methylation during conjugation. A knockout of any of the Ezl1p complex components causes disassociation of the complex and loss of H3K9 methylation along with aberrant H3K27 methylation, which implicates the Ezl1p complex in both H3K9me3 and H3K27me3 during conjugation (S.D. Taverna *et al.*, unpublished data) [158]. Knockouts of the EZL1 complex also result in an increased accumulation of scnRNAs and ncRNAs produced in the developing macronucleus from the M IES, which indicates the existence of a feedback mechanism controlling both scnRNA and ncRNA production throughout the cell during conjugation (S.D. Taverna *et al.*, unpublished data). DCL1, TWI1, and EZL1 complex knockouts also form aberrant DNA elimination bodies, which contain a number of proteins including the chromodomain proteins, Pdd1p, and Pdd3p (S.D. Taverna *et al.*, unpublished data) [158]. Like other components of RNAi-directed DNA elimination, knockouts of the Ezl1p complex caused failure of DNA elimination (S.D. Taverna *et al.*, unpublished data). In the case of EZL1 knockouts, a failure to complete conjugation has also been observed [158].

Methylation of H3K9 and H3K27 by the Ezl1p complex is an integral part of the RNAi-dependent DNA elimination process [31, 155, 158]. Indeed, the inhibition of this methylation by the Ezl1p complex through knockout of any component of RNAi-directed DNA elimination upstream or mutation of histone 3 itself is sufficient to block binding of the chromodomain proteins, Pdd1p and Pdd3p, and its association with other proteins to form DNA elimination bodies necessary for DNA elimination [75, 117, 118, 155, 158]. Mutation of H3K9Q directly blocks the site from methylation, while mutations of H3S10E and H3S28E created an artificially phospho-switch, which naturally prevents methylation of the lysine directly downstream. All of these histone 3 mutations prevent Pdd1p and Pdd3p association with IESs [155, 158].

4.6.2 Protein Binding of Modified Chromatin, Protein Aggregate Formation, and DNA Elimination

The role of chromodomain proteins in RNAi-directed heterochromatin formation and heterochromatin formation in general in eukaryotes is well documented [111, 181–183]. Once H3K9me3 and H3K27me3 modification occurs on histones associated with IESs, the aforementioned chromodomain proteins, Pdd1p and Pdd3p, are able to bind the IES chromatin which, along with other associated proteins, condenses the approximately 6000 IES loci into a handful of cellular foci referred to as *DNA elimination bodies* [14, 31, 158, 159, 161–163, 184–186]. In these DNA elimination bodies a domesticated piggyBac transposase, Tbp2p, directs the endonucleolytic cleavage of IESs at the IES boundaries, excising the IES [14]. Although these double-strand breaks are thought to be repaired through one of the dsDNA break repair pathways, it is currently unknown which pathway is responsible for this repair in *T. thermophila.*

Chromodomain proteins are pivotal heterochromatin histone readers. Knockouts of chromodomain proteins cause derepression of heterochromatin [183, 187]; likewise, knockouts of PDD1 also see a decrease in heterochromatin formation [31, 158]. This implies that the establishment of H3K9me3 and H3K27me3, and the binding of the two chromodomain proteins (Pdd1p and Pdd3p) to these marks, are interconnected in DNA elimination body formation and DNA elimination (see Fig. 7a,b) [31, 158]. Pdd1p and Pdd3p, along with Pdd2p, were discovered by the isolation of proteins enriched in developing zygotic macronuclei late during conjugation, and were the first identified proteins shown to play a role in DNA elimination [159, 161, 184, 185]. Pdd1p contains two chromodomains, and is capable of binding either H3K9me3 or H3K27me3 peptides *in vitro*, and to colocalize with H3K9me3, H3K27me3-modified chromatin and IESs late in conjugation [31, 158, 159, 188].

Pdd1p may play multiple roles during development, as it has been shown to localize within crescent micronuclei early during meiosis, within parental macronuclei and developing zygotic macronuclei, as well as in a cytoplasmic body known as the *conjusome* [159, 184, 189, 190]. The biological roles of Pdd1p in the crescent micronucleus and parental macronucleus are unknown, although a loss of expression during the early developmental stages is sufficient to block DNA elimination, thereby indicating that such Pdd1p localization is biologically relevant [190]. The localization of Pdd1p in the conjusome is thought to reflect the conjusome's role as a distribution center for the parental and developing macronuclei, or as a staging ground for Pdd1p transition from the parental macronuclei into the developing zygotic macronuclei later in conjugation [189]. Other proteins that are known to localize to the developing zygotic macronucleus later in conjugation, such as Lia1p, Lia3p, and Lia5p, also appear in the conjusome [162, 163]. In order to signal a transition from the parental macronucleus to the conjusome and the developing zygotic macronucleus, Pdd1p is phosphorylated up to four times [159]; this phosphorylation is lost as the conjugation proceeds, however, which may trigger DNA elimination body formation. The colocalization of Pdd1p with H3K9me3, H3K27me3, and IESs occurs in the developing zygotic macronucleus [31, 158, 159, 188]. Initially, the localization of Pdd1p is diffuse throughout the entire nucleus, but as the developing zygotic macronucleus matures the Pdd1p is concentrated into approximately 10 foci of average size 1 μm, termed *DNA elimination bodies* [184]. These Pdd1p-containing DNA elimination

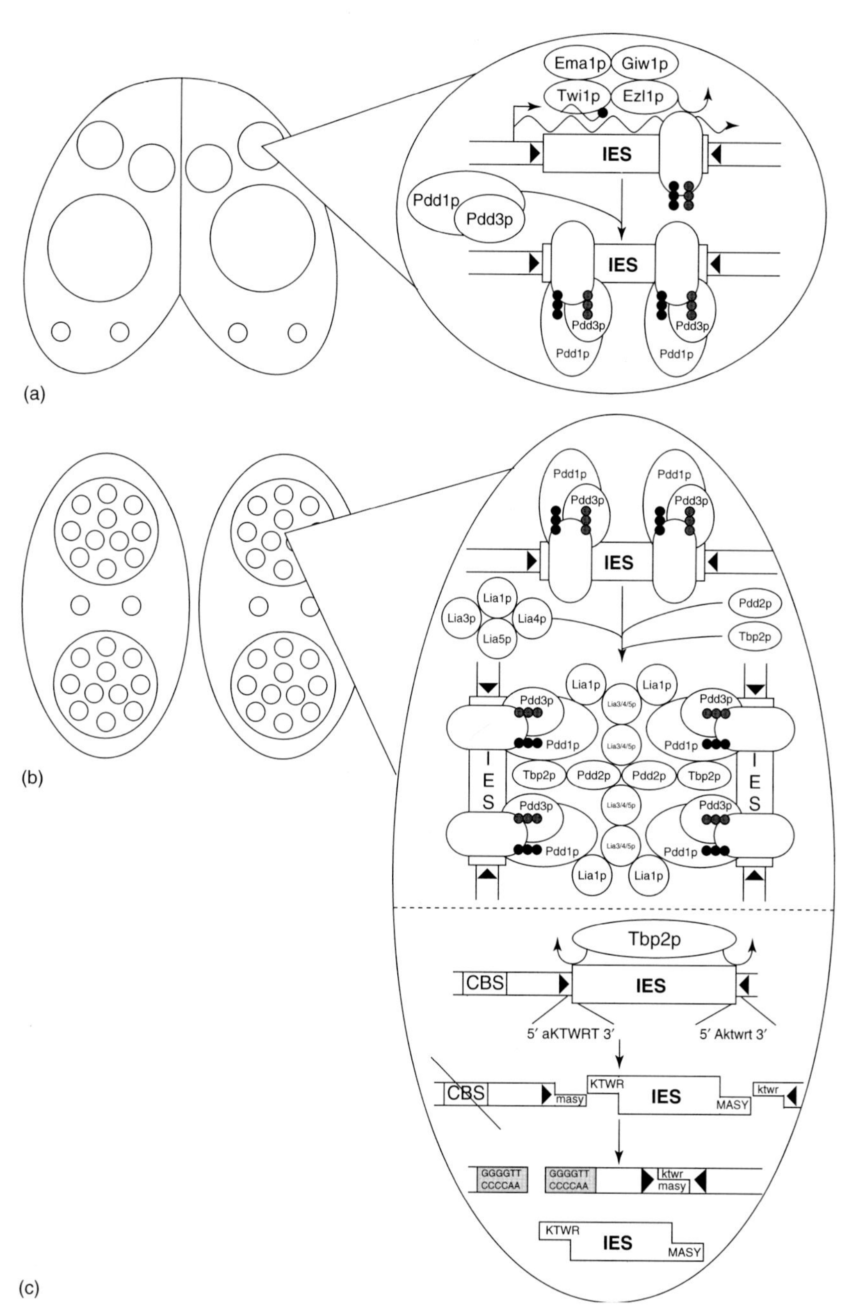

Fig. 7 RNAi-directed histone methylation on internal eliminated sequences (IESs) leads to their assembly into DNA elimination bodies and excision by a domesticated transposase, Tpb2p.

bodies also contain a number of other proteins including Pdd2p, Pdd3p, Lia1p, Lia3p, Lia4p, Lia5p, Tbp2p, and the Ezl1p complex (see Fig. 7b) (S.D. Taverna *et al.*, unpublished data) [14, 158, 161–163, 188]. Double knockouts of WAG1 and CNJB, LIA1 knockouts, EZL1 complex knockouts, PDD1 somatic knockouts, and PDD2 somatic knockouts each disrupt DNA elimination body formation (S.D. Taverna *et al.*, unpublished data) [149, 190, 191]. The tethering of Pdd1 to an artificial IES with no native histone methylation is also sufficient to direct DNA elimination, indicating that Pdd1p itself is sufficient to recruit its accessory proteins such as Tbp2p and to trigger DNA elimination [31].

The third chromodomain protein, Pdd3p, has been shown to bind strongly to H3K9me3, but not to H3K27me3, *in vitro* [31, 158]. Pdd3p localization is limited to the developing zygotic macronucleus where, like Pdd1p, it is initially diffuse but later condenses into the DNA elimination bodies [161]. The second programmed DNA degradation protein, Pdd2p, has no known homology [185] but demonstrates a localization that differs slightly from that of Pdd1p, by localizing only to the parental and developing macronuclei [185, 191]. Like Pdd1p and Pdd3p, the localization of Pdd2p in the developing zygotic macronucleus is initially diffuse until DNA elimination body formation. In a similar manner to Pdd1p, Pdd2p is phosphorylated once during transition from the parental macronucleus to the developing zygotic macronucleus [188]; again, this phosphorylation is removed immediately prior to DNA elimination body formation. PDD2 somatic knockouts are sufficient to cause the failure of cells to undergo DNA elimination and to complete conjugation which, like PDD1 somatic knockouts, may indicate a vital role for early localization in the parental macronucleus [191].

Other proteins have been found to influence DNA elimination body formation. For example, a diverse group of proteins that participated in this process were identified by their localization specifically to differentiating macronuclei, and thus were named localization in macronuclear anlagen (Lia) proteins [162, 163]. Lia1p, Lia4p, and Lia5p each play a role in DNA elimination body formation; typically, Lia5p contains a plant homeodomain (PHD) Zn Finger, while Lia4p contains a putative chromo shadow domain; otherwise, these proteins show no obvious homology to other known proteins. Of the Lia proteins, Lia1p is the best characterized, and localizes to both the conjusome and developing zygotic macronucleus [162]. Late in conjugation Lia1p is found in association with Pdd1p and IESs in DNA elimination bodies. Knockouts of LIA1 fail to eliminate IESs and complete conjugation, much like many other proteins in RNAi-directed DNA elimination. The preliminary characterization of Lia3p, Lia4p, and Lia5p has shown a diffuse localization early in the developing zygotic macronucleus, and later localization in DNA elimination bodies [163]. LIA3, LIA4, and LIA5 knockouts also fail to undergo DNA elimination and complete conjugation (A.W.-Y. Shieh *et al.*, unpublished data). While the role of these non-chromodomain proteins in RNAi-directed DNA elimination is not clear, it is possible that these proteins form a scaffold through which Pdd1p and Pdd3p, by interacting with specific classes of IESs, can be brought together to form the foci necessary for DNA elimination by the domesticated piggyBac transposase, Tbp2p.

Domesticated transposases have been shown to play an important role in a variety of eukaryotic organisms, for example, RAG1/RAG2 recombinase in VDJ (variable, diverse, and joining) recombination in the human immune system [192, 193]. Ciliates appear to have domesticated transposases in order to facilitate the removal of repetitive sequences and IESs during conjugation [12–14]. In *T. thermophila*, Tbp2p – a piggyBac transposase homolog – is essential for removing IESs during conjugation (see Fig. 7c) [14]. An analysis of the TBP2 ORF shows homology with, and preservation of, the catalytic DDD motif in the domesticated piggyBac transposase in *P. tetraurelia*, PGM, and other piggyBac transposases in *H. sapiens*, *Xenopus* spp., and the moth, *Trichoplusia ni*. Tbp2p colocalizes with H3K9m3, H3K27me3, and Pdd1p in the developing zygotic macronucleus, before and after DNA elimination body formation. The knockdown of TBP2 using RNA hairpins does not inhibit Pdd1p association with H3K9me3 and H3K27me3 [14, 70]; however, TBP2 knockdown does inhibit DNA elimination body formation, IES removal, and completion of conjugation, thus implying an essential function downstream of Pdd1p and Pdd3p binding [14]. An *in vitro* analysis of the catalytic DDD motif of Tbp2p has shown that it is capable of cutting the consensus piggyBac cleavage sequence, 5′-TTAA-3′, as well as a variety of divergent sequences (see Fig. 7c) [14, 194, 195]. As noted above, Tbp2p cleavage produces a 4 bp 5′ overhang, which is not observed in mutants of the Tbp2p DDD catalytic motif [14, 195].

4.7 Chromosome Breakage in the Developing Somatic Nucleus

The epigenetic RNAi-directed DNA elimination process in *T. thermophila* is only a part of the global genome rearrangement that occurs in the developing zygotic macronucleus during conjugation. Chromosome breakage and differential chromosome amplification must also take place for this process to be complete [105–110, 196–199] (for a review, see Ref. [5]). This epigenomic process differs between *P. tetraurelia* and *T. thermophila*; in the former species the process seems to depend on RNAi-dependent DNA elimination machinery, whereas in *T. thermophila* chromosome breakage during conjugation is prompted by a conserved DNA sequence called the chromosome breakage sequence (CBS) [12, 104, 200]. Chromosome breakage and differential chromosome amplification have been shown to be essential for completion of conjugation, and are linked to RNAi-directed DNA elimination [33, 117, 149, 158, 191]. The conserved 15 bp CBS sequence is sufficient and necessary for chromosome breakage and telomere addition, which is blocked in CBS mutants (see Fig. 7c) [200–203]. Genomic analysis of the *T. thermophila* genome has shown that, with little variation, the CBS is present at all sites of chromosome breakage [107, 108]. Like IES excision, chromosome breakage appears to be dependent on the piggyBac transposase, Tbp2p [14].

5 Chromosome Fragmentation and Elimination of DNA during Conjugation in *Oxytricha*

The studies of DNA elimination in *P. tetraurelia* and *T. thermophila*, as described above, have revealed the role of sRNAs and long ncRNAs in remodeling genomes during development. They have also hinted to the possible mechanisms that allow phenotypic traits to be propagated to the next generation. Whilst DNA elimination and chromosome fragmentation occur throughout the entire ciliate clade [5], it remains unclear whether RNAs play a similar role in more distantly related ciliates. However, recent investigations on these processes in a subgroup of ciliates known as *stichotrichs* has provided a definitive answer to this question [76]. Whilst the stichotrichs – which include the genera *Oxytricha* and *Stylonichia* – undergo DNA elimination and chromosome breakage, these processes are much more extreme and result in the elimination of more than 95% of the genome and of gene-sized mini-chromosomes of approximately 2 kb in size [204–207] (for a review, see Ref. [5]). A further complication in the understanding of these processes in *Oxytricha* and *Stylonichia* was the discovery of scrambled genes in the micronucleus [208–214]. Recent data acquired from *Oxytricha trifallax* have indicated that parental macronuclear ncRNA is able to direct the unscrambling of genes, DNA elimination, and chromosome breakage [76].

5.1 Gene Unscrambling and Domesticated Transposases in DNA Elimination and Chromosome Breakage

As in other ciliates, it seems likely that in stichotrichs DNA elimination – and, by extension, gene scrambling – in the micronucleus represent ways to prevent active transposons from appearing in the somatic macronuclear genome [5]. Yet, by scrambling the macronuclear-destined sequences (MDSs) of genes in the germline micronucleus, the stichotrichs ensure that DNA elimination must occur during sexual reproduction, in order to generate intact coding regions if progeny are to be viable. In this case, gene scrambling takes several forms, with some MDSs having undergone permutation in linear order, while others are even inverted with respect to the other MDSs to complicate the unscrambling process further (see Fig. 8b) [208–214].

To date, the scrambled genes discovered have included actin I, α telomere-binding protein (αTBP) and DNA polymerase α, with many more likely waiting to be discovered. Similar to *P. tetraurelia*, the MDSs of *O. trifallax* are bordered by short repeats (termed *pointers*) that may help direct gene unscrambling and DNA elimination [215] although, unfortunately, these repeats are too short to unambiguously accomplish this task. The discovery of parental macronuclear ncRNA during conjugation, and its role in gene unscrambling and DNA elimination, illuminates how these processes occur in *O. trifallax* and possibly in stichotrichs in general [76]. Subsequent RT-PCR analyses of RNA isolated from conjugating *O. trifallax* early and late in conjugation detected the presence of both sense and anti-sense ncRNAs. These ncRNAs, which are longer than mRNAs and contain telomeres, imply that the general transcription of all mini-chromosomes is initiated at the telomere sequence early during conjugation. RNAi against these ncRNAs during conjugation was sufficient to block the rearrangement of the target genes in the developing macronucleus. In order to validate the role of the parental macronucleus in producing these ncRNAs, Landweber and coworkers injected (into either the macronucleus or the cytoplasm) artificial DNA and RNA transcripts to a known gene (telomere-end-binding protein-β; TEBPβ), which contained different permutations of the MDSs. Upon the completion of conjugation, some TEBPβ genes containing

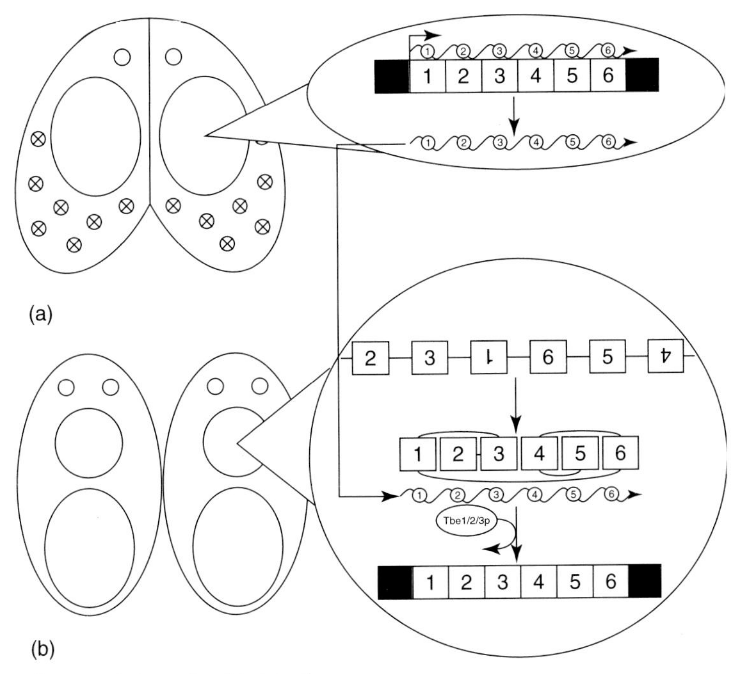

Fig. 8 Unscrambling of genes that are jumbled in the germ line genome of *O. trifallax* is guided by maternally produced template RNAs.

the alternative MDS order were found in the developing macronucleus, thus verifying the ability of artificial DNA in the parental macronucleus to produce ncRNA transcripts and to alter DNA elimination in the developing macronucleus.

Similar to *P. tetraurelia* and *T. thermophila*, a family of domesticated transposases has been found to play a role in gene unscrambling and DNA elimination in *O. trifallax* [12–14]. In this case, the transposases, termed telomere-bearing element 1 (TBE1), TBE2, and TBE3, belong to the TBE family of transposons and are not retained in the macronucleus after DNA elimination and chromosome breakage [13, 216, 217]. The triple knockdown of these transposases is sufficient to cause aberrant gene unscrambling and DNA elimination [13].

Taken together, these data have led to the proposal of a model (see Fig. 8) for gene unscrambling, DNA elimination, and chromosome breakage in *O. trifallax* [76]. At an early stage in conjugation, the bidirectional transcription of all mini-chromosomes in the parental macronucleus produces ncRNA. Following its appearance, the latter is transported to the developing macronucleus later during conjugation, where it directs gene unscrambling (if necessary) and the DNA elimination of IESs via a family of domesticated transposases (TBE1, TBE2, and TBE3) to produce a functional minichromosome in the developing macronucleus [13, 76]. Although the presence of sRNAs, a Piwi homolog, and heterochromatin marks have each been found in the stichotrich, *Stylonichia*, it remains to be seen whether any of these play a role in gene unscrambling and DNA elimination in *Oxytricha*, *Stylonichia*, and in other stichotrichs, similar to *P. tetraurelia* and *T. thermophila* [33, 66, 85, 155, 158, 218]. It should be noted here that one point remains consistent in DNA elimination in all ciliates, namely the presence of ncRNA.

6 Perspective

Ciliates have long held the fascination of scientists, as they were among some of the first microorganisms to be studied in detail. Indeed, it was while developing *Paramecium* as a genetic model that Sonneborn first realized that many traits did not follow simple Mendelian rules of inheritance, and instead proposed that the cytoplasm might play a role in regulating the development of stable phenotypes. Although, today, molecular explanations for many of Sonneborn's observations have still not been provided, ciliates have nevertheless emerged as an important study system when investigating epigenetic mechanisms. Notably, their nuclear dimorphism has provided an informative biological context within which to uncover the mechanisms responsible for the differential regulation of homologous sequences. Ultimately, many of the mechanisms identified were shown to be common regulatory schemes used widely among eukaryotes. As an example, studies conducted in *Tetrahymena* provided the key data to show that transcriptional regulators acted by modifying chromatin [28, 29].

More recently, studies with ciliates have helped to reveal important roles for both long and short ncRNAs in mediating epigenetic regulation [33, 65, 66, 69, 76, 117, 118]. The majority of these new insights have resulted from studies aimed at elucidating the mechanisms that these organisms employ to remodel their somatic genomes during nuclear differentiation. An important paradigm that is now emerging from investigations of somatic nuclear differentiation of *Paramecium* and *Tetrahymena*, is that DNA rearrangement provides a means of genome surveillance, serving to remove the repetitive DNA from the transcriptionally active somatic nucleus, so that any potentially deleterious elements (e.g., transposons) which are silent in the germline cannot be spread. The ciliates identify this "junk" DNA by making an RNA copy of their germline genome during meiosis, thus processing bidirectional transcripts into an abundant class of sRNAs (scnRNAs) that can be used as the specificity factors to recognize germline-limited sequences [33, 66, 69, 74, 117, 118]. DNA rearrangement can be considered an innovative endpoint in the ciliate version of the piRNA pathway. In metazoans, the piRNA pathway serves to protect the germline from transposable elements via RNAi-directed silencing [120, 123, 125, 126, 219, 220] (see also Refs [63, 221]). In ciliates, the silencing of these sequences is permanent in the somatic genome, as they are eliminated during differentiation. It is clear, therefore, that a piRNA-mediated genome defense can serve as an evolutionary ancient mechanism.

The mechanistic connection between epigenetic silencing and DNA elimination is quite direct, as evidenced in *Tetrahymena*, where the germline-derived scnRNAs guide DNA rearrangements by directing heterochromatic modifications to the IESs. As noted in Sect. 4.3, both histone H3K9 and H3K27 methylation are established on IES chromatin at the start of differentiation of the somatic macronucleus [31, 158], a discovery that was made immediately after RNAi was found to direct heterochromatin modification to silent genomic domains in *S. pombe* [112]. Taken together, the results of investigations in these unicellular models confirmed that RNAi-directed heterochromatin formation could provide a common means of instituting transcriptional gene silencing at homologous loci. Although the exact details of how sRNAs can direct chromatin modifications to specific sequences remain rather unclear, future studies in *Tetrahymena* and in other model systems will surely provide more detailed insights into these fundamental mechanisms.

What has become increasingly apparent is that ciliates have ways to communicate homologous sequence information between the germline and somatic genomes, from one generation to the next. As first revealed in studies of d48 *Paramecium* strains, the simple absence or presence of a DNA sequence in the parental somatic nucleus can "template" the same genome structure after DNA rearrangement of the new copy in the zygotic somatic genome [51, 49]. Evidence acquired from both *Paramecium* and *Tetrahymena* has indicated that this comparison of genome content is mediated by an interaction between scnRNAs and longer ncRNAs (see Figs 5 and 6), produced from the different nuclei [69, 75]. The syntheses and sites of action of these different ncRNAs exhibit both temporal and spatial separations, which allows the ncRNAs created in the parental somatic nucleus to block the action of scnRNAs, whereas those in the developing zygotic macronucleus will help to guide DNA elimination by interacting with the remaining scnRNA pool. Indeed, it is quite likely that these RNA-mediated genome comparisons that occur during development are responsible for some of the enigmatic examples of non-Mendelian inheritance, as originally described by Sonneborn.

The control of gene unscrambling in *Oxytricha*, via ncRNAs produced from the parental somatic genome, is perhaps the most intriguing phenomenon yet discovered [76]. As illustrated in Fig. 8, these ncRNAs are proposed to interact directly at the scrambled loci derived from the germline genome, and to guide the correct ordering of the mixed-up and inverted gene segments to ensure the assembly of a functional ORF. Whilst it is rather remarkable to consider that RNA could dramatically restructure the DNA of an organism, recent data acquired from this group of ciliates has further revealed that the copy number of the putative ncRNA templates can epigenetically regulate the copy number of the homologous chromosomes in the next generation [3, 4]. While the detailed mechanisms underlying these phenomena remain to be elucidated, these observations reveal nonetheless that homologous RNAs have a much-underappreciated capacity to influence gene expression and genome organization. Today, with much biology still awaiting illumination, the ciliated protozoa are clearly an important group of eukaryotes that are capable of revealing surprising modes of epigenetic regulation.

References

1 Sonneborn, T.M. (1937) Sex, sex inheritance and sex determination in *Paramecium aurelia*. *Proc. Natl Acad. Sci. USA*, **23** (7), 378–385.

2 Philippe, H., Germot, A., Moreira, D. (2000) The new phylogeny of eukaryotes. *Curr. Opin. Genet. Dev.*, **10** (6), 596–601.

3 Nowacki, M., Haye, J.E., Fang, W., Vijayan, V., Landweber, L.F. (2010) RNA-mediated

epigenetic regulation of DNA copy number. *Proc. Natl Acad. Sci. USA*, **107** (51), 22140–22144.
4 Heyse, G., Jonsson, F., Chang, W.J., Lipps, H.J. (2010) RNA-dependent control of gene amplification. *Proc. Natl Acad. Sci. USA*, **107** (51), 22134–22139.
5 Prescott, D.M. (1994) The DNA of ciliated protozoa. *Microbiol. Rev.*, **58** (2), 233–267.
6 Betermier, M. (2004) Large-scale genome remodelling by the developmentally programmed elimination of germ line sequences in the ciliate *Paramecium*. *Res. Microbiol.*, **155** (5), 399–408.
7 Yao, M.C., Gorovsky, M.A. (1974) Comparison of the sequences of macro- and micronuclear DNA of *Tetrahymena pyriformis*. *Chromosoma*, **48** (1), 1–18.
8 Klobutcher, L.A., Herrick, G. (1995) Consensus inverted terminal repeat sequence of *Paramecium* IESs: resemblance to termini of Tc1-related and Euplotes Tec transposons. *Nucleic Acids Res.*, **23** (11), 2006–2013.
9 Herrick, G., Cartinhour, S., Dawson, D., Ang, D., Sheets, R., Lee, A., Williams, K. (1985) Mobile elements bounded by C4A4 telomeric repeats in *Oxytricha fallax*. *Cell*, **43**, 759–768.
10 Ribas-Aparicio, R.M., Sparkowski, J.J., Proulx, A.E., Mitchell, J.D., Klobutcher, L.A. (1987) Nucleic acid splicing events occur frequently during macronuclear development in the protozoan *Oxytricha nova* and involve the elimination of unique DNA. *Genes Dev.*, **1**, 323–336.
11 Jahn, C.L., Klobutcher, L.A. (2002) Genome remodeling in ciliated protozoa. *Annu. Rev. Microbiol.*, **56**, 489–520.
12 Baudry, C., Malinsky, S., Restituito, M., Kapusta, A., Rosa, S., Meyer, E., Betermier, M. (2009) PiggyMac, a domesticated piggyBac transposase involved in programmed genome rearrangements in the ciliate *Paramecium tetraurelia*. *Genes Dev.*, **23** (21), 2478–2483.
13 Nowacki, M., Higgins, B.P., Maquilan, G.M., Swart, E.C., Doak, T.G., Landweber, L.F. (2009) A functional role for transposases in a large eukaryotic genome. *Science*, **324** (5929), 935–938.
14 Cheng, C.Y., Vogt, A., Mochizuki, K., Yao, M.C. (2010) A domesticated piggyBac transposase plays key roles in heterochromatin dynamics and DNA cleavage during programmed DNA deletion in *Tetrahymena thermophila*. *Mol. Biol. Cell*, **21** (10), 1753–1762.
15 Hannon, G.J. (2002) RNA interference. *Nature*, **418** (6894), 244–251.
16 Allis, C.D., Glover, C.V., Bowen, J.K., Gorovsky, M.A. (1980) Histone variants specific to the transcriptionally active, amitotically dividing macronucleus of the unicellular eucaryote, *Tetrahymena thermophila*. *Cell*, **20** (3), 609–617.
17 Hayashi, T., Hayashi, H., Fusauchi, Y., Iwai, K. (1984) *Tetrahymena* histone H3. Purification and two variant sequences. *J. Biochem. (Tokyo)*, **95** (6), 1741–1749.
18 Liu, X., Li, B., Gorovsky, M.A. (1996) Essential and nonessential histone H2A variants in *Tetrahymena thermophila*. *Mol. Cell. Biol.*, **16** (8), 4305–4311.
19 Sugai, T., Hiwatashi, K. (1974) Cytological and autoradiographic studies of the micronucleus at meiotic prophase in *Tetrahymena pyriformis*. *J. Protozool.*, **21**, 542–548.
20 Martindale, D.W., Allis, C.D., Bruns, P.J. (1985) RNA and protein synthesis during meiotic prophase in *Tetrahymena thermophila*. *J. Protozool.*, **32** (4), 644–649.
21 Stargell, L.A., Bowen, J., Dadd, C.A., Dedon, P.C., Davis, M., Cook, R.G., Allis, C.D., Gorovsky, M.A. (1993) Temporal and spatial association of histone H2A variant hv1 with transcriptionally competent chromatin during nuclear development in *Tetrahymena thermophila*. *Genes Dev.*, **7**, 2641–2651.
22 Yu, L., Gorovsky, M.A. (1997) Constitutive expression, not a particular primary sequence, is the important feature of the H3 replacement variant hv2 in *Tetrahymena thermophila*. *Mol. Cell. Biol.*, **17** (11), 6303–6310.
23 Shen, X., Yu, L., Weir, J.W., Gorovsky, M.A. (1995) Linker histones are not essential and affect chromatin condensation in vivo. *Cell*, **82** (1), 47–56.
24 Shen, X., Gorovsky, M.A. (1996) Linker histone H1 regulates specific gene expression but not global transcription in vivo. *Cell*, **86** (3), 475–483.

25 Allfrey, V.G., Faulkner, R., Mirsky, A.E. (1964) Acetylation and methylation of histones and their possible role in the regulation of RNA synthesis. *Proc. Natl Acad. Sci. USA*, **51**, 786–794.

26 Ren, Q., Gorovsky, M.A. (2001) Histone H2A.Z acetylation modulates an essential charge patch. *Mol. Cell*, **7** (6), 1329–1335.

27 Ren, Q., Gorovsky, M.A. (2003) The nonessential H2A N-terminal tail can function as an essential charge patch on the H2A.Z variant N-terminal tail. *Mol. Cell. Biol.*, **23** (8), 2778–2789.

28 Brownell, J.E., Allis, C.D. (1995) An activity gel assay detects a single, catalytically active histone acetyltransferase subunit in *Tetrahymena* macronuclei. *Proc. Natl Acad. Sci. USA*, **92** (14), 6364–6368.

29 Brownell, J.E., Zhou, J., Ranalli, T., Kobayashi, R., Edmondson, D.G., Roth, S.Y., Allis, C.D. (1996) Tetrahymena histone acetyltransferase A: a homolog to yeast Gcn5p linking histone acetylation to gene activation. *Cell*, **84** (6), 843–851.

30 Strahl, B.D., Ohba, R., Cook, R.G., Allis, C.D. (1999) Methylation of histone H3 at lysine 4 is highly conserved and correlates with transcriptionally active nuclei in *Tetrahymena*. *Proc. Natl Acad. Sci. USA*, **96** (26), 14967–14972.

31 Taverna, S.D., Coyne, R.S., Allis, C.D. (2002) Methylation of histone h3 at lysine 9 targets programmed DNA elimination in *Tetrahymena*. *Cell*, **110** (6), 701–711.

32 Nakayama, J., Rice, J.C., Strahl, B.D., Allis, C.D., Grewal, S.I. (2001) Role of histone H3 lysine 9 methylation in epigenetic control of heterochromatin assembly. *Science*, **292** (5514), 110–113.

33 Mochizuki, K., Fine, N.A., Fujisawa, T., Gorovsky, M.A. (2002) Analysis of a piwi-related gene implicates small RNAs in genome rearrangement in *Tetrahymena*. *Cell*, **110** (6), 689–699.

34 Wei, Y., Mizzen, C.A., Cook, R.G., Gorovsky, M.A., Allis, C.D. (1998) Phosphorylation of histone H3 at serine 10 is correlated with chromosome condensation during mitosis and meiosis in *Tetrahymena*. *Proc. Natl Acad. Sci. USA*, **95** (13), 7480–7484.

35 Wei, Y., Yu, L., Bowen, J., Gorovsky, M.A., Allis, C.D. (1999) Phosphorylation of histone H3 is required for proper chromosome condensation and segregation. *Cell*, **97** (1), 99–109.

36 Sonneborn, T.M. (1963) Does Pre-Formed Cell Structure Play an Essential Role in Cell Heredity? in: Allen, J.M. (Ed.) *The Nature of Biological Diversity*, McGraw-Hill, New York, pp. 165–221.

37 Beisson, J., Sonneborn, T.M. (1965) Cytoplasmic inheritance of the organization of the cell cortex in *Paramecium aurelia*. *Proc. Natl Acad. Sci. USA*, **53**, 275–282.

38 Sonneborn, T.M. (1977) Genetics of cellular differentiation: stable nuclear differentiation in eucaryotic unicells. *Annu. Rev. Genet.*, **11**, 349–367.

39 Sonneborn, T.M. (1974) Paramecium aurelia, in: King, R.C. (Ed.) *Handbook of Genetics: Plants, Plant Viruses and Protists*, Plenum Press, New York, pp. 469–594.

40 Sonneborn, T.M. (1947) Recent advances in the genetics of *Paramecium* and *Euplotes*. *Adv. Genet.*, **1**, 263–358.

41 Nanney, D.L. (1957) Mating-type inheritance at conjugation in variety 4 of *Paramecium aurelia*. *J. Protozool.*, **4**, 89–95.

42 Butzel, H.M. (1973) Abnormalities in nuclear behavior and mating type determination in cytoplasmically bridged exconjugants of doublet *Paramecium aurelia*. *J. Eukaryot. Microbiol.*, **20** (1), 140–142.

43 Koizumi, S. (1971) The cytoplasmic factor that fixes macronuclear mating type determination in *Paramecium aurelia*, syngen 4. *Genetics*, **68** (Suppl.), 34.

44 Sonneborn, T.M. (1948) The determination of hereditary antigenic differences in genically identical *Paramecium* cells. *Proc. Natl Acad. Sci. USA*, **34** (8), 413–418.

45 Brygoo, Y., Keller, A.-M. (1981) Genetic analysis of mating type differentiation in *Paramecium tetraurelia*. III. A mutation restricted to mating type E and affecting the determination of mating type. *Dev. Genet.*, **2**, 13–22.

46 Meyer, E., Keller, A.-M. (1996) A Mendelian mutation affecting mating-type determination also affects developmental genomic rearrangements in *Paramecium tetraurelia*. *Genetics*, **143**, 191–202.

47 Duharcourt, S., Butler, A., Meyer, E. (1995) Epigenetic self-regulation of developmental excision of an internal eliminated sequence in *Paramecium tetraurelia*. *Genes Dev.*, **9**, 2065–2077.

48 Epstein, L.M., Forney, J.D. (1984) Mendelian and non-Mendelian mutations affecting surface antigen expression in *Paramecium tetraurelia*. *Mol. Cell. Biol.*, **4**, 1583–1590.
49 Preer, L.B., Hamilton, G., Preer, J.R. (1992) Micronuclear DNA from *Paramecium tetraurelia*: serotype 51A gene has internally eliminated sequences. *J. Protozool.*, **39**, 678–682.
50 Godiska, R., Aufderheide, K.J., Gilley, D., Hendrie, P., Fitzwater, T., Preer, L.B., Polisky, B., Preer, J.R.J. (1987) Transformation of *Paramecium* by microinjection of a cloned serotype gene. *Proc. Natl Acad. Sci. USA*, **84** (21), 7590–7594.
51 Koizumi, S., Kobayashi, S. (1989) Microinjection of plasmid DNA encoding the A surface antigen of *Paramecium tetraurelia* restores the ability to regenerate a wild-type macronucleus. *Mol. Cell. Biol.*, **9** (10), 4398–4401.
52 You, Y., Aufderheide, K., Morand, J., Rodkey, K., Forney, J. (1991) Macronuclear transformation with specific DNA fragments controls the content of the new macronuclear genome in *Paramecium tetraurelia*. *Mol. Cell. Biol.*, **11** (2), 1133–1137.
53 Jessop-Murray, H., Martin, L.D., Gilley, D., Preer, J.R., Jr, Polisky, B. (1991) Permanent rescue of a non-Mendelian mutation of *Paramecium* by microinjection of specific DNA sequences. *Genetics*, **129** (3), 727–734.
54 Scott, J.M., Leeck, C.L., Forney, J.D. (1994) Analysis of the micronuclear B type surface protein gene in *Paramecium tetraurelia*. *Nucleic Acids Res.*, **22** (23), 5079–5084.
55 Scott, J.M., Mikami, K., Leeck, C.L., Forney, J.D. (1994) Non-Mendelian inheritance of macronuclear mutations is gene specific in *Paramecium tetraurelia*. *Mol. Cell. Biol.*, **14** (4), 2479–2484.
56 Kim, C.S., Preer, J.R. Jr, Polisky, B. (1994) Identification of DNA segments capable of rescuing a non-Mendelian mutant in *Paramecium*. *Genetics*, **136** (4), 1325–1328.
57 Forney, J.D., Yantiri, F., Mikami, K. (1996) Developmentally controlled rearrangement of surface protein genes in *Paramecium tetraurelia*. *J. Eukaryot. Microbiol.*, **43** (6), 462–467.
58 Napoli, C., Lemieux, C., Jorgensen, R. (1990) Introduction of a chimeric chalcone synthase gene into *Petunia* results in reversible co-suppression of homologous genes in trans. *Plant Cell*, **2** (4), 279–289.
59 Romano, N., Macino, G. (1992) Quelling: transient inactivation of gene expression in *Neurospora crassa* by transformation with homologous sequences. *Mol. Microbiol.*, **6** (22), 3343–3353.
60 Ruiz, F., Vayssié, L., Klotz, C., Sperling, L., Madeddu, L. (1998) Homology-dependent gene silencing in *Paramecium*. *Mol. Biol. Cell*, **9** (4), 931–943.
61 Galvani, A., Sperling, L. (2001) Transgene-mediated post-transcriptional gene silencing is inhibited by 3' non-coding sequences in *Paramecium*. *Nucleic Acids Res.*, **29** (21), 4387–4394.
62 Lejeune, E., Allshire, R.C. (2011) Common ground: small RNA programming and chromatin modifications. *Curr. Opin. Cell Biol.*, **23** (3), 258–265.
63 Malone, C.D., Hannon, G.J. (2009) Small RNAs as guardians of the genome. *Cell*, **136** (4), 656–668.
64 Carthew, R.W., Sontheimer, E.J. (2009) Origins and mechanisms of miRNAs and siRNAs. *Cell*, **136** (4), 642–655.
65 Lee, S.R., Collins, K. (2006) Two classes of endogenous small RNAs in *Tetrahymena thermophila*. *Genes Dev.*, **20** (1), 28–33.
66 Lepere, G., Nowacki, M., Serrano, V., Gout, J.F., Guglielmi, G., Duharcourt, S., Meyer, E. (2009) Silencing-associated and meiosis-specific small RNA pathways in *Paramecium tetraurelia*. *Nucleic Acids Res.*, **37** (3), 903–915.
67 Galvani, A., Sperling, L. (2002) RNA interference by feeding in *Paramecium*. *Trends Genet.*, **18** (1), 11–12.
68 Garnier, O., Serrano, V., Duharcourt, S., Meyer, E. (2004) RNA-mediated programming of developmental genome rearrangements in *Paramecium tetraurelia*. *Mol. Cell. Biol.*, **24** (17), 7370–7379.
69 Lepere, G., Betermier, M., Meyer, E., Duharcourt, S. (2008) Maternal noncoding transcripts antagonize the targeting of DNA elimination by scanRNAs in *Paramecium tetraurelia*. *Genes Dev.*, **22** (11), 1501–1512.
70 Howard-Till, R.A., Yao, M.C. (2006) Induction of gene silencing by hairpin RNA

expression in *Tetrahymena thermophila* reveals a second small RNA pathway. *Mol. Cell. Biol.*, **26** (23), 8731–8742.

71 Hamilton, A.J., Baulcombe, D.C. (1999) A species of small antisense RNA in posttranscriptional gene silencing in plants. *Science*, **286** (5441), 950–952.

72 Lee, S.R., Collins, K. (2007) Physical and functional coupling of RNA-dependent RNA polymerase and Dicer in the biogenesis of endogenous siRNAs. *Nat. Struct. Mol. Biol.*, **14** (7), 604–610.

73 Marker, S., Le Mouel, A., Meyer, E., Simon, M. (2010) Distinct RNA-dependent RNA polymerases are required for RNAi triggered by double-stranded RNA versus truncated transgenes in *Paramecium tetraurelia*. *Nucleic Acids Res.*, **38** (12), 4092–4107.

74 Chalker, D.L., Yao, M.C. (2001) Nongenic, bidirectional transcription precedes and may promote developmental DNA deletion in *Tetrahymena thermophila*. *Genes Dev.*, **15** (10), 1287–1298.

75 Aronica, L., Bednenko, J., Noto, T., DeSouza, L.V., Siu, K.W., Loidl, J., Pearlman, R.E., Gorovsky, M.A., Mochizuki, K. (2008) Study of an RNA helicase implicates small RNA-noncoding RNA interactions in programmed DNA elimination in *Tetrahymena*. *Genes Dev.*, **22** (16), 2228–2241.

76 Nowacki, M., Vijayan, V., Zhou, Y., Schotanus, K., Doak, T.G., Landweber, L.F. (2008) RNA-mediated epigenetic programming of a genome-rearrangement pathway. *Nature*, **451** (7175), 153–158.

77 Yao, M.C., Fuller, P., Xi, X. (2003) Programmed DNA deletion as an RNA-guided system of genome defense. *Science*, **300** (5625), 1581–1584.

78 Duharcourt, S., Keller, A., Meyer, E. (1998) Homology-dependent maternal inhibition of developmental excision of internal eliminated sequences in *Paramecium tetraurelia*. *Mol. Cell Biol.*, **18** (12), 7075–7085.

79 Aury, J.M., Jaillon, O., Duret, L., Noel, B., Jubin, C.,, Porcel, B.M., Segurens, B., Daubin, V., Anthouard, V., Aiach, N., Arnaiz, O., Billaut, A., Beisson, J., Blanc, I., Bouhouche, K., Camara, F., Duharcourt, S., Guigo, R., Gogendeau, D., Katinka, M., Keller, A.M., Kissmehl, R., Klotz, C., Koll, F., Le Mouel, A., Lepere, G., Malinsky, S., Nowacki, M., Nowak, J.K., Plattner, H., Poulain, J., Ruiz, F., Serrano, V., Zagulski, M., Dessen, P., Betermier, M., Weissenbach, J., Scarpelli, C., Schachter, V., Sperling, L., Meyer, E., Cohen, J., Wincker, P. (2006) Global trends of whole-genome duplications revealed by the ciliate *Paramecium tetraurelia*. *Nature*, **444** (7116), 171–178.

80 Dippell, R.V. (1954) A preliminary report on the chromosomal constitution of certain variety 4 races of *Paramecium aurelia*. *Caryologia*, **6**, 1109–1111.

81 Blaszczyk, J., Tropea, J.E., Bubunenko, M., Routzahn, K.M., Waugh, D.S., Court, D.L., Ji, X. (2001) Crystallographic and modeling studies of RNase III suggest a mechanism for double-stranded RNA cleavage. *Structure*, **9** (12), 1225–1236.

82 Han, J., Lee, Y., Yeom, K.H., Kim, Y.K., Jin, H., Kim, V.N. (2004) The Drosha-DGCR8 complex in primary microRNA processing. *Genes Dev.*, **18** (24), 3016–3027.

83 Zhang, H., Kolb, F.A., Jaskiewicz, L., Westhof, E., Filipowicz, W. (2004) Single processing center models for human Dicer and bacterial RNase III. *Cell*, **118** (1), 57–68.

84 Kim, V.N., Han, J., Siomi, M.C. (2009) Biogenesis of small RNAs in animals. *Nat. Rev. Mol. Cell Biol.*, **10** (2), 126–139.

85 Bouhouche, K., Gout, J.-F., Kapusta, A., Bétermier, M., Meyer, E. (2011) Functional specialization of Piwi proteins in *Paramecium tetraurelia* from post-transcriptional gene silencing to genome remodelling. *Nucleic Acids Res.*, **39** (10), 4249–4264.

86 Nowacki, M., Zagorski-Ostoja, W., Meyer, E. (2005) Nowa1p and Nowa2p: novel putative RNA binding proteins involved in trans-nuclear crosstalk in *Paramecium tetraurelia*. *Curr. Biol.*, **15** (18), 1616–1628.

87 Meller, V.H., Wu, K.H., Roman, G., Kuroda, M.I., Davis, R.L. (1997) roX1 RNA paints the X chromosome of male *Drosophila* and is regulated by the dosage compensation system. *Cell*, **88**, 445–457.

88 Amrein, H., Axel, R. (1997) Genes expressed in neurons of adult male *Drosophila*. *Cell*, **88**, 459–469.

89 Brown, C.J., Hendrich, B.D., Rupert, J.L., Lafreniere, R.G., Xing, Y., Lawrence, J., Willard, H.F. (1992) The human XIST gene: analysis of a 17 kb inactive X-specific RNA that contains conserved repeats and is highly localized within the nucleus. *Cell*, **71**, 527–542.

90 Brockdorff, N., Ashworth, A., Kay, G.F., McCabe, V.M., Norris, D.P., Cooper, P.J., Swift, S., Rastan, S. (1992) The product of the mouse Xist gene is a 15 kb inactive X-specific transcript containing no conserved ORF and located in the nucleus. *Cell*, **71**, 515–526.

91 Brown, C.J., Ballabio, A., Rupert, J.L., Lafreniere, R.G., Grompe, M., Tonlorenzi, R., Willard, H.F. (1991) A gene from the region of the human X inactivation centre is expressed exclusively from the inactive X chromosome. *Nature*, **349** (6304), 38–44.

92 Bartolomei, M.S., Zemel, S., Tilghman, S.M. (1991) Parental imprinting of the mouse H19 gene. *Nature*, **351** (6322), 153–155.

93 Ponting, C.P., Oliver, P.L., Reik, W. (2009) Evolution and functions of long noncoding RNAs. *Cell*, **136** (4), 629–641.

94 Klobutcher, L.A., Herrick, G. (1997) Developmental genome reorganization in ciliated protozoa: the transposon link. *Prog. Nucleic Acid Res. Mol. Biol.*, **56**, 1–62.

95 Gratias, A., Betermier, M. (2003) Processing of double-strand breaks is involved in the precise excision of *Paramecium* internal eliminated sequences. *Mol. Cell Biol.*, **23** (20), 7152–7162.

96 Mayer, K.M., Mikami, K., Forney, J.D. (1998) A mutation in *Paramecium tetraurelia* reveals functional and structural features of developmentally excised DNA elements. *Genetics*, **148**, 139–149.

97 Mayer, K., Forney, J. (1999) A mutation in the flanking 5′-TA-3′ dinucleotide prevents excision of an internal eliminated sequence from the *Paramecium tetraurelia* genome. *Genetics*, **151** (2), 597–604.

98 Gratias, A., Lepere, G., Garnier, O., Rosa, S., Duharcourt, S., Malinsky, S., Meyer, E., Betermier, M. (2008) Developmentally programmed DNA splicing in *Paramecium* reveals short-distance crosstalk between DNA cleavage sites. *Nucleic Acids Res.*, **36** (10), 3244–3251.

99 Ruiz, F., Krzywicka, A., Klotz, C., Keller, A., Cohen, J., Koll, F., Balavoine, G., Beisson, J. (2000) The SM19 gene, required for duplication of basal bodies in *Paramecium*, encodes a novel tubulin, eta-tubulin. *Curr. Biol.*, **10** (22), 1451–1454.

100 Elick, T.A., Bauser, C.A., Fraser, M.J. (1996) Excision of the piggyBac transposable element in vitro is a precise event that is enhanced by the expression of its encoded transposase. *Genetica*, **98** (1), 33–41.

101 Mitra, R., Fain-Thornton, J., Craig, N.L. (2008) piggyBac can bypass DNA synthesis during cut and paste transposition. *EMBO J.*, **27** (7), 1097–1109.

102 Kapusta, A., Matsuda, A., Marmignon, A., Ku, M., Silve, A., Meyer, E., Forney, J.D., Malinsky, S., Betermier, M. (2011) Highly precise and developmentally programmed genome assembly in *Paramecium* requires ligase IV-dependent end joining. *PLoS Genet.*, **7** (4), e1002049.

103 Duharcourt, S., Lepere, G., Meyer, E. (2009) Developmental genome rearrangements in ciliates: a natural genomic subtraction mediated by non-coding transcripts. *Trends Genet.*, **25** (8), 344–350.

104 Le Mouel, A., Butler, A., Caron, F., Meyer, E. (2003) Developmentally regulated chromosome fragmentation linked to imprecise elimination of repeated sequences in paramecia. *Eukaryot. Cell*, **2** (5), 1076–1090.

105 Altschuler, M.I., Yao, M.C. (1985) Macronuclear DNA of *Tetrahymena thermophila* exists as defined subchromosomal-sized molecules. *Nucleic Acids Res.*, **13** (16), 5817–5831.

106 Woodard, J., Kaneshiro, E., Gorovsky, M.A. (1972) Cytochemical studies on the problem of macronuclear subnuclei in *Tetrahymena*. *Genetics*, **70** (2), 251–260.

107 Cassidy-Hanley, D., Bisharyan, Y., Fridman, V., Gerber, J., Lin, C., Orias, E., Orias, J.D., Ryder, H., Vong, L., Hamilton, E.P. (2005) Genome-wide characterization of *Tetrahymena thermophila* chromosome breakage sites. II. Physical and genetic mapping. *Genetics*, **170** (4), 1623–1631.

108 Hamilton, E., Bruns, P., Lin, C., Merriam, V., Orias, E., Vong, L., Cassidy-Hanley, D. (2005) Genome-wide characterization of *Tetrahymena thermophila* chromosome breakage sites. I. Cloning and identification of functional sites. *Genetics*, **170** (4), 1611–1621.

109 Doerder, F.P., Deak, J.C., Lief, J.H. (1992) Rate of phenotypic assortment in *Tetrahymena thermophila*. *Dev. Genet.*, **13** (2), 126–132.

110 Conover, R.K., Brunk, C.F. (1986) Macronuclear DNA molecules of *Tetrahymena thermophila*. *Mol. Cell. Biol.*, **6** (3), 900–905.

111 Allis, C.D., Jenuwein, T., Reinberg, D., Caparros, M.-L.A.E. (Eds) (2007) *Epigenetics*, Cold Spring Harbor Press, Cold Spring Harbor.

112 Volpe, T.A., Kidner, C., Hall, I.M., Teng, G., Grewal, S.I., Martienssen, R.A. (2002) Regulation of heterochromatic silencing and histone H3 lysine-9 methylation by RNAi. *Science*, **297** (5588), 1833–1837.

113 Reinhart, B.J., Bartel, D.P. (2002) Small RNAs correspond to centromere heterochromatic repeats. *Science*, **297** (5588), 1831.

114 Wassenegger, M., Heimes, S., Riedel, L., Sanger, H.L. (1994) RNA-directed de novo methylation of genomic sequences in plants. *Cell*, **76**, 567–576.

115 Brown, D.D., Dawid, I.B. (1968) Specific gene amplification in oocytes. *Science*, **160**, 272–280.

116 Fire, A., Xu, S., Montgomery, M.K., Kostas, S.A., Driver, S.E., Mello, C.C. (1998) Potent and specific genetic interference by double-stranded RNA in *Caenorhabditis elegans*. *Nature*, **391**, 806–811.

117 Malone, C.D., Anderson, A.M., Motl, J.A., Rexer, C.H., Chalker, D.L. (2005) Germ line transcripts are processed by a Dicer-like protein that is essential for developmentally programmed genome rearrangements of *Tetrahymena thermophila*. *Mol. Cell. Biol.*, **25** (20), 9151–9164.

118 Mochizuki, K., Gorovsky, M.A. (2005) A Dicer-like protein in *Tetrahymena* has distinct functions in genome rearrangement, chromosome segregation, and meiotic prophase. *Genes Dev.*, **19** (1), 77–89.

119 Aravin, A.A., Hannon, G.J., Brennecke, J. (2007) The Piwi-piRNA pathway provides an adaptive defense in the transposon arms race. *Science*, **318** (5851), 761–764.

120 Brennecke, J., Aravin, A.A., Stark, A., Dus, M., Kellis, M., Sachidanandam, R., Hannon, G.J. (2007) Discrete small RNA-generating loci as master regulators of transposon activity in *Drosophila*. *Cell*, **128** (6), 1089–1103.

121 Carmell, M.A., Girard, A., van de Kant, H.J., Bourc'his, D., Bestor, T.H., de Rooij, D.G., Hannon, G.J. (2007) MIWI2 is essential for spermatogenesis and repression of transposons in the mouse male germline. *Dev. Cell*, **12** (4), 503–514.

122 Das, P.P., Bagijn, M.P., Goldstein, L.D., Woolford, J.R., Lehrbach, N.J., Sapetschnig, A., Buhecha, H.R., Gilchrist, M.J., Howe, K.L., Stark, R., Matthews, N., Berezikov, E., Ketting, R.F., Tavare, S., Miska, E.A. (2008) Piwi and piRNAs act upstream of an endogenous siRNA pathway to suppress Tc3 transposon mobility in the *Caenorhabditis elegans* germline. *Mol. Cell*, **31** (1), 79–90.

123 Gunawardane, L.S., Saito, K., Nishida, K.M., Miyoshi, K., Kawamura, Y., Nagami, T., Siomi, H., Siomi, M.C. (2007) A slicer-mediated mechanism for repeat-associated siRNA 5′ end formation in *Drosophila*. *Science*, **315** (5818), 1587–1590.

124 Kuramochi-Miyagawa, S., Watanabe, T., Gotoh, K., Totoki, Y., Toyoda, A., Ikawa, M., Asada, N., Kojima, K., Yamaguchi, Y., Ijiri, T.W., Hata, K., Li, E., Matsuda, Y., Kimura, T., Okabe, M., Sakaki, Y., Sasaki, H., Nakano, T. (2008) DNA methylation of retrotransposon genes is regulated by Piwi family members MILI and MIWI2 in murine fetal testes. *Genes Dev.*, **22** (7), 908–917.

125 Saito, K., Nishida, K.M., Mori, T., Kawamura, Y., Miyoshi, K., Nagami, T., Siomi, H., Siomi, M.C. (2006) Specific association of Piwi with rasiRNAs derived from retrotransposon and heterochromatic regions in the *Drosophila* genome. *Genes Dev.*, **20** (16), 2214–2222.

126 Vagin, V.V., Sigova, A., Li, C., Seitz, H., Gvozdev, V., Zamore, P.D. (2006) A distinct small RNA pathway silences selfish genetic elements in the germline. *Science*, **313** (5785), 320–324.

127 Martindale, D.W., Allis, C.D., Bruns, P. (1982) Conjugation in *Tetrahymena thermophila*: a temporal analysis of cytological stages. *Exp. Cell Res.*, **140**, 227–236.

128 Ray, C. (1956) Meiosis and nuclear behavior in *Tetrahymena pyriformis*. *J. Eukaryot. Microbiol.*, **3** (2), 88–96.

129 Wenkert, D., Allis, C.D. (1984) Timing of the appearance of macronuclear-specific histone variant hv1 and gene expression in developing new macronuclei of *Tetrahymena thermophila*. *J. Cell Biol.*, **98** (6), 2107–2117.

130 Stargell, L.A., Gorovsky, M.A. (1994) TATA-binding protein and nuclear differentiation in *Tetrahymena thermophila*. *Mol. Cell. Biol.*, **14** (1), 723–734.

131 Mochizuki, K., Gorovsky, M.A. (2004) RNA polymerase II localizes in *Tetrahymena thermophila* meiotic micronuclei when micronuclear transcription associated with genome rearrangement occurs. *Eukaryot. Cell*, **3** (5), 1233–1240.

132 Henderson, I.R., Zhang, X., Lu, C., Johnson, L., Meyers, B.C., Green, P.J., Jacobsen, S.E. (2006) Dissecting *Arabidopsis thaliana* DICER function in small RNA processing, gene silencing and DNA methylation patterning. *Nat. Genet.*, **38** (6), 721–725.

133 Elbashir, S.M., Lendeckel, W., Tuschl, T. (2001) RNA interference is mediated by 21- and 22-nucleotide RNAs. *Genes Dev.*, **15** (2), 188–200.

134 Robertson, H.D., Dunn, J.J. (1975) Ribonucleic acid processing activity of *Escherichia coli* ribonuclease III. *J. Biol. Chem.*, **250** (8), 3050–3056.

135 Schweitz, H., Ebel, J.P. (1971) A study of the mechanism of action of *E. coli* ribonuclease 3. *Biochimie*, **53** (5), 585–593.

136 Crouch, R.J. (1974) Ribonuclease 3 does not degrade deoxyribonucleic acid-ribonucleic acid hybrids. *J. Biol. Chem.*, **249** (4), 1314–1316.

137 Mochizuki, K., Gorovsky, M.A. (2004) Conjugation-specific small RNAs in *Tetrahymena* have predicted properties of scan (scn) RNAs involved in genome rearrangement. *Genes Dev.*, **18** (17), 2068–2073.

138 Hammond, S.M., Bernstein, E., Beach, D., Hannon, G.J. (2000) An RNA-directed nuclease mediates post-transcriptional gene silencing in *Drosophila* cells. *Nature*, **404** (6775), 293–296.

139 Noto, T., Kurth, H.M., Kataoka, K., Aronica, L., DeSouza, L.V., Siu, K.W., Pearlman, R.E., Gorovsky, M.A., Mochizuki, K. (2010) The *Tetrahymena* argonaute-binding protein Giw1p directs a mature argonaute-siRNA complex to the nucleus. *Cell*, **140** (5), 692–703.

140 Kurth, H.M., Mochizuki, K. (2009) 2′-O-methylation stabilizes Piwi-associated small RNAs and ensures DNA elimination in *Tetrahymena*. *RNA*, **15** (4), 675–685.

141 Horwich, M.D., Li, C., Matranga, C., Vagin, V., Farley, G., Wang, P., Zamore, P.D. (2007) The *Drosophila* RNA methyltransferase, DmHen1, modifies germline piRNAs and single-stranded siRNAs in RISC. *Curr. Biol.*, **17** (14), 1265–1272.

142 Saito, K., Sakaguchi, Y., Suzuki, T., Suzuki, T., Siomi, H., Siomi, M.C. (2007) Pimet, the *Drosophila* homolog of HEN1, mediates 2′-O-methylation of Piwi- interacting RNAs at their 3′ ends. *Genes Dev.*, **21** (13), 1603–1608.

143 Yu, B., Yang, Z., Li, J., Minakhina, S., Yang, M., Padgett, R.W., Steward, R., Chen, X. (2005) Methylation as a crucial step in plant microRNA biogenesis. *Science*, **307** (5711), 932–935.

144 Kirino, Y., Mourelatos, Z. (2007) The mouse homolog of HEN1 is a potential methylase for Piwi-interacting RNAs. *RNA*, **13** (9), 1397–1401.

145 Kirino, Y., Vourekas, A., Kim, N., de Lima Alves, F., Rappsilber, J., Klein, P.S., Jongens, T.A., Mourelatos, Z. (2010) Arginine methylation of vasa protein is conserved across phyla. *J. Biol. Chem.*, **285** (11), 8148–8154.

146 Kuramochi-Miyagawa, S., Kimura, T., Ijiri, T.W., Isobe, T., Asada, N., Fujita, Y., Ikawa, M., Iwai, N., Okabe, M., Deng, W., Lin, H., Matsuda, Y., Nakano, T. (2004) Mili, a mammalian member of piwi family gene, is essential for spermatogenesis. *Development*, **131** (4), 839–849.

147 Thomson, T., Liu, N., Arkov, A., Lehmann, R., Lasko, P. (2008) Isolation of new polar granule components in *Drosophila* reveals P body and ER associated proteins. *Mech. Dev.*, **125** (9-10), 865–873.

148 Arkov, A.L., Ramos, A. (2010) Building RNA-protein granules: insight from the germline. *Trends Cell Biol.*, **20** (8), 482–490.

149 Bednenko, J., Noto, T., DeSouza, L.V., Siu, K.W., Pearlman, R.E., Mochizuki, K., Gorovsky, M.A. (2009) Two GW repeat proteins interact with *Tetrahymena thermophila* argonaute and promote genome rearrangement. *Mol. Cell. Biol.*, **29** (18), 5020–5030.

150 He, X.J., Hsu, Y.F., Zhu, S., Wierzbicki, A.T., Pontes, O., Pikaard, C.S., Liu, H.L., Wang, C.S., Jin, H., Zhu, J.K. (2009) An effector of RNA-directed DNA methylation in *Arabidopsis* is an ARGONAUTE

4- and RNA-binding protein. *Cell*, **137** (3), 498–508.

151 Chekulaeva, M., Parker, R., Filipowicz, W. (2010) The GW/WG repeats of *Drosophila* GW182 function as effector motifs for miRNA-mediated repression. *Nucleic Acids Res.*, **38** (19), 6673–6683.

152 Behm-Ansmant, I., Rehwinkel, J., Doerks, T., Stark, A., Bork, P., Izaurralde, E. (2006) mRNA degradation by miRNAs and GW182 requires both CCR4:NOT deadenylase and DCP1:DCP2 decapping complexes. *Genes Dev.*, **20** (14), 1885–1898.

153 Chalker, D.L., Fuller, P., Yao, M.C. (2005) Communication between parental and developing genomes during *Tetrahymena* nuclear differentiation is likely mediated by homologous RNAs. *Genetics*, **169** (1), 149–160.

154 Chalker, D.L., Yao, M.C. (1996) Non-Mendelian, heritable blocks to DNA rearrangement are induced by loading the somatic nucleus of *Tetrahymena thermophila* with germ line-limited DNA. *Mol. Cell. Biol.*, **16** (7), 3658–3667.

155 Liu, Y., Mochizuki, K., Gorovsky, M.A. (2004) Histone H3 lysine 9 methylation is required for DNA elimination in developing macronuclei in *Tetrahymena*. *Proc. Natl Acad. Sci. USA*, **101** (6), 1679–1684.

156 Onodera, Y., Haag, J.R., Ream, T., Nunes, P.C., Pontes, O., Pikaard, C.S. (2005) Plant nuclear RNA polymerase IV mediates siRNA and DNA methylation-dependent heterochromatin formation. *Cell*, **120** (5), 613–622.

157 Kanno, T., Huettel, B., Mette, M.F., Aufsatz, W., Jaligot, E., Daxinger, L., Kreil, D.P., Matzke, M., Matzke, A.J. (2005) Atypical RNA polymerase subunits required for RNA-directed DNA methylation. *Nat. Genet.*, **37** (7), 761–765.

158 Liu, Y., Taverna, S.D., Muratore, T.L., Shabanowitz, J., Hunt, D.F., Allis, C.D. (2007) RNAi-dependent H3K27 methylation is required for heterochromatin formation and DNA elimination in *Tetrahymena*. *Genes Dev.*, **21** (12), 1530–1545.

159 Madireddi, M.T., Coyne, R.S., Smothers, J.F., Mickey, K.M., Yao, M.C., Allis, C.D. (1996) Pdd1p, a novel chromodomain-containing protein, links heterochromatin assembly and DNA elimination in *Tetrahymena*. *Cell*, **87** (1), 75–84.

160 Coyne, R., Nikiforov, M.A., Smothers, J.F., Allis, C.D., Yao, M.C. (1999) Parental expression of the chromodomain protein Pdd1p is required for completion of programmed DNA elimination and nuclear differentiation. *Mol. Cell*, **4** (5), 865–872.

161 Nikiforov, M.A., Gorovsky, M.A., Allis, C.D. (2000) A novel chromodomain protein, pdd3p, associates with internal eliminated sequences during macronuclear development in *Tetrahymena thermophila*. *Mol. Cell. Biol.*, **20** (11), 4128–4134.

162 Rexer, C.H., Chalker, D.L. (2007) Lia1p, a novel protein required during nuclear differentiation for genome-wide DNA rearrangements in *Tetrahymena thermophila*. *Eukaryot. Cell*, **6** (8), 1320–1329.

163 Yao, M.C., Yao, C.H., Halasz, L.M., Fuller, P., Rexer, C.H., Wang, S.H., Jain, R., Coyne, R.S., Chalker, D.L. (2007) Identification of novel chromatin-associated proteins involved in programmed genome rearrangements in *Tetrahymena*. *J. Cell Sci.*, **120** (Pt 12), 1978–1989.

164 Cam, H.P., Sugiyama, T., Chen, E.S., Chen, X., FitzGerald, P.C., Grewal, S.I. (2005) Comprehensive analysis of heterochromatin- and RNAi-mediated epigenetic control of the fission yeast genome. *Nat. Genet.*, **37** (8), 809–819.

165 Rea, S., Eisenhaber, F., O'Carroll, D., Strahl, B.D., Sun, Z.W., Schmid, M., Opravil, S., Mechtler, K., Ponting, C.P., Allis, C.D., Jenuwein, T. (2000) Regulation of chromatin structure by site-specific histone H3 methyltransferases. *Nature*, **406** (6796), 593–599.

166 Ivanova, A.V., Bonaduce, M.J., Ivanov, S.V., Klar, A.J.S. (1998) The chromo and SET domains of the Clr4 protein are essential for silencing in fission yeast. *Nat. Genet.*, **19** (2), 192–195.

167 Jackson, J.P., Lindroth, A.M., Cao, X., Jacobsen, S.E. (2002) Control of CpNpG DNA methylation by the KRYPTONITE histone H3 methyltransferase. *Nature*, **416** (6880), 556–560.

168 Goodrich, J., Puangsomlee, P., Martin, M., Long, D., Meyerowitz, E.M., Coupland, G. (1997) A Polycomb-group gene regulates homeotic gene expression in *Arabidopsis*. *Nature*, **386** (6620), 44–51.

169 Fong, Y., Bender, L., Wang, W., Strome, S. (2002) Regulation of the different chromatin states of autosomes and X chromosomes in the germ line of *C. elegans*. *Science*, **296** (5576), 2235–2238.

170 van der Lugt, N.M., Domen, J., Linders, K., van Roon, M., Robanus-Maandag, E., te Riele, H., van der Valk, M., Deschamps, J., Sofroniew, M., van Lohuizen, M., Berns, A. (1994) Posterior transformation, neurological abnormalities, and severe hematopoietic defects in mice with a targeted deletion of the bmi-1 proto-oncogene. *Genes Dev.*, **8** (7), 757–769.

171 Cao, R., Tsukada, Y., Zhang, Y. (2005) Role of Bmi-1 and Ring1A in H2A ubiquitylation and Hox gene silencing. *Mol. Cell*, **20** (6), 845–854.

172 Wang, H., Wang, L., Erdjument-Bromage, H., Vidal, M., Tempst, P., Jones, R.S., Zhang, Y. (2004) Role of histone H2A ubiquitination in Polycomb silencing. *Nature*, **431** (7010), 873–878.

173 Lagarou, A., Mohd-Sarip, A., Moshkin, Y.M., Chalkley, G.E., Bezstarosti, K., Demmers, J.A., Verrijzer, C.P. (2008) dKDM2 couples histone H2A ubiquitylation to histone H3 demethylation during Polycomb group silencing. *Genes Dev.*, **22** (20), 2799–2810.

174 Gearhart, M.D., Corcoran, C.M., Wamstad, J.A., Bardwell, V.J. (2006) Polycomb group and SCF ubiquitin ligases are found in a novel BCOR complex that is recruited to BCL6 targets. *Mol. Cell. Biol.*, **26** (18), 6880–6889.

175 Sanchez, C., Sanchez, I., Demmers, J.A., Rodriguez, P., Strouboulis, J., Vidal, M. (2007) Proteomics analysis of Ring1B/Rnf2 interactors identifies a novel complex with the Fbxl10/Jhdm1B histone demethylase and the Bcl6 interacting corepressor. *Mol. Cell. Proteomics*, **6** (5), 820–834.

176 Simon, J.A., Kingston, R.E. (2009) Mechanisms of polycomb gene silencing: knowns and unknowns. *Nat. Rev. Mol. Cell Biol.*, **10** (10), 697–708.

177 Cao, R., Wang, L., Wang, H., Xia, L., Erdjument-Bromage, H., Tempst, P., Jones, R.S., Zhang, Y. (2002) Role of histone H3 lysine 27 methylation in Polycomb-group silencing. *Science*, **298** (5595), 1039–1043.

178 Czermin, B., Melfi, R., McCabe, D., Seitz, V., Imhof, A., Pirrotta, V. (2002) Drosophila enhancer of Zeste/ESC complexes have a histone H3 methyltransferase activity that marks chromosomal Polycomb sites. *Cell*, **111** (2), 185–196.

179 Muller, J., Hart, C.M., Francis, N.J., Vargas, M.L., Sengupta, A., Wild, B., Miller, E.L., O'Connor, M.B., Kingston, R.E., Simon, J.A. (2002) Histone methyltransferase activity of a *Drosophila* Polycomb group repressor complex. *Cell*, **111** (2), 197–208.

180 Kuzmichev, A., Nishioka, K., Erdjument-Bromage, H., Tempst, P., Reinberg, D. (2002) Histone methyltransferase activity associated with a human multiprotein complex containing the Enhancer of Zeste protein. *Genes Dev.*, **16** (22), 2893–2905.

181 Bannister, A.J., Zegerman, P., Partridge, J.F., Miska, E.A., Thomas, J.O., Allshire, R.C., Kouzarides, T. (2001) Selective recognition of methylated lysine 9 on histone H3 by the HP1 chromo domain. *Nature*, **410** (6824), 120–124.

182 Ekwall, K., Javerzat, J.P., Lorentz, A., Schmidt, H., Cranston, G., Allshire, R. (1995) The chromodomain protein Swi6: a key component at fission yeast centromeres. *Science*, **269** (5229), 1429–1431.

183 Lorentz, A., Ostermann, K., Fleck, O., Schmidt, H. (1994) Switching gene swi6, involved in repression of silent mating-type loci in fission yeast, encodes a homologue of chromatin-associated proteins from *Drosophila* and mammals. *Gene*, **143** (1), 139–143.

184 Madireddi, M.T., Davis, M.C., Allis, C.D. (1994) Identification of a novel polypeptide involved in the formation of DNA-containing vesicles during macronuclear development in *Tetrahymena*. *Dev. Biol.*, **165** (2), 418–431.

185 Smothers, J.F., Mizzen, C.A., Tubbert, M.M., Cook, R.G., Allis, C.D. (1997) Pdd1p associates with germline-restricted chromatin and a second novel anlagen-enriched protein in developmentally programmed DNA elimination structures. *Development*, **124** (22), 4537–4545.

186 Yao, M.C., Choi, J., Yokoyama, S., Austerberry, C.F., Yao, C.H. (1984) DNA elimination in *Tetrahymena*: a developmental process involving extensive

breakage and rejoining of DNA at defined sites. *Cell*, **36** (2), 433–440.

187 Eissenberg, J.C., James, T.C., Foster-Hartnett, D.M., Hartnett, T., Ngan, V., Elgin, S.C. (1990) Mutation in a heterochromatin-specific chromosomal protein is associated with suppression of position-effect variegation in *Drosophila melanogaster*. *Proc. Natl Acad. Sci. USA*, **87** (24), 9923–9927.

188 Smothers, J.F., Madireddi, M.T., Warner, F.D., Allis, C.D. (1997) Programmed DNA degradation and nucleolar biogenesis occur in distinct organelles during macronuclear development in *Tetrahymena*. *J. Euk. Microbiol.*, **44**, 79–88.

189 Janetopoulos, C., Cole, E., Smothers, J.F., Allis, C.D., Aufderheide, K.J. (1999) The conjusome: a novel structure in *Tetrahymena* found only during sexual reorganization. *J. Cell Sci.*, **112** (Pt 7), 1003–1011.

190 Coyne, R.S., Nikiforov, M.A., Smothers, J.F., Allis, C.D., Yao, M.C. (1999) Parental expression of the chromodomain protein Pdd1p is required for completion of programmed DNA elimination and nuclear differentiation. *Mol. Cell*, **4** (5), 865–872.

191 Nikiforov, M., Smothers, J., Gorovsky, M., Allis, C. (1999) Excision of micronuclear-specific DNA requires parental expression of Pdd2p and occurs independently from DNA replication in *Tetrahymena thermophila*. *Genes Dev.*, **13** (21), 2852–2862.

192 Agrawal, A., Eastman, Q.M., Schatz, D.G. (1998) Transposition mediated by RAG1 and RAG2 and its implications for the evolution of the immune system. *Nature*, **394** (6695), 744–751.

193 Hiom, K., Melek, M., Gellert, M. (1998) DNA transposition by the RAG1 and RAG2 proteins: a possible source of oncogenic translocations. *Cell*, **94** (4), 463–470.

194 Cary, L.C., Goebel, M., Corsaro, B.G., Wang, H.G., Rosen, E., Fraser, M.J. (1989) Transposon mutagenesis of baculoviruses: analysis of *Trichoplusia ni* transposon IFP2 insertions within the FP-locus of nuclear polyhedrosis viruses. *Virology*, **172** (1), 156–169.

195 Saveliev, S.V., Cox, M.M. (1996) Developmentally programmed DNA deletion in *Tetrahymena thermophila* by a transposition-like reaction pathway. *EMBO J.*, **15** (11), 2858–2869.

196 Doerder, F.P., Debault, L.E. (1975) Cytofluorimetric analysis of nuclear DNA during meiosis, fertilization and macronuclear development in the ciliate *Tetrahymena pyriformis*, syngen 1. *J. Cell Sci.*, **17** (3), 471–493.

197 Yao, M.C. (1981) Ribosomal RNA gene amplification in *Tetrahymena* may be associated with chromosome breakage and DNA elimination. *Cell*, **24** (3), 765–774.

198 Yao, M.C., Yao, C.H. (1981) Repeated hexanucleotide C-C-C-C-A-A is present near free ends of macronuclear DNA of *Tetrahymena*. *Proc. Natl Acad. Sci. USA*, **78** (12), 7436–7439.

199 Yao, M.C., Kimmel, A.R., Gorovsky, M.A. (1974) A small number of cistrons for ribosomal RNA in the germinal nucleus of a eukaryote, *Tetrahymena pyriformis*. *Proc. Natl Acad. Sci. USA*, **71** (8), 3082–3086.

200 Yao, M.C., Zheng, K., Yao, C.H. (1987) A conserved nucleotide sequence at the sites of developmentally regulated chromosomal breakage in *Tetrahymena*. *Cell*, **48** (5), 779–788.

201 Fan, Q., Yao, M.-C. (1996) New telomere formation coupled with site-specific chromosome breakage in *Tetrahymena thermophila*. *Mol. Cell. Biol.*, **16** (3), 1267–1274.

202 Kapler, G.M., Blackburn, E.H. (1994) A weak germ-line excision mutation blocks developmentally controlled amplification of the rDNA minichromosome of *Tetrahymena thermophila*. *Genes Dev.*, **8** (1), 84–95.

203 Fan, Q., Yao, M.C. (2000) A long stringent sequence signal for programmed chromosome breakage in *Tetrahymena thermophila*. *Nucleic Acids Res.*, **28** (4), 895–900.

204 Swanton, M.T., Heumann, J.M., Prescott, D.M. (1980) Gene-sized DNA molecules of the macronuclei in three species of hypotrichs: size distributions and absence of nicks. DNA of ciliated protozoa. VIII. *Chromosoma*, **77** (2), 217–227.

205 Swanton, M.T., Greslin, A.F., Prescott, D.M. (1980) Arrangement of coding and non-coding sequences in the DNA molecules coding for rRNAs in *Oxytricha* sp. DNA of ciliated protozoa. VII. *Chromosoma*, **77** (2), 203–215.

206 Wunning, I.U., Lipps, H.J. (1983) A transformation system for the hypotrichous ciliate *Stylonychia mytilus*. *EMBO J.*, **2** (10), 1753–1757.

207 Steinbruck, G. (1983) Over-amplification of genes in macronuclei of hypotrichous ciliates. *Chromosoma*, **88**, 156–163.

208 Greslin, A.F., Loukin, S.H., Oka, Y., Prescott, D.M. (1988) An analysis of the macronuclear actin genes of *Oxytricha*. *DNA*, **7** (8), 529–536.

209 Greslin, A.F., Prescott, D.M., Oka, Y., Loukin, S.H., Chappell, J.C. (1989) Reordering of nine exons is necessary to form a functional actin gene in *Oxytricha nova*. *Proc. Natl Acad. Sci. USA*, **86** (16), 6264–6268.

210 Mitcham, J.L., Lynn, A.J., Prescott, D.M. (1992) Analysis of a scrambled gene: the gene encoding alpha-telomere-binding protein in *Oxytricha nova*. *Genes Dev.*, **6** (5), 788–800.

211 Prescott, D.M., Greslin, A.F. (1992) Scrambled actin I gene in the micronucleus of *Oxytricha nova*. *Dev. Genet.*, **13** (1), 66–74.

212 DuBois, M., Prescott, D.M. (1995) Scrambling of the actin I gene in two *Oxytricha* species. *Proc. Natl Acad. Sci. USA*, **92** (9), 3888–3892.

213 Hoffman, D.C., Prescott, D.M. (1997) Evolution of internal eliminated segments and scrambling in the micronuclear gene encoding DNA polymerase alpha in two *Oxytricha* species. *Nucleic Acids Res.*, **25** (10), 1883–1889.

214 Prescott, J.D., DuBois, M.L., Prescott, D.M. (1998) Evolution of the scrambled germline gene encoding alpha-telomere binding protein in three hypotrichous ciliates. *Chromosoma*, **107** (5), 293–303.

215 Landweber, L.F., Kuo, T.-C., Curtis, E.A. (2000) Evolution and assembly of an extremely scrambled gene. *Proc. Natl Acad. Sci. USA*, **97** (7), 3298–3303.

216 Doak, T.G., Doerder, F.P., Jahn, C.L., Herrick, G. (1994) A proposed superfamily of transposase genes: transposon-like elements in ciliated protozoa and a common "D35E" motif. *Proc. Natl Acad. Sci. USA*, **91** (3), 942–946.

217 Hunter, D.J., Williams, K., Cartinhour, S., Herrick, G. (1989) Precise excision of telomere-bearing transposons during *Oxytricha fallax* macronuclear development. *Genes Dev.*, **3** (12B), 2101–2112.

218 Juranek, S.A., Rupprecht, S., Postberg, J., Lipps, H.J. (2005) snRNA and heterochromatin formation are involved in DNA excision during macronuclear development in stichotrichous ciliates. *Eukaryot. Cell*, **4** (11), 1934–1941.

219 Bourc'his, D., Bestor, T.H. (2004) Meiotic catastrophe and retrotransposon reactivation in male germ cells lacking Dnmt3L. *Nature*, **431** (7004), 96–99.

220 Kato, Y., Kaneda, M., Hata, K., Kumaki, K., Hisano, M., Kohara, Y., Okano, M., Li, E., Nozaki, M., Sasaki, H. (2007) Role of the Dnmt3 family in de novo methylation of imprinted and repetitive sequences during male germ cell development in the mouse. *Hum. Mol. Genet.*, **16** (19), 2272–2280.

221 Siomi, M.C., Sato, K., Pezic, D., Aravin, A.A. (2011) PIWI-interacting small RNAs: the vanguard of genome defence. *Nat. Rev. Mol. Cell Biol.*, **12** (4), 246–258.

Index

Epigenetic Regulation and Epigenomics: Advances in Molecular Biology and Medicine, First Edition. Edited by Robert A. Meyers.

d

h

i

n